PACEMAKER®

Geometry

TEACHER'S ANSWER EDITION

GLOBE FEARON
Pearson Learning Group

Contents

Pacemaker® Geometry, First Edition

We thank the following educators, who provided valuable comments and suggestions during the development of this book:

CONSULTANTS
Martha C. Beech, Center for Performance Technology, Florida State University, Tallahassee, Florida; **Kay McClain**, Department of Teaching and Learning, Vanderbilt University, Nashville, Tennessee

REVIEWERS
Chapters 1–4: **Donna Hambrick**, formerly of Woodham High School, Pensacola, Florida; **Joseph H. Bean**, Wythe County Public Schools, Wytheville, Virginia
Chapters 5–7: **Jack Ray Whittemore**, Olympic High School, Charlotte, North Carolina; **Martha (Marty) Penn**, Monte Vista Christian School, Watsonville, California
Chapters 8–10: **Judy Ann Mock**, Detroit Public Schools, Detroit, Michigan; **Marguerite L. Hart**, MSD Washington Township, Indianapolis, Indiana
Chapters 11–13: **Charlene Ekrut**, Wichita Unified School District #259, Wichita, Kansas; **Dr. Judith C. Branch-Boyd**, Chicago Public Schools, Chicago, Illinois

PROJECT STAFF
Art and Design: Evelyn Bauer, Susan Brorein, Joan Jacobus, Jen Visco *Editorial:* Jane Books, Danielle Camaleri, Phyllis Dunsay, Amy Feldman, Elizabeth Fernald, Mary Ellen Gilbert, Justian Kelly, Dena R. Pollak *Manufacturing:* Mark Cirillo, Thomas Dunne *Marketing:* Clare Harrison *Production:* Karen Edmonds, Roxanne Knoll, Jill Kuhfuss *Publishing Operations:* Carolyn Coyle, Tom Daning, Richetta Lobban

Photo Credits appear on page 453.

ISBN: 0-130-23838-4

Printed in the United States of America
5 6 7 8 9 10 05 04

1-800-321-3106
www.pearsonlearning.com

Pacemaker® Geometry, First Edition

The Proven Solution for Today's Classrooms from Globe Fearon

Students and educators alike face increased challenges in the twenty-first century. Students need preparation for the rigors of more difficult standards and proficiency tests as well as the ability to successfully apply learning to daily challenges—including the challenges of the workplace. Educators must meet the needs of diverse classrooms, keep learning up to date and relevant, and create supportive learning environments for a range of learning styles.

Globe Fearon's *Pacemaker® Geometry* supplies educators and students with materials and techniques that are accessible, predictable, age-appropriate, and relevant. This new program provides a solid, well-balanced approach to teaching geometry.

Designed to help students gain access to required content, *Pacemaker Geometry* features

- Concise, manageable chapters.
- Concepts presented in a predictable format.
- Frequent opportunities for review.
- Pretaught vocabulary and a controlled reading level.
- High-interest, age-appropriate design.
- Solid teacher support.
- A full-color Student Edition that engages students.
- Margin notes that provide support.
- Visuals that address a variety of learning styles.
- A Workbook that parallels each chapter in the Student Edition.
- A Teacher's Answer Edition that provides answers on the student pages.
- A Classroom Resource Binder that has an abundant variety of teacher support.
- A Solution Key that provides solutions to the exercises in the Student Edition.

▶ Pacemaker® Geometry, First Edition

The STUDENT EDITION includes

- Unit openers that list chapters for planning.
- Chapter openers that preteach vocabulary, state chapter objectives, and provide a chapter project for portfolio assessment.
- Content that is divided into manageable lessons.
- Stepped-out examples.
- Ample review for reinforcement.
- Relevant features such as *Math Connection, On-the-Job Math,* and *Math In Your Life.*
- Chapter reviews that promote vocabulary and content retention.
- Unit reviews for standardized test practice.

The TEACHER'S ANSWER EDITION includes

- Instruction on how to use the program.
- ESL/ELL notes for each chapter.
- Cross-references for all components.
- Answers at point-of-use.

The WORKBOOK includes

- Complete practice for every chapter.
- Skill practice worksheets that provide essential practice and encourage student mastery of content.
- Critical Thinking questions to encourage students to investigate geometric concepts individually and in groups.

The CLASSROOM RESOURCE BINDER includes

- A complete set of reproducibles that review, reinforce, and enrich key skills and concepts.
- A variety of assessment opportunities.
- Organizational and planning charts for ease of classroom management.
- Paragraph proofs and two-column proofs to build confidence and empower students in logical thinking.

The SOLUTION KEY includes

- Answers to the Workbook and the Classroom Resource Binder.
- Stepped-out solutions for the Student Edition.

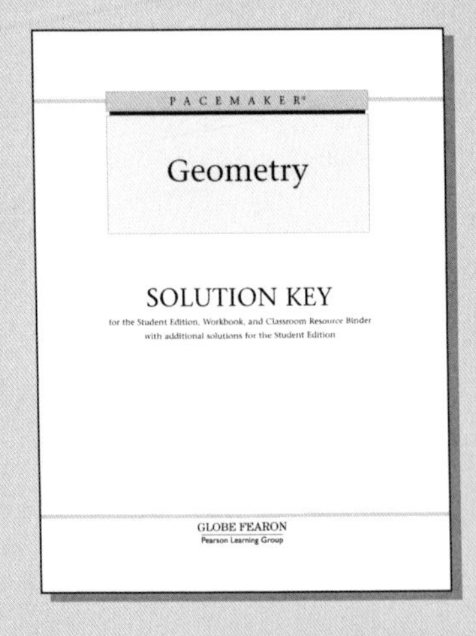

Each unit has an organized opener and practical review.

Supports easy planning and provides a quick reference point with the **Unit Opener** chapter listing.

Unit One

1

Ensure students' understanding and retention of key concepts with the **Unit Review.**

Unit 1 Review

Write the letter of the correct answer.

1. $\overline{AB}$ is 18 cm long. Point M is the midpoint of $\overline{AB}$. Find AM.
A. 6 cm
B. 9 cm
C. 12 cm
D. 36 cm

2. $\angle A$ and $\angle B$ are complementary angles. $\angle A$ is 35°. Find m$\angle B$.
A. 20°
B. 35°
C. 55°
D. 145°

3. $\angle 1$ and $\angle 3$ are vertical angles. $\angle 1$ is 80°. Find m$\angle 3$.
A. 10°
B. 40°
C. 80°
D. 100°

4. Which of the following is the Reflexive Property of Equality?
A. If $a = b$, then $b = a$.
B. If $a = b$, then $a + 1 = b + 1$.
C. If $a = b$, then $3a = 3b$.
D. $a = a$

Use the diagram to answer Items 5 and 6. In this diagram, $p \parallel q$.

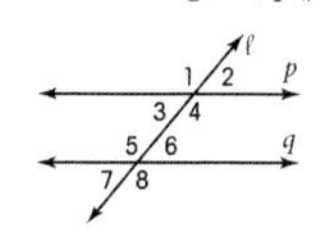

5. Which of the following is NOT true?
A. $\angle 1 \cong \angle 4$
B. $\angle 4 \cong \angle 6$
C. $\angle 3 \cong \angle 6$
D. $\angle 4 \cong \angle 8$

6. $\angle 3$ is 50°. Find m$\angle 5$.
A. 25°
B. 40°
C. 50°
D. 130°

Critical Thinking
In the diagram below, $\overleftrightarrow{AB} \parallel \overleftrightarrow{DE}$. Find m$\angle FCG$.

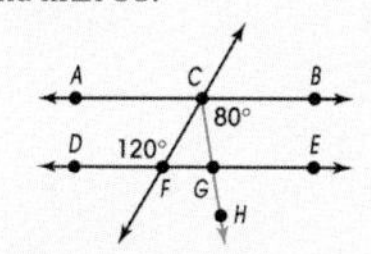

CHALLENGE Which angle is congruent to $\angle GCB$? Explain.

120 Unit 1 • Review

Foster independent thinking and writing with open-ended critical thinking questions.

Further assess students' understanding with **Challenge questions.**

***Pacemaker Geometry* Student Edition, pages 1, 120**

▸ Chapter Openers draw students into chapter content.

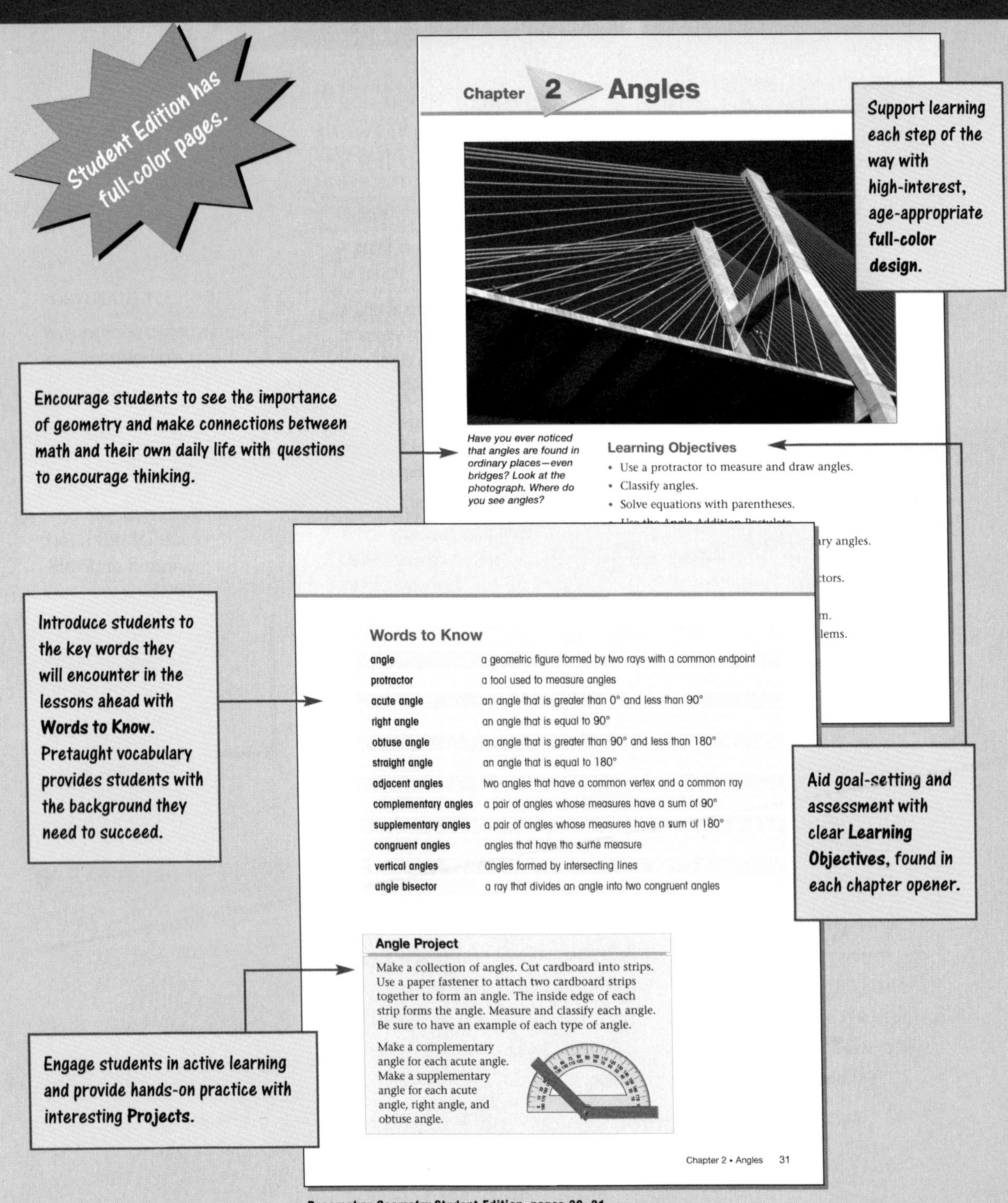

Student Edition has full-color pages.

Support learning each step of the way with high-interest, age-appropriate **full-color design.**

Chapter 2 Angles

Have you ever noticed that angles are found in ordinary places—even bridges? Look at the photograph. Where do you see angles?

Encourage students to see the importance of geometry and make connections between math and their own daily life with questions to encourage thinking.

Learning Objectives

- Use a protractor to measure and draw angles.
- Classify angles.
- Solve equations with parentheses.
- ...ry angles.
- ...tors.
- ...m.
- ...lems.

Aid goal-setting and assessment with clear **Learning Objectives,** found in each chapter opener.

Words to Know

angle	a geometric figure formed by two rays with a common endpoint
protractor	a tool used to measure angles
acute angle	an angle that is greater than 0° and less than 90°
right angle	an angle that is equal to 90°
obtuse angle	an angle that is greater than 90° and less than 180°
straight angle	an angle that is equal to 180°
adjacent angles	two angles that have a common vertex and a common ray
complementary angles	a pair of angles whose measures have a sum of 90°
supplementary angles	a pair of angles whose measures have a sum of 180°
congruent angles	angles that have the same measure
vertical angles	angles formed by intersecting lines
angle bisector	a ray that divides an angle into two congruent angles

Introduce students to the key words they will encounter in the lessons ahead with **Words to Know.** Pretaught vocabulary provides students with the background they need to succeed.

Angle Project

Make a collection of angles. Cut cardboard into strips. Use a paper fastener to attach two cardboard strips together to form an angle. The inside edge of each strip forms the angle. Measure and classify each angle. Be sure to have an example of each type of angle.

Make a complementary angle for each acute angle. Make a supplementary angle for each acute angle, right angle, and obtuse angle.

Engage students in active learning and provide hands-on practice with interesting **Projects.**

Chapter 2 • Angles 31

Pacemaker Geometry **Student Edition, pages 30, 31**

Manageable lessons teach single concepts.

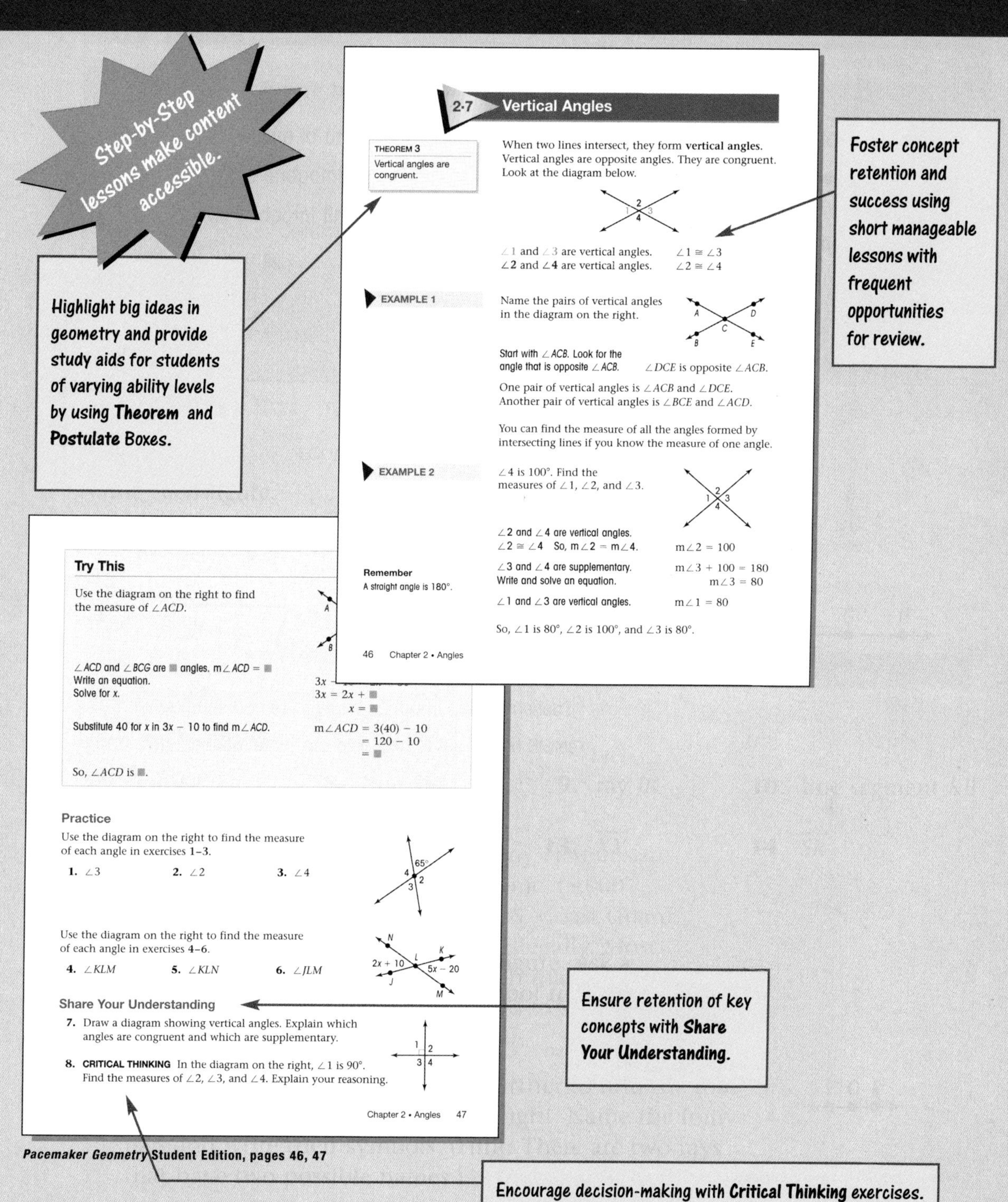

2·7 Vertical Angles

THEOREM 3
Vertical angles are congruent.

When two lines intersect, they form **vertical angles**. Vertical angles are opposite angles. They are congruent. Look at the diagram below.

$\angle 1$ and $\angle 3$ are vertical angles. $\angle 1 \cong \angle 3$
$\angle 2$ and $\angle 4$ are vertical angles. $\angle 2 \cong \angle 4$

EXAMPLE 1 Name the pairs of vertical angles in the diagram on the right.

Start with $\angle ACB$. Look for the angle that is opposite $\angle ACB$. $\angle DCE$ is opposite $\angle ACB$.

One pair of vertical angles is $\angle ACB$ and $\angle DCE$.
Another pair of vertical angles is $\angle BCE$ and $\angle ACD$.

You can find the measure of all the angles formed by intersecting lines if you know the measure of one angle.

EXAMPLE 2 $\angle 4$ is 100°. Find the measures of $\angle 1$, $\angle 2$, and $\angle 3$.

$\angle 2$ and $\angle 4$ are vertical angles.
$\angle 2 \cong \angle 4$ So, $m\angle 2 = m\angle 4$. $m\angle 2 = 100$

Remember
A straight angle is 180°.

$\angle 3$ and $\angle 4$ are supplementary. Write and solve an equation. $m\angle 3 + 100 = 180$
$m\angle 3 = 80$

$\angle 1$ and $\angle 3$ are vertical angles. $m\angle 1 = 80$

So, $\angle 1$ is 80°, $\angle 2$ is 100°, and $\angle 3$ is 80°.

46 Chapter 2 • Angles

Try This

Use the diagram on the right to find the measure of $\angle ACD$.

$\angle ACD$ and $\angle BCG$ are ■ angles. $m\angle ACD =$ ■
Write an equation. $3x$ –
Solve for x. $3x = 2x +$ ■
$x =$ ■

Substitute 40 for x in $3x - 10$ to find $m\angle ACD$. $m\angle ACD = 3(40) - 10$
$= 120 - 10$
$=$ ■

So, $\angle ACD$ is ■.

Practice

Use the diagram on the right to find the measure of each angle in exercises 1–3.

1. $\angle 3$ **2.** $\angle 2$ **3.** $\angle 4$

Use the diagram on the right to find the measure of each angle in exercises 4–6.

4. $\angle KLM$ **5.** $\angle KLN$ **6.** $\angle JLM$

Share Your Understanding

7. Draw a diagram showing vertical angles. Explain which angles are congruent and which are supplementary.

8. CRITICAL THINKING In the diagram on the right, $\angle 1$ is 90°. Find the measures of $\angle 2$, $\angle 3$, and $\angle 4$. Explain your reasoning.

Chapter 2 • Angles 47

Pacemaker Geometry Student Edition, pages 46, 47

Features provide relevant connections to real life.

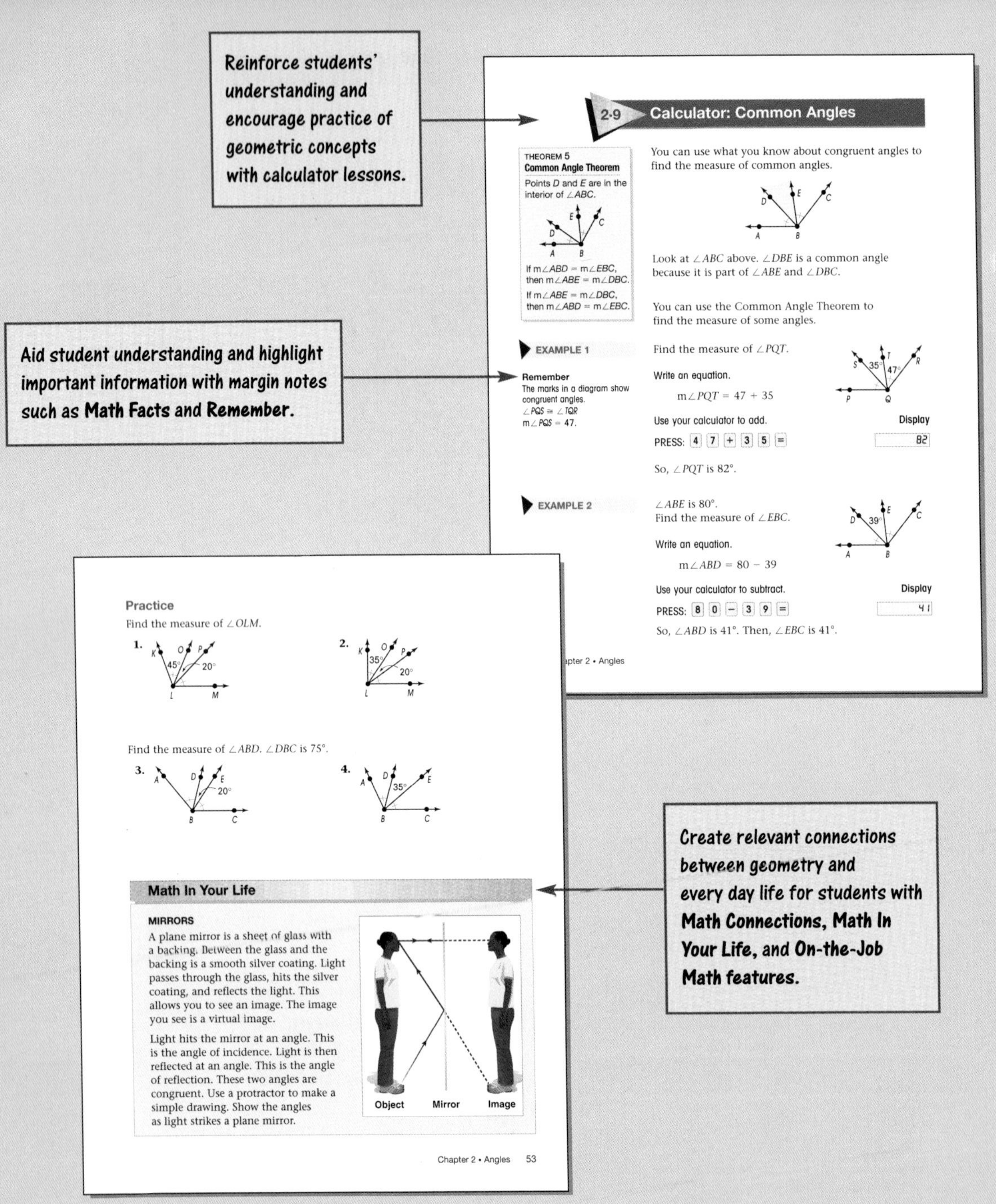

Reinforce students' understanding and encourage practice of geometric concepts with calculator lessons.

Aid student understanding and highlight important information with margin notes such as **Math Facts** and **Remember**.

Create relevant connections between geometry and every day life for students with **Math Connections, Math In Your Life, and On-the-Job Math features.**

2·9 Calculator: Common Angles

THEOREM 5
Common Angle Theorem
Points D and E are in the interior of $\angle ABC$.

If $m\angle ABD = m\angle EBC$, then $m\angle ABE = m\angle DBC$.
If $m\angle ABE = m\angle DBC$, then $m\angle ABD = m\angle EBC$.

You can use what you know about congruent angles to find the measure of common angles.

Look at $\angle ABC$ above. $\angle DBE$ is a common angle because it is part of $\angle ABE$ and $\angle DBC$.

You can use the Common Angle Theorem to find the measure of some angles.

EXAMPLE 1

Find the measure of $\angle PQT$.

Remember
The marks in a diagram show congruent angles.
$\angle PQS \cong \angle TQR$
$m\angle PQS = 47$.

Write an equation.

$m\angle PQT = 47 + 35$

Use your calculator to add.

PRESS: 4 7 + 3 5 = | Display: 82

So, $\angle PQT$ is 82°.

EXAMPLE 2

$\angle ABE$ is 80°.
Find the measure of $\angle EBC$.

Write an equation.

$m\angle ABD = 80 - 39$

Use your calculator to subtract.

PRESS: 8 0 − 3 9 = | Display: 41

So, $\angle ABD$ is 41°. Then, $\angle EBC$ is 41°.

...apter 2 • Angles

Practice

Find the measure of $\angle OLM$.

1. 45°, 20°

2. 35°, 20°

Find the measure of $\angle ABD$. $\angle DBC$ is 75°.

3. 20°

4. 35°

Math In Your Life

MIRRORS

A plane mirror is a sheet of glass with a backing. Between the glass and the backing is a smooth silver coating. Light passes through the glass, hits the silver coating, and reflects the light. This allows you to see an image. The image you see is a virtual image.

Light hits the mirror at an angle. This is the angle of incidence. Light is then reflected at an angle. This is the angle of reflection. These two angles are congruent. Use a protractor to make a simple drawing. Show the angles as light strikes a plane mirror.

Chapter 2 • Angles 53

Pacemaker Geometry **Student Edition, pages 52, 53**

Problem solving is emphasized in every chapter.

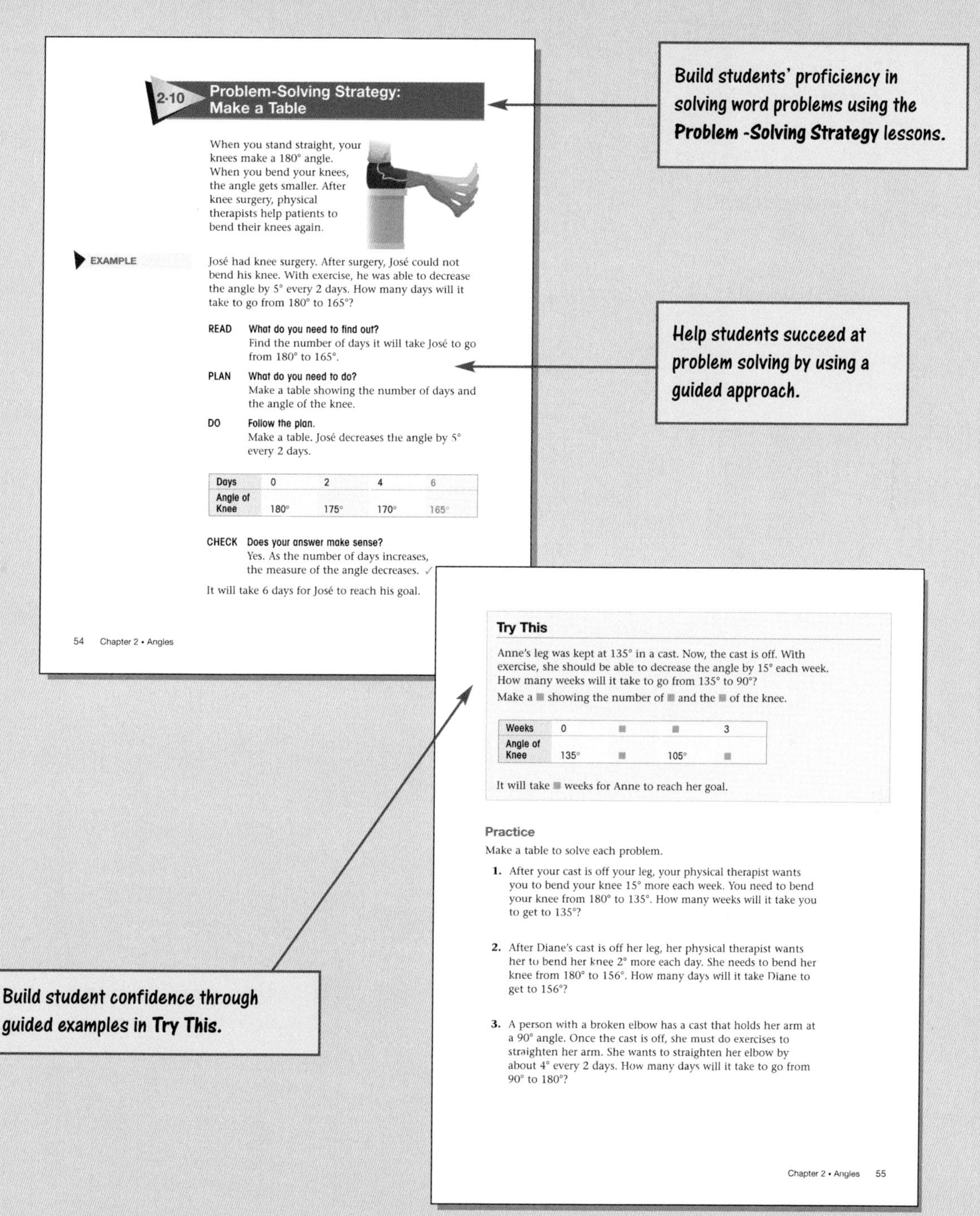

Build students' proficiency in solving word problems using the **Problem -Solving Strategy** lessons.

2·10 Problem-Solving Strategy: Make a Table

When you stand straight, your knees make a 180° angle. When you bend your knees, the angle gets smaller. After knee surgery, physical therapists help patients to bend their knees again.

EXAMPLE

José had knee surgery. After surgery, José could not bend his knee. With exercise, he was able to decrease the angle by 5° every 2 days. How many days will it take to go from 180° to 165°?

READ **What do you need to find out?**
Find the number of days it will take José to go from 180° to 165°.

PLAN **What do you need to do?**
Make a table showing the number of days and the angle of the knee.

DO **Follow the plan.**
Make a table. José decreases the angle by 5° every 2 days.

Days	0	2	4	6
Angle of Knee	180°	175°	170°	165°

CHECK **Does your answer make sense?**
Yes. As the number of days increases, the measure of the angle decreases. ✓

It will take 6 days for José to reach his goal.

54 Chapter 2 • Angles

Help students succeed at problem solving by using a guided approach.

Try This

Anne's leg was kept at 135° in a cast. Now, the cast is off. With exercise, she should be able to decrease the angle by 15° each week. How many weeks will it take to go from 135° to 90°?

Make a ■ showing the number of ■ and the ■ of the knee.

Weeks	0	■	■	3
Angle of Knee	135°	■	105°	■

It will take ■ weeks for Anne to reach her goal.

Practice

Make a table to solve each problem.

1. After your cast is off your leg, your physical therapist wants you to bend your knee 15° more each week. You need to bend your knee from 180° to 135°. How many weeks will it take you to get to 135°?
2. After Diane's cast is off her leg, her physical therapist wants her to bend her knee 2° more each day. She needs to bend her knee from 180° to 156°. How many days will it take Diane to get to 156°?
3. A person with a broken elbow has a cast that holds her arm at a 90° angle. Once the cast is off, she must do exercises to straighten her arm. She wants to straighten her elbow by about 4° every 2 days. How many days will it take to go from 90° to 180°?

Chapter 2 • Angles 55

Build student confidence through guided examples in **Try This.**

***Pacemaker Geometry* Student Edition, pages 54, 55**

Problem-solving applications make math relevant.

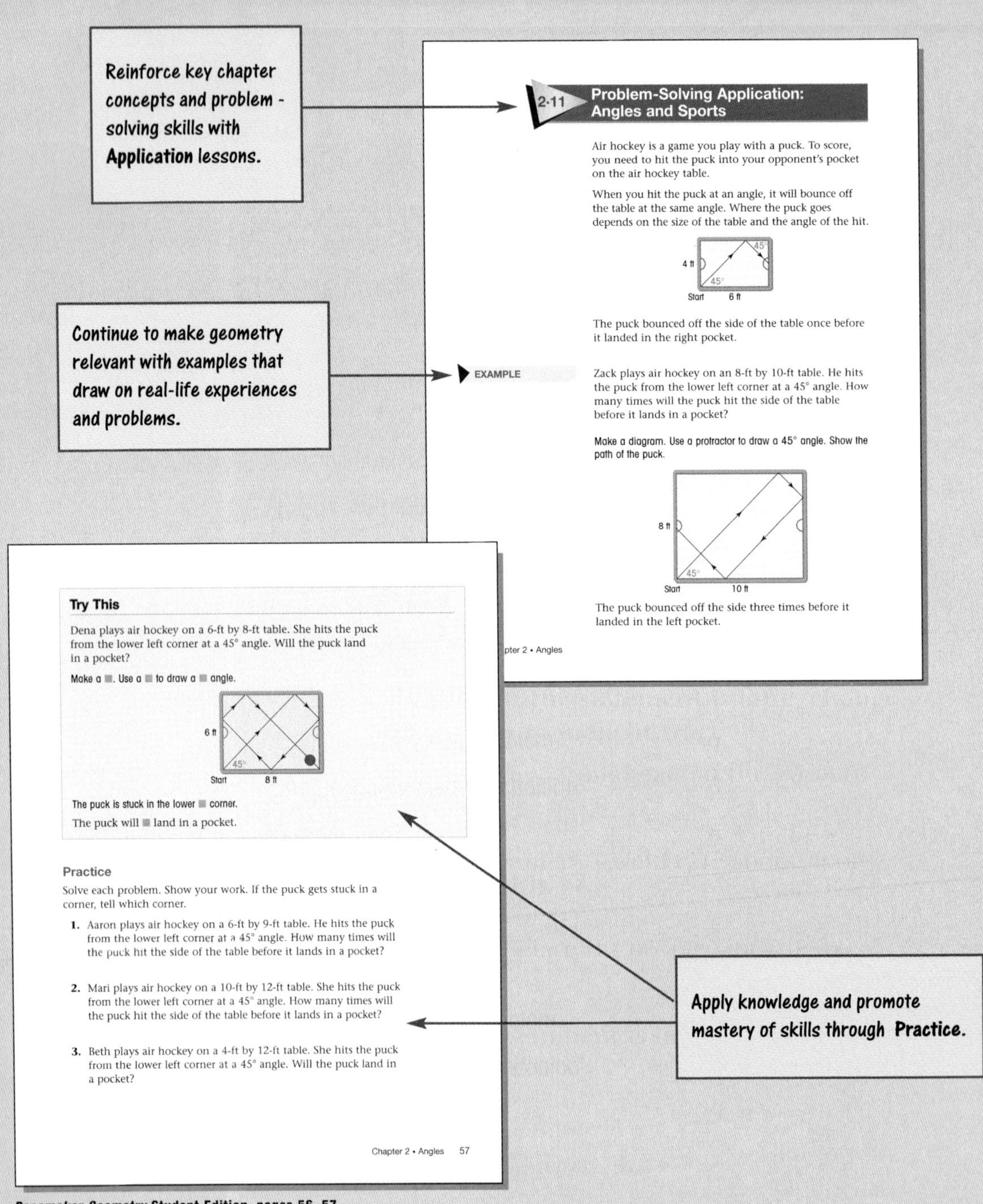

Pacemaker Geometry Student Edition, pages 56, 57

Complete Chapter Review assesses understanding.

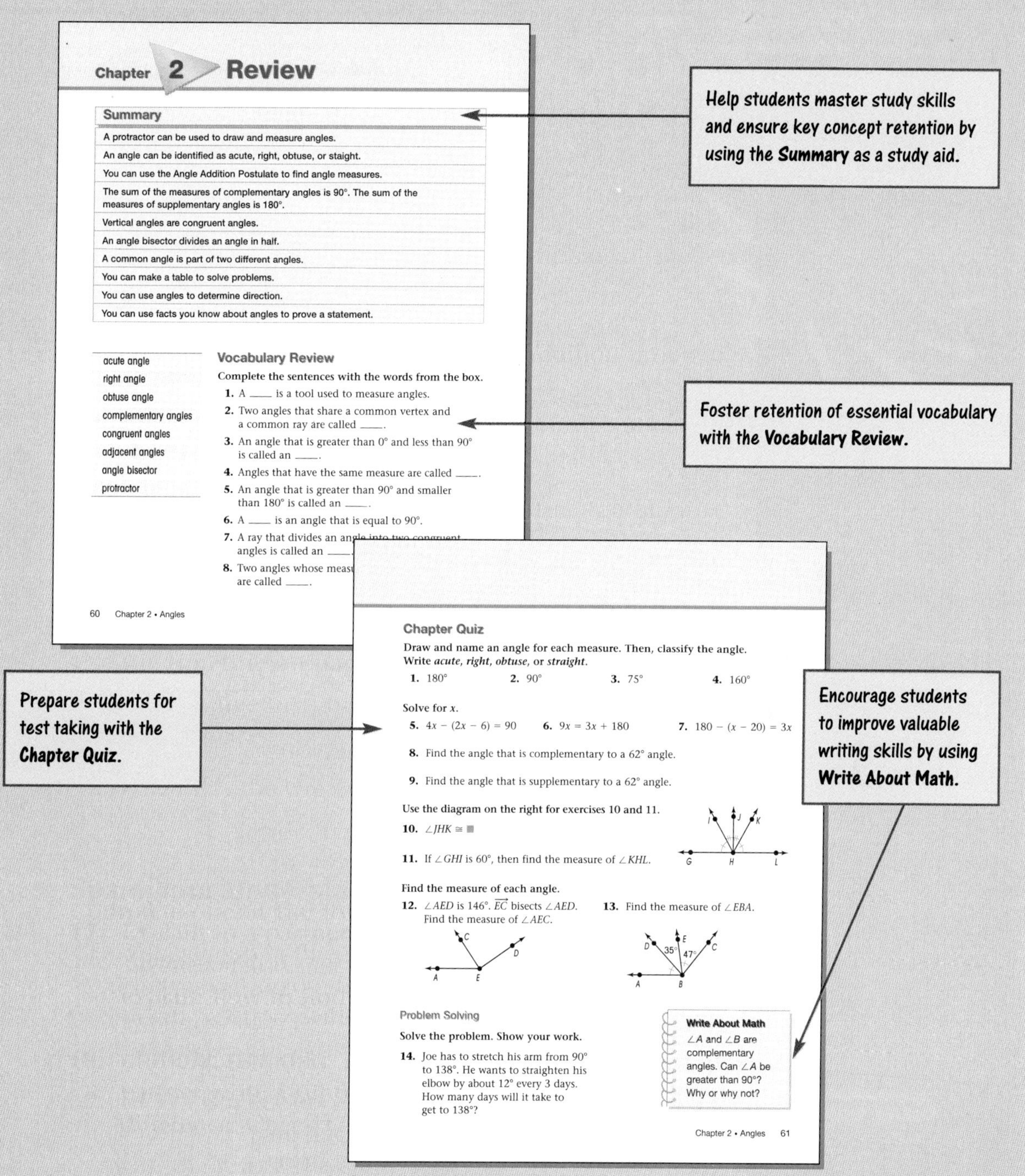

Chapter 2 Review

Summary

A protractor can be used to draw and measure angles.
An angle can be identified as acute, right, obtuse, or staight.
You can use the Angle Addition Postulate to find angle measures.
The sum of the measures of complementary angles is 90°. The sum of the measures of supplementary angles is 180°.
Vertical angles are congruent angles.
An angle bisector divides an angle in half.
A common angle is part of two different angles.
You can make a table to solve problems.
You can use angles to determine direction.
You can use facts you know about angles to prove a statement.

acute angle
right angle
obtuse angle
complementary angles
congruent angles
adjacent angles
angle bisector
protractor

Vocabulary Review

Complete the sentences with the words from the box.

1. A ____ is a tool used to measure angles.
2. Two angles that share a common vertex and a common ray are called ____.
3. An angle that is greater than 0° and less than 90° is called an ____.
4. Angles that have the same measure are called ____.
5. An angle that is greater than 90° and smaller than 180° is called an ____.
6. A ____ is an angle that is equal to 90°.
7. A ray that divides an an[gle into two congruent] angles is called an ____
8. Two angles whose meas[...] are called ____.

60 Chapter 2 • Angles

Chapter Quiz

Draw and name an angle for each measure. Then, classify the angle. Write *acute*, *right*, *obtuse*, or *straight*.

1. 180° 2. 90° 3. 75° 4. 160°

Solve for x.

5. $4x - (2x - 6) = 90$ 6. $9x = 3x + 180$ 7. $180 - (x - 20) = 3x$

8. Find the angle that is complementary to a 62° angle.

9. Find the angle that is supplementary to a 62° angle.

Use the diagram on the right for exercises 10 and 11.

10. $\angle JHK \cong$ ■

11. If $\angle GHI$ is 60°, then find the measure of $\angle KHL$.

Find the measure of each angle.

12. $\angle AED$ is 146°. $\overrightarrow{EC}$ bisects $\angle AED$. Find the measure of $\angle AEC$.

13. Find the measure of $\angle EBA$.

Problem Solving

Solve the problem. Show your work.

14. Joe has to stretch his arm from 90° to 138°. He wants to straighten his elbow by about 12° every 3 days. How many days will it take to get to 138°?

Write About Math
$\angle A$ and $\angle B$ are complementary angles. Can $\angle A$ be greater than 90°? Why or why not?

Chapter 2 • Angles 61

Help students master study skills and ensure key concept retention by using the **Summary** as a study aid.

Foster retention of essential vocabulary with the **Vocabulary Review.**

Prepare students for test taking with the **Chapter Quiz.**

Encourage students to improve valuable writing skills by using **Write About Math.**

Pacemaker Geometry Student Edition pages 60, 61

Teacher's Answer Edition has point-of-use answers.

Component reference annotation correlates the lesson to the Workbook page and to the Classroom Resource Binder page.

Point-of-use answers ease classroom management.

More practice is provided in the Workbook, page 10, and in the Classroom Resource Binder, page 16.

2·7 Vertical Angles

Getting Started Have students draw intersecting lines and measure the angles.

THEOREM 3
Vertical angles are congruent.

When two lines intersect, they form **vertical angles.** Vertical angles are opposite angles. They are congruent. Look at the diagram below.

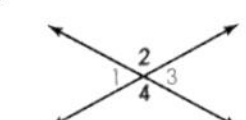

$\angle 1$ and $\angle 3$ are vertical angles. $\angle 1 \cong \angle 3$
$\angle 2$ and $\angle 4$ are vertical angles. $\angle 2 \cong \angle 4$

EXAMPLE 1

Name the pairs of vertical angles in the diagram on the right.

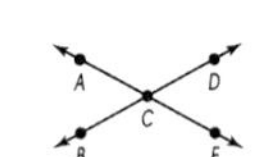

Start with $\angle ACB$. Look for the angle that is opposite $\angle ACB$. $\angle DCE$ is opposite $\angle ACB$.

One pair of vertical angles is $\angle ACB$ and $\angle DCE$.
Another pair of vertical angles is $\angle BCE$ and $\angle ACD$.

You can find the measure of all the angles formed by intersecting lines if you know the measure of one angle.

EXAMPLE 2

$\angle 4$ is 100°. Find the measures of $\angle 1$, $\angle 2$, and $\angle 3$.

$\angle 2$ and $\angle 4$ are vertical angles.
$\angle 2 \cong \angle 4$ So, $m\angle 2 = m\angle 4$. $m\angle 2 = 100$

$\angle 3$ and $\angle 4$ are supplementary. Write and solve an equation. $m\angle 3 + 100 = 180$, $m\angle 3 = 80$

$\angle 1$ and $\angle 3$ are vertical angles. $m\angle 1 = 80$

So, $\angle 1$ is 80°, $\angle 2$ is 100°, and $\angle 3$ is 80°.

...er
...angle is 180°.

...apter 2 • Angles

Try This

Use the diagram on the right to find the measure of $\angle ACD$.

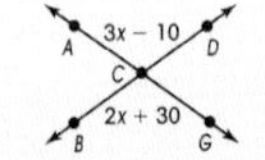

$\angle ACD$ and $\angle BCG$ are ■ angles. vertical $m\angle ACD =$ ■ $m\angle BCG$
Write an equation. $3x - 10 = 2x + 30$
Solve for x. $3x = 2x +$ ■ 40
$x =$ ■ 40

Substitute 40 for x in $3x - 10$ to find $m\angle ACD$. $m\angle ACD = 3(40) - 10 = 120 - 10 =$ ■ 110

So, $\angle ACD$ is ■. 110°

Practice

Use the diagram on the right to find the measure of each angle in exercises **1–3**.

1. $\angle 3$ 65° **2.** $\angle 2$ 115° **3.** $\angle 4$ 115°

Use the diagram on the right to find the measure of each angle in exercises **4–6**.

4. $\angle KLM$ 30° **5.** $\angle KLN$ 150° **6.** $\angle JLM$ 150°

Share Your Understanding

7. Draw a diagram showing vertical angles. Explain which angles are congruent and which are supplementary.
Check students' work.

8. **CRITICAL THINKING** In the diagram on the right, $\angle 1$ is 90°. Find the measures of $\angle 2$, $\angle 3$, and $\angle 4$. Explain your reasoning.
$\angle 2 = 90°$; $\angle 3 = 90°$; $\angle 4 = 90°$; Possible answer: $\angle 1$ and $\angle 2$ form a 180° angle. Because $\angle 1 = 90°$, $\angle 2 = 90°$. $\angle 1$ and $\angle 4$, and $\angle 2$ and $\angle 3$ are vertical angles. They are congruent. So, $\angle 3 = 90°$ and $\angle 4 = 90°$.

Chapter 2 • Angles 47

Annotations offer answer support for efficient classroom management.

***Pacemaker Geometry* Teacher's Answer Edition, pages 46, 47**

Workbook provides practice for every chapter.

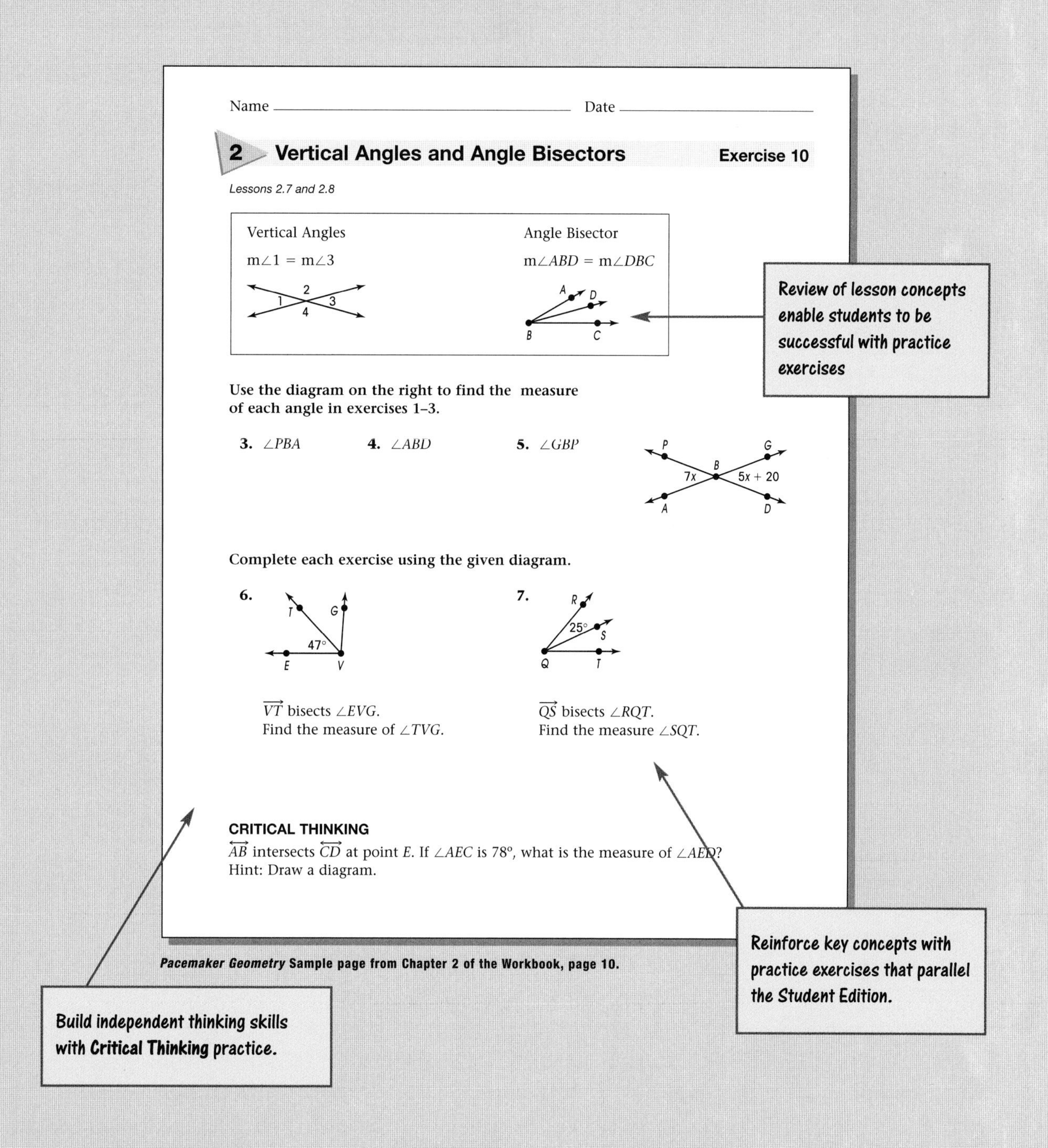

Name ______________________ Date ______________

2 Vertical Angles and Angle Bisectors — Exercise 10

Lessons 2.7 and 2.8

Vertical Angles

$m\angle 1 = m\angle 3$

Angle Bisector

$m\angle ABD = m\angle DBC$

Use the diagram on the right to find the measure of each angle in exercises 1–3.

3. $\angle PBA$ **4.** $\angle ABD$ **5.** $\angle GBP$

Complete each exercise using the given diagram.

6. $\overrightarrow{VT}$ bisects $\angle EVG$.
Find the measure of $\angle TVG$.

7. $\overrightarrow{QS}$ bisects $\angle RQT$.
Find the measure $\angle SQT$.

CRITICAL THINKING

$\overleftrightarrow{AB}$ intersects $\overleftrightarrow{CD}$ at point E. If $\angle AEC$ is 78°, what is the measure of $\angle AED$?
Hint: Draw a diagram.

Pacemaker Geometry **Sample page from Chapter 2 of the Workbook, page 10.**

Classroom Resource Binder provides review and

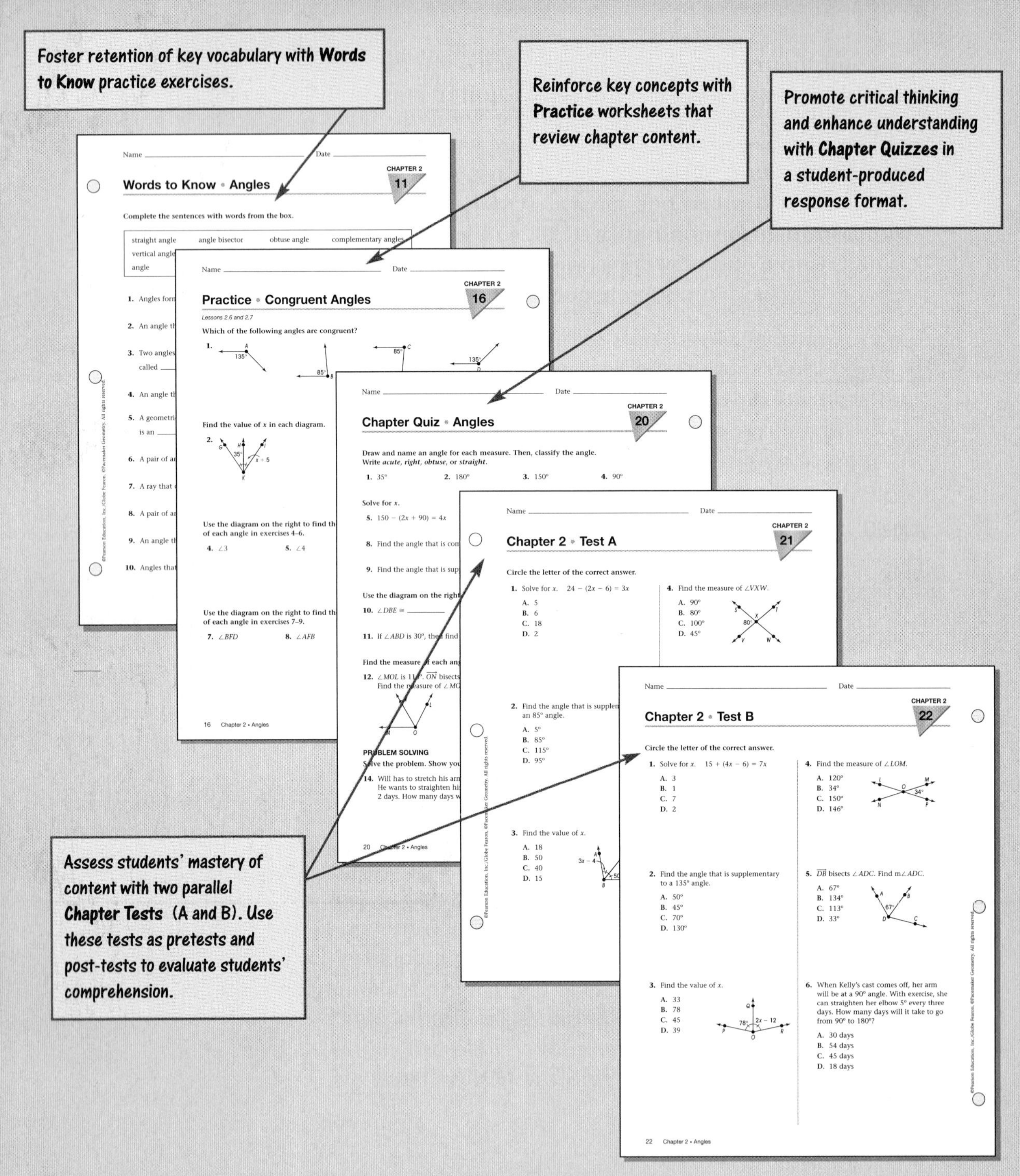

practice to meet the needs of all your students.

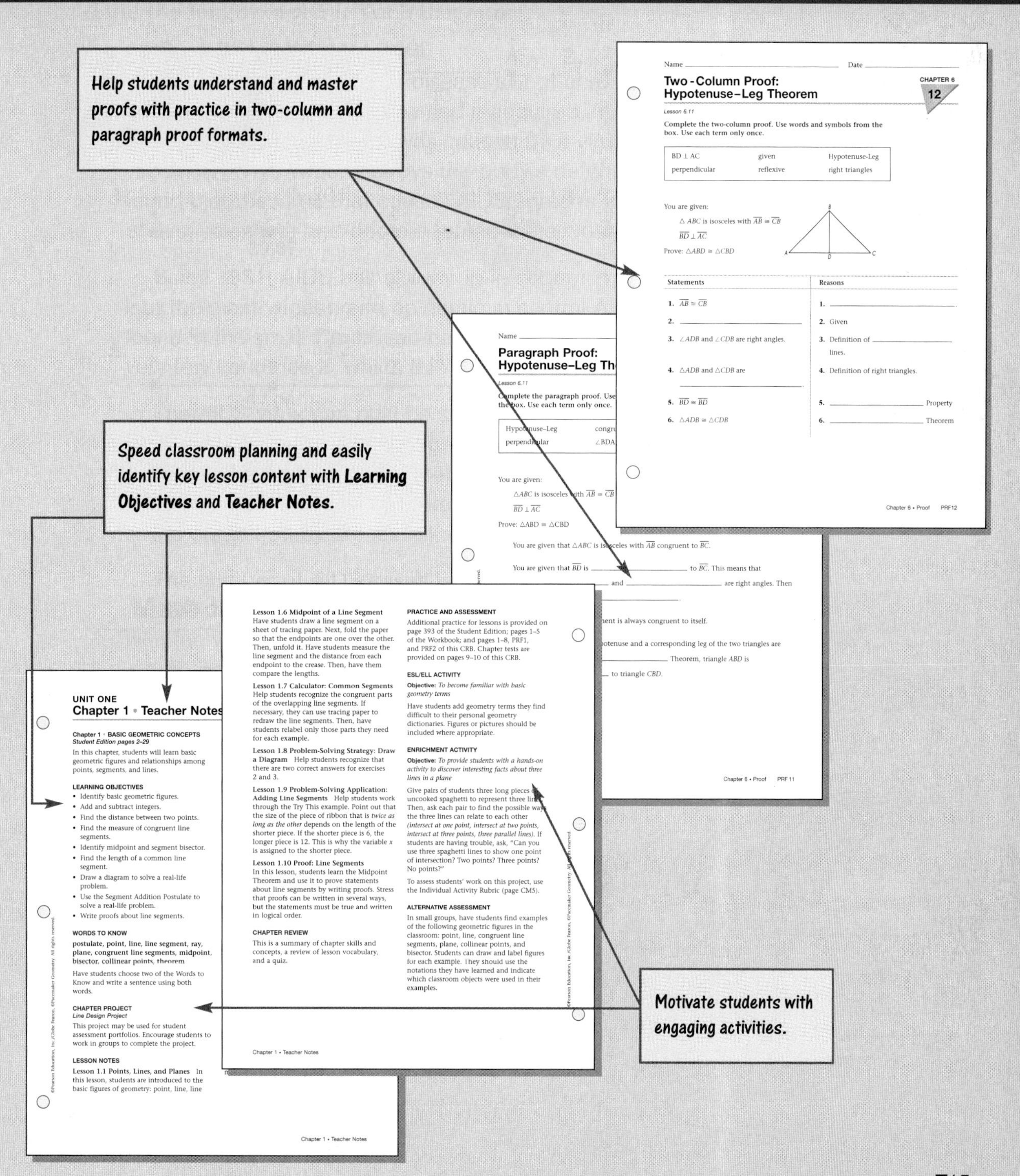

▶ A variety of Globe Fearon programs offer support.

Use these valuable Globe Fearon programs and worktexts with your students of all ability levels for extra practice, reinforcement, reteaching, enrichment, and test preparation.

Pacemaker® Pre-Algebra

Presents concepts and skills students need to get ready to take on algebra.

Pacemaker® Algebra 1

Encourages student success in algebra by providing the essential concepts and skills of a first-year algebra course.

Pacemaker® Basic Math

Guides students from understanding whole numbers and basic arithmetic operations to mastery of simple geometry and algebra.

Writing in Mathematics

Shows your entire class how to communicate clearly and effectively in mathematics.

Math for Proficiency

Helps students develop the mastery and confidence they need to succeed.

Success in Math

Presents core math concepts in short, manageable lessons that require little background knowledge.

Globe Fearon, an imprint of Pearson Learning Group, features a wide array of materials to help tailor instruction to meet the diverse needs of your students.

PACEMAKER®

Geometry

GLOBE FEARON
Pearson Learning Group

Pacemaker® Geometry, First Edition

We thank the following educators, who provided valuable comments and suggestions during the development of this book:

CONSULTANTS

Martha C. Beech, Center for Performance Technology, Florida State University, Tallahassee, Florida
Kay McClain, Department of Teaching and Learning, Vanderbilt University, Nashville, Tennessee

REVIEWERS

Chapters 1–4: **Donna Hambrick**, formerly of Woodham High School, Pensacola, Florida; **Joseph H. Bean**, Wythe County Public Schools, Wytheville, Virginia
Chapters 5–7: **Jack Ray Whittemore**, Olympic High School, Charlotte, North Carolina; **Martha (Marty) Penn**, Monte Vista Christian School, Watsonville, California
Chapters 8–10: **Judy Ann Mock**, Detroit Public Schools, Detroit, Michigan; **Marguerite L. Hart**, MSD Washington Township, Indianapolis, Indiana
Chapters 11–13: **Charlene Ekrut**, Wichita Unified School District #259, Wichita, Kansas; **Dr. Judith C. Branch-Boyd**, Chicago Public Schools, Chicago, Illinois

PROJECT STAFF

Art and Design: Evelyn Bauer, Susan Brorein, Joan Jacobus, Jen Visco *Editorial:* Jane Books, Danielle Camaleri, Phyllis Dunsay, Amy Feldman, Elizabeth Fernald, Mary Ellen Gilbert, Justian Kelly, Dena R. Pollak *Manufacturing:* Mark Cirillo *Marketing:* Clare Harrison *Production:* Karen Edmonds, Roxanne Knoll, Jill Kuhfuss *Publishing Operations:* Carolyn Coyle, Tom Daning, Richetta Lobban

Photo Credits appear on page 453.

ISBN: 0-130-23837-6

Printed in the United States of America
5 6 7 8 9 10 05 04

1-800-321-3106
www.pearsonlearning.com

Contents

Chapter 3 Reasoning and Proofs 62

Chapter 4 Perpendicular and Parallel Lines 88

Chapter 13 Right Triangle Trigonometry 368

Student Handbook 391

A Note to the Student

Welcome to geometry! Geometry is about shapes—such as lines, points, circles, and squares—and the properties of these shapes. Algebra is about numbers, symbols, and calculations. You will use what you learned in algebra in geometry. The neat thing about geometry is that you can draw a lot of the problems on paper!

Each lesson is set up the same way. The first page tells about the lesson and shows you how to do the math. Then, you can try your skills in **Try This**, followed by **Practice**. From lesson to lesson, you can share what you learned with a partner in **Share Your Understanding**. Margin notes will give you helpful hints, such as math facts, postulates, and theorems.

Postulates and theorems are special features of geometry. They help us to show how shapes relate to each other. Postulates are like math facts for geometry. Theorems are statements that can only be proved by using postulates. This is like saying, "I did my homework," and then proving it by showing your teacher what you did to get the work done. Understanding postulates and theorems will help you succeed in geometry.

Application lessons show you how to apply your new knowledge to different subjects. **Problem-Solving** lessons show you different ways to work out problems by using what you have learned. **Calculator** lessons give you another tool to solve geometry problems.

Math Connections, **On-the-Job Math**, and **Math In Your Life** are features that include interesting information about how we use math every day. Other features help you to organize your study of geometry. In the beginning of each chapter, you will find **Learning Objectives** that will help you focus on the important points in the chapter. You will also find important vocabulary in **Words to Know**. At the end of each chapter, there is a **Chapter Review**. At the end of each unit, there is a **Unit Review**.

Geometry is probably not like the math you have done before. So, if you can, work with a friend. Just give geometry a chance. You might find you like it!

Unit One

Chapter 1
Basic Geometric Concepts

Chapter 2
Angles

Chapter 3
Reasoning and Proofs

Chapter 4
Perpendicular and Parallel Lines

Chapter 1

Basic Geometric Concepts

Look at the photograph. These lines are reflections from a prism. Where do the lines start? Where do they end?

Learning Objectives

- Identify basic geometric figures.
- Add and subtract integers.
- Find the distance between two points.
- Find the measure of congruent line segments.
- Identify midpoint and segment bisector.
- Find the length of a common line segment.
- Draw a diagram to solve a real-life problem.
- Use the Segment Addition Postulate to solve a real-life problem.
- Write proofs about line segments.

ESL/ELL Note Have students create a vocabulary list for every chapter. They should write the vocabulary word, draw a picture when possible, and write the page number where they can find that concept.

Words to Know

postulate	a statement that is accepted without proof
point	a location in space
line	a geometric figure made up of infinitely many points; it extends endlessly in two directions
line segment	a part of a line; it has two endpoints
ray	a part of a line; it has one endpoint
plane	a flat surface; it has no thickness; it extends endlessly in all directions
congruent line segments	line segments that have the same length
midpoint	the point that divides a line segment into two congruent parts
bisector	a line that intersects a line segment at its midpoint
collinear points	points on the same line
theorem	a statement that can be proved

Line Design Project

You can make a curve using only straight lines. On graph paper, mark off points along two lines as shown. Connect these points with straight lines.

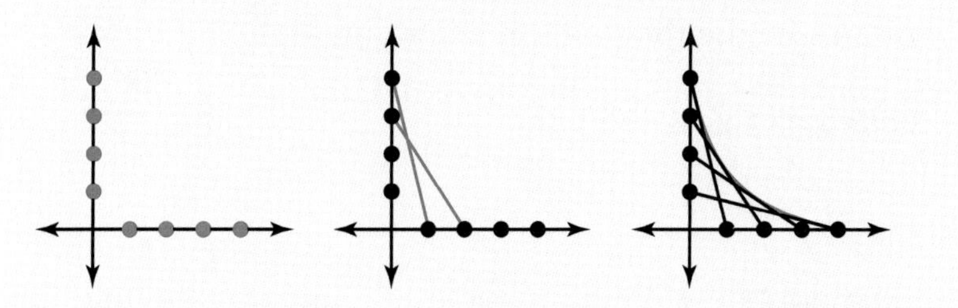

Measure each line segment. Which line segments are congruent?

Project
Have students use colored string and cardboard to make a string design. See the Classroom Resource Binder for a scoring rubric to assess this project.

More practice is provided in the Workbook, page 1, and in the Classroom Resource Binder, page 2.

1·1 Points, Lines, and Planes

Getting Started
Draw a line on the board. Have students place points anywhere on the line. To demonstrate a ray or a line segment, take some points off and/or erase parts of the line.

A geometric figure is a set of points. The chart below shows the basic geometric figures. The **postulates** on the left describe two of these figures.

POSTULATE 1

Two points determine exactly one line.

POSTULATE 2

Three points not on a single line determine exactly one plane.

Figure	Name	Symbol
A **point** is a location in space.	point *A*	
A **line** is made up of infinitely many points. It extends endlessly in two directions.	line *AB* or line *BA*	$\overleftrightarrow{AB}$ or $\overleftrightarrow{BA}$
A **line segment** is part of a line. It has two endpoints.	line segment *AB* or line segment *BA*	$\overline{AB}$ or $\overline{BA}$
A **ray** is part of a line. It has one endpoint. When you name a ray, name the endpoint first.	ray *BA*	$\overrightarrow{BA}$
A **plane** is a flat surface. It has no thickness. It extends endlessly in all directions.	plane *ABC*	

You can use the chart to identify geometric figures.

EXAMPLE

Name three of the basic geometric figures in the diagram on the right.

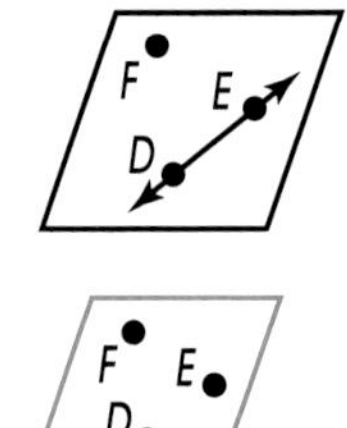

Copy each figure.

line *DE* point *F* plane *DEF*

Three figures are line *DE*, point *F*, and plane *DEF*. Try to find other figures in the diagram.

Try These

1. Name the figure below.

Match the figure to one in the chart.

There is one endpoint. The figure is a ■. ray

Place the endpoint first in the name.

This is ray ■. *HG*

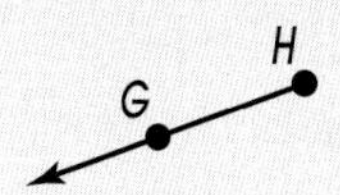

2. Draw $\overleftrightarrow{MN}$.

The symbol ⟷ is used for a ■. line

The letters *M* and *N* are two ■ on the line. points

The drawing of $\overleftrightarrow{MN}$ is below. Label the points.

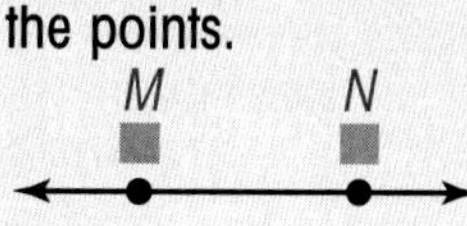

Practice

Name each figure. Some answers may vary.

1. C B ray *CB*

2. 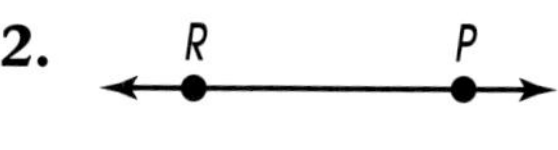

line *RP*

3. •D point *D*

4. 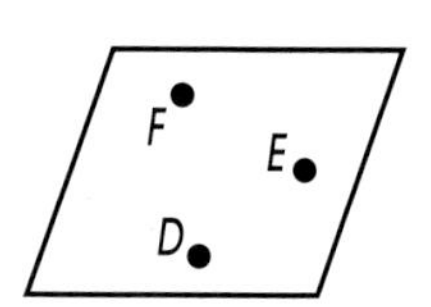

plane *DEF*

5. 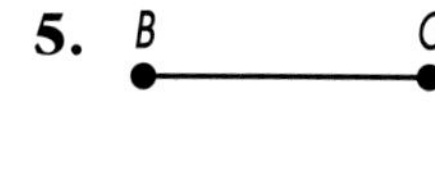

line segment *BC*

6. 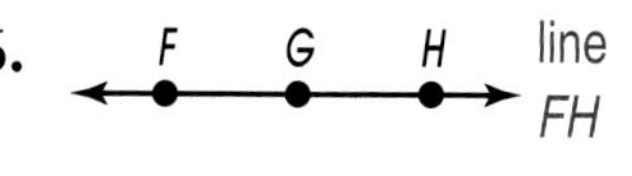

line *FH*

Draw each figure.

7. point *C* •C

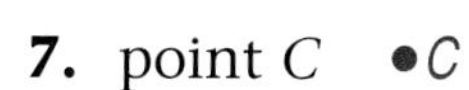

8. line *SM* (S M)

9. ray *JK* (J K)

10. line segment *KB* (K B)

11. $\overrightarrow{WJ}$ (W J)

12. $\overline{BC}$ (B C)

13. $\overleftrightarrow{XY}$ (X Y)

14. $\overleftrightarrow{SR}$ (S R)

Share Your Understanding

15. Write the symbol for a geometric figure. Ask a partner to draw the figure the symbol represents. Check your partner's work. Answers will vary.

16. **CRITICAL THINKING** Work with a partner to find the four different rays in the figure on the right. Name the four rays and write their symbols. (Hint: There are two rays that have two possible names.) $\overrightarrow{PQ}$ or $\overrightarrow{PR}$, $\overrightarrow{QP}$, $\overrightarrow{QR}$, $\overrightarrow{RQ}$ or $\overrightarrow{RP}$

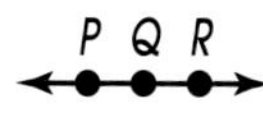

More practice is provided in the Classroom Resource Binder, page 3.

1·2 Algebra Review: Adding Integers

Getting Started
Have students draw and label their own integer number lines to use with the lesson.

Integers are the numbers ..., $^-2$, $^-1$, 0, $^+1$, $^+2$, ...
Integers describe distance and direction.
The integer $^-3$ is 3 spaces to the left of 0.

Math Fact
Addends are the integers you add.

The absolute value of an integer describes distance from 0. The symbol $|^-3|$ means "the absolute value of $^-3$."

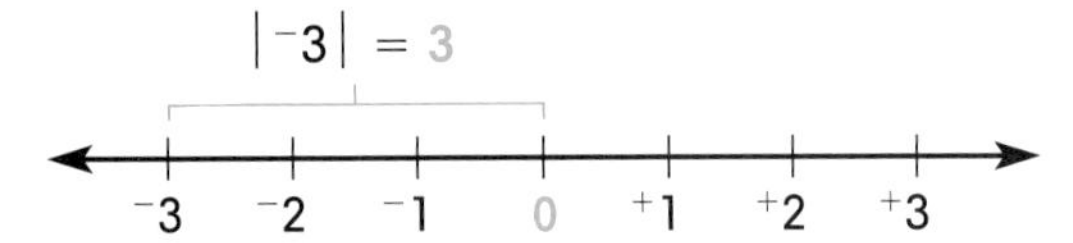

You can use absolute value to add integers.

When the addends have the same sign, you find the sum of the absolute values.

EXAMPLE

Add. $^-5 + (^-7)$

Find the absolute value of each integer. $|^-5| = 5$ $|^-7| = 7$

Notice $^-5$ and $^-7$ have the same sign. Find the sum of the absolute values. $5 + 7 = 12$

$^-5$ and $^-7$ are negative. Use a negative sign in the answer. $^-5 + (^-7) = {}^-12$

The sum of $^-5$ and $^-7$ is $^-12$.

When the addends have different signs, you find the difference of the absolute values.

Try This

Add. $^-5 + 7$

Find the absolute value of each integer. $|^-5| = ■$ 5 $|7| = ■$ 7

Notice $^-5$ and 7 have different signs.
Find the difference of the absolute values. $7 ■ 5 = 2$ –

7 has the larger absolute value and 7 is positive.
Use a positive sign in the answer. $^-5 + 7 = ■$ 2

The sum of $^-5$ and 7 is ■. 2

Practice

Find the absolute value of each integer.

1. $|^-8|$ 8 **2.** $|29|$ 29 **3.** $|^-500|$ 500 **4.** $|8|$ 8

Add.

5. $^-3 + 4$ 1 **6.** $^-7 + (^-8)$ $^-15$ **7.** $^-9 + (^-8)$ $^-17$ **8.** $^-6 + 5$ $^-1$

9. $^-6 + 6$ 0 **10.** $^-6 + (^-6)$ $^-12$ **11.** $8 + (^-3)$ 5 **12.** $8 + (^-16)$ $^-8$

13. $15 + 11$ 26 **14.** $^-1 + (^-9)$ $^-10$ **15.** $9 + (^-1)$ 8 **16.** $^-10 + 1$ $^-9$

17. First, find the absolute value of each integer. Find the difference of the absolute values. Use the sign of the integer with the larger absolute value in the answer. In $7 + (^-8)$, use a negative sign for the sum, $^-1$. In $^-7 + 8$, use a positive sign for the sum, 1.

Share Your Understanding

17. Explain to a partner how to add $7 + (^-8)$. Use the words *absolute value, sign, difference,* and *larger* in your explanation. Have your partner explain how to add $^-7 + 8$. See above.

18. **CRITICAL THINKING** Pick an integer from $^-10$ to 10. Have a partner write an addition problem with that integer as the sum. Check your partner's work. Answers will vary.

More practice is provided in the Classroom Resource Binder, page 3.

1·3 Algebra Review: Subtracting Integers

Getting Started
Review addition of integers. Have students find each sum below.

$7 + 4$	[11]
$^-7 + 4$	[$^-3$]
$7 + (^-4)$	[3]
$^-7 + (^-4)$	[$^-11$]

Every integer except zero has an opposite.

The opposite of $^+3$ is $^-3$.

The opposite of $^-2$ is $^+2$.

You use opposites to subtract integers. To subtract an integer, you add its opposite.

$$\text{Subtract } ^-3 \longrightarrow \text{Add } ^+3$$
$$5 - (^-3) = 5 + (^+3) = 8$$

EXAMPLE 1

Subtract. $4 - (^-6)$

Write the subtraction problem.	$4 - (^-6)$
Change the problem to addition.	$4 + (^+6)$
Find the sum.	$4 + 6 = 10$

The answer to $4 - (^-6)$ is 10.

EXAMPLE 2

Subtract. $^-6 - 1$

Write the subtraction problem.	$^-6 - (^+1)$
Change the problem to addition.	$^-6 + (^-1)$
Find the sum.	$^-6 + (^-1) = {}^-7$

The answer to $^-6 - 1$ is $^-7$.

EXAMPLE 3

Subtract. $^-9 - (^-5)$

Write the subtraction problem.	$^-9 - (^-5)$
Change the problem to addition.	$^-9 + (^+5)$
Find the sum.	$^-9 + 5 = {}^-4$

The answer to $^-9 - (^-5)$ is $^-4$.

Try This

Subtract. $^-3 - (^-8)$

Write the subtraction problem.	$^-3$ $^-8$ ■ − ■
Change the problem to ■. addition	$^-3 +$ ■ $^+8$
Find the sum.	$^-3 + 8 =$ ■ 5

The answer to $^-3 - (^-8)$ is ■. 5

Practice

Rewrite each subtraction problem as an addition problem.

1. $5 - 4$
$5 + (^-4)$

2. $4 - (^-6)$
$4 + (^+6)$

3. $^-2 - 5$
$^-2 + (^-5)$

4. $^-6 - (^-1)$
$^-6 + (^+1)$

Subtract.

5. $5 - 4$ 1

6. $4 - (^-6)$ 10

7. $^-2 - 5$ $^-7$

8. $^-6 - (^-1)$ $^-5$

9. $7 - 8$ $^-1$

10. $3 - (^-1)$ 4

11. $^-2 - 2$ $^-4$

12. $^-5 - (^-5)$ 0

13. $^-3 - 0$ $^-3$

14. $^-9 - 8$ $^-17$

15. $^-10 - (^-11)$ 1

16. $20 - (^-20)$ 40

Share Your Understanding

17. Explain to a partner how to find $7 - 8$. Use the words *subtraction* and *sum* in your explanation. Have your partner explain how to find $7 - (^-8)$. First, write the subtraction problem. Change $7 - 8$ to $7 + (^-8)$. Find the sum, $^-1$. Change $7 - (^-8)$ to $7 + 8$. Find the sum, 15.

18. **CRITICAL THINKING** Pick an integer from $^-10$ to 10. Have a partner write a subtraction problem with that integer as the difference. Check your partner's work. Answers will vary.

More practice is provided in the Workbook, page 1, and in the Classroom Resource Binder, page 3.

1·4 Distance Between Two Points

Getting Started
Review absolute value as the distance between a point and zero on the number line. Have students find $|^-4|$ on a number line. Ask for a volunteer to give and explain his or her answer. [4]

You can find the distance between two points on a number line. You can count the number of spaces between the two points.

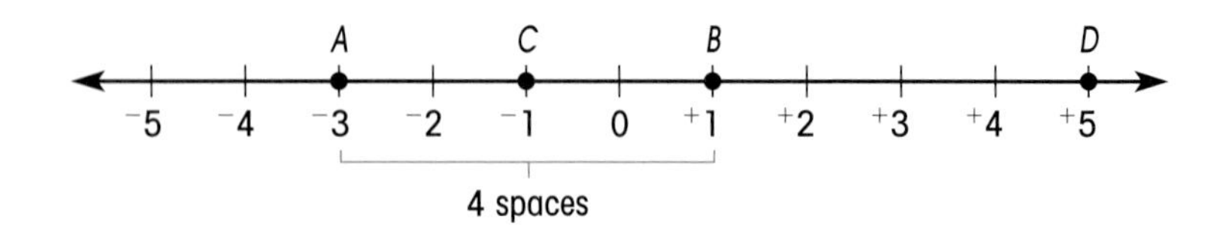

Look at the number line above. The distance between points *A* and *B* is 4.

You do not need to count spaces to find distance. You can use absolute values.

EXAMPLE

Use the number line above to find the distance between points *C* and *D*.

> **POSTULATE 3**
> The distance between points *C* and *D* is $|c - d|$ or $|d - c|$.

Match point *C* with an integer on the number line.	Point *C* is at $^-1$.
Match point *D* with an integer on the number line.	Point *D* is at 5.
Subtract the integers.	$^-1 - 5 = ^-1 + (^-5)$ $= ^-6$
Find the absolute value of the difference.	$|^-6| = 6$

The distance between points *C* and *D* is 6.

Math Fact
CD means the length of $\overline{CD}$.

The length of a line segment is the distance between the endpoints. The length of $\overline{CD}$ is 6. You write $CD = 6$.

Try This

Use the number line below to find the length of $\overline{EF}$.

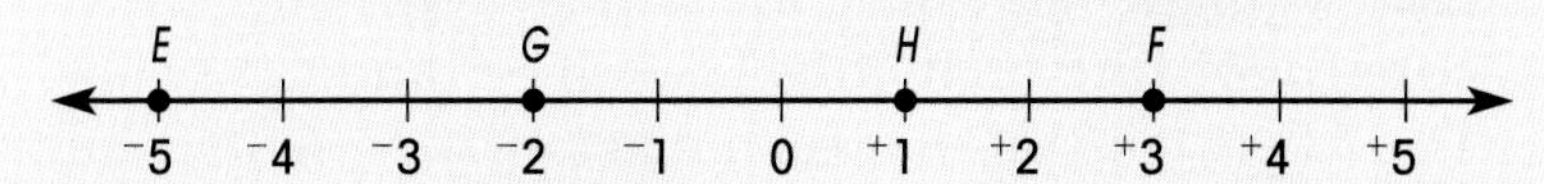

Match point ■ with an integer. E — Point E is at ⁻5.

Match point ■ with an integer. F — Point F is at ■. 3

■ the integers. subtract — ⁻5 − ■ = ⁻5 + ■ = ■ (3; (⁻3); ⁻8)

Find the ■ of the difference. absolute value — |■| = ■ (⁻8; 8)

EF = ■ 8

The length of $\overline{EF}$ is ■. 8

Practice

Use the number line below for exercises **1–7**.

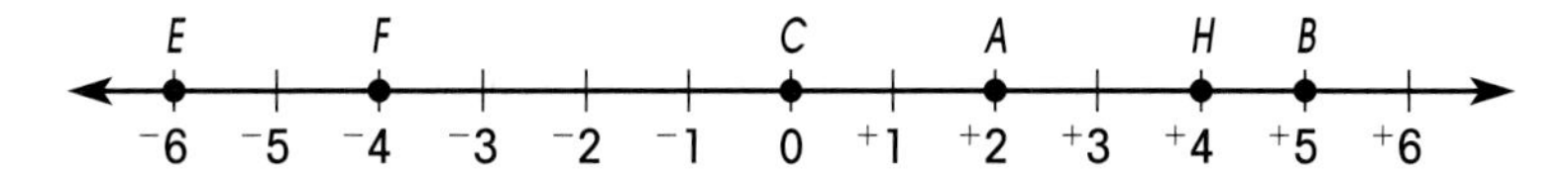

Find the distance between each set of points.

1. Points H and A 2

2. Points F and A 6

3. Points C and B 5

Find the length of each line segment.

4. $\overline{AB}$ 3

5. $\overline{AC}$ 2

6. $\overline{EC}$ 6

7. $\overline{EH}$ 10

Share Your Understanding

8. Choose one negative point and one positive point on a number line. Explain how to find the distance between these two points. To find the distance between two points, match each point with an integer. Subtract the integers. Find the absolute value of the difference.

9. **CRITICAL THINKING** Look at the number line on the right. Point A is at ⁻1. The distance between points A and B is 2. Where is point B on the number line? (Hint: There are two possible answers.) ⁺1 or ⁻3

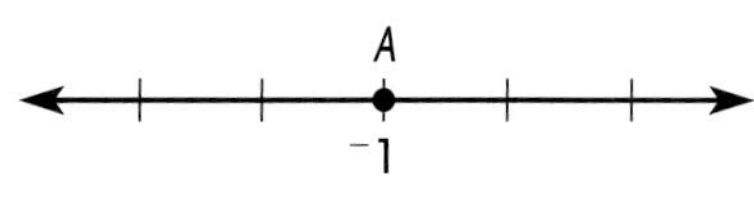

More practice is provided in the Workbook, page 2, and in the Classroom Resource Binder, page 4.

1·5 Congruent Line Segments

Getting Started
Have students use a ruler to measure the length of objects in the classroom. Discuss units of measure.

Line segments that have the same lengths are **congruent line segments**. Look at $\overline{AB}$ and $\overline{CD}$ below. The labels tell you that each diagram represents a line segment that is 2 centimeters long.

Math Fact
If $AB = CD$,
then $\overline{AB} \cong \overline{CD}$.

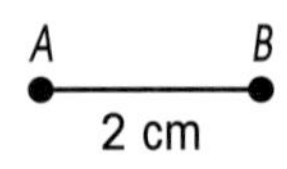

C D
2 cm

$\overline{AB}$ is congruent to $\overline{CD}$. The symbol $\cong$ means "is congruent to." You can write $\overline{AB} \cong \overline{CD}$.

A small mark on a diagram tells you that those line segments are congruent.

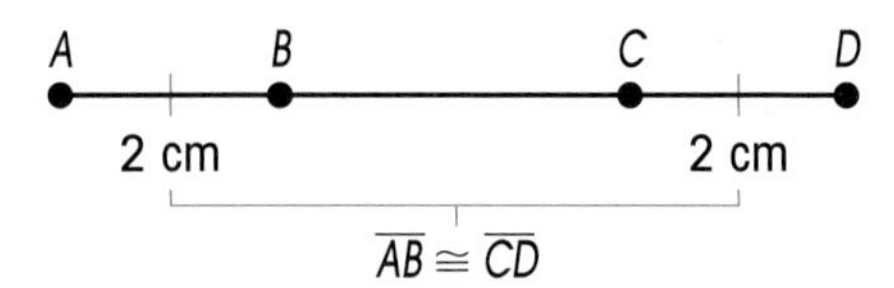

You can use what you know about congruent line segments to find an unknown value.

EXAMPLE

Find the value of x in the diagram.

J K L M
$x + 1$ 8

Math Fact
If $\overline{JK} \cong \overline{LM}$,
then $JK = LM$.

The marks on the diagram show that $\overline{JK} \cong \overline{LM}$. Write an equation.	$JK = LM$ $x + 1 = 8$
Subtract 1 from both sides of the equation.	$x + 1 - 1 = 8 - 1$ $x = 7$

Because $\overline{JK}$ is congruent to $\overline{LM}$, x is equal to 7.

Try This

Find the value of x in the diagram.

P Q R S
$2x - 3$ 15

The diagram shows $\overline{PQ} \cong$ ■. $\overline{RS}$

Then, $PQ =$ ■ RS

Write an equation.

$2x - 3 =$ ■ 15

Solve for x. Add 3 to both sides.

$2x - 3 +$ ■ $= 15 +$ ■ 3, 3

Divide both sides of the equation by 2.

$$2x = 18$$

$$\frac{2x}{2} = \frac{18}{2}$$

$x =$ ■ 9

Because $\overline{PQ}$ is ■ congruent to $\overline{RS}$, x is ■. 9

Practice

1. Identify the congruent line segments.
(Hint: Be sure the units are the same.) $\overline{AB}$ and $\overline{GH}$

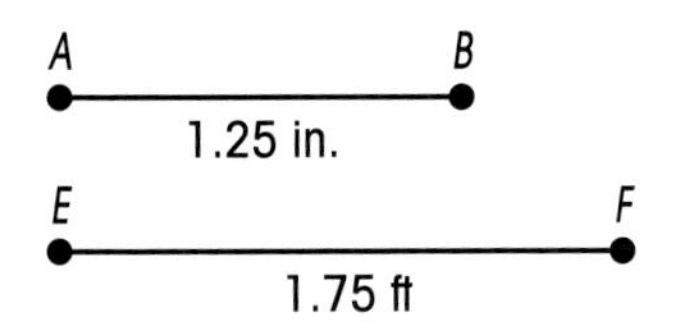

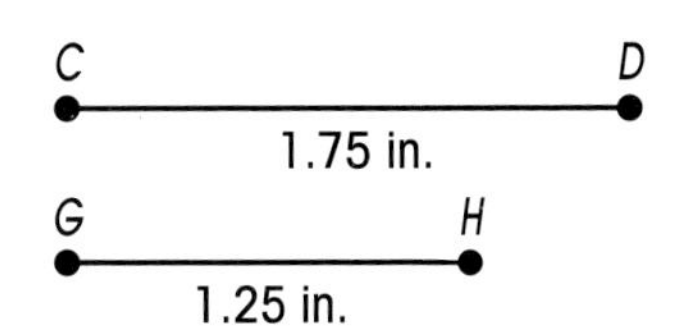

Find the value of x in each diagram.

2.

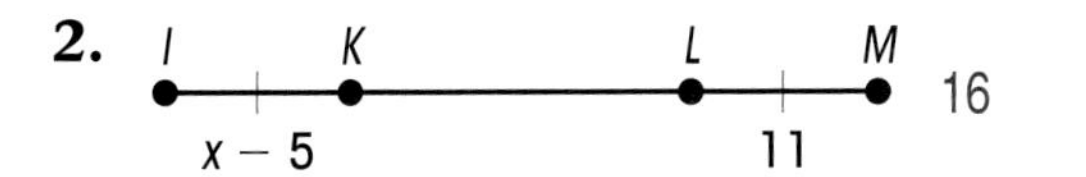

16

3.

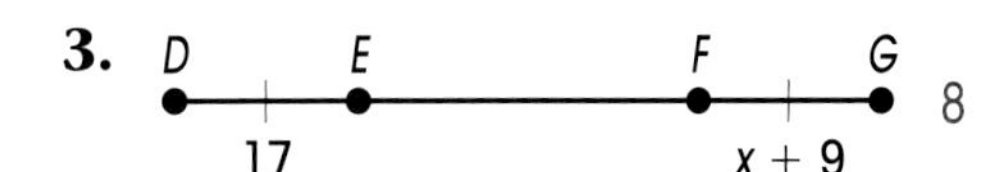

8

4.

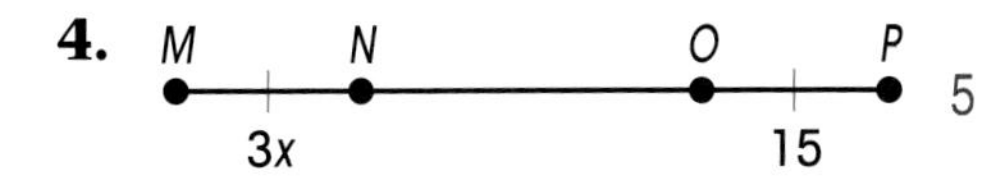

5

5.

3

6. A B C D
21 $5x - 9$ 6

6

7.

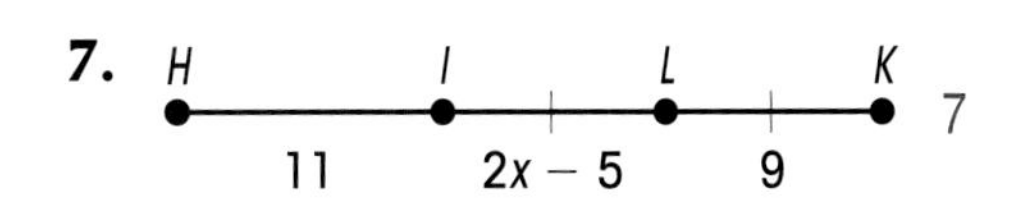

7

Try This

Find the length of $\overline{EF}$.

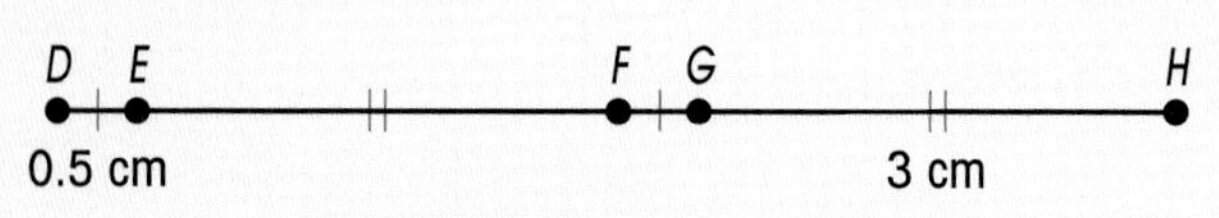

Find the segment that has the same number of marks as $\overline{EF}$.

$\overline{EF}$ has 2 marks.

$\overline{GH}$ ■ has 2 marks.

The same number of marks on the diagram means $\overline{EF} \cong$ ■. $\overline{GH}$

$EF =$ ■ GH

GH is 3 cm, then EF is ■. 3 cm

The length of $\overline{EF}$ is ■. 3 cm

Practice

Use the diagram below to complete exercises **8–11**.

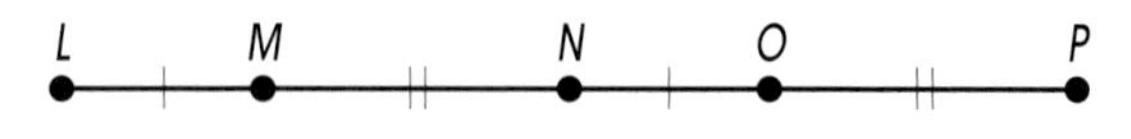

8. $\overline{LM} \cong$ ■ $\overline{NO}$ **9.** $\overline{MN} \cong$ ■ $\overline{OP}$ **10.** $\overline{NO} \cong$ ■ $\overline{LM}$ **11.** $\overline{OP} \cong$ ■ $\overline{MN}$

Use the diagram on the right to complete exercises **12** and **13**.

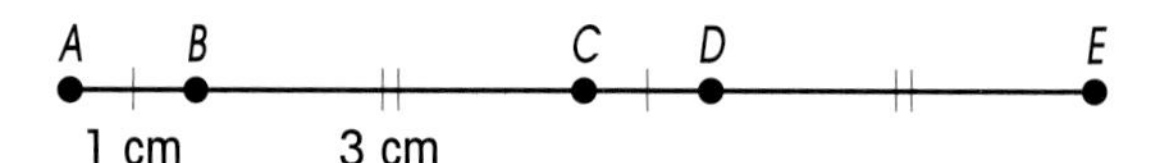

12. Find the length of $\overline{CD}$. 1 cm

13. Find the length of $\overline{DE}$. 3 cm

Share Your Understanding

14. Explain to a partner how you know two line segments are congruent. Use the words *same* and *lengths* in your explanation. Congruent line segments have the same lengths.

15. **CRITICAL THINKING** Look at $\overline{PQ}$ below. Find the length of $\overline{QR}$. (Hint: Write an equation and then solve for x.) 15

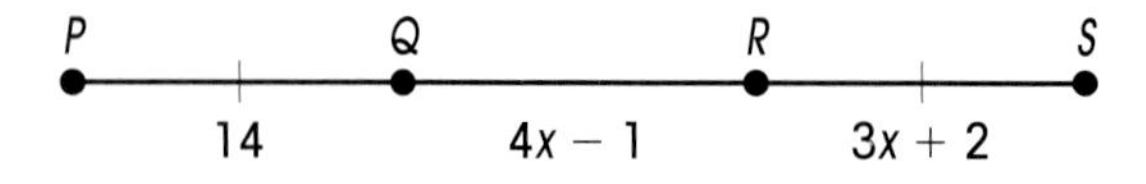

CONSTRUCTION
A Congruent Line Segment

You can use a compass to copy a line segment. The new line segment will be congruent to the given line segment. Follow the steps below to construct $\overline{DE}$ congruent to $\overline{AB}$.

STEP 1 Use a straightedge to draw a **ray** with **endpoint *D***.

STEP 2 Place one end of the compass on **point *A***. Open the compass until the pencil end is over **point *B***. Do not change this compass position.

STEP 3 Take the end of the compass that was on point A. Place it on **point *D***.

STEP 4 Use the pencil end of the compass to **draw an arc** that intersects the ray. An arc is part of a circle.

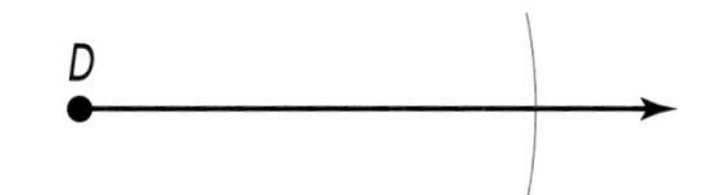

STEP 5 Label the point where the arc and the ray meet as **point *E***.

$\overline{DE} \cong \overline{AB}$

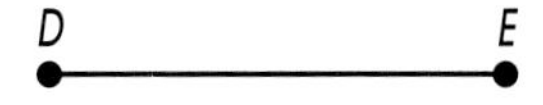

More practice is provided in the Workbook, page 3, and in the Classroom Resource Binder, page 5.

Getting Started
Write *midday*, *midway*, and *midterm* on the board and ask students what they mean. Then ask students what all the words have in common (*mid* meaning middle).

1·6 Midpoint of a Line Segment

The **midpoint** of a line segment is the point that divides the line segment into two congruent parts.

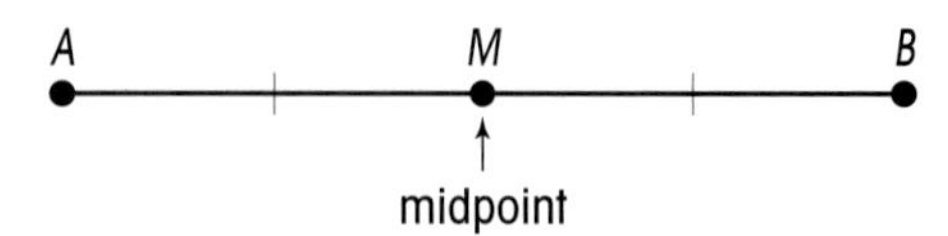

Look at $\overline{AB}$ above. Point M is the midpoint of $\overline{AB}$. $\overline{AM}$ and $\overline{MB}$ are congruent. $\overline{AM} \cong \overline{MB}$

You can use what you know about midpoints to find the length of line segments.

EXAMPLE 1

Point A is the midpoint of $\overline{DF}$. Find the length of $\overline{AF}$.

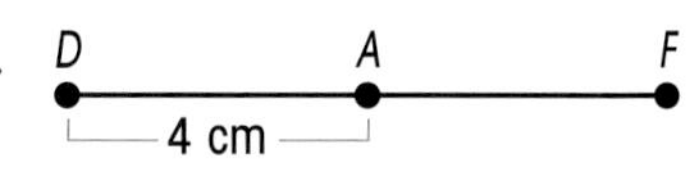

Because point A is the midpoint of $\overline{DF}$, $\overline{DA} \cong \overline{AF}$.

DA is 4 cm, so AF is 4 cm.

The length of $\overline{AF}$ is 4 cm.

Math Fact
A bisector can be a line, a line segment, or a ray.

Any line that intersects a segment at its midpoint is called a **bisector**.

EXAMPLE 2

Line n bisects $\overline{TR}$ at point S. Find the length of $\overline{TS}$.

Because line n bisects $\overline{TR}$ at point S, point S is the midpoint of $\overline{TR}$. $\quad TS = \frac{1}{2}TR$

Math Fact
You can use one letter to name a line, such as line n.

Look at the diagram to find TR. $\quad TR = 10$

Find $\frac{1}{2}$ of TR by dividing it by 2. $\quad TS = \frac{1}{2} \cdot 10$

$= 10 \div 2 = 5$

The length of $\overline{TS}$ is 5 in.

Try This

Point F is the midpoint of $\overline{EG}$.
Find the length of $\overline{EG}$.

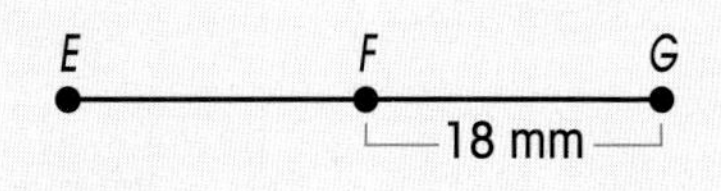

Because point F is the ■ of $\overline{EG}$, midpoint
$EG = 2 \cdot FG$. Look at the diagram to find FG. FG is ■. 18 mm

$EG = 2 \cdot FG$
$EG = ■ \cdot 18 = ■$ 2, 36

The length of $\overline{EG}$ is ■. 36 mm

Practice

Decide if point M is the midpoint of each line segment. Write *midpoint* or *not the midpoint.*

1. A M B — 2 cm, 2 cm
midpoint

2. R M S — 1 in., 3 in.
not the midpoint

3. P M Q — 2 mm, 6 mm
not the midpoint

Point B is the midpoint of $\overline{AC}$. Find the length of $\overline{AB}$ in each segment below.

4.
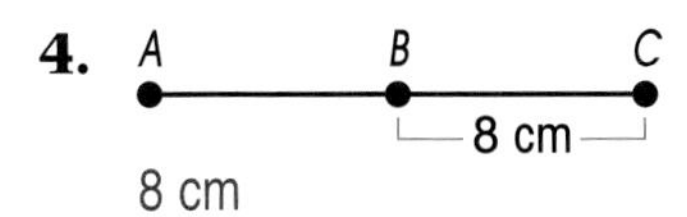

8 cm

5. A B C — 12 mm
12 mm

6.
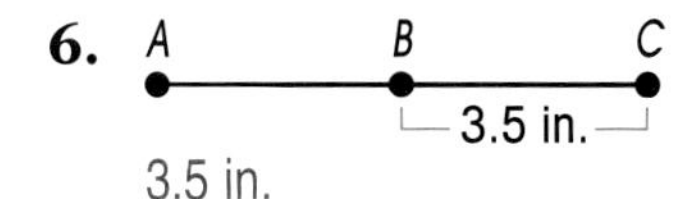

3.5 in.

7. A B C — 8 cm
4 cm

8.
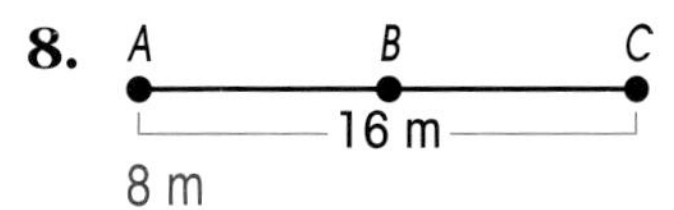

8 m

9.
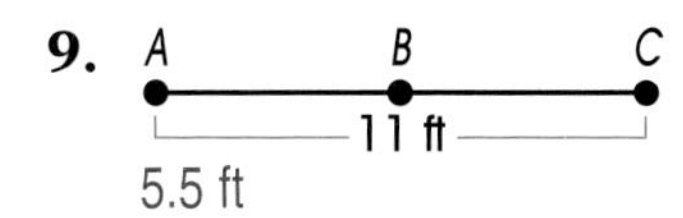

5.5 ft

Line a bisects $\overline{KM}$ at point L. Find the length of $\overline{KM}$ in each segment below.

10. 10 cm
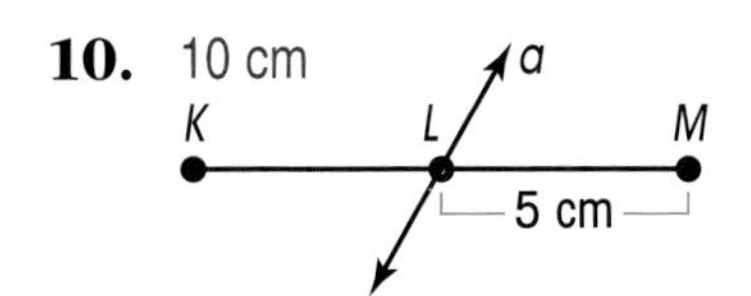

11. 32 in.
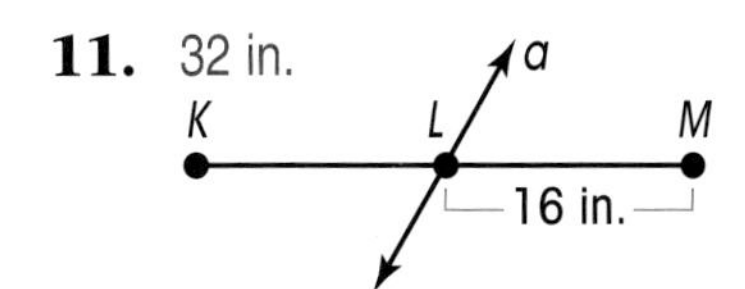

12. 8.4 mm
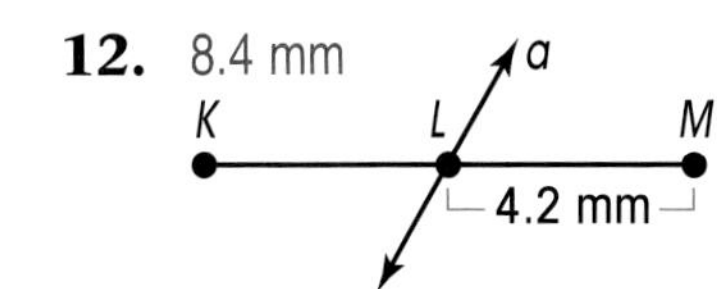

Try This

Point B is the midpoint of $\overline{AD}$.
Point C is the midpoint of $\overline{BD}$.
Find the length of $\overline{BC}$.

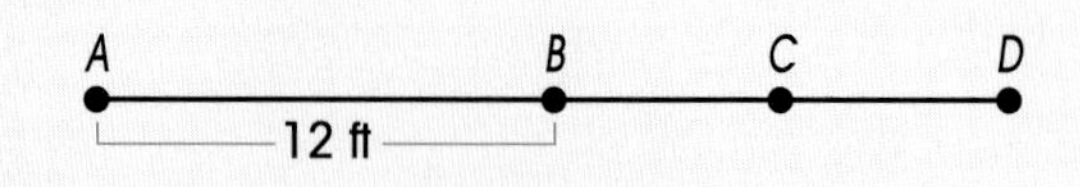

Because point B is the ■ of $\overline{AD}$, midpoint
$AB = BD$.

$AB = 12$
$BD = $ ■ 12

Because point C is the midpoint of $\overline{BD}$,
BC is $\frac{1}{2}$ of BD. Divide the length of $\overline{BD}$ by ■. 2

$BC = \frac{1}{2} \cdot$ ■ BD
$=$ ■ $\div 2 =$ ■ 6
12

The length of $\overline{BC}$ is ■. 6 ft

Practice

Point Q is the midpoint of $\overline{PS}$. Point R is the midpoint of $\overline{QS}$. Use the diagram on the right to complete exercises **13–16**.

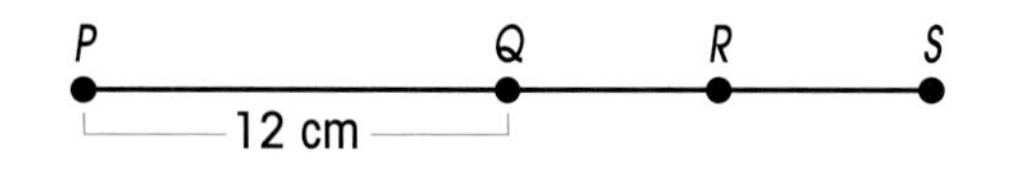

13. $QS =$ ■ 12 cm **14.** $QR =$ ■ 6 cm **15.** $PR =$ ■ 18 cm **16.** $PS =$ ■ 24 cm

Point L is the midpoint of $\overline{KM}$. Point K is the midpoint of $\overline{JM}$. Use the diagram on the right to complete exercises **17–20**.

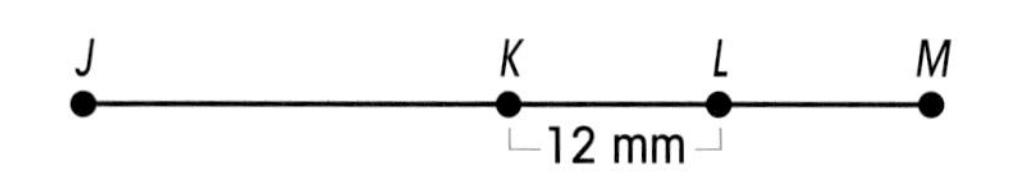

17. $LM =$ ■ 12 mm **18.** $KM =$ ■ 24 mm **19.** $JK =$ ■ 24 mm **20.** $JL =$ ■ 36 mm

Share Your Understanding

21. Explain to a partner how to find the midpoint of a line segment if you are given the length of the whole line segment. Use the words *length, whole,* and *divide* in your explanation.
Divide the length of the whole line segment by 2. Count that many units from an endpoint to locate the midpoint.

22. **CRITICAL THINKING** A line bisects $\overline{JK}$ at point P. The length of $\overline{JP}$ is 9 cm. Find the length of $\overline{JK}$. (Hint: Draw a picture first.) 18 cm

CONSTRUCTION
A Bisector to a Line Segment

Follow the steps below to construct line ℓ so that it bisects $\overline{DF}$ at point E.

STEP 1 Use a straightedge to draw a line segment. Label the line segment $\overline{DF}$.

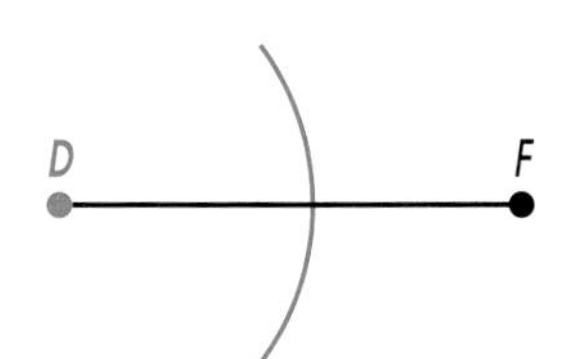

STEP 2 Place one end of the compass on point D. Open the compass until the pencil end is past the midpoint of $\overline{DF}$. Draw a large arc that intersects $\overline{DF}$. Do not change this compass position.

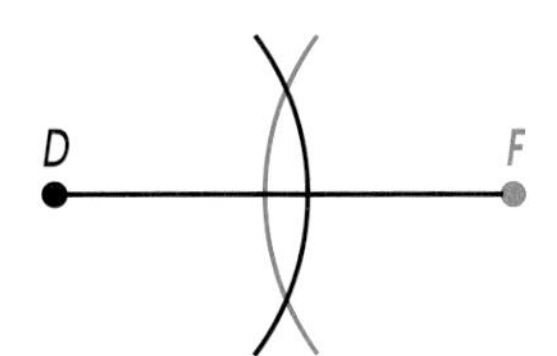

STEP 3 Take the end of the compass that was on point D. Place it on point F. Draw another large arc that intersects $\overline{DF}$. Make sure that the second arc intersects the first arc both above and below the line segment.

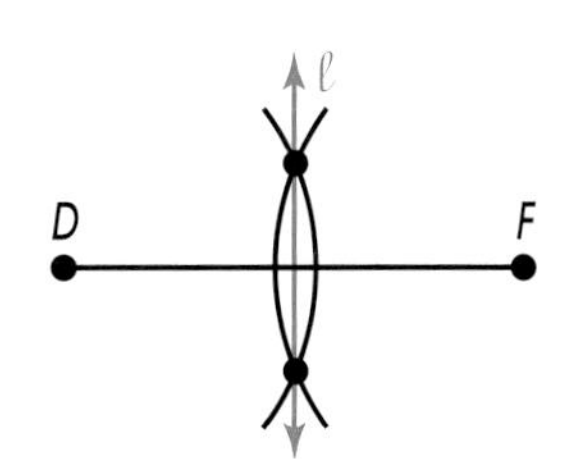

STEP 4 Use a straightedge to draw a line that connects the points where the two arcs intersect above and below the line segment. Label the line ℓ.

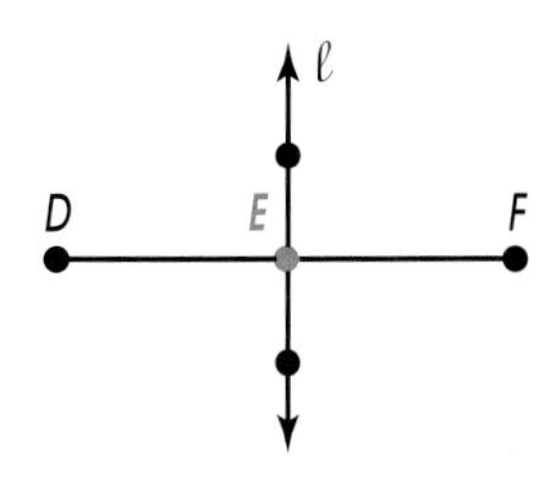

STEP 5 Label the point where line ℓ intersects $\overline{DF}$ as point E.

Line ℓ is the bisector of $\overline{DF}$.

1-7 Calculator: Common Segments

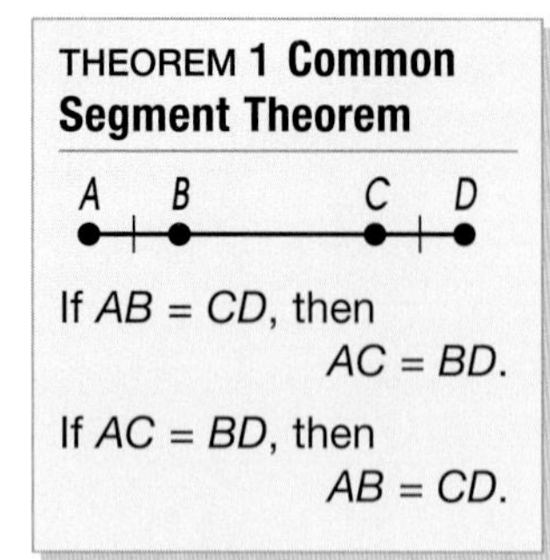

You can use what you know about congruent line segments to find the length of line segments.

Look at $\overline{AD}$ above. $\overline{BC}$ is a common segment because it is part of $\overline{AC}$ and part of $\overline{BD}$.

EXAMPLE 1

Find the length of $\overline{AC}$.

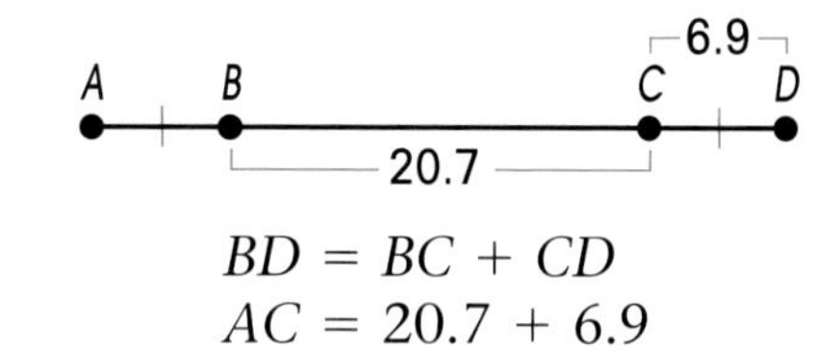

Write an equation to find the length of $\overline{AC}$. Since $AB = CD$, $AC = BD$.

$$BD = BC + CD$$
$$AC = 20.7 + 6.9$$

Use your calculator to add.

PRESS: 6 . 9 + 2 0 . 7 =

Display 27.6

The length of $\overline{BD}$ is 27.6. So the length of $\overline{AC}$ is 27.6.

EXAMPLE 2

Find the length of $\overline{LM}$.

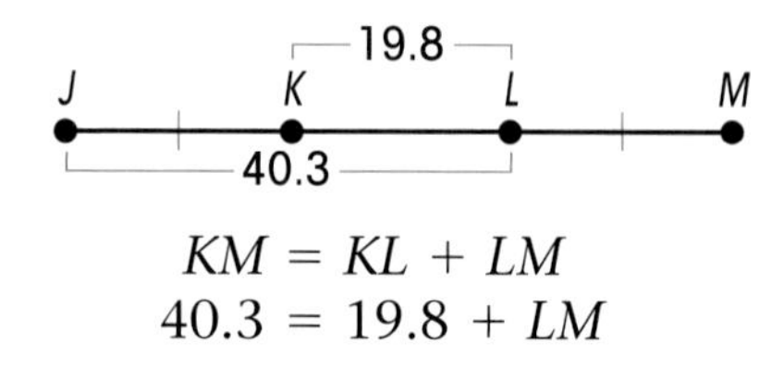

Write an equation to find the length of $\overline{LM}$. Since $JK = LM$, $JL = KM$.

$$KM = KL + LM$$
$$40.3 = 19.8 + LM$$

Subtract 19.8 from both sides to find LM.

$$40.3 - 19.8 = 19.8 - 19.8 + LM$$

Use your calculator to subtract.

PRESS: 4 0 . 3 − 1 9 . 8 =

Display 20.5

The length of $\overline{LM}$ is 20.5.

Practice

Find the length of $\overline{AC}$ in each diagram.

1.

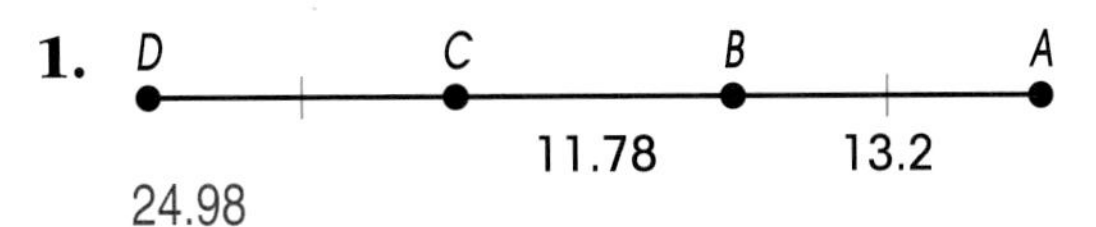

2.

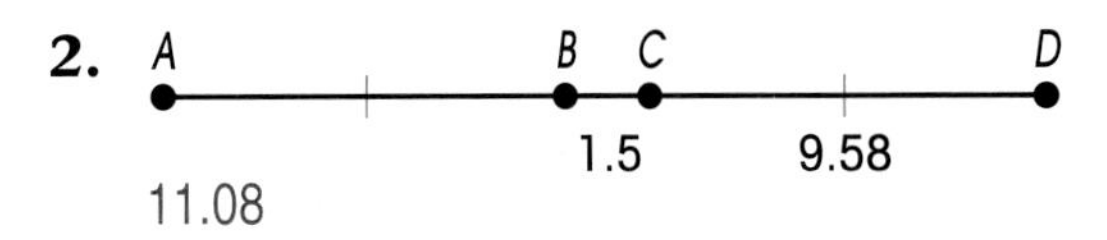

3.

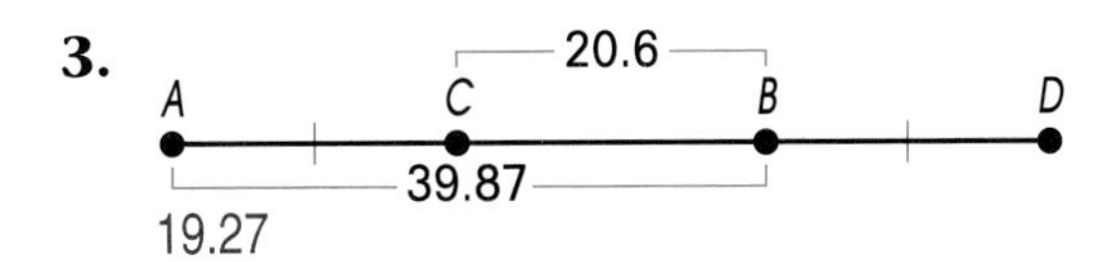

4.

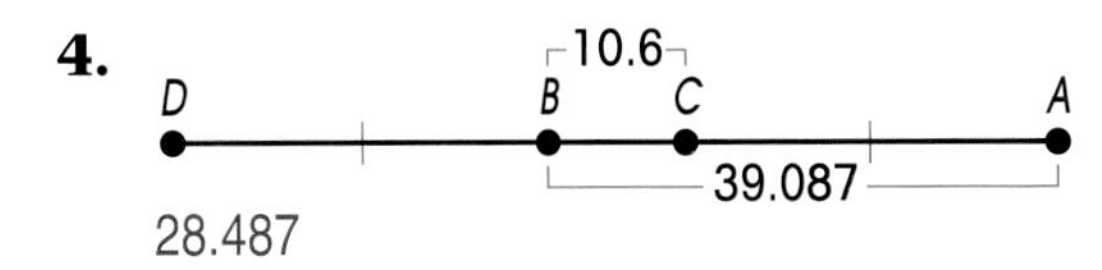

5.

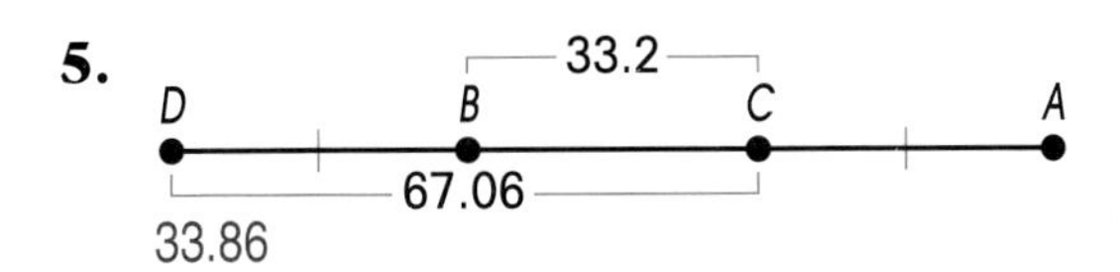

6.

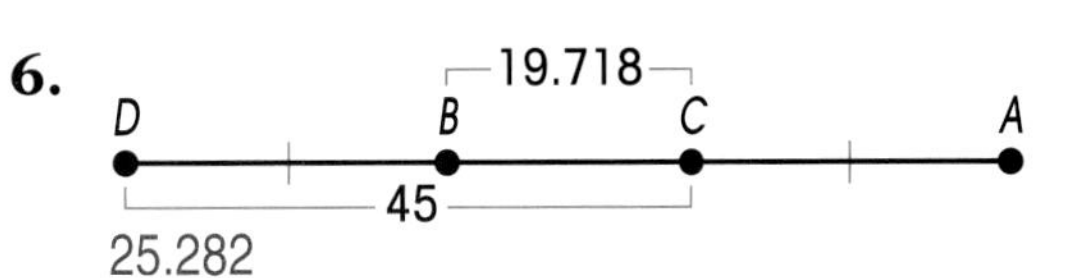

Math In Your Life

LINES OF SYMMETRY

A line of symmetry bisects a figure into two congruent parts. Many things in the world around you have a line of symmetry. Look at the bug on the right. There is a vertical line of symmetry.

Trace the bug. Cut out the tracing. Then, fold it along the line of symmetry. Each side should match.

Vertical Line of Symmetry

Some things have one line of symmetry like the bug. Other things have more than one line of symmetry. Draw, trace, or photograph things around you that have lines of symmetry. Write about the lines of symmetry you find.

More practice is provided in the Workbook, page 4, and in the Classroom Resource Binder, page 6.

1-8 Problem-Solving Strategy: Draw a Diagram

Getting Started
Have three students stand in a line an arm's length apart. Describe each student as a point on a line. Have students decide which point is between the other two points. Students may rearrange themselves and repeat the activity.

Points that are on the same line are collinear. Points A, B, and C are **collinear points**.

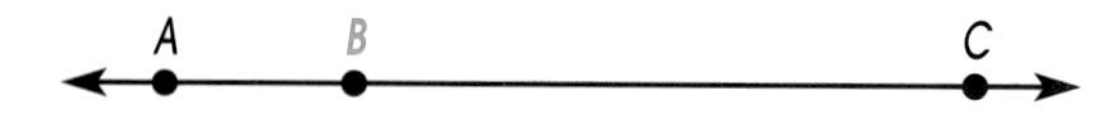

Point B is between points A and C because

Point B is on $\overleftrightarrow{AC}$ and $AB + BC = AC$.

EXAMPLE

Jo delivers mail to points L, R, and T along a straight road. The points are not in that order. The distance from L to T is 15 miles. The distance from L to R is 40 miles. Find the distance Jo drives to deliver the mail. There are two possible answers.

READ **What do you need to find out?**
Find the distance Jo drives.

PLAN **What do you need to do?**
Decide where points L, R, and T are on the road. Draw a line with points L, R, and T.

DO **Follow the plan.**
Draw a line for each of the possible answers.

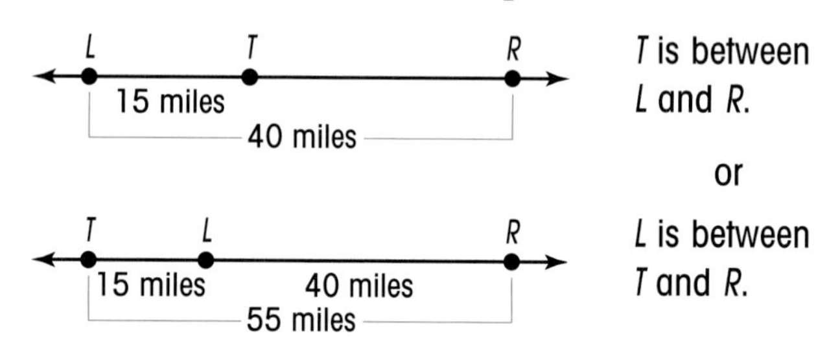

Jo could drive 40 miles or 55 miles.

CHECK **Does your answer make sense?**
Does the information in the diagram match the information in the word problem? Yes. ✓

Try This

A bus travels along a straight road. It makes stops at points *C*, *K*, *M*, and *P*. The stops are not in that order. Point *C* is between *M* and *K*. Point *P* is between *C* and *K*. Which point is the third stop along the route? There are two possible answers.

Draw a diagram of the bus route. First, draw point C between points *M* and *K*. There are two possible diagrams.

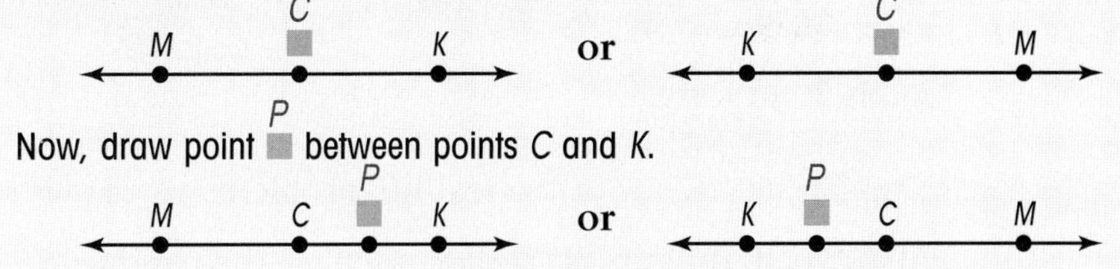

Now, draw point P between points *C* and *K*.

Find the third point in each diagram. The third point is the third bus stop.

The third stop on the bus route could be point P or point *C*.

Practice

Solve each problem. Be sure to draw a diagram.

1. A bus makes stops at points *G*, *N*, and S along a straight road. The points are not in that order. Point *G* is between points *N* and *S*. It is 20 miles from point N to point *S*. The distance from point *G* to point *S* is 14 miles. What is the distance from point *N* to point *G*? 6 miles

2. Sam delivers mail to points *B*, M, and *P* along a straight road. The points are not in that order. The distance from point *B* to point *P* is 5 miles. The distance from point *M* to point *P* is 12 miles. Find the distance from point *M* to point *B*. There are two possible answers. 7 miles or 17 miles

3. A bus travels along a straight road. It makes stops at points *D*, *J*, *Q*, and *V*. The stops are not in that order. Point *J* is between points *D* and *Q*. Point *V* is between points *J* and *Q*. Which point is the second stop along the route? There are two possible answers. point *J* or point *V*

More practice is provided in the Workbook, page 5, and in the Classroom Resource Binder, page 7.

1·9 Problem-Solving Application: Adding Line Segments

Getting Started Remind students that they can add to find the total length. They can subtract to find a part of a line segment.

> POSTULATE 4 **Segment Addition Postulate**
>
> If three points are on the same straight line and point B is between points A and C, then $AB + BC = AC$.

You know that a line segment can be broken into pieces. The length of each piece can be added together to get the total length of the complete line segment.

Look at $\overline{AC}$.

A B C
2 cm 5 cm

$AB + BC = 2 + 5$

$\overline{AC}$ is 7 cm long.

You can use this to solve real-life problems.

EXAMPLE 1

Al drove 30 miles from Lodi to Atco on the parkway. After lunch he drove 36 miles from Atco to Rio on the Parkway. How many miles did he drive from Lodi to Rio?

Draw and label a diagram with the given information.

Lodi Atco Rio
30 miles 36 miles

To find the total distance, add the two short distances.

Lodi to Atco = 30
Atco to Rio = 36
$30 + 36 = 66$

Al drove 66 miles from Lodi to Rio on the parkway.

EXAMPLE 2

A plumber needs to cut one piece of pipe into two pieces. The total length of the pipe is 36 inches. He needs to cut off 24 inches. How much pipe is left?

Draw and label a diagram with the given information.

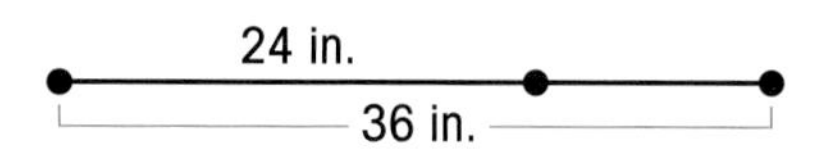

Find the length of pipe that is left. Subtract the length of the piece that was cut from the total length.

Total length = 36 inches
Piece of pipe = 24 inches
$36 - 24 = 12$

There are 12 inches of pipe left.

Try This

Hilda has 51 inches of ribbon. She needs to cut this ribbon into two pieces. One piece of ribbon should be twice as long as the other. How many inches will the longer piece of ribbon be?

Draw and label a diagram.

x — $2x$

Let x = the ■ shorter piece of ribbon

Let $2x$ = ■ 2 times the shorter piece of ribbon

The total length of both pieces of ribbon is 51 in.

Write an equation. $x + 2x = 51$

Solve for x. $3x = 51$

$\frac{3x}{3} = \frac{■}{3}$ 51

$x =$ ■ 17

The shorter piece of ribbon is 17 in. Multiply this by 2 to find the length of the longer piece of ribbon. ■ 2 $\cdot$ 17 = 34

The longer piece of ribbon is ■ 34 in. long.

Practice

Solve each problem. Show your work.

1. Altogether, Alicia has to drive 104 miles on the turnpike. Before lunch, she drives 74 miles. How many more miles does she have to drive on the turnpike? 30 miles

2. A woodcutter has to cut a piece of wood into two pieces. One piece has to be twice as long as the other piece. The total piece of wood is 45 cm long. How long will the shorter piece of wood be? 15 cm

3. A carpenter cuts a piece of lumber into two pieces. One piece is twice the size of the other. If the smaller piece is 16 in. long, how long was the whole piece of lumber before it was cut? 48 in.

An alternate two-column proof lesson is provided on page 406 of the student book.

Getting Started
Review the definition of *midpoint* with the students.

1•10 Proof: Line Segments

You can use information you know is true to prove a **theorem**. You begin with a fact that is given. Then, you use only what you know is true to prove the statement. Sometimes this takes several steps.

THEOREM 2
Midpoint Theorem
The midpoint M of $\overline{AB}$ divides $\overline{AB}$ in half so that $AM = \frac{1}{2}AB$.

You can use what you know about midpoints to prove the Midpoint Theorem. You can write a proof in paragraph form.

EXAMPLE

You are given:
M is the midpoint of $\overline{AB}$.

A M B

Prove: $AM = \frac{1}{2}AB$

Begin with what you are given. Then, step by step use reasoning to reach the statement you need to prove.

You are given that M is the midpoint of $\overline{AB}$. You know that the midpoint divides a line segment into two congruent parts. Because $AM = MB$, you can write $AB = AM + AM$. You can then combine the like terms on the right side of this equation. So, $AB = 2AM$. Divide both sides of this equation by 2. Then, $\frac{1}{2}AB = AM$ or $AM = \frac{1}{2}AB$.

So, you prove that $AM = \frac{1}{2}AB$. ✓

Math Fact
Substitute means "to replace."

There may be more than one way to write this proof. Only use information you know is true. The statements you write should be in logical order.

Try This

Copy and complete the proof.

You are given: $\overline{AC} \cong \overline{BD}$

Prove: $\overline{AB} \cong \overline{CD}$

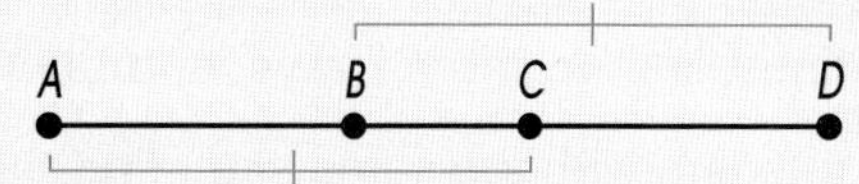

Look at the diagram. Look at what you need to prove. To prove $\overline{AB} \cong \overline{CD}$, you need to take away the common segment. Begin with the given.

You are given that $\overline{AC} \cong$ ■. The measures of these congruent line segments are equal. $AC =$ ■. You can rename AC and rename BD. $AC = AB + BC$ and $BD = BC + CD$. So, $AB + BC = BC + CD$. Now subtract BC from both sides of this equation. Then, $AB =$ ■. This means that $\overline{AB} \cong$ ■.

So, you prove that ■. ✓

$\overline{BD}$ BD CD $\overline{CD}$ $\overline{AB} \cong \overline{CD}$

Practice

Copy and complete the proof.

You are given:

Line t bisects $\overline{MN}$ at point P.

Prove: $MN = 2MP$

You are given that line t __bisects__ __MN__ at point P. Line t divides MN into __two__ congruent line segments. Then, $MN = 2$ __MP__.

So, you prove that _____. ✓ $MN = 2MP$

More Practice is provided in the Classroom Resource Binder.

Chapter 1 Review

Summary

Points, lines, line segments, rays, and planes are basic geometric figures.
To add integers with the same sign, find the sum of the absolute values. Then, use the sign of the addends.
To add integers with different signs, find the difference of the absolute values. Use the sign of the addend with the greater absolute value.
To subtract an integer, add its opposite.
To find the distance between two points on a number line, find the absolute value of the difference between the two points.
Congruent line segments have the same length.
A midpoint and a bisector divide a segment in half.
A common segment is part of two different line segments.
You can draw a diagram to solve distance problems.
You can use segment addition to solve measurement problems.
You can use facts you know about line segments to prove a statement.

More vocabulary review is provided in the Classroom Resource Binder, page 1.

point
line segment
ray
congruent line segments
midpoint
collinear points
bisector

Vocabulary Review

Complete the sentences with words from the box.

1. Points on the same line are ____. collinear points
2. A ____ is part of a line that has two endpoints. line segment
3. Line segments with the same length are ____. congruent line segments
4. A point is a location in space.
5. A ____ is part of a line that has one endpoint. ray
6. A point that separates a line segment into two congruent line segments is called a ____. midpoint
7. A line that intersects a segment at its midpoint is a ____. bisector

Chapter Quiz **More assessment** is provided in the Classroom Resource Binder.

Name each figure.

1. M L — ray ML

2. J C — line segment JC

3. 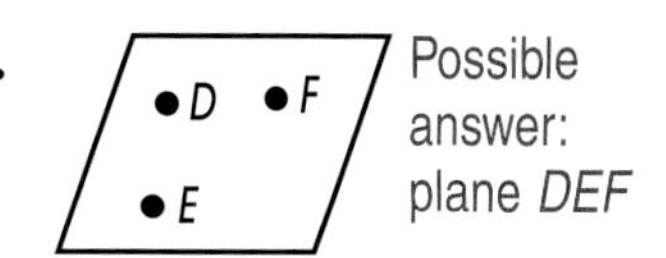

Possible answer: plane DEF

Add or subtract.

4. $^-9 + (^-3)$ $^-12$ **5.** $19 + (^-4)$ 15 **6.** $7 - (^-5)$ 12 **7.** $^-6 - 8$ $^-14$

Find the distance between each set of points.

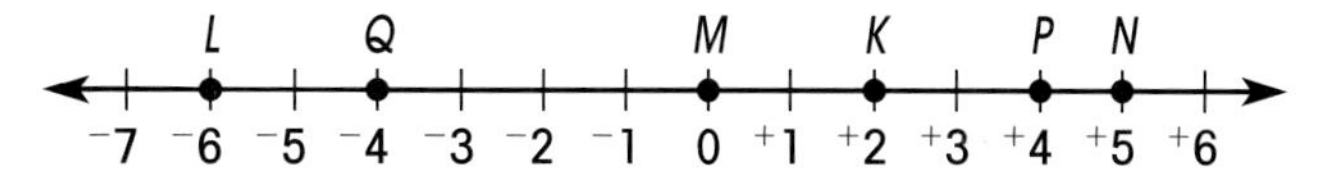

8. Points Q and L 2

9. Points N and M 5

Use the diagrams on the right for exercises 10–12.

10. Solve for x. 3

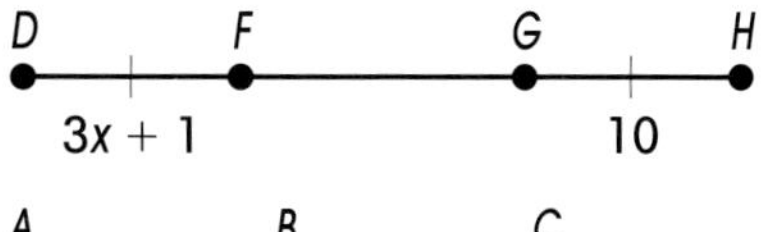

11. Point B is the midpoint of $\overline{AC}$. Find the length of $\overline{AB}$. 5 m

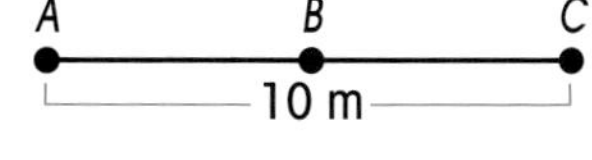

12. Find the length of $\overline{LM}$. 6 cm

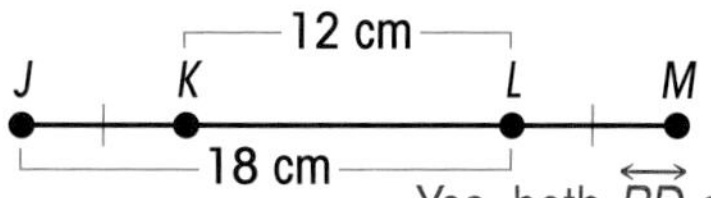

Problem Solving

Solve each problem. Show your work.

13. Points E, P, and S are collinear. The points are not in that order. The distance from E to S is 10 m. The distance from P to S is 24 m. Find the distance from P to E. 14 m or 34 m

14. A wire 90 cm long is cut into two pieces. One piece is twice as long as the other piece. How long is the shorter piece of wire? 30 cm

Write About Math

Do symbols $\overleftrightarrow{PD}$ and $\overleftrightarrow{DP}$ name the geometric figure below? Explain why or why not.

P D

Yes, both $\overleftrightarrow{PD}$ and $\overleftrightarrow{DP}$ name the line. The order of the points when naming a line does not matter.

Additional Practice for this chapter is provided on page 393 of the student book.

Chapter 2 Angles

Have you ever noticed that angles are found in ordinary places—even bridges? Look at the photograph. Where do you see angles?

Learning Objectives

- Use a protractor to measure and draw angles.
- Classify angles.
- Solve equations with parentheses.
- Use the Angle Addition Postulate.
- Find complementary and supplementary angles.
- Find the measure of congruent angles.
- Identify vertical angles and angle bisectors.
- Find the measure of a common angle.
- Make a table to solve a real-life problem.
- Use angles and direction to solve problems.
- Write proofs about angles.

ESL/ELL Note Have students make flash cards for acute, right, and obtuse angles. The front should have two examples of the angle with measures. The back should have the type of angle and its definition.

Words to Know

angle	a geometric figure formed by two rays with a common endpoint
protractor	a tool used to measure angles
acute angle	an angle that is greater than 0° and less than 90°
right angle	an angle that is equal to 90°
obtuse angle	an angle that is greater than 90° and less than 180°
straight angle	an angle that is equal to 180°
adjacent angles	two angles that have a common vertex and a common ray
complementary angles	a pair of angles whose measures have a sum of 90°
supplementary angles	a pair of angles whose measures have a sum of 180°
congruent angles	angles that have the same measure
vertical angles	angles formed by intersecting lines
angle bisector	a ray that divides an angle into two congruent angles

Angle Project

Make a collection of angles. Cut cardboard into strips. Use a paper fastener to attach two cardboard strips together to form an angle. The inside edge of each strip forms the angle. Measure and classify each angle. Be sure to have an example of each type of angle.

Make a complementary angle for each acute angle. Make a supplementary angle for each acute angle, right angle, and obtuse angle.

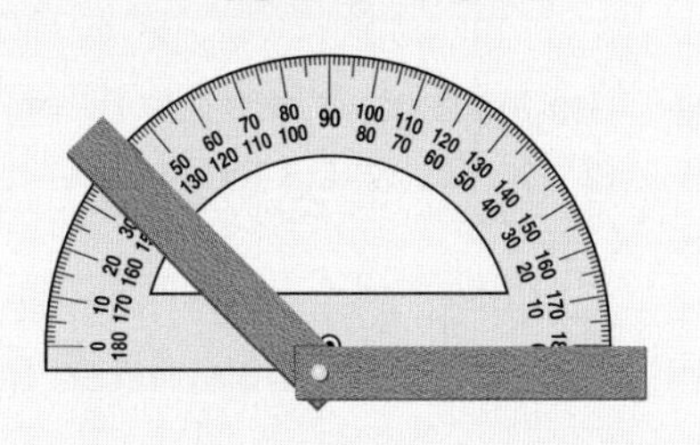

Project
Students can begin work on the project after completing Lesson 2.5. See the Classroom Resource Binder for a scoring rubric to assess this project.

More practice is provided in the Workbook, page 6, and in the Classroom Resource Binder, page 12.

2·1 Using a Protractor

Getting Started Point out the two scales on a protractor. Have students practice placing the protractor on angles in different positions.

An **angle** is formed by two rays with a common endpoint. The endpoint is called the vertex of the angle.

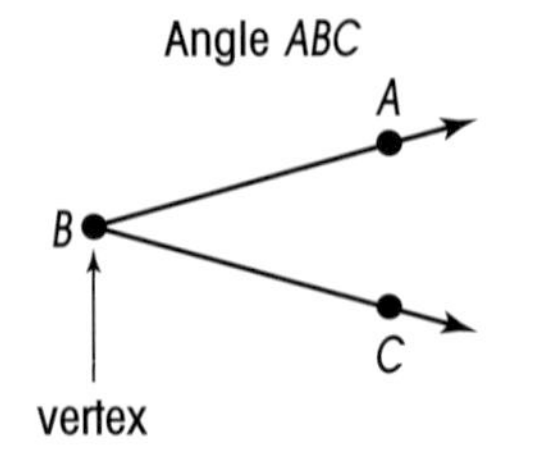

You can use the symbol $\angle$ to name an angle. The angle on the left is $\angle ABC$ or $\angle CBA$. The middle letter of the name is the vertex. You can also use the vertex alone to name this angle as $\angle B$.

You can measure the space between the two rays of an angle with a **protractor**. An angle is measured in degrees. The symbol for degrees is °.

EXAMPLE

Find the measure of $\angle DEF$.

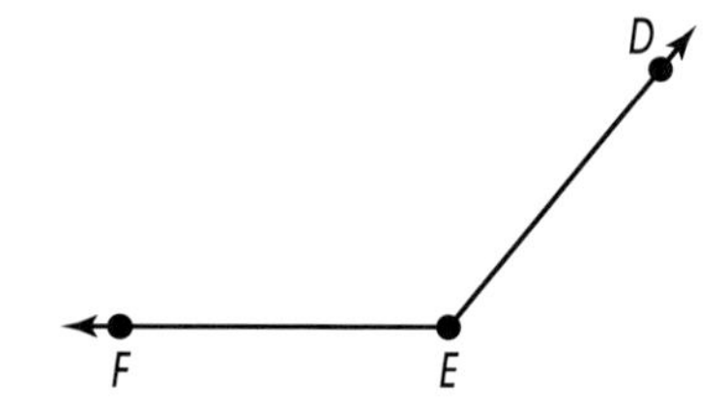

Place the center of the protractor's straight edge on the vertex. Be sure you can see the vertex through the hole in the center of the protractor. One ray must pass through 0° on the protractor.

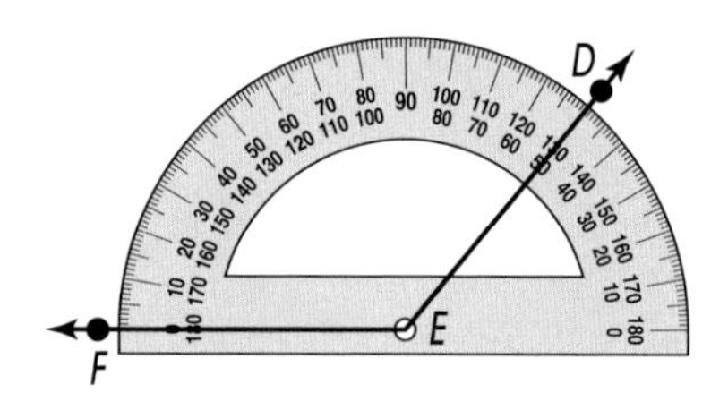

Use the scale that reads 0° on the first ray. Read the number of degrees where the second ray crosses the protractor.

$\angle DEF$ is 130°. The measure of $\angle DEF$ is 130°. This can also be written as $m\angle DEF = 130$.

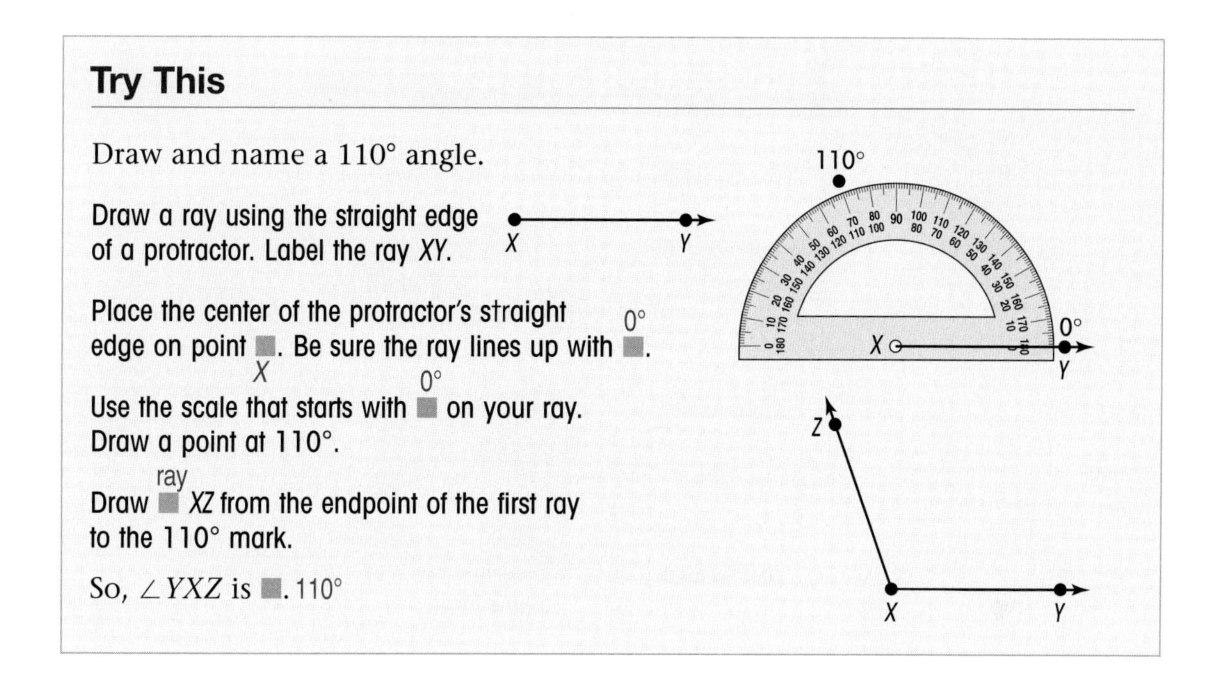

Try This

Draw and name a 110° angle.

Draw a ray using the straight edge of a protractor. Label the ray XY.

Place the center of the protractor's straight edge on point ■ X. Be sure the ray lines up with ■ 0°.

Use the scale that starts with ■ 0° on your ray. Draw a point at 110°.

Draw ■ ray XZ from the endpoint of the first ray to the 110° mark.

So, $\angle YXZ$ is ■. 110°

Practice

Look at the protractor on the right.
Find the measure of each angle.

1. $\angle EAD$ 30°

2. $\angle BAD$ 150°

3. $\angle EAB$ 180°

4. $\angle EAC$ 90°

5. $\angle BAF$ 25°

6. $\angle EAF$ 155°

Draw and name each angle. Check students' work.

7. 165°

8. 60°

9. 90°

Share Your Understanding

10. Explain to a partner how you know which scale to read on a protractor when you measure an angle. Use *scale, ray,* and *0°* in your explanation. Use the scale on the protractor that starts with 0° on your ray.

11. **CRITICAL THINKING** Draw an angle. Have a partner measure the angle. Then, draw an angle that is 5° larger. Answers will vary.

More practice is provided in the Workbook, page 6, and in the Classroom Resource Binder, page 12.

Classifying Angles

Getting Started
Students can place the corner of an index card on an angle to determine if the angle is less than, equal to, or greater than 90°.

Math Fact
When you see the symbol ⌐ in an angle, it means that the angle is a right angle.

The chart below shows the four basic types of angles.

Type of Angle	Size of Angle	Example of Angle
Acute Angle	Greater than 0° and less than 90°	D, E, F
Right Angle	Equal to 90°	E, G, F
Obtuse Angle	Greater than 90° and less than 180°	H, I, J
Straight Angle	Equal to 180°	K, L, M

To classify an angle, find its measure. Then, find it in the chart above.

EXAMPLE

Classify $\angle JKL$ as acute, right, obtuse, or straight.

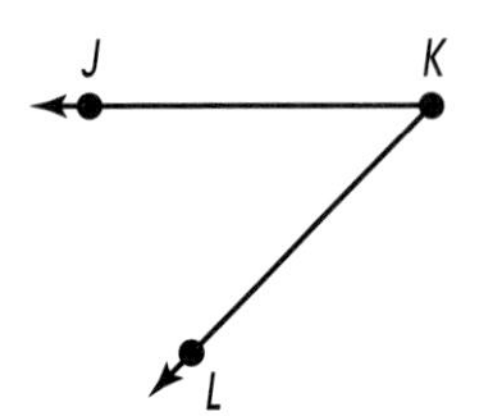

First, measure the angle.

The angle is 45°.
45° is greater than 0° and less than 90°.

So, $\angle JKL$ is an acute angle.

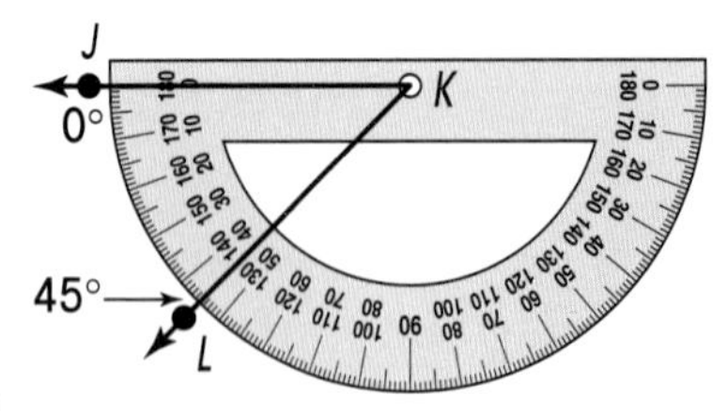

Try This

Draw an obtuse angle. Then, name the angle.

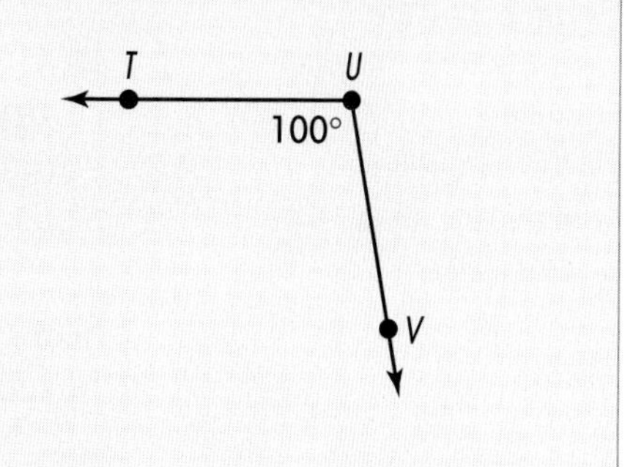

Decide on a measure for your obtuse angle.
An obtuse angle is greater than ■ [90°] and less than ■ [180°].

Suppose you choose an angle that is 100°.

Use the protractor to draw your obtuse ■. [angle]
Label the angle with three ■. [points]

∠TUV is ■ [100°]. So, ∠TUV is an ■ [obtuse] angle.

Practice

Classify each angle. Write *acute, right, obtuse,* or *straight.*

1. 110° obtuse

2. right

3. 50° acute

4. 15° acute

5. 180° straight

6. 105° obtuse

Draw each angle. Then, name the angle. Check students' work.

7. acute **8.** obtuse **9.** right **10.** straight

Share Your Understanding

11. Explain to a partner how you know an angle is obtuse. Use the words *greater than* and *less than* in your explanation.
An obtuse angle is greater than 90° and less than 180°.

12. **CRITICAL THINKING** The measure of an obtuse angle is twice the measure of an acute angle. What could be the measure of the acute angle? (Hint: There is more than one possible answer.) greater than 45° but less than 90°

More practice is provided in the Classroom Resource Binder, page 13.

2-3 Algebra Review: Solving Equations With Parentheses

Getting Started
Give students the following equations to solve. Review inverse operations.

$x - 1 = 24$ [25]
$4x = 24$ [6]
$6 + x = 24$ [18]

In Chapter 1, you solved equations to find the length of line segments. In this chapter, you will solve equations to find the measure of angles.

Sometimes you need to solve equations that have parentheses.

EXAMPLE 1

Solve for x. $2x - (x - 24) = 90$

Math Fact
Like terms are terms that have the same variable with the same exponent. $2x$ and x are like terms.

Undo the parentheses. $2x - (x - 24) = 90$

Combine like terms. $2x - x + 24 = 90$

Subtract 24 from both sides of the equation. $x + 24 - 24 = 90 - 24$

$x = 66$

The solution is $x = 66$.

Sometimes you need to solve equations that have a variable on both sides of the equal sign.

EXAMPLE 2

Solve for x. $4x = 80 + 2x$

Subtract $2x$ from both sides. $4x - 2x = 80 + 2x - 2x$

Divide both sides by 2. $\frac{2x}{2} = \frac{80}{2}$

$x = 40$

The solution is $x = 40$.

Try This

Solve for x. $90 - (x + 10) = 3x$

Undo the ■. parentheses $90 - x - 10 = 3x$

Combine the numbers. $80 - x = 3x$

Add x to both sides of the equation. $80 - x + x = 3x + x$

Simplify. ■ $= 4x$ 80

Divide both sides by ■. 4 $\frac{80}{4} = \frac{4x}{4}$

20 ■ $= x$

The solution is $x =$ ■. 20

Practice

Solve for x.

1. $4x + (x + 25) = 90$ 13 **2.** $2x + (x - 24) = 90$ 38 **3.** $3x - (x + 14) = 180$ 97

4. $90 = 4x - (2x - 6)$ 42 **5.** $4x = 72 - 2x$ 12 **6.** $14x = x + 169$ 13

7. $90 + 2x = 4x$ 45 **8.** $3x + 140 = 4x$ 140 **9.** $70 - (2x + 5) = 3x$ 13

10. $64 - (3x - 4) = x$ 17 **11.** $20 + (4x - 6) = 6x$ 7 **12.** $4x = 52 - (x + 7)$ 9

Share Your Understanding

13. Explain to a partner how to solve for x in the equation $14x + (15 + x) = 180$. Use the words *parentheses, like terms, subtract,* and *divide* in your explanation. Undo the parentheses. Combine the like terms. Subtract 15 from both sides. Divide both sides by 15 to find that x is 11.

14. **CRITICAL THINKING** Give an example of an equation with a variable on each side of the equal sign. Tell how to solve it. Check students' work.

More practice is provided in the Workbook, page 7, and in the Classroom Resource Binder, page 14.

Adding and Subtracting Angle Measures

Getting Started Review the Segment Addition Postulate.

POSTULATE 5
Angle Addition Postulate

If point D is in the interior of $\angle ABC$, then $m\angle ABD + m\angle DBC = m\angle ABC$.

Adjacent angles are two angles with a common vertex and a common ray. $\angle ABD$ and $\angle DBC$ are adjacent angles. The combined angle is $\angle ABC$.

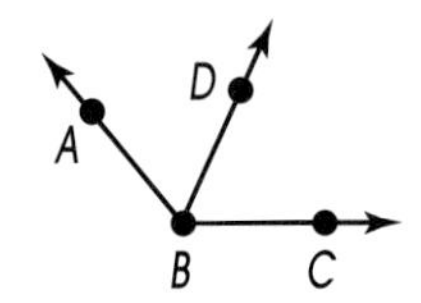

You can find the measure of a combined angle if you know the measures of the adjacent angles.

EXAMPLE 1

Find the measure of $\angle EFG$.

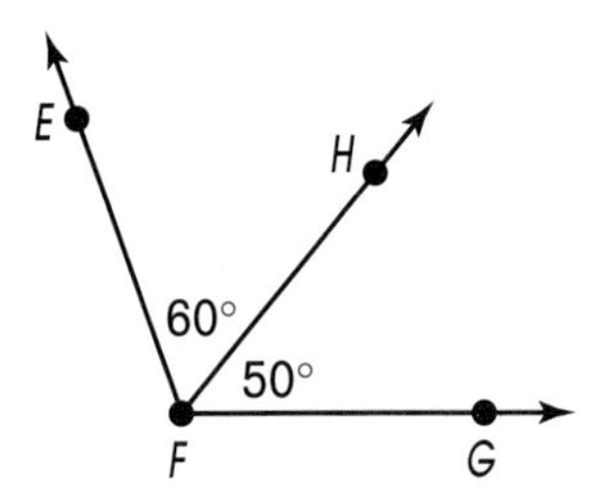

Write an equation. $m\angle EFH + m\angle HFG = m\angle EFG$

Substitute the given values. $60 + 50 = m\angle EFG$

Then, add. $110 = m\angle EFG$

So, $\angle EFG$ is 110°.

EXAMPLE 2

$\angle MKL$ is 55°. Find the measure of $\angle JKM$.

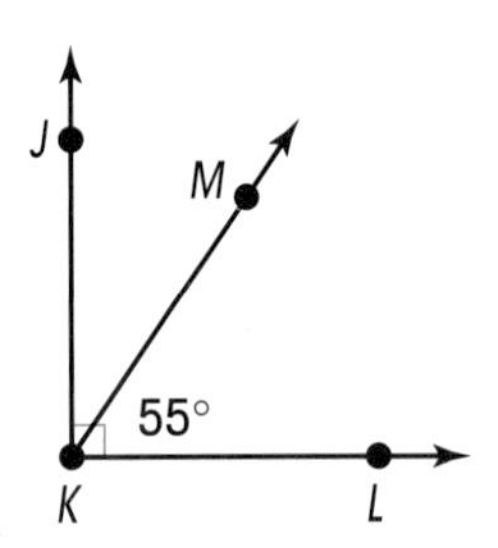

Write an equation. $m\angle JKM + m\angle MKL = m\angle JKL$

Substitute the given values. $m\angle JKM + 55 = 90$

Subtract. $m\angle JKM + 55 - 55 = 90 - 55$

$m\angle JKM = 35$

So, $\angle JKM$ is 35°.

Remember
A right angle is 90°.

Try This

Draw a diagram and find the measure of $\angle UST$.

$\angle RSU$ is adjacent to $\angle UST$.
$\angle RST$ is a straight angle.
$\angle RSU$ is 115°.

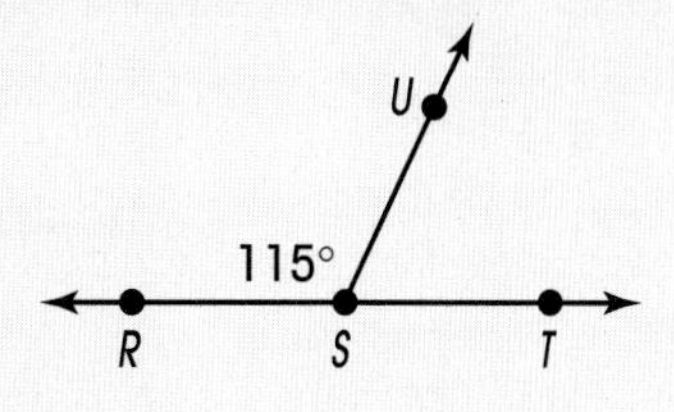

Write an equation. $m\angle RSU + m\angle UST = \blacksquare$ m∠RST

Substitute. Notice that ∠RST is a ■ angle. straight $115 + m\angle UST = 180$

Subtract 115 from both sides. $115 + m\angle UST - 115 = 180 - \blacksquare$ 115

$m\angle UST = \blacksquare$ 65

So, $\angle UST$ is ■. 65°

Practice

Draw a diagram. Find the measure of each angle.

1. $\angle CBE$ is adjacent to $\angle EBA$.
$\angle CBE$ is 40° and $\angle EBA$ is 80°.
Find the measure of $\angle CBA$. 120°

2. $\angle JKL$ is adjacent to $\angle LKM$.
$\angle JKL$ is 15° and $\angle LKM$ is 35°.
Find the measure of $\angle JKM$. 50°

3. $\angle DEH$ is adjacent to $\angle HEG$.
$\angle DEG$ is 106° and $\angle HEG$ is 55°.
Find the measure of $\angle DEH$. 51°

4. $\angle PQR$ is adjacent to $\angle RQS$.
$\angle PQS$ is 110° and $\angle RQS$ is 34°.
Find the measure of $\angle PQR$. 76°

5. Adjacent angles are two angles with a common vertex and a common ray. These two angles combine to form a third angle.

Share Your Understanding

5. Write about adjacent angles. Describe the three angles that are formed. See above.

6. **CRITICAL THINKING** $\angle EFG$ is 110°. Write and solve an equation to find the measure of $\angle EFH$ and $\angle HFG$. $x + x + 20 = 110$; ∠EFH is 45°; ∠HFG is 65°

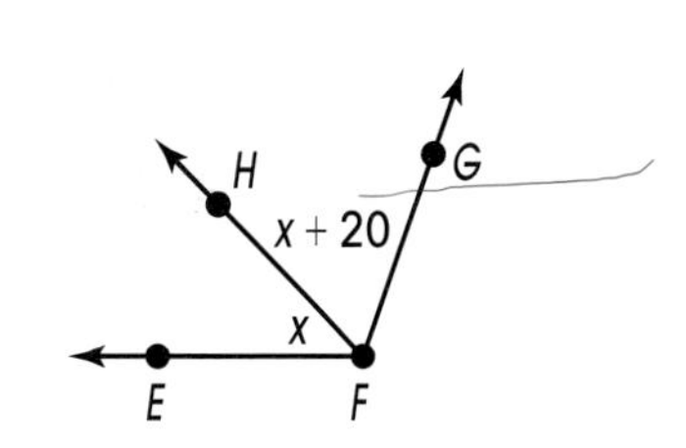

More practice is provided in the Workbook, page 8, and in the Classroom Resource Binder, page 15.

2·5 Complementary and Supplementary Angles

Getting Started
Have students fold the corner of a rectangular sheet of paper. Have them measure each angle formed and find the sum. Repeat for each corner. Have them vary the size of the angles created by each fold.

If the sum of the measures of two angles is 90° or 180°, the angle pair has a special name.

A pair of angles whose measures have a sum of 90° are **complementary angles**. If these angles are adjacent, they will form a right angle.

EXAMPLE 1

Decide if $\angle A$ and $\angle B$ are complementary angles.

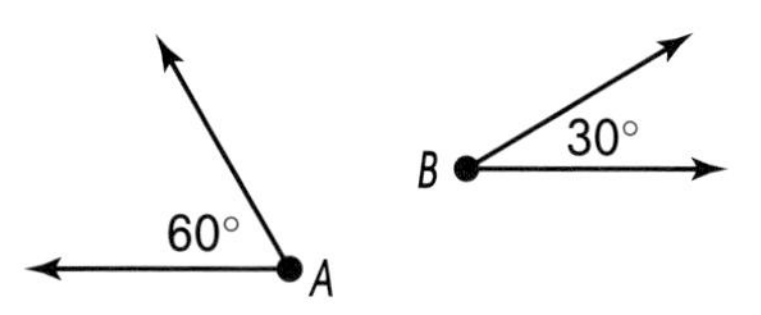

Remember
An angle can be named by using the letter of the vertex.

Find the sum of the angles.
$\angle A$ is 60° and $\angle B$ is 30°. $60 + 30 = 90$

Does the sum equal 90°? Yes.

Yes, $\angle A$ and $\angle B$ are complementary angles.

A pair of angles whose measures have a sum of 180° are **supplementary angles**. If these angles are adjacent, they will form a straight angle.

EXAMPLE 2

Decide if $\angle FGH$ and $\angle HGI$ are supplementary angles.

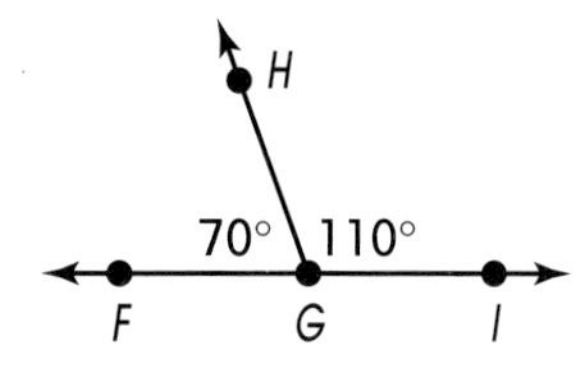

Find the sum of the angles. $70 + 110 = 180$

Does the sum equal 180°? Yes.

Yes, $\angle FGH$ and $\angle HGI$ are supplementary angles.

Try This

$\angle K$ is 55°. Find the angle that is complementary to $\angle K$.

The pair of angles is ■ complementary. The sum of the measures of the two angles is ■ 90. Write an equation. $55 + x = 90$

Subtract 55 from both sides. $55 + x - 55 = 90 - ■$ 55

Simplify. $x = ■$ 35

The angle complementary to a 55° angle is a ■ 35° angle.

Practice

1. Decide if $\angle S$ and $\angle T$ are complementary or supplementary angles. Explain your answer
Supplementary, because the sum of their measures equals 180°.

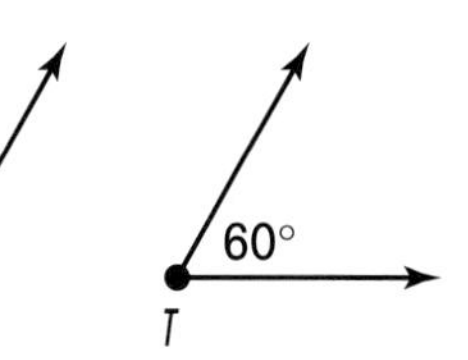

Find the angle that is complementary to each angle.

2. 20° 70° **3.** 50° 40° **4.** 10° 80° **5.** 15° 75°

6. 65° 25° **7.** 45° 45° **8.** 42° 48° **9.** $x°$ $90° - x°$

Find the angle that is supplementary to each angle.

10. 20° 160° **11.** 130° 50° **12.** 60° 120° **13.** 150° 30°

14. 35° 145° **15.** 145° 35° **16.** 115° 65° **17.** $x°$ $180° - x°$

Share Your Understanding

18. Explain to a partner how to decide if a pair of angles is complementary. Use *sum*, *angles*, and *90°* in your explanation.
If the sum of the measures of two angles is 90°, then the angles are complementary.

19. **CRITICAL THINKING** An angle is 16° more than its complementary angle. Find the measure of each angle. (Hint: Write an equation first. Let x represent the measure of the complementary angle.)
The angle is 53° and its complement is 37°.

More practice is provided in the Workbook, page 9, and in the Classroom Resource Binder, page 16.

Congruent Angles

Getting Started
Have students name the congruent line segments in the following diagram.
Have them explain their choice. [$\overline{AB} \cong \overline{CD}$]

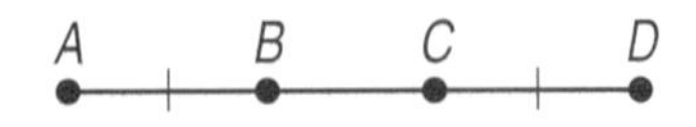

In Chapter 1, you learned that congruent line segments are segments that have the same length. **Congruent angles** are angles that have the same measure.

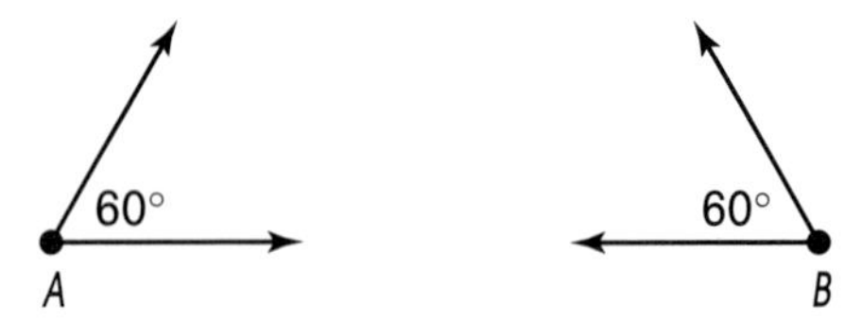

Math Fact
If $m\angle A = m\angle B$,
then $\angle A \cong \angle B$.

The measures of $\angle A$ and $\angle B$ are equal. You can write $m\angle A = m\angle B$. This means that $\angle A$ is congruent to $\angle B$. You can write $\angle A \cong \angle B$.

A small mark on a diagram tells you that the angles are congruent.

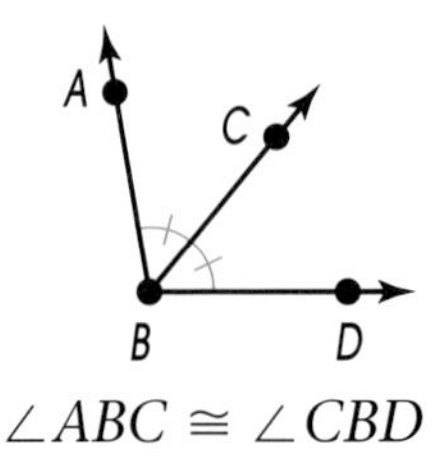

$\angle ABC \cong \angle CBD$

You can use what you know about congruent angles to find an unknown value.

EXAMPLE

Find the value of x in the diagram on the right.

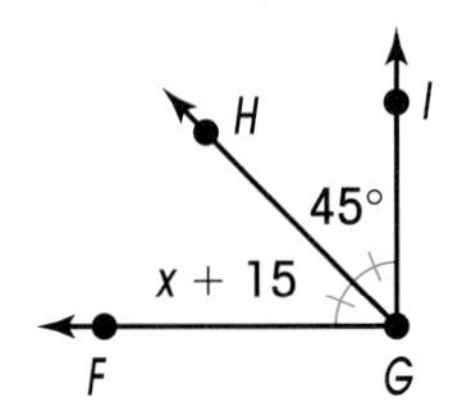

Math Fact
If $\angle A \cong \angle B$,
then $m\angle A = m\angle B$.

The marks on the diagram show that $\angle FGH \cong \angle HGI$.

$$m\angle FGH = m\angle HGI$$

Write an equation.

$$x + 15 = 45$$

Solve for x. Subtract 15 from both sides of the equation.

$$x + 15 - 15 = 45 - 15$$
$$x = 30$$

Because $\angle FGH$ is congruent to $\angle HGI$, x is equal to 30.

Try This

Find the value of x in the diagram on the right.

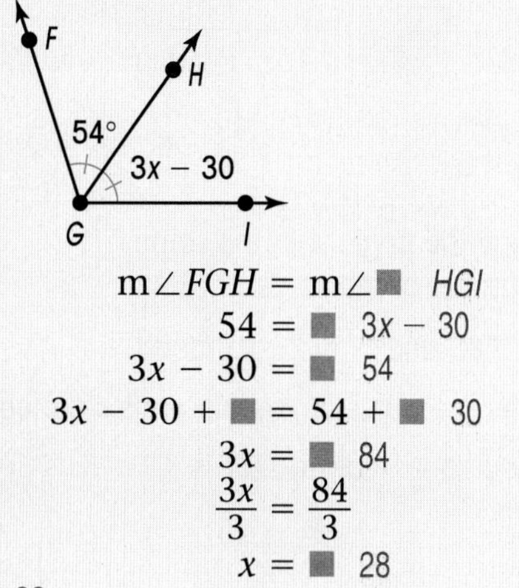

The drawing shows $\angle FGH \cong$ ■. $\angle HGI$
Write an equation. Rewrite the equation so that x is on the left of the equal sign.
Solve for x. Add 30 to both sides.

Divide both sides of the equation by 3.

$$m\angle FGH = m\angle ■ \quad HGI$$
$$54 = ■ \quad 3x - 30$$
$$3x - 30 = ■ \quad 54$$
$$3x - 30 + ■ = 54 + ■ \quad 30$$
$$3x = ■ \quad 84$$
$$\frac{3x}{3} = \frac{84}{3}$$
$$x = ■ \quad 28$$

Because $\angle FGH$ is congruent to ■ ($\angle HGI$), x is equal to ■ (28).

Practice

1. Which of the following angles are congruent? $\angle T$ and $\angle V$; $\angle S$ and $\angle R$

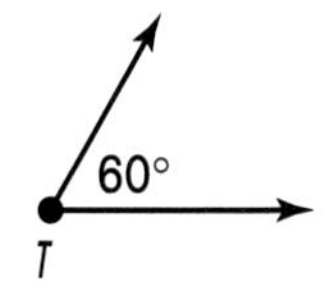

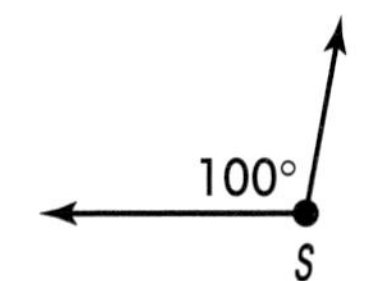

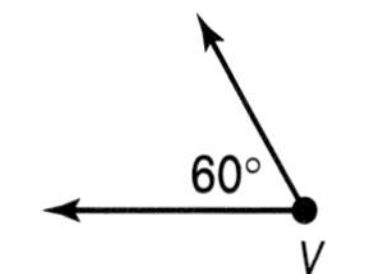

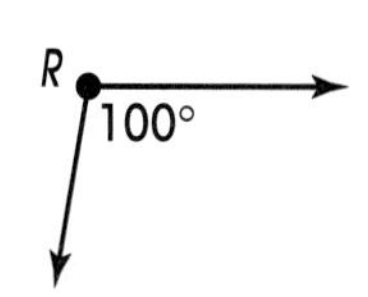

Find the value of x in each diagram.

2. 82

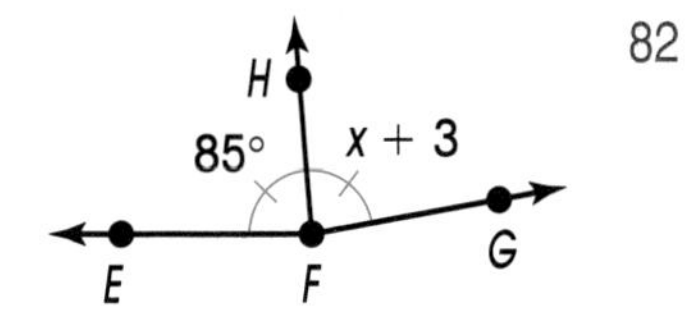

3. 20

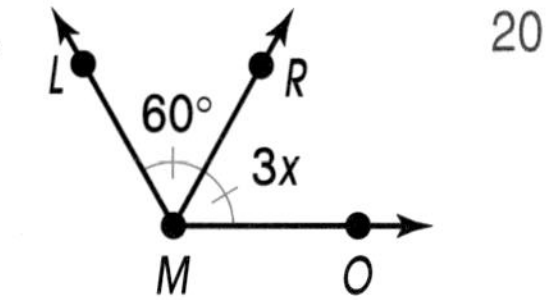

4. 3

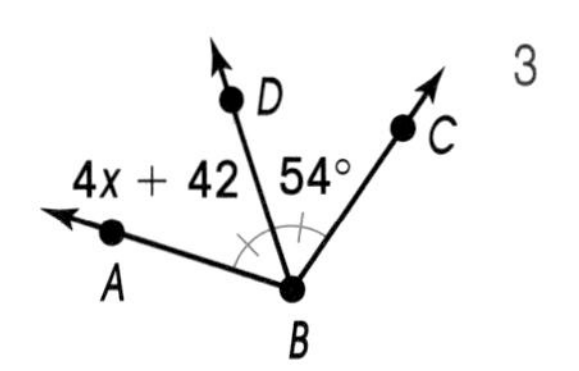

5. 35

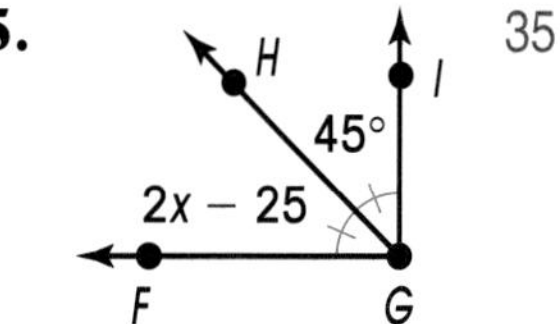

Try This

In the diagram, $\angle EBF$ is 25° and $\angle EBD$ is 45°. Find $m\angle ABC$.

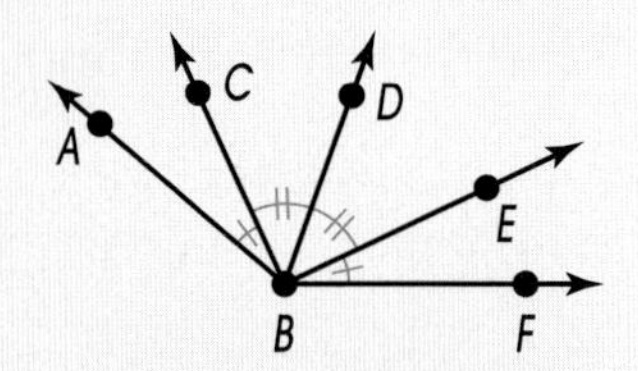

Find the angle with the same number of marks as $\angle ABC$.	$\angle ABC$ has 1 mark. $\angle EBF$ ■ has 1 mark.
$\angle ABC$ and ■ $\angle EBF$ are congruent.	$\angle ABC \cong$ ■ $\angle EBF$
The measures of the two angles are equal.	$m\angle ABC = m\angle$■ EBF
You know that $m\angle EBF$ is ■. 25	$m\angle ABC =$ ■ 25
So, $\angle ABC$ is ■. 25°	

Practice

Use the diagram on the right for exercises **6–13**.

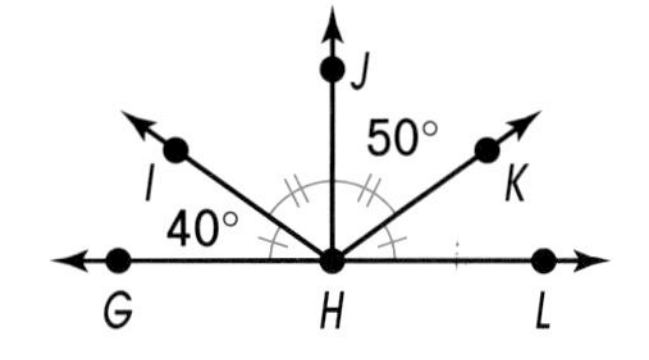

Name each congruent angle.

6. $\angle GHI \cong$ ■ $\angle KHL$ **7.** $\angle JHK \cong$ ■ $\angle JHI$

Find the measure of each angle.

8. $\angle JHI =$ ■ 50° **9.** $\angle KHL =$ ■ 40° **10.** $\angle GHJ =$ ■ 90°

11. $\angle IHK =$ ■ 100° **12.** $\angle GHK =$ ■ 140° **13.** $\angle IHL =$ ■ 140°

14. $\angle ABC$ is congruent to $\angle CBD$; Write the equation $x + 16 = 62$. Solve for x. Subtract 16 from both sides; $x = 46$ and $\angle ABD = 124°$

Share Your Understanding

14. Explain to a partner how to solve for x in the diagram on the right. Use the words *congruent, equation, subtract,* and *both sides* in your explanation. Have your partner find $m\angle ABD$.
See above.

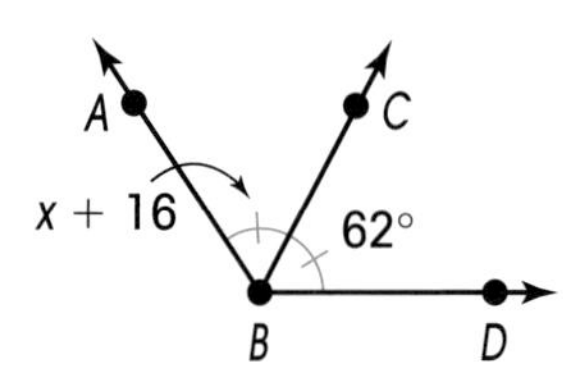

15. **CRITICAL THINKING** Find $m\angle ABC$. Use the information given in the diagram on the right. Explain your reasoning.
45°; $\angle ABD$ is a right angle. $\angle ABC \cong \angle CBD$ So, the measures of these angles are equal. Then, $90 \div 2 = 45$.

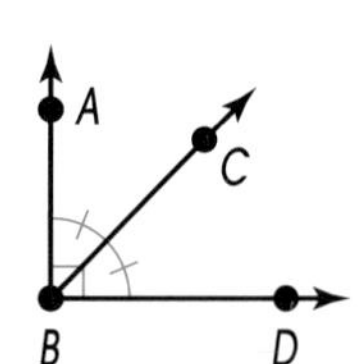

CONSTRUCTION
A Congruent Angle

You already know how to copy a line segment. Follow the steps below to construct $\angle EDG$ so that it is congruent to $\angle ABC$.

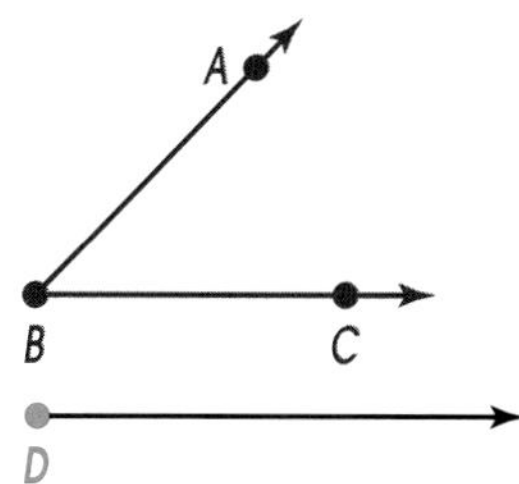

STEP 1 Draw a ray. Label its endpoint *D*.

STEP 2 Trace $\angle ABC$. Place one end of a compass on point *B*. Draw an arc that intersects both sides of the angle. Do not change this compass position.

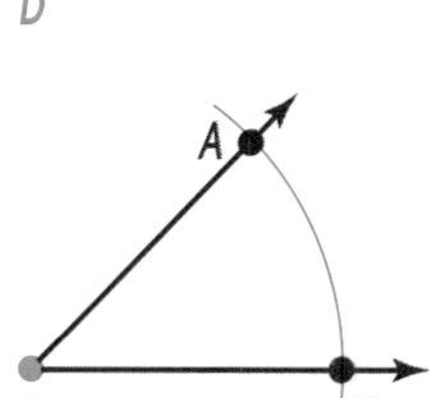

STEP 3 Take the compass end that was on point *B* and place it on point *D*. Draw a large arc to intersect your ray. Label this point *G*.

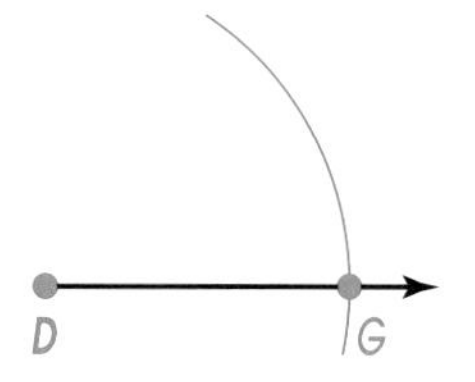

STEP 4 Go back to $\angle ABC$. Place one end of the compass on point *C*. Move the pencil end until it is on point *A*. Do not change this compass position.

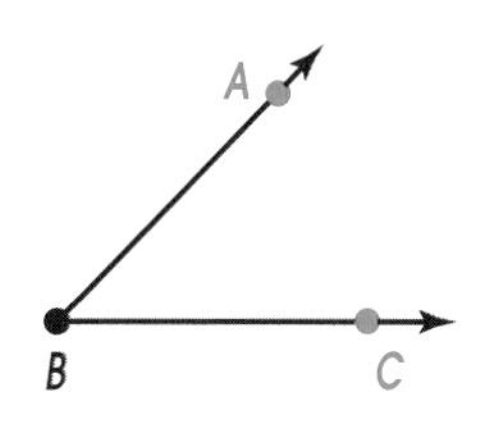

STEP 5 Take the compass end that was on point *C* and place it on point *G*. Draw a small arc to intersect the arc that you drew in STEP 3. Label this point *E*.

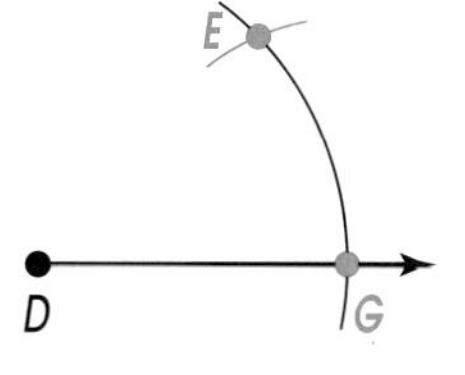

STEP 6 Draw a ray from point *D* to point *E* to complete $\angle EDG$.

$\angle ABC \cong \angle EDG$

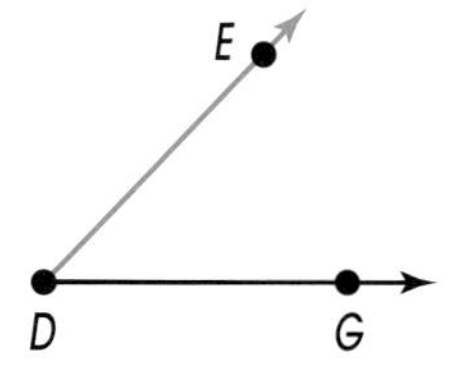

More practice is provided in the Workbook, page 10, and in the Classroom Resource Binder, page 16.

2-7 Vertical Angles

Getting Started Have students draw intersecting lines and measure the angles.

THEOREM 3

Vertical angles are congruent.

When two lines intersect, they form **vertical angles**. Vertical angles are opposite angles. They are congruent. Look at the diagram below.

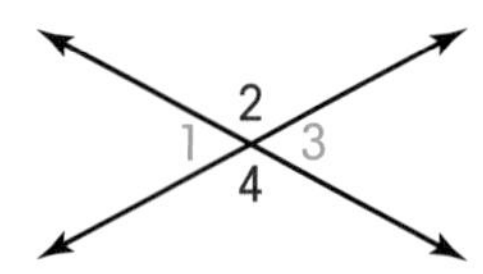

$\angle 1$ and $\angle 3$ are vertical angles. $\angle 1 \cong \angle 3$
$\angle 2$ and $\angle 4$ are vertical angles. $\angle 2 \cong \angle 4$

EXAMPLE 1

Name the pairs of vertical angles in the diagram on the right.

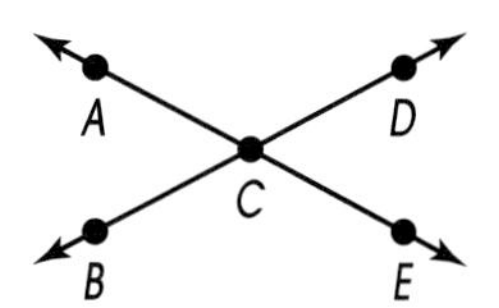

Start with $\angle ACB$. Look for the angle that is opposite $\angle ACB$. $\angle DCE$ is opposite $\angle ACB$.

One pair of vertical angles is $\angle ACB$ and $\angle DCE$.
Another pair of vertical angles is $\angle BCE$ and $\angle ACD$.

You can find the measure of all the angles formed by intersecting lines if you know the measure of one angle.

EXAMPLE 2

$\angle 4$ is 100°. Find the measures of $\angle 1$, $\angle 2$, and $\angle 3$.

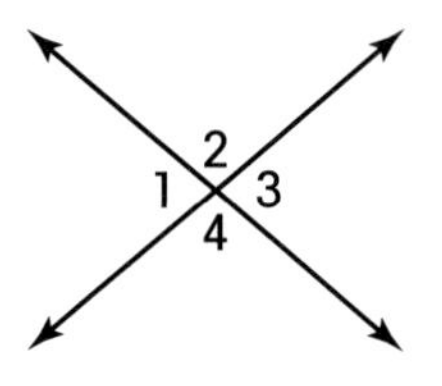

$\angle 2$ and $\angle 4$ are vertical angles.
$\angle 2 \cong \angle 4$ So, $m\angle 2 = m\angle 4$. $m\angle 2 = 100$

$\angle 3$ and $\angle 4$ are supplementary.
Write and solve an equation. $m\angle 3 + 100 = 180$
$m\angle 3 = 80$

$\angle 1$ and $\angle 3$ are vertical angles. $m\angle 1 = 80$

Remember
A straight angle is 180°.

So, $\angle 1$ is 80°, $\angle 2$ is 100°, and $\angle 3$ is 80°.

Try This

Use the diagram on the right to find the measure of $\angle ACD$.

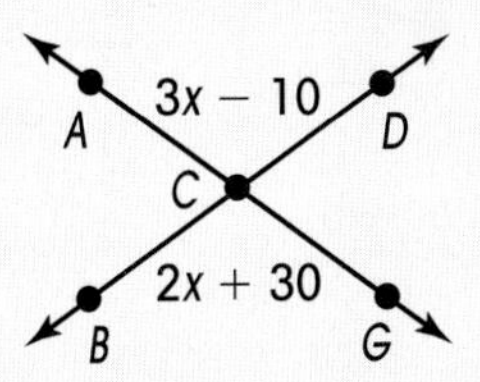

$\angle ACD$ and $\angle BCG$ are ■ vertical angles. $m\angle ACD =$ ■ $m\angle BCG$

Write an equation. $3x - 10 = 2x + 30$

Solve for x. $3x = 2x +$ ■ 40

$x =$ ■ 40

Substitute 40 for x in $3x - 10$ to find $m\angle ACD$. $m\angle ACD = 3(40) - 10$

$= 120 - 10$

$=$ ■ 110

So, $\angle ACD$ is ■. 110°

Practice

Use the diagram on the right to find the measure of each angle in exercises **1–3**.

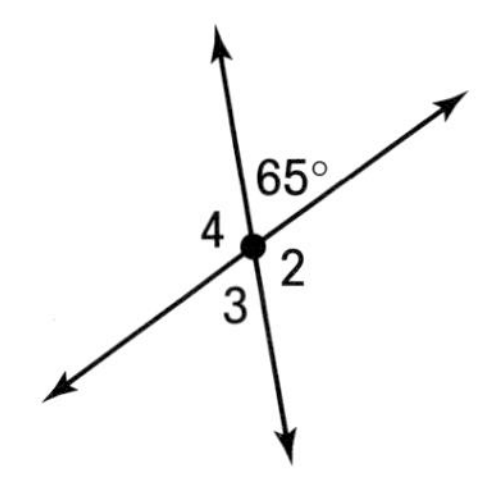

1. $\angle 3$ 65° **2.** $\angle 2$ 115° **3.** $\angle 4$ 115°

Use the diagram on the right to find the measure of each angle in exercises **4–6**.

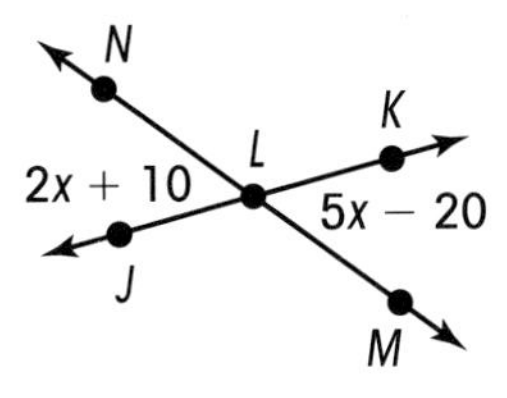

4. $\angle KLM$ 30° **5.** $\angle KLN$ 150° **6.** $\angle JLM$ 150°

Share Your Understanding

7. Draw a diagram showing vertical angles. Explain which angles are congruent and which are supplementary.
Check students' work.

8. **CRITICAL THINKING** In the diagram on the right, $\angle 1$ is 90°. Find the measures of $\angle 2$, $\angle 3$, and $\angle 4$. Explain your reasoning.

1 2
3 4

$\angle 2 = 90°$; $\angle 3 = 90°$; $\angle 4 = 90°$; Possible answer: $\angle 1$ and $\angle 2$ form a 180° angle. Because $\angle 1 = 90°$, $\angle 2 = 90°$. $\angle 1$ and $\angle 4$, and $\angle 2$ and $\angle 3$ are vertical angles. They are congruent. So, $\angle 3 = 90°$ and $\angle 4 = 90°$.

More practice is provided in the Workbook, page 10, and in the Classroom Resource Binder, page 17.

2·8 Angle Bisectors

Getting Started Review bisector of a line segment.

> THEOREM 4
>
> **Angle Bisector Theorem**
>
> The bisector $\overrightarrow{BD}$ of $\angle ABC$ divides $\angle ABC$ in half so that $m\angle ABD = \frac{1}{2}m\angle ABC$.

An **angle bisector** is a ray that divides an angle into two congruent angles.

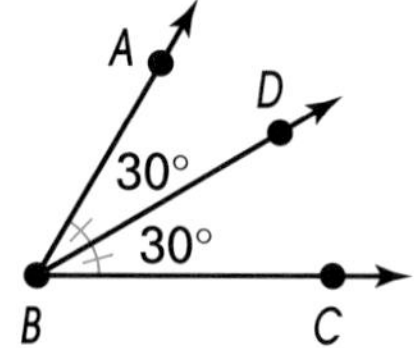

$\overrightarrow{BD}$ divides $\angle ABC$ into two congruent angles.

$\overrightarrow{BD}$ is the angle bisector of $\angle ABC$.

You can say $\overrightarrow{BD}$ bisects $\angle ABC$.

EXAMPLE 1

Decide if $\overrightarrow{KM}$ bisects $\angle JKL$.

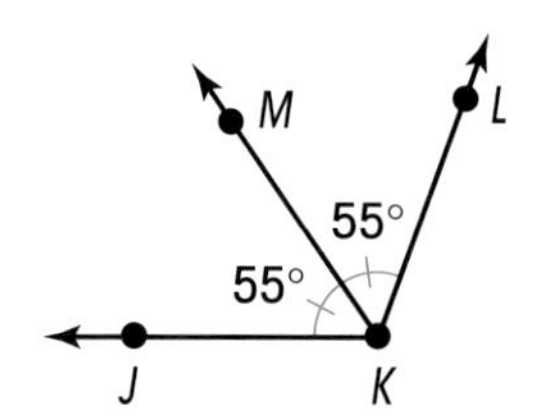

Are $\angle JKM$ and $\angle MKL$ congruent? Yes, $m\angle JKM = m\angle MKL$.

So, $\overrightarrow{KM}$ bisects $\angle JKL$.

Knowing that an angle is bisected helps you to find the angle measures.

EXAMPLE 2

$\overrightarrow{QS}$ bisects $\angle PQR$. $\angle PQR$ is 80°. Find the measures of $\angle PQS$ and $\angle SQR$.

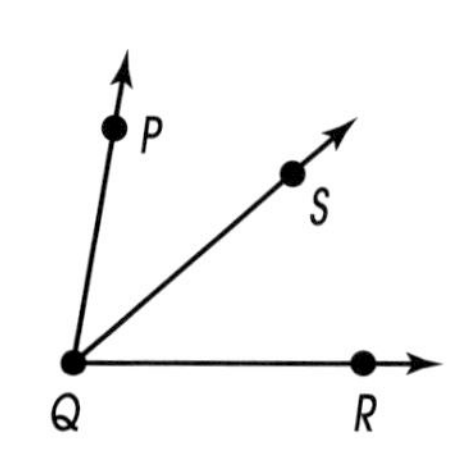

$\overrightarrow{QS}$ bisects $\angle PQR$. $\overrightarrow{QS}$ divides $\angle PQR$ into two congruent angles. $m\angle PQS = m\angle SQR$

Divide the measure of $\angle PQR$ by 2. $80 \div 2 = 40$

So, $\angle PQS$ is 40° and $\angle SQR$ is 40°.

Try This

$\overrightarrow{AD}$ bisects $\angle BAC$.
Find the measure of $\angle BAC$.

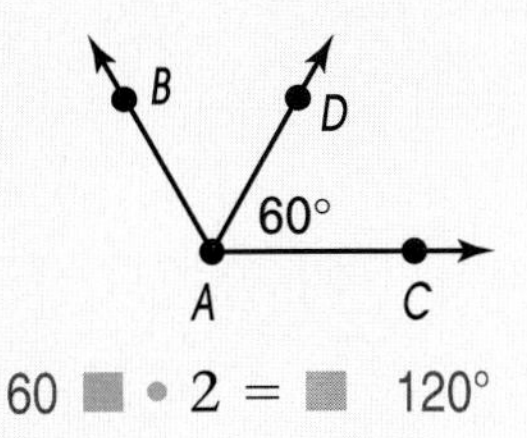

$\overrightarrow{AD}$ ■ $\angle BAC$. $\overrightarrow{AD}$ divides $\angle BAC$ into two ■ angles. (bisects; congruent)
Multiply $m\angle DAC$ by ■. 2

60 ■ • 2 = ■ 120°

So, $\angle BAC$ is ■. 120°

Practice

Find the measure of each angle.

1. 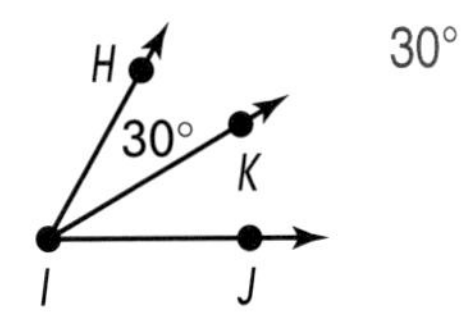

30°

$\overrightarrow{IK}$ bisects $\angle HIJ$.
Find the measure of $\angle KIJ$.

2. 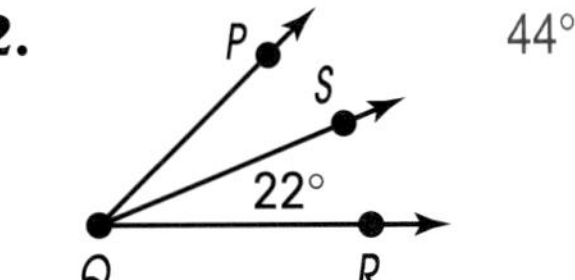

44°

$\overrightarrow{QS}$ bisects $\angle PQR$.
Find the measure of $\angle PQR$.

3. 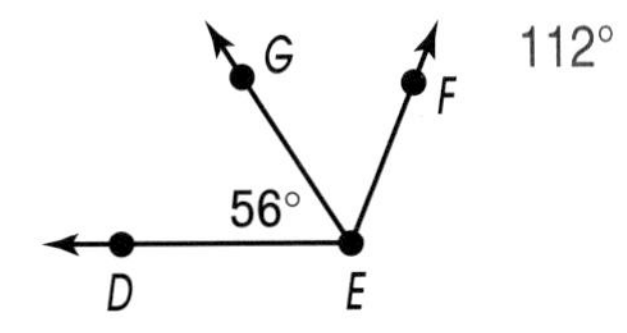

112°

$\overrightarrow{EG}$ bisects $\angle DEF$.
Find the measure of $\angle DEF$.

4. 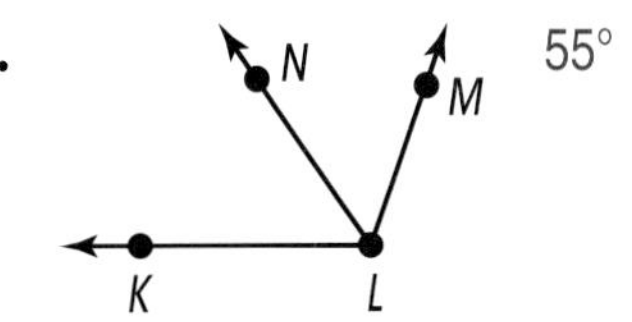

55°

$\overrightarrow{LN}$ bisects $\angle KLM$. $\angle KLM$ is 110°.
Find the measure of $\angle KLN$.

5. 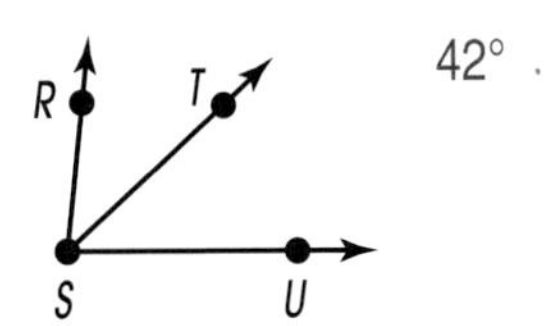

42°

$\overrightarrow{ST}$ bisects $\angle RSU$. $\angle RSU$ is 84°.
Find the measure of $\angle TSU$.

6. 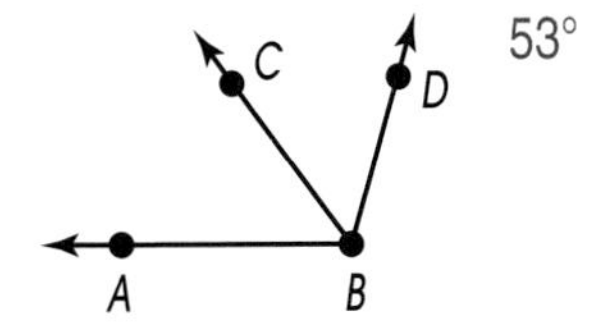

53°

$\overrightarrow{BC}$ bisects $\angle ABD$. $\angle ABD$ is 106°.
Find the measure of $\angle CBD$.

Try This

$\overrightarrow{BC}$ bisects $\angle ABD$.
Find the value of x.

$\overrightarrow{BC}$ ■ bisects $\angle ABD$. $m\angle ABC =$ ■ $m\angle CBD$

Write an equation. $55 = x + 22$

Solve for x. 33 ■ $= x$

The value of x is ■. 33

Practice

$\overrightarrow{ST}$ bisects $\angle RSU$. Find the value of x in each diagram.

7. 19

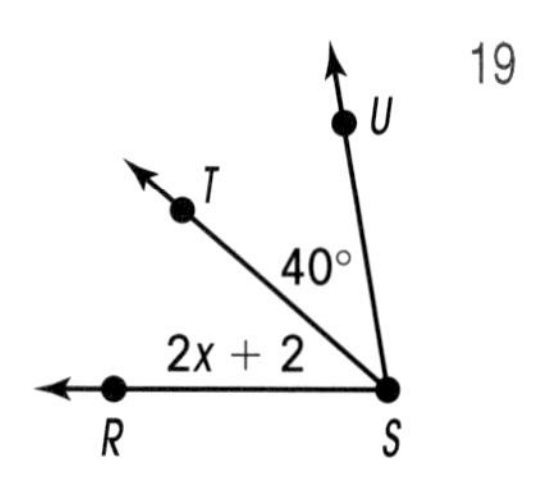

8. 10

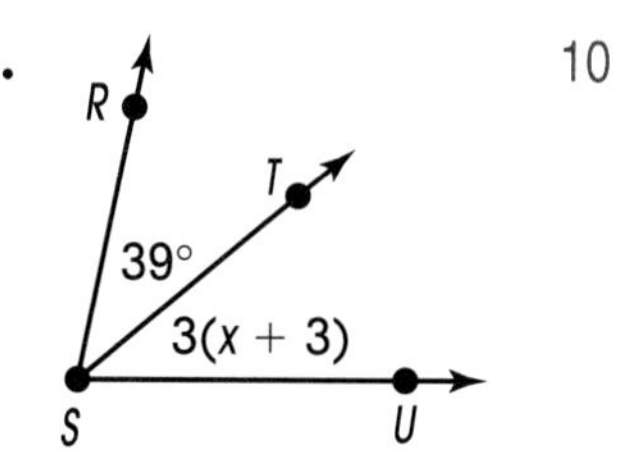

9. 20

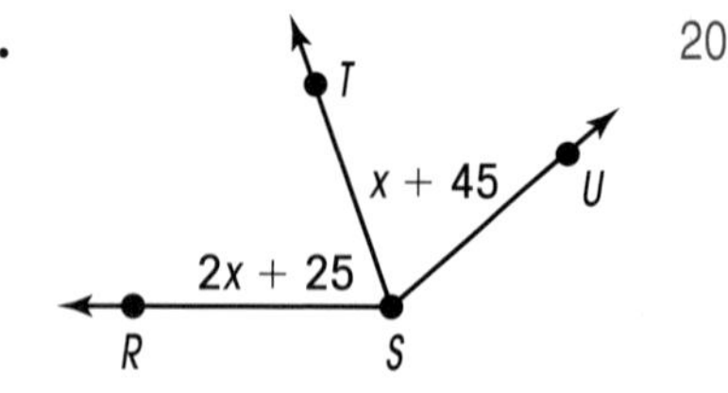

10. 10

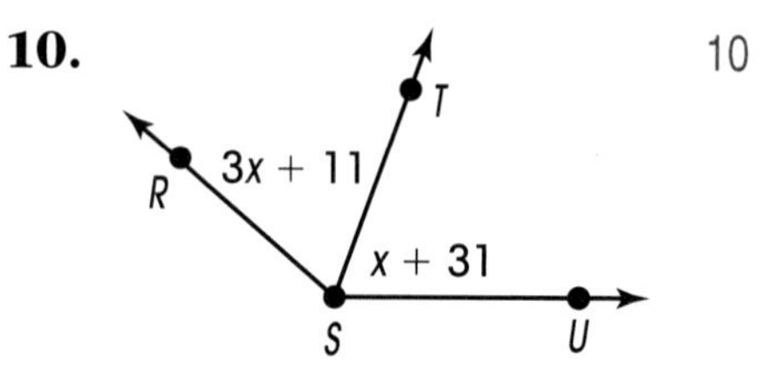

Share Your Understanding

11. Explain to a partner how you know if a ray is an angle bisector. Use *congruent* in your explanation. An angle bisector divides an angle into two congruent angles. So, if the two angles are congruent, then the ray is an angle bisector.

12. **CRITICAL THINKING** Take turns with a partner. Draw an angle on a sheet of paper. Measure the angle. Then, tell your partner to find where the bisector of the angle should be drawn. Check students' work.

CONSTRUCTION
An Angle Bisector

Follow the steps below to construct $\overrightarrow{BD}$ so that it bisects $\angle B$.

On a sheet of paper, copy $\angle B$.

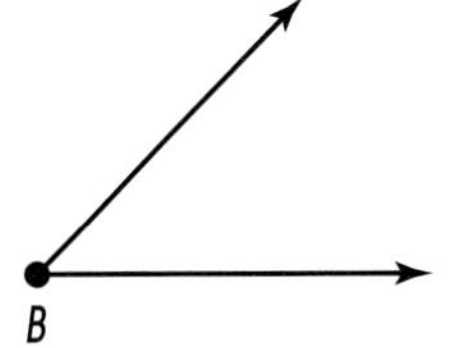

STEP 1 Place one end of the compass on point B. Draw an arc through both sides of the angle.

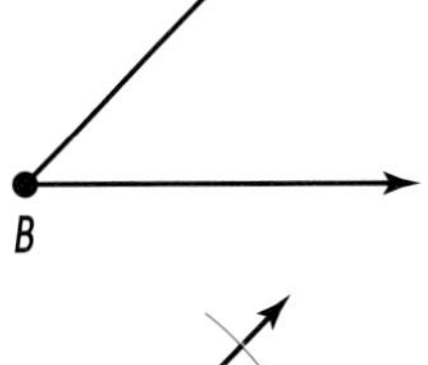

STEP 2 Place points where the arc intersects each side. Name these points A and C.

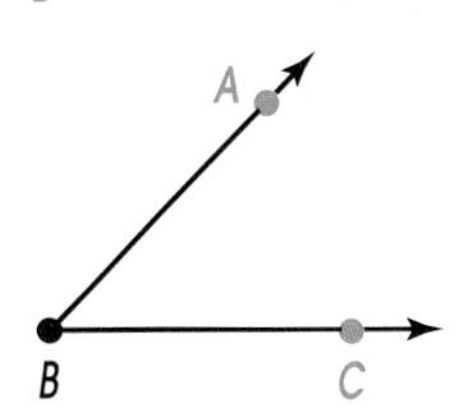

STEP 3 Place one end of the compass on point C. Draw a small arc inside $\angle B$. The arc should be more than halfway between A and C. Do not move the compass position.

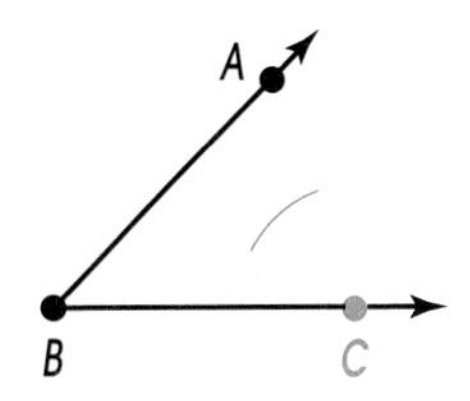

STEP 4 Take the compass end that was on point C and place it on point A. Draw a small arc. This arc should intersect the small arc from STEP 3 at exactly one point. Label this point D.

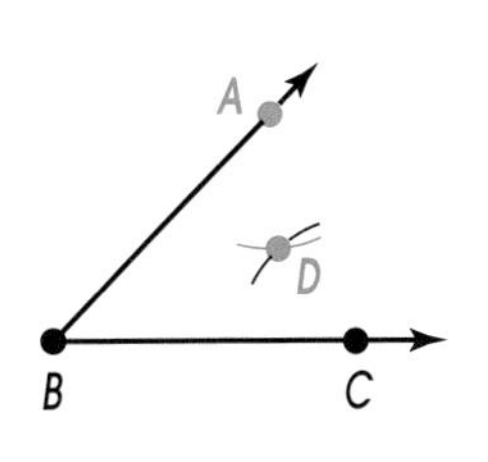

STEP 5 Draw a ray from point B through point D. This is $\overrightarrow{BD}$.

$\overrightarrow{BD}$ is the bisector of $\angle B$. This angle can now be named $\angle ABC$.

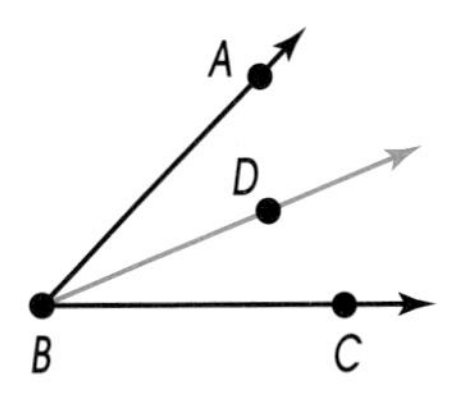

2·9 Calculator: Common Angles

Getting Started Review common segments.

You can use what you know about congruent angles to find the measure of common angles.

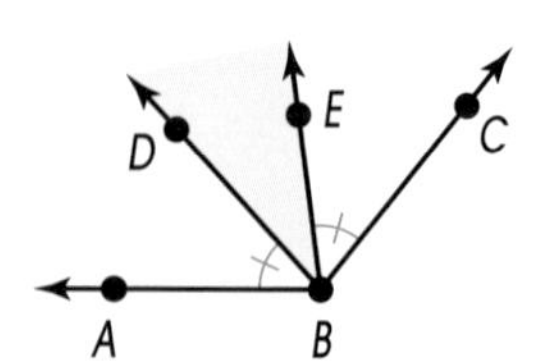

Look at $\angle ABC$ above. $\angle DBE$ is a common angle because it is part of $\angle ABE$ and $\angle DBC$.

THEOREM 5
Common Angle Theorem

Points D and E are in the interior of $\angle ABC$.

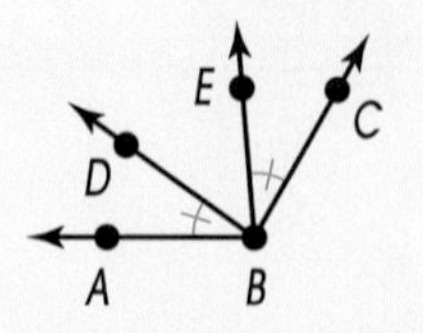

If $m\angle ABD = m\angle EBC$, then $m\angle ABE = m\angle DBC$.

If $m\angle ABE = m\angle DBC$, then $m\angle ABD = m\angle EBC$.

You can use the Common Angle Theorem to find the measure of some angles.

EXAMPLE 1

Remember
The marks in a diagram show congruent angles.
$\angle PQS \cong \angle TQR$
$m\angle PQS = 47$.

Find the measure of $\angle PQT$.

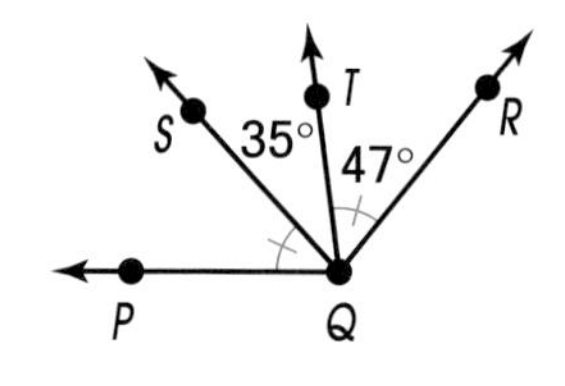

Write an equation.

$$m\angle PQT = 47 + 35$$

Use your calculator to add.

PRESS: 4 7 + 3 5 =

Display

82

So, $\angle PQT$ is 82°.

EXAMPLE 2

$\angle ABE$ is 80°.
Find the measure of $\angle EBC$.

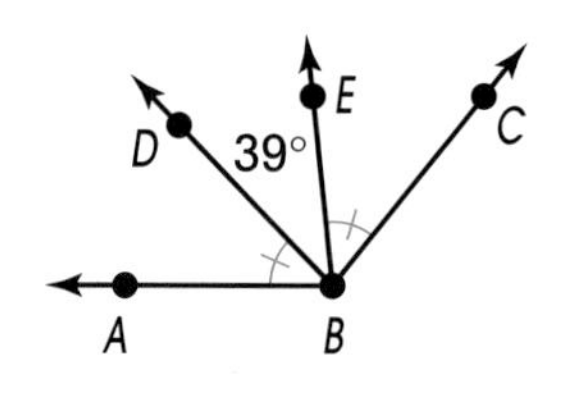

Write an equation.

$$m\angle ABD = 80 - 39$$

Use your calculator to subtract.

PRESS: 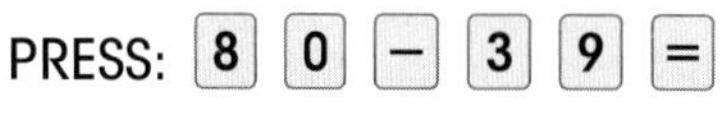

Display

So, $\angle ABD$ is 41°. Then, $\angle EBC$ is 41°.

Practice

Find the measure of $\angle OLM$.

1. 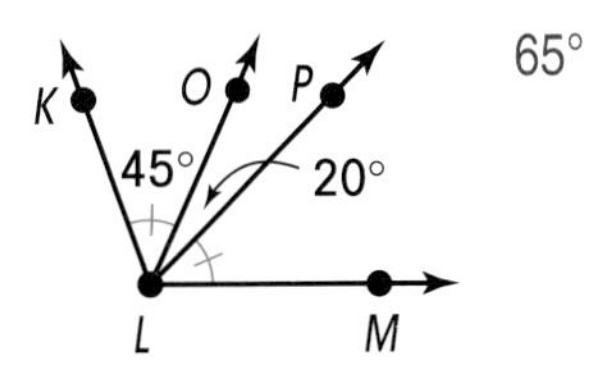

65°

2. 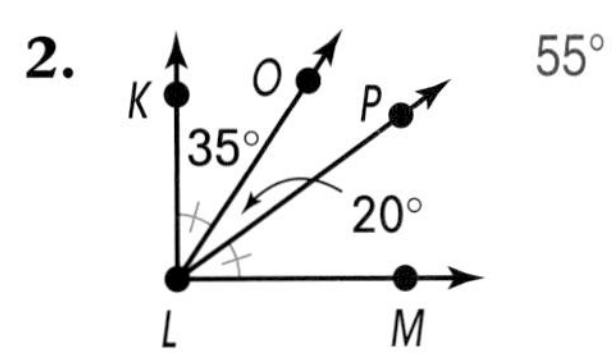

55°

Find the measure of $\angle ABD$. $\angle DBC$ is 75°.

3. 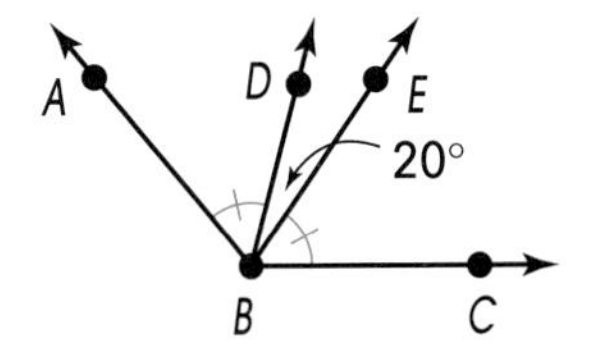

55°

4. 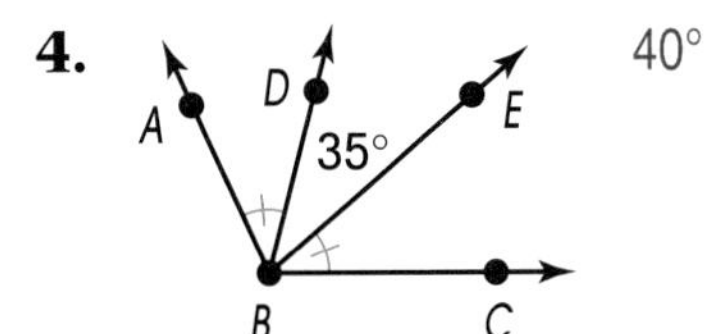

40°

Math In Your Life

MIRRORS

A plane mirror is a sheet of glass with a backing. Between the glass and the backing is a smooth silver coating. Light passes through the glass, hits the silver coating, and reflects the light. This allows you to see an image. The image you see is a virtual image.

Light hits the mirror at an angle. This is the angle of incidence. Light is then reflected at an angle. This is the angle of reflection. These two angles are congruent. Use a protractor to make a simple drawing. Show the angles as light strikes a plane mirror.

Check students' work.

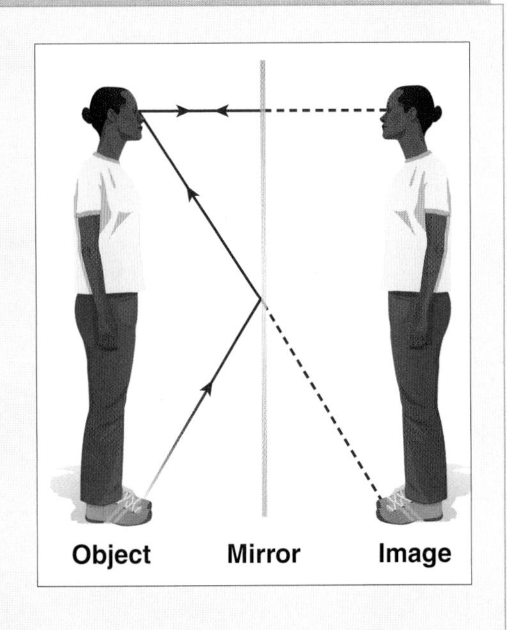

More practice is provided in the Workbook, page 11, and in the Classroom Resource Binder, page 18.

2·10 Problem-Solving Strategy: Make a Table

Getting Started
Discuss what happens when a person has an arm or a leg in a cast. When the cast is removed, a person needs to exercise in order to regain mobility in the limb.

When you stand straight, your knees make a 180° angle. When you bend your knees, the angle gets smaller. After knee surgery, physical therapists help patients to bend their knees again.

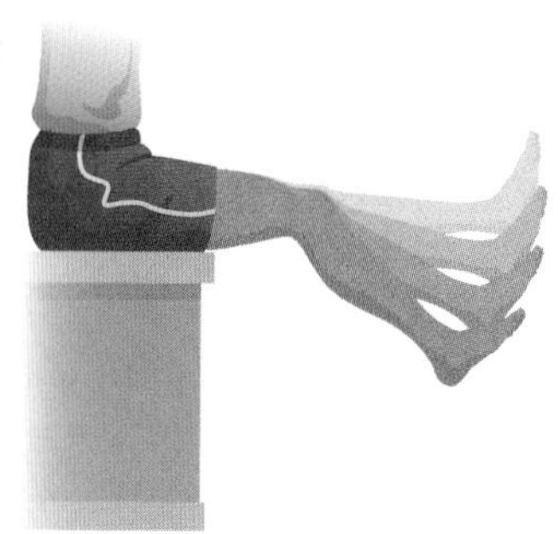

EXAMPLE

José had knee surgery. After surgery, José could not bend his knee. With exercise, he was able to decrease the angle by 5° every 2 days. How many days will it take to go from 180° to 165°?

READ **What do you need to find out?**
Find the number of days it will take José to go from 180° to 165°.

PLAN **What do you need to do?**
Make a table showing the number of days and the angle of the knee.

DO **Follow the plan.**
Make a table. José decreases the angle by 5° every 2 days.

Days	0	2	4	6
Angle of Knee	180°	175°	170°	165°

CHECK **Does your answer make sense?**
Yes. As the number of days increases, the measure of the angle decreases. ✓

It will take 6 days for José to reach his goal.

Try This

Anne's leg was kept at 135° in a cast. Now, the cast is off. With exercise, she should be able to decrease the angle by 15° each week. How many weeks will it take to go from 135° to 90°?

Make a ■ showing the number of ■ and the ■ of the knee.
table weeks angle

Weeks	0	■ 1	■ 2	3
Angle of Knee	135°	■ 120°	105°	■ 90°

3
It will take ■ weeks for Anne to reach her goal.

Practice

Make a table to solve each problem. Check students' work.

1. After your cast is off your leg, your physical therapist wants you to bend your knee 15° more each week. You need to bend your knee from 180° to 135°. How many weeks will it take you to get to 135°? 3 weeks

2. After Diane's cast is off her leg, her physical therapist wants her to bend her knee 2° more each day. She needs to bend her knee from 180° to 156°. How many days will it take Diane to get to 156°? 12 days

3. A person with a broken elbow has a cast that holds her arm at a 90° angle. Once the cast is off, she must do exercises to straighten her arm. She wants to straighten her elbow by about 4° every 2 days. How many days will it take to go from 90° to 180°? 45 days

More practice is provided in the Workbook, page 12, and in the Classroom Resource Binder, page 19.

2·11 Problem-Solving Application: Angles and Sports

Getting Started
Have students draw air hockey tables on graph paper to use with the lesson. Make sure students use a scale, such as 1 box = 1 foot.

Air hockey is a game you play with a puck. To score, you need to hit the puck into your opponent's pocket on the air hockey table.

When you hit the puck at an angle, it will bounce off the table at the same angle. Where the puck goes depends on the size of the table and the angle of the hit.

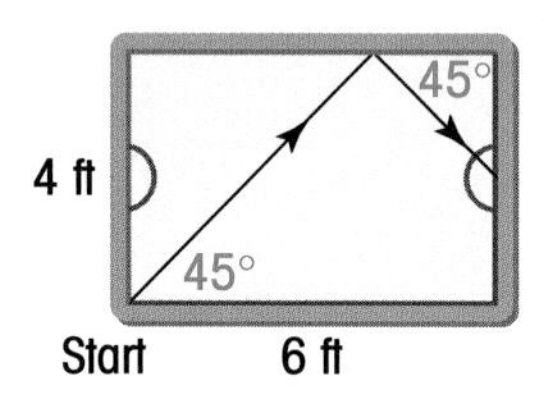

The puck bounced off the side of the table once before it landed in the right pocket.

EXAMPLE

Zack plays air hockey on an 8-ft by 10-ft table. He hits the puck from the lower left corner at a 45° angle. How many times will the puck hit the side of the table before it lands in a pocket?

Make a diagram. Use a protractor to draw a 45° angle. Show the path of the puck.

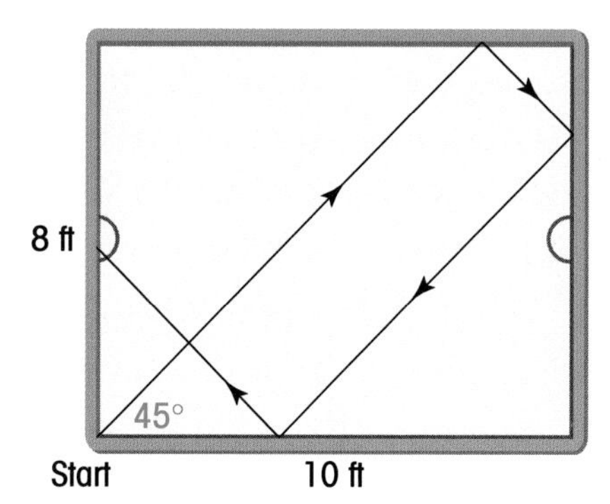

The puck bounced off the side three times before it landed in the left pocket.

Try This

Dena plays air hockey on a 6-ft by 8-ft table. She hits the puck from the lower left corner at a 45° angle. Will the puck land in a pocket?

Make a ■ (diagram). Use a ■ (protractor) to draw a ■ (45°) angle.

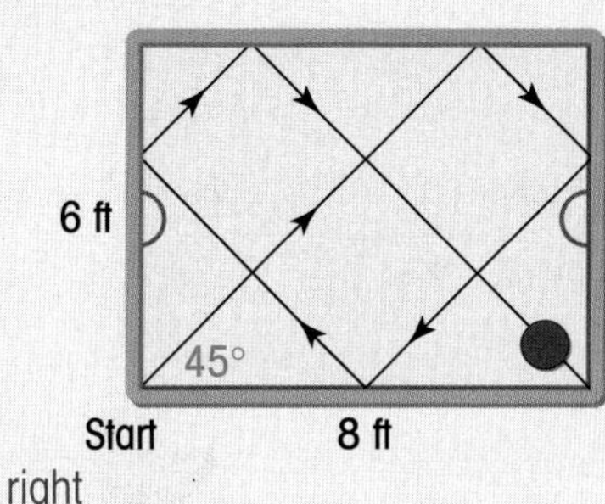

The puck is stuck in the lower ■ (right) corner.

The puck will ■ (not) land in a pocket.

Practice

Solve each problem. Show your work. If the puck gets stuck in a corner, tell which corner.

1. Aaron plays air hockey on a 6-ft by 9-ft table. He hits the puck from the lower left corner at a 45° angle. How many times will the puck hit the side of the table before it lands in a pocket?
   1 time

2. Mari plays air hockey on a 4-ft by 9-ft table. She hits the puck from the lower left corner at a 45° angle. How many times will the puck hit the side of the table before it lands in a corner?
   9 times

3. Beth plays air hockey on a 4-ft by 12-ft table. She hits the puck from the lower left corner at a 45° angle. Will the puck land in a pocket?
   The puck will not land in a pocket. The puck gets stuck in the upper right corner.

An alternate two-column proof lesson is provided on page 407 of the student book.

Proof: Angles

Getting Started Review the proofs in Chapter 1.

A theorem is a statement you can prove is true by using facts that you know to be true. You learned the Angle Bisector Theorem. Now, you can use information you know about angles to prove this theorem. You can write the proof in paragraph form.

> THEOREM 4
> **Angle Bisector Theorem**
> The bisector $\overrightarrow{BD}$ of $\angle ABC$ divides $\angle ABC$ in half so that $m\angle ABD = \frac{1}{2} m\angle ABC$.

EXAMPLE

You are given:

$\overrightarrow{BD}$ is the bisector of $\angle ABC$.

Prove: $m\angle ABD = \frac{1}{2}m\angle ABC$

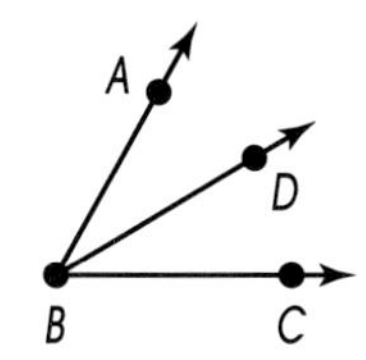

Begin with what you are given. Then, step by step, use logical reasoning to reach the statement you wish to prove.

You are given that $\overrightarrow{BD}$ bisects $\angle ABC$. This means that $m\angle ABD = m\angle DBC$. You can use the Angle Additon Postulate to write $m\angle ABC = m\angle ABD + m\angle DBC$.

You can substitute $m\angle ABD$ for $m\angle DBC$ in this equation. Then, $m\angle ABC = m\angle ABD + m\angle ABD$ or $m\angle ABC = 2m\angle ABD$. Now, divide both sides of this equation by 2. Then, $m\angle ABD = \frac{1}{2}m\angle ABC$.

So, you prove that $m\angle ABD = \frac{1}{2}m\angle ABC$. ✓

This proof shows that an angle bisector divides an angle in half.

Try This

Copy and complete the proof.

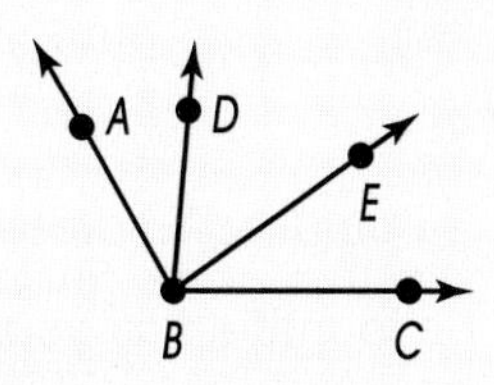

You are given:

$\angle ABE \cong \angle DBC$

Prove: $\angle ABD \cong \angle EBC$

Look at the diagram. To prove $\angle ABD \cong \angle EBC$, you need to take away the common angle.

You are ■ that $\angle ABE \cong \angle DBC$. This means that $m\angle ABE = m\angle DBC$. given

You can use the Angle ■ Postulate to rename $m\angle ABE$ and rename $m\angle DBC$. $m\angle ABE = m\angle ABD + m\angle DBE$ and $m\angle DBC = m\angle DBE + m\angle EBC$. Addition

Use substitution to rewrite the equation $m\angle ABE = m\angle DBC$ as $m\angle ABD + m\angle DBE = m\angle DBE + m\angle EBC$. Notice that $m\angle DBE$ is on both sides of this equation. Subtract ■ from both sides of the equation. Then, $m\angle ABD = m\angle EBC$ and $\angle ABD \cong \angle EBC$. $m\angle DBE$

So, you prove that ■. ✓ $\angle ABD \cong \angle EBC$

Practice

Copy and complete the proof.

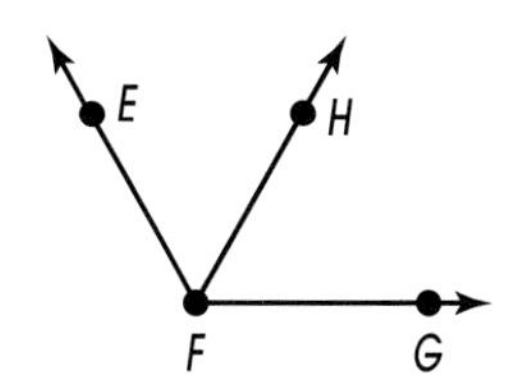

You are given: $\overrightarrow{FH}$ bisects $\angle EFG$.
Prove: $m\angle EFG = 2m\angle EFH$

You are given *that $\overrightarrow{FH}$* bisects *$\angle EFG$. $\angle EFH \cong$* $\angle HFG$.
Then, $m\angle EFH =$ $m\angle HFG$.

You can use the Angle Addition *Postulate to write $m\angle EFG =$* $m\angle EFH + m\angle HFG$.
Now, substitute $m\angle EFH$ for $m\angle HFG$ in the equation.

Then, $m\angle EFG = m\angle EFH +$ ____. *You can add the terms.* $m\angle EFH$
$m\angle EFG =$ $2m\angle EFH$

So, you prove that ____. ✓ $m\angle EFG = 2m\angle EFH$

More practice is provided in the Classroom Resource Binder.

Chapter 2 Review

Summary

A protractor can be used to draw and measure angles.
An angle can be identified as acute, right, obtuse, or staight.
You can use the Angle Addition Postulate to find angle measures.
The sum of the measures of complementary angles is 90°. The sum of the measures of supplementary angles is 180°.
Vertical angles are congruent angles.
An angle bisector divides an angle in half.
A common angle is part of two different angles.
You can make a table to solve problems.
You can use angles to determine direction.
You can use facts you know about angles to prove a statement.

More vocabulary review is provided in the Classroom Resource Binder, page 11.

acute angle
right angle
obtuse angle
complementary angles
congruent angles
adjacent angles
angle bisector
protractor

Vocabulary Review

Complete the sentences with the words from the box.

1. A ____ is a tool used to measure angles. protractor
2. Two angles that share a common vertex and a common ray are called ____. adjacent angles
3. An angle that is greater than 0° and less than 90° is called an ____. acute angle
4. Angles that have the same measure are called ____. congruent angles
5. An angle that is greater than 90° and smaller than 180° is called an ____. obtuse angle
6. A ____ is an angle that is equal to 90°. right angle
7. A ray that divides an angle into two congruent angles is called an ____. angle bisector
8. Two angles whose measures have a sum of 90° are called ____. complementary angles

Write About Math No. The sum of the measures of the two angles must equal 90°. If one angle is greater than 90°, the sum would be greater than 90°.

Chapter Quiz

More assessment is provided in the Classroom Resource Binder.

Draw and name an angle for each measure. Then, classify the angle. Write *acute, right, obtuse,* or *straight.* Check students' work.

1. 180° straight **2.** 90° right **3.** 75° acute **4.** 160° obtuse

Solve for x.

5. $4x - (2x - 6) = 90$ 42 **6.** $9x = 3x + 180$ 30 **7.** $180 - (x - 20) = 3x$ 50

8. Find the angle that is complementary to a 62° angle. 28°

9. Find the angle that is supplementary to a 62° angle. 118°

Use the diagram on the right for exercises 10 and 11.

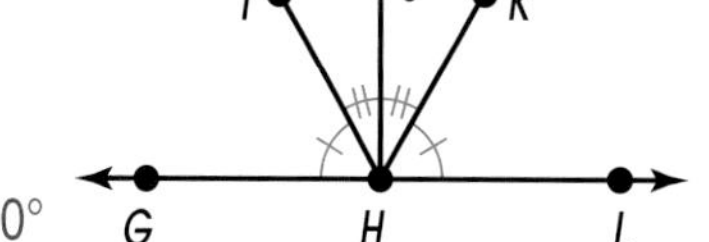

10. $\angle JHK \cong$ ■ $\angle JHI$ or $\angle IHJ$

11. If $\angle GHI$ is 60°, then find the measure of $\angle KHL$. 60°

Find the measure of each angle.

12. $\angle AED$ is 146°. $\overrightarrow{EC}$ bisects $\angle AED$. Find the measure of $\angle AEC$. 73°

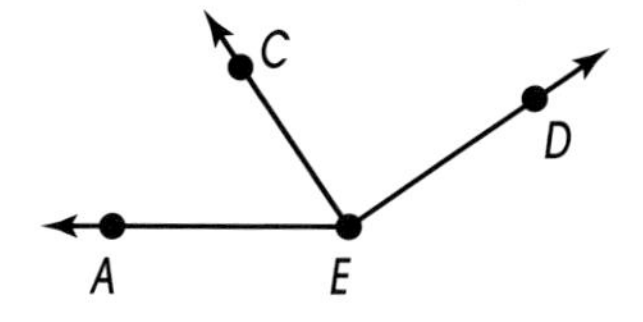

13. Find the measure of $\angle EBA$. 82°

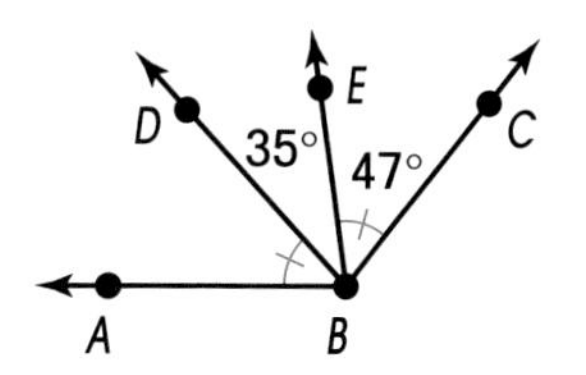

Problem Solving

Solve the problem. Show your work.

14. Joe has to stretch his arm from 90° to 138°. He wants to straighten his elbow by about 12° every 3 days. How many days will it take to get to 138°? 12 days

Write About Math

$\angle A$ and $\angle B$ are complementary angles. Can $\angle A$ be greater than 90°? Why or why not?

See top of page.

Additional Practice for this chapter is provided on page 394 of the student text.

Chapter 3 Reasoning and Proofs

The game of chess requires reasoning. You need to plan your moves. What other games require reasoning?

Learning Objectives

- Use inductive reasoning to continue a pattern.
- Use deductive reasoning in direct and indirect proofs.
- Complete conditional statements.
- Use the Properties of Equality and Congruence.
- Write proofs as a paragraph proof, two-column proof, or flow proof.
- Find a counterexample.
- Write proofs about angles and line segments.

ESL/ELL Note Help students to distinguish among the different forms a proof can take: paragraph proof, two-column proof, or flow proof.

Words to Know

inductive reasoning	a way to reach a conclusion based on a pattern
deductive reasoning	a way to reach a conclusion based on known facts
conditional statement	a statement that uses the words "if" and "then"
hypothesis	the "if" part of a conditional statement
conclusion	the "then" part of a conditional statement
properties of equality	properties used to solve equations
properties of congruence	properties used to prove statements about geometric figures, such as line segments and angles
proof	a way to reach a conclusion using reasoning
counterexample	an example that shows a statement is false
indirect proof	a proof that can be used when there are only two possibilities; if one possibility is false, the other must be true

Famous Trial Project

A lawyer uses logical reasoning about the evidence to prove a person's guilt or innocence. Write about a famous trial. Tell about the crime, the defendant, one or more of the lawyers, and the jury decision.

Project
See the Classroom Resource Binder for a scoring rubric to assess this project.

More practice is provided in the Workbook, page 13, and in the Classroom Resource Binder, page 24.

3·1 Inductive Reasoning

Getting Started
Discuss a train or bus schedule. Elicit that based on experience, you can predict where a train or a bus will be at a given time.

You use **inductive reasoning** when you reach a conclusion based on a pattern. Sometimes you can find a pattern in a sequence of numbers. You can then use the pattern to find more numbers in the sequence.

EXAMPLE 1

Find a pattern in the sequence of numbers below.

1, 3, 5, 7, 9, ...

You can add 2 to each number to find the next number.

1, 3, 5, 7, 9, 11, 13, ...

The pattern is add 2.

Sometimes the pattern is visual.

EXAMPLE 2

Continue the pattern.

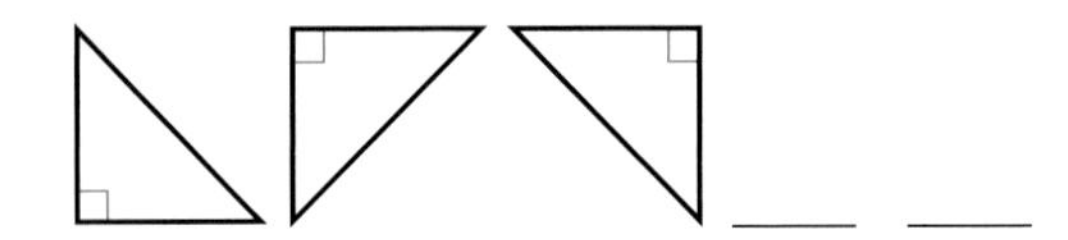

Each figure shows a $\frac{1}{4}$-turn clockwise.

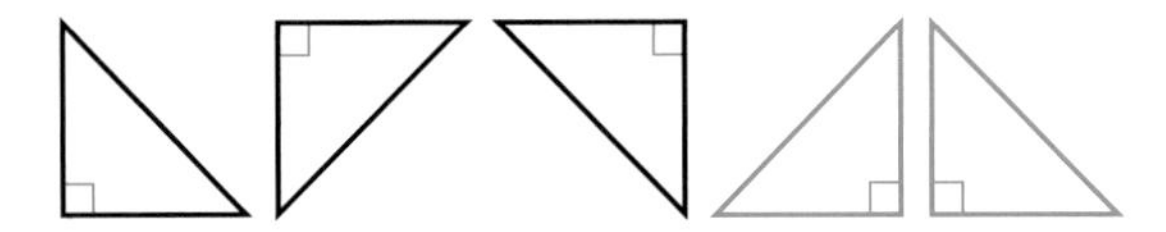

You use inductive reasoning when you reach a conclusion based on experience. The conclusion you reach may be faulty.

EXAMPLE 3

Al noticed that it rained every Monday for 6 weeks. He reasons that it will rain next Monday. Is Al correct? Will it rain next Monday?

On any given day, it may rain or it may not rain.

Al's reasoning is faulty. It may not rain next Monday.

Try This

How many line segments are formed using 6 collinear points?
Use a pattern.

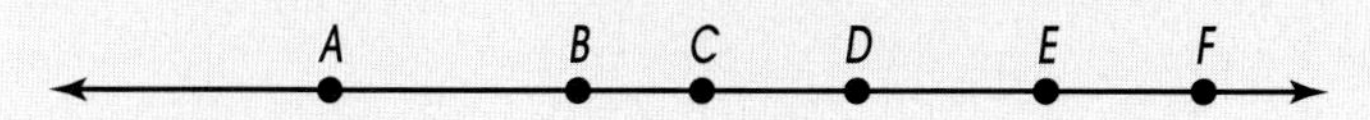

Count the number of line segments formed by 1 point, 2 points, 3 points, and 4 points.

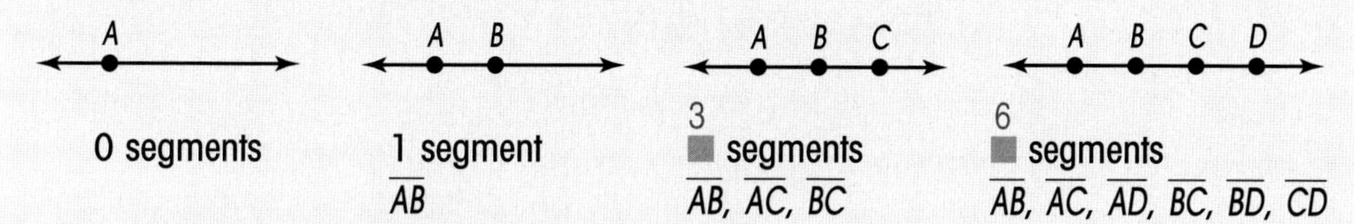

0 segments	1 segment $\overline{AB}$	3 ■ segments $\overline{AB}, \overline{AC}, \overline{BC}$	6 ■ segments $\overline{AB}, \overline{AC}, \overline{AD}, \overline{BC}, \overline{BD}, \overline{CD}$

Put this information into a table. Look for a pattern in the second row.

Number of Points (n)	1	2	3	4	5	6
Number of Line Segments	0	1	3	6	■ 10	■ 15

Pattern +1 +2 +3 +4 ■ +5

5 points form ■ line segments and 6 points form ■ line segments.
10 15

Practice

Continue each pattern. Explain the pattern. Check students' explanations.

1. 50, 45, 48, 43, 46, ____, ____ 41, 44

2. 0, 1, 3, 6, 10, ____, ____ 15, 21

3. 2, 3, 2, 3, ____, ____ 2, 3

4. 3, 4, 2, 3, 1, ____, ____ 2, 0

5.

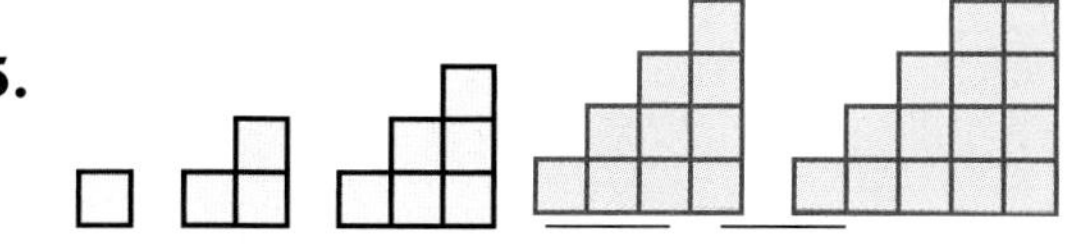

6.

Share Your Understanding

7. Write a number pattern for your partner to continue. Then, have your partner write a number pattern for you to continue.
Answers will vary. Check students' work.

8. **CRITICAL THINKING** Look at Try This. Find the number of line segments for 10 points. 45 line segments

More practice is provided in the Workbook, page 13, and in the Classroom Resource Binder, page 24.

3·2 Deductive Reasoning

Getting Started
Horses like sugar cubes. Ed is a horse. What can you say about Ed? [Ed likes sugar cubes.]

When you reach a conclusion from known facts, you use **deductive reasoning**. You can use what you know about line segments and angles to reach a conclusion.

EXAMPLE 1

Use deductive reasoning. Write a conclusion for the following statement.

F is between E and G.

Math Fact
A postulate is a statement that is accepted as true.

Draw a diagram.

E F G

Write a fact you know. $EF + FG = EG$

Name the fact you used. Segment Addition Postulate

Use the fact to write a conclusion.

If F is between E and G, then $EF + FG = EG$.

The conclusion is "then $EF + FG = EG$."

Sometimes you can use deductive reasoning to reach a conclusion from a diagram.

EXAMPLE 2

Use deductive reasoning. Write a conclusion about the diagram on the left.

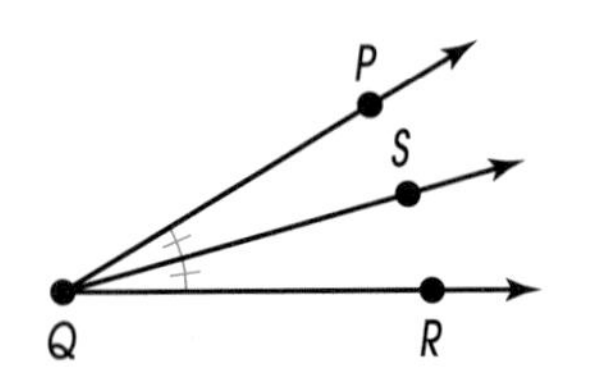

What does the diagram show? $\angle PQS \cong \angle SQR$

List facts you can conclude. $m\angle PQS = m\angle SQR$
$\overrightarrow{QS}$ is the angle bisector.

Use one fact to write a conclusion.

One conclusion is "$\overrightarrow{QS}$ is the angle bisector."

Try This

Use deductive reasoning. Write a conclusion for the following statement.

M is the midpoint of $\overline{JK}$.

Draw and label a diagram.

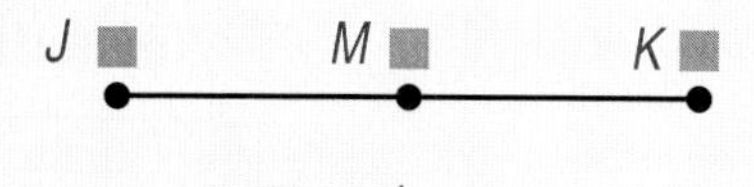

What fact do you know about midpoints?

A midpoint divides a line segment into two ■ segments. congruent

Use the fact to write a conclusion.

One conclusion is $\overline{JM} \cong$ ■. $\overline{MK}$

Practice

Use deductive reasoning. Write a conclusion for each statement.

1. $\angle N$ is a straight angle. $m\angle N = 180$
2. M is the midpoint of $\overline{CD}$. $\overline{CM} \cong \overline{MD}$
3. $\angle Y$ is an acute angle. Answers may vary. The angle should be between 0° and 90°.
4. $\overrightarrow{MN}$ bisects $\angle OMP$. $\angle OMN \cong \angle NMP$
5. Point A lies between point O and point P, and $\overline{OA} \cong \overline{AP}$. A is the midpoint of $\overline{OP}$
6. $\angle C$ and $\angle H$ are complementary angles. $m\angle C + m\angle H = 90$

Use the diagram on the right for exercises **7–9**. Some answers may vary.

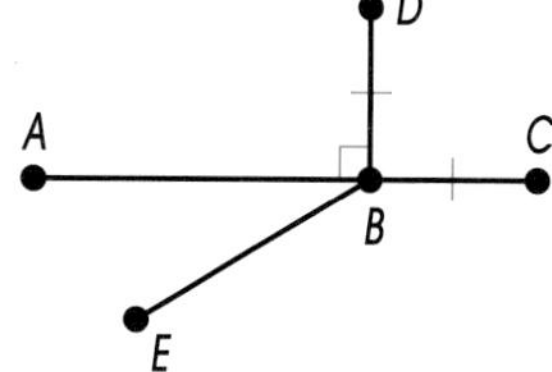

7. Write a conclusion about $\angle ABD$ and $\angle DBC$. See above.

7. $\angle ABD$ and $\angle DBC$ are right angles

8. Write a conclusion about $\overline{BD}$ and $\overline{BC}$. $\overline{BD} \cong \overline{BC}$
9. Write a conclusion about $\angle ABE$ and $\angle EBC$.
   The angles are supplementary.

Share Your Understanding

10. Write a statement about line segments or angles. Ask your partner to write a conclusion. Answers will vary.

11. **CRITICAL THINKING** Decide if the following is inductive reasoning or deductive reasoning. The temperature was 32°F for 10 days. The temperature will be 32°F on the eleventh day.
    Inductive reasoning; a conclusion was reached based on a pattern.

More practice is provided in the Workbook, page 13, and in the Classroom Resource Binder, page 24.

Getting Started Discuss each statement.

1. If two segments are congruent, then they have equal measures. [True; this is the definition]
2. If $\angle K$ is obtuse, then $m\angle K = 150$. [False; $m\angle K$ can be any angle measure between 90 and 180.]

3-3 Conditional Statements

A **conditional statement** uses the words if and then.

If you stand in the rain, then you will get wet.

hypothesis: you stand in the rain — conclusion: you will get wet

The if part of the statement is the **hypothesis**. The then part of the statement is the **conclusion**.

EXAMPLE 1

Look at the statement below. Underline the hypothesis and circle the conclusion.

If M is the midpoint of $\overline{AB}$, then $\overline{AM} \cong \overline{MB}$.

The hypothesis follows if. M is the midpoint of $\overline{AB}$

The conclusion follows then. $\overline{AM} \cong \overline{MB}$

Underline the hypothesis and circle the conclusion.

If <u>M is the midpoint of $\overline{AB}$</u>, then ($\overline{AM} \cong \overline{MB}$.)

Sometimes the hypothesis is at the end of a statement. To find the hypothesis, look for the word *if*.

EXAMPLE 2

Look at the statement below. Underline the hypothesis and circle the conclusion.

$\angle P$ and $\angle Q$ are supplementary angles if the sum of $m\angle P$ and $m\angle Q$ is 180.

The hypothesis follows if.

the sum of $m\angle P$ and $m\angle Q$ is 180

The conclusion begins the statement.

$\angle P$ and $\angle Q$ are supplementary angles

Underline the hypothesis and circle the conclusion.

($\angle P$ and $\angle Q$ are supplementary angles) if <u>the sum of $m\angle P$ and $m\angle Q$ is 180</u>.

Try This

Write a conclusion to complete the following statement.
Tell what fact you used.

If $BA = 4$ and $DC = 4$, then ____.

The lengths of BA and ■ are equal. The lengths of congruent line segments are equal. You can conclude that $\overline{BA} \cong$ ■. Now, complete the statement above.

DC

$\overline{DC}$

If $BA = 4$ and $DC = 4$, then $\overline{BA}$ ■ $\overline{DC}$.

$\cong$

State the fact you used to complete the statement.

congruent

The definition of ■ line segments was used to reach this conclusion.

Practice

Write each conditional statement on a sheet of paper. Underline the hypothesis. Circle the conclusion.

1. If you work for pay, then you will earn money.
2. Vertical angles are formed if two lines intersect.
3. If two angles are congruent, then their measures are equal.
4. If point C is between point A and point B, then $AC + CB = AB$.

Write a conclusion for each conditional statement.

5. If $\angle R$ and $\angle S$ are vertical angles, then ____. $\angle R \cong \angle S$
6. If $\overline{OP}$ is the angle bisector of $\angle AOL$, then ____. $\angle AOP \cong \angle POL$
7. If point T is the midpoint of $\overline{AB}$, then ____. $\overline{AT} \cong \overline{TB}$
8. If the measure of $\angle Q$ is 90, then ____. $\angle Q$ is a right angle.

Share Your Understanding

9. Write a conditional statement for this fact. You can add vitamin D to your diet by drinking milk.
   If you drink milk, then you add vitamin D to your diet.
10. **CRITICAL THINKING** Give two possible hypotheses for this conditional statement. If ____, then they are congruent.
    two angles have equal measures; two line segments have equal lengths

More practice is provided in the Workbook, page 14, and in the Classroom Resource Binder, page 25.

3-4 Algebra Review: Properties of Equality

Getting Started
Have students explain how to solve $x + 7 = 20$. Discuss the idea of keeping both sides of the equation balanced.

One of the tools you can use in reasoning in geometry are the **properties of equality** from algebra.

Properties of Equality	
Addition	If $x = y$, then $x + 2 = y + 2$.
Subtraction	If $x = y$, then $x - 7 = y - 7$.
Multiplication	If $x = y$, then $x \bullet 5 = y \bullet 5$.
Division	If $x = y$, then $x \div 8 = y \div 8$.

EXAMPLE 1

Name the property used to reach the conclusion below.

If $m\angle B + 30 = 180$, then $m\angle B = 150$.

Look at each equation. You subtract 30 from both sides of the equation.

$$m\angle B + 30 = 180$$
$$m\angle B = 150$$

You can use the Subtraction Property of Equality.

Here are four more properties of equality.

Properties of Equality	
Reflexive	$x = x$
Symmetric	If $x = y$, then $y = x$.
Transitive	If $x = y$ and $y = z$, then $x = z$.
Substitution	If $x = 3$, then you can use 3 for x. So, $x + y = 3 + y$.

EXAMPLE 2

Name the property used to reach the conclusion below.

If $m\angle C = m\angle D$ and $m\angle D = 30$, then $m\angle C = 30$.

Look for the property that links three things together.

The Transitive Property of Equality shows that $m\angle C = 30$.

Try These

Name the property of equality you can use to reach each conclusion.

1. $AB = AB$ — Reflexive ■ of Equality Property
2. If $m\angle C - 50 = 90$, then $m\angle C = 140$. — ■ Property of Equality Addition
3. If $2AM = AB$, then $AM = \frac{1}{2}AB$. — ■ Property of Equality Division or Multiplication

Practice

Name the property of equality you can use to reach each conclusion.

1. If $x = 3$, then $x \bullet 5 = 3 \bullet 5$. substitution
2. If $EF = AB$ and $AB = 3$, then $EF = 3$. transitive
3. If $2m\angle ABD = m\angle ABC$, then $m\angle ABD = \frac{1}{2}m\angle ABC$. division (or multiplication)
4. If $AB = CD$, then $CD = AB$. symmetric
5. If $EF + 10 = 15$, then $EF = 5$. subtraction
6. If $m\angle HIJ = 35$, then $2m\angle HIJ = 70$. multiplication
7. $m\angle PQR = m\angle PQR$ reflexive
8. If $m\angle 1 + m\angle 2 = 180$ and $m\angle 1 = m\angle 3$, then $m\angle 3 + m\angle 2 = 180$. substitution

Share Your Understanding

9. You are given $m\angle 2 = 50$ and $m\angle 1 + m\angle 2 = 90$. Explain to a partner how you can use substitution to rewrite the second equation. Ask your partner to find $m\angle 1$.
   Answers will vary. Check the students' work.

10. **CRITICAL THINKING** Look at the diagram on the right. Find y. Name the property of equality you use for each step of your solution.

$2y + 30 = 50$ definition of congruent angles
$2y = 20$ subtraction property
$y = 10$ division property

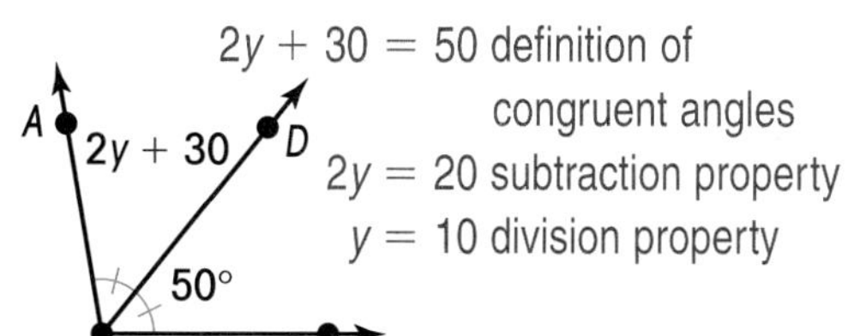

More practice is provided in the Workbook, page 14, and in the Classroom Resource Binder, page 25.

3-5 Properties of Congruence

Getting Started
Have students complete each statement to review the properties of equality.

1. Reflexive $a =$ ____
2. Symmetric
   If $a = b$ then ____.
3. Transitive
   If $a = b$ and $b = c$, then ____.

[a; $b = a$; $a = c$]

When you work with line segments and angles, you use the **properties of congruence**. These are properties of geometric figures.

Properties of Congruence	
Reflexive	$\overline{CD} \cong \overline{CD}$
	$\angle A \cong \angle A$
Symmetric	If $\overline{CD} \cong \overline{EF}$, then $\overline{EF} \cong \overline{CD}$.
	If $\angle A \cong \angle B$, then $\angle B \cong \angle A$.
Transitive	If $\overline{CD} \cong \overline{EF}$ and $\overline{EF} \cong \overline{GH}$, then $\overline{CD} \cong \overline{GH}$.
	If $\angle A \cong \angle B$ and $\angle B \cong \angle C$, then $\angle A \cong \angle C$.

You can use the properties of congruence to write a conclusion about line segments and angles.

EXAMPLE 1

Use the Transitive Property to write a conclusion.

If $\overline{RS} \cong \overline{YZ}$ and $\overline{YZ} \cong \overline{JK}$, then ______.

Look at the statements. $\overline{YZ}$ links $\overline{RS}$ and $\overline{JK}$. Because $\overline{YZ}$ is congruent to both $\overline{RS}$ and $\overline{JK}$, you can write $\overline{RS} \cong \overline{JK}$.

If $\overline{RS} \cong \overline{YZ}$ and $\overline{YZ} \cong \overline{JK}$, then $\overline{RS} \cong \overline{JK}$.

EXAMPLE 2

Use the Symmetric Property to write a conclusion.

If $\angle P \cong \angle Q$, then ____.

The symmetric property changes the order.
To change the order, begin with $\angle Q$.

If $\angle P \cong \angle Q$, then $\angle Q \cong \angle P$.

Try These

Name the property of congruence you can use to reach each conclusion.

1. If $\overline{JK} \cong \overline{MN}$, then $\overline{MN} \cong \overline{JK}$. Symmetric ■ Property of Congruence
2. If $\angle R \cong \angle S$ and $\angle S \cong \angle T$, then $\angle R \cong \angle T$. Transitive ■ Property of Congruence
3. $\overline{PQ} \cong \overline{PQ}$ Reflexive ■ Property of Congruence

Practice

Name the property of congruence you can use to reach each conclusion.

1. If $\angle DEF \cong \angle PQR$, then $\angle PQR \cong \angle DEF$. Symmetric
2. $\angle 1 \cong \angle 1$ Reflexive
3. If $\overline{LM} \cong \overline{PQ}$ and $\overline{PQ} \cong \overline{RS}$, then $\overline{LM} \cong \overline{RS}$. Transitive
4. If $\overline{AB} \cong \overline{HI}$, then $\overline{HI} \cong \overline{AB}$. Symmetric

Use a property of congruence or a property of equality to write each conclusion. Name the property you used. Some answers may vary.

5. If $x = 5$ and $5 = y$, then ____. $x = y$; Transitive
6. $\overline{AB} \cong$ ____ $\overline{AB}$; Reflexive
7. If $m\angle ABC + 30 = 90$, then ____. $m\angle ABC = 60$; Subtraction property of equality
8. If $AM = \frac{1}{2}AB$, then ____. $2AM = AB$; Multiplication property of equality

Share Your Understanding

9. Choose a property of congruence. Use the property to write a statement about angles. Ask your partner to name the property you used.
Check students' work.

10. **CRITICAL THINKING** In the diagram on the right, $\overline{BD}$ bisects $\overline{AC}$ and $\overline{DC} \cong \overline{BD}$. Use the transitive property to show that $\overline{AD} \cong \overline{BD}$.
Because $\overline{BD}$ bisects $\overline{AC}$, $\overline{AD} \cong \overline{DC}$. You are given $\overline{DC} \cong \overline{BD}$. By the transitive property $\overline{AD} \cong \overline{BD}$.

B
A D C

More practice is provided in the Workbook, page 15, and in the Classroom Resource Binder, page 26.

3·6 Paragraph Proof

Getting Started
Have students use 'prove' or 'proof' in a sentence. Discuss the everyday meanings of these words.

You began your work with proofs in Chapter 1 and in Chapter 2. A **proof** uses logical reasoning to establish the truth of a statement. To write a proof, you use only information you know is true.

You can write a proof in paragraph form. This is a paragraph proof. Begin with the information you are given. Then tell how you can reach the statement you wish to prove.

EXAMPLE

You are given:
M is the midpoint of $\overline{AB}$.

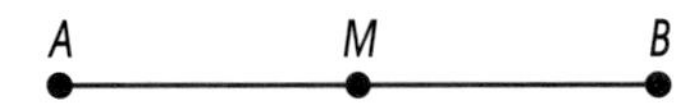

Prove: $AB = 2AM$

You are given that M is the midpoint of $\overline{AB}$. Because the midpoint divides a line segment into two congruent parts, you can write $AM = AB$. Then, $AB = AM + AM$. You can combine the like terms on the right side of this equation. The equation becomes $AB = 2AM$.

So, you prove that $AB = 2AM$. ✓

You can write this proof in more than one way. Be sure that the statements you write are true. Be sure that they follow one after the other in logical order.

Try This

Copy and complete the paragraph proof.

You are given: $\overline{AB} \cong \overline{CD}$

Prove: $\overline{AC} \cong \overline{BD}$

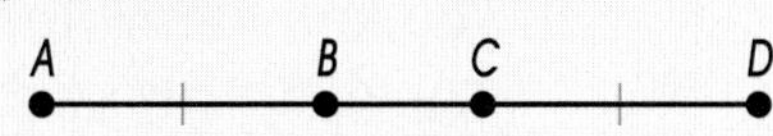

Look at the diagram. Look at what you need to prove. To prove $\overline{AC} \cong \overline{BD}$, you need to add the common segment. Begin with the given.

You are given that $\overline{AB} \cong$ ■. Because the measures of congruent line segments are ■, $AB = CD$. You can ■ BC to both sides of this equation. So, $AB + BC = CD +$ ■. You can rewrite this equation as $AB + BC = BC + CD$.

Now, you can use the Segment ■ Postulate to write $AC = BD$. By definition of ■ line segments, $\overline{AC} \cong \overline{BD}$.

So, you prove that ■.

CD; equal; add; BC; Addition; congruent; ✓ $\overline{AC} \cong \overline{BD}$

Practice

Write a paragraph proof for each of the following. See Additional Answers in the back of this book.

1. You are given:
$\angle JKM \cong \angle NKL$

Prove: $\angle JKN \cong \angle MKL$

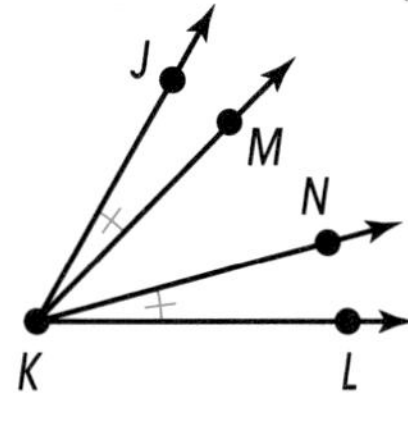

2. You are given:
$\angle PQT \cong \angle SQR$

Prove: $\angle PQS \cong \angle TQR$

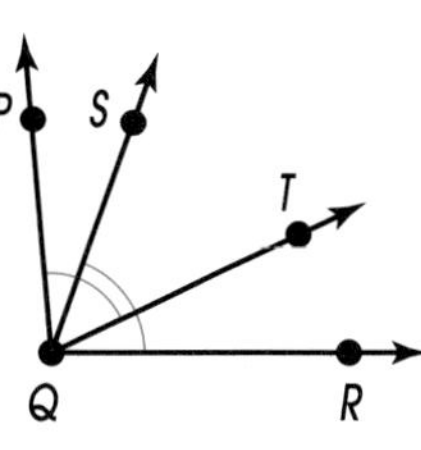

Share Your Understanding

3. Work with a partner. Compare the definition of a midpoint to the Midpoint Theorem. How do these two statements differ?

3. Possible answer: The definition of a midpoint says that the midpoint divides the segment into two congruent segments. The theorem says that if M is a midpoint, then it divides AB in half so that $AM = \frac{1}{2}AB$.

4. **CRITICAL THINKING** How many bisectors does a line segment have? Use a diagram to help explain your reasoning.
Check students' diagrams. A line segment has an infinite number of bisectors, all of which pass through the midpoint.

More practice is provided in the Workbook, page 16, and in the Classroom Resource Binder, page 27.

Two-Column Proof

Getting Started
Discuss the reasoning used in a deductive proof. Point out that the paragraph proof is but one form used to write a deductive proof.

You know how to write a proof in paragraph form. You can also write proofs in two-column form. One column is for statements and the other column is for reasons. The reasons tell you why the statement is true.

You can use what is given, definitions, properties, postulates, and theorems for reasons. Here is the proof you learned in two-column form.

EXAMPLE

You are given:
M is the midpoint of $\overline{AB}$.

A M B

Prove: $AB = 2AM$

Statements	Reasons
1. M is the midpoint of $\overline{AB}$.	1. Given
2. $\overline{AM} \cong \overline{MB}$	2. Definition of a midpoint
3. $AM = MB$	3. Definition of congruent line segments
4. $AB = AM + MB$	4. Segment Addition Postulate
5. $AB = AM + AM$	5. Substitution Property of Equality
6. $AB = 2AM$	6. Combine like terms

So, you prove that $AB = 2AM$. ✓

Compare the two-column proof with the paragraph proof in Lesson 3.6. In a paragraph proof, you can give a reason before or after a statement. The writing style is up to you.

In a two-column proof, you write the statement on the left and the reason on the right. These are your two columns.

Try This

Copy and complete the two-column proof.

You are given: $\overline{AB} \cong \overline{CD}$
Prove: $\overline{AC} \cong \overline{BD}$

Statements	Reasons
1. $\overline{AB} \cong \overline{CD}$	1. ■ Given
2. ■ $AB = CD$	2. Definition of congruent line segments
3. $BC = BC$	3. ■ Property of Equality Reflexive
4. $AB + BC = CD + BC$	4. ■ Property of Equality Addition
5. $AB + BC = BC + CD$	5. Commutative ■ property
6. $AC = BD$	6. Segment ■ Postulate Addition
7. $\overline{AC} \cong \overline{BD}$	7. ■ of congruent line segments Definition

Practice

Write a two-column proof for each of the following. See Additional Answers in the back of this book.

1. You are given:
$\angle JKM \cong \angle NKL$

Prove: $\angle JKN \cong \angle MKL$

2. You are given:
$\angle PQT \cong \angle SQR$

Prove: $\angle PQS \cong \angle TQR$

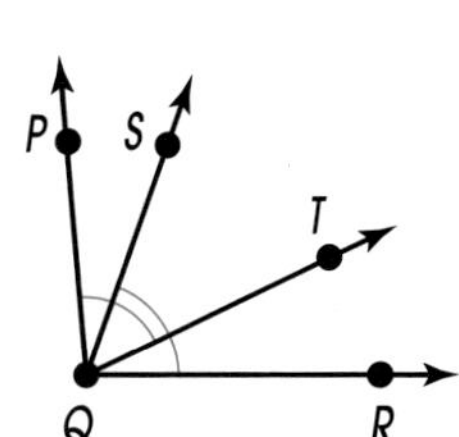

Share Your Understanding

3. **CRITICAL THINKING** Choose the type of proof you like to use, paragraph proof or two-column proof. Explain why.
Answers will vary.

3·8 Calculator: Finding a Counterexample

Getting Started
Ask students to decide if the following statement is true or false. Have students explain their reasoning.

Any number multiplied by 3 is an odd number.

[False; 2 • 3 = 6; 6 is an even number.]

Scientists conduct experiments to try to discover facts. They reach a conclusion based on the results of many experiments. This is inductive reasoning.

When you use inductive reasoning, you need to be careful. Sometimes you can find a **counterexample**. This shows that the reasoning is false.

EXAMPLE

Find a counterexample to prove that the following statement is false.

$n^2 + n + 5$ will always be a prime number.

Remember
A prime number is only divisible by itself and 1. Some prime numbers are: 2, 3, 5, 7, 11, 13, and 17.

Use your calculator to evaluate $n^2 + n + 5$ for $n = 1$. **Display**

PRESS: 1 × 1 + 1 + 5 = 7

7 is a prime number. The statement is true for $n = 1$.

Use your calculator to evaluate $n^2 + n + 5$ for $n = 2$. **Display**

PRESS: 2 × 2 + 2 + 5 = 11

11 is a prime number. The statement is true for $n = 2$.

Use your calculator to evaluate $n^2 + n + 5$ for $n = 3$. **Display**

PRESS: 3 × 3 + 3 + 5 = 17

17 is a prime number. The statement is true for $n = 3$.

Use your calculator to evaluate $n^2 + n + 5$ for $n = 4$. **Display**

PRESS: 4 × 4 + 4 + 5 = 25

25 is not a prime number.
The statement above is not true for $n = 4$.

So, $n^2 + n + 5$ will not always be a prime number.
A counterexample is $n = 4$.

Practice

Find a counterexample to prove each statement is false. Use your calculator. (Hint: Make an organized list.)

1. $n^2 + n + 11$ is a prime number. $n = 11$

2. $n^2 + n + 17$ is a prime number. $n = 17$

3. $(a + b)^2 = a^2 + b^2$ $a = 3$ and $b = 2$

4. If $x^2 = y^2$, then $x = y$. $x = 1$ and $y = -1$

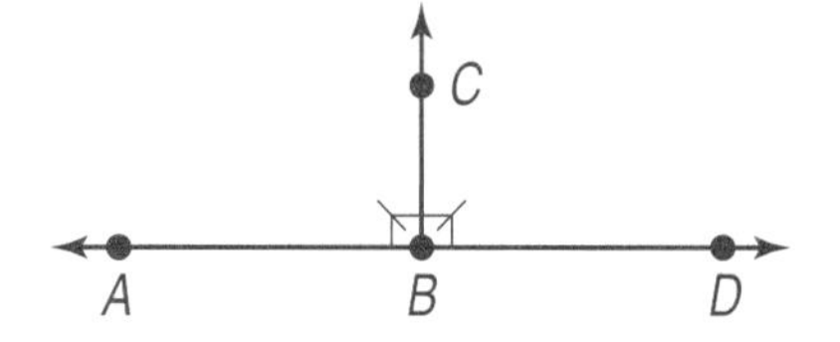

5. If two angles are congruent, they must be vertical angles. (Hint: Draw a diagram.) False. See above for a possible diagram.

Math Connection

FRACTALS

A fractal is a figure that repeats itself following a rule. Follow the directions below to create this fractal.

Take a large sheet of paper and a ruler. Draw a triangle like the one on the right. All of the sides of the triangle are congruent.

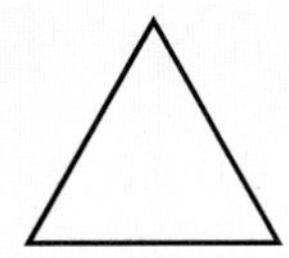

Find the midpoint of each side. Then, connect the midpoints to form another triangle. Shade this new triangle.

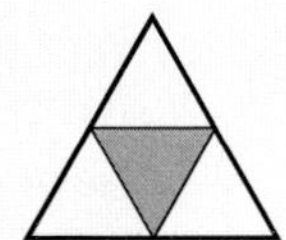

Repeat the activity in the three triangles that are not shaded. Each smaller triangle you form is similar to the first triangle.

The word *fractal* was first used by mathematician Benoit Mandelbrot.

More practice is provided in the Workbook, page 17, and in the Classroom Resource Binder, page 28.

3·9 Problem-Solving Skill: Indirect Proof

Getting Started
Discuss the word *contradict*. Write the following on the board:
$\angle A$ is acute. $\angle A$ is 120°.
Point out that the second statement contradicts the first.

The proofs you have done so far are direct proofs. Sometimes it is easier to use an **indirect proof**. Assume that the opposite of what you want to prove is true. Prove that this assumption is false. Then, what you want to prove must be true.

EXAMPLE 1

You are given: $AB \neq BD$
B is between point A and point D.

Prove: B is not the midpoint of AD.

Math Fact
The symbol $\neq$ means "is not equal to."

Assume the opposite of the prove statement is true.

B is the midpoint of AD.

Prove this assumption is false.

If B is the midpoint, then $AB = BD$.
This contradicts the given fact that $AB \neq BD$.

Thus, what you assume is false. The prove statement must be true.

So, you prove that B is not the midpoint. ✓

EXAMPLE 2

You are given: Point A and Point B

Prove: There is one line between A and B.

Assume the opposite of the prove statement is true.

More than one line can be drawn between point A and point B.

Prove the assumption is false. Draw a diagram.

You can only draw one line between point A and point B. Anything else you draw is a curve.

So, you prove that there is one line between A and B. ✓

Try These

Write the opposite of each statement.

1. $\overrightarrow{AD}$ bisects $\angle CAB$. $\overrightarrow{AD}$ does ■ bisect $\angle CAB$. not
2. $\overline{PQ} \cong \overline{ST}$ $\overline{PQ}$ ■ $\overline{ST}$ $\ncong$
3. $m\angle 1 = m\angle 2$ $m\angle 1$ ■ $m\angle 2$ $\neq$

Practice

Write the opposite of each statement.

1. $\overleftrightarrow{JK}$ does not bisect $\overline{GH}$ $\overleftrightarrow{JK}$ bisects $\overline{GH}$.
2. $AB = CD$ $AB \neq CD$
3. P is the midpoint of $\overline{MN}$ P is not the midpoint of $\overline{MN}$.
4. $m\angle 5 = m\angle 7$ $m\angle 5 \neq m\angle 7$
5. $\angle A \cong \angle B$ $\angle A \ncong \angle B$
6. $\angle E$ and $\angle F$ are complementary angles. $\angle E$ and $\angle F$ are not complementary angles.
7. Copy and complete the following indirect proof.

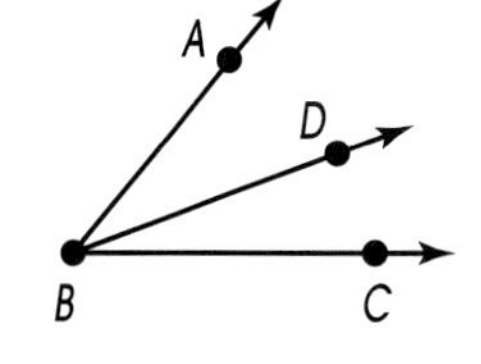

You are given: $m\angle ABD = m\angle DBC$.
$\overrightarrow{BD}$ is between $\overrightarrow{BA}$ and $\overrightarrow{BC}$.
Prove: $\overrightarrow{BD}$ is the angle bisector.
Assume: $\overrightarrow{BD}$ is not the angle bisector

If $\overrightarrow{BD}$ is not the angle bisector, $m\angle ABD \neq m\angle DBC$. This contradicts the given fact that $m\angle ABD = m\angle DBC$. Thus, what you assume is false.
The prove statement must be true.
So, you prove that $\overrightarrow{BD}$ is the angle bisector.

Share Your Understanding

8. Write a statement you want to prove. Have your partner write the opposite of this statement. Check students' work.

9. **CRITICAL THINKING** Write an indirect proof to prove: *Two straight lines intersect at one point.* (Hint: Draw a diagram.) See Additional Answers in the back of this book.

More practice is provided in the Workbook, page 18, and in the Classroom Resource Binder, page 29.

3-10 Problem-Solving Application: Flow Proof

Getting Started
Have students use a flow chart to show the steps they take every school day, from the time they get out of bed in the morning until they arrive in their first class.

You learned to write a paragraph proof and a two-column proof. Another way to show the reasoning in a proof is the flow proof. In a flow proof, you place each statement and its reason in a box. You use arrows to show how each statement leads to another.

EXAMPLE

You are given:
M is the midpoint of $\overline{AB}$.

Prove: $AB = 2AM$

Write each statement and reason in a box. Use arrows to show how each statement leads to another.

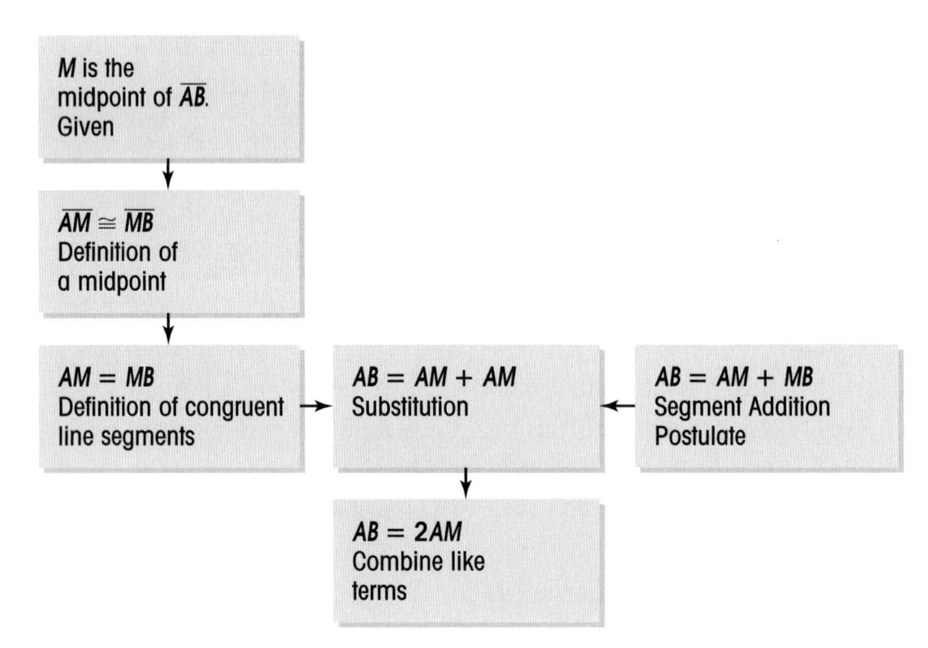

So, you prove that $AB = 2AM$. ✓

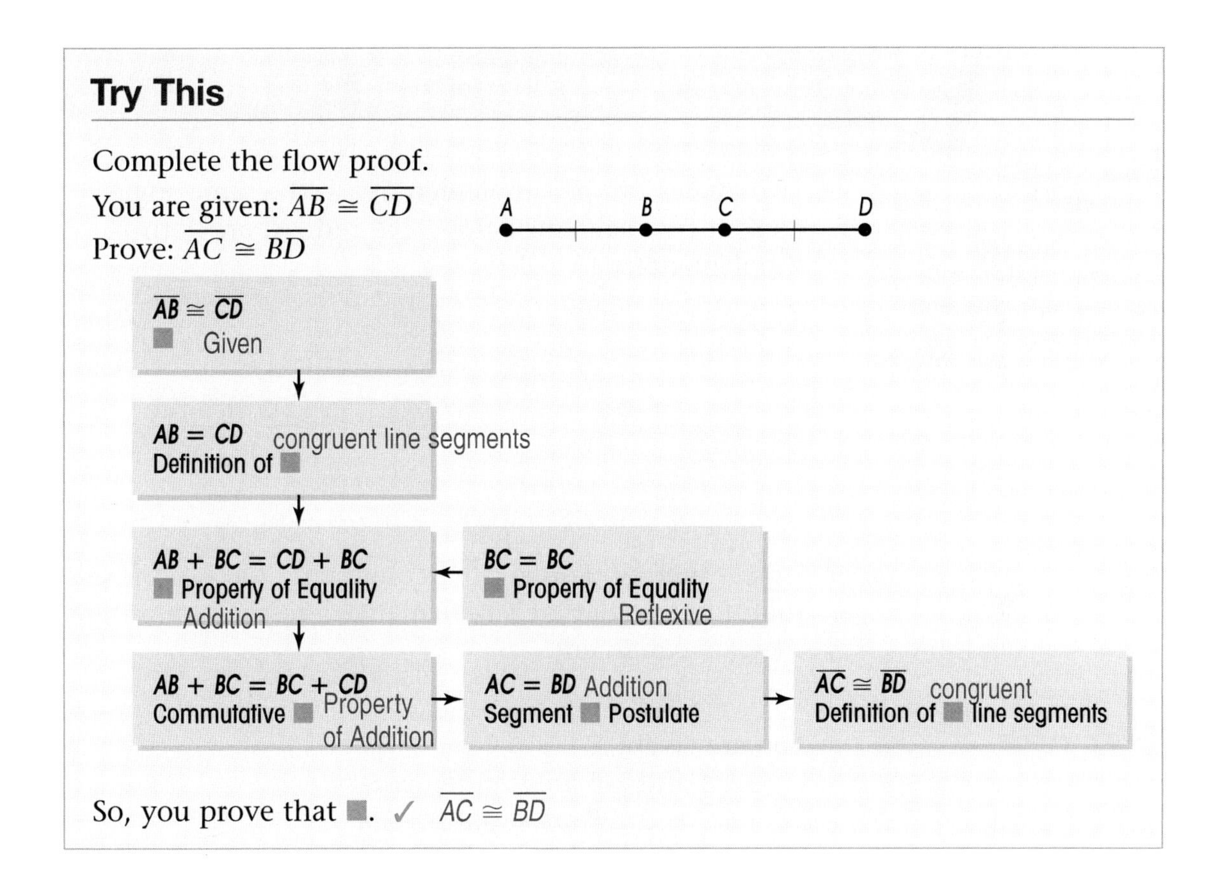

Practice

Write a flow proof for each of the following. See Additional Answers in the back of this book.

1. You are given:
$\angle JKM \cong \angle NKL$

Prove: $\angle JKN \cong \angle MKL$

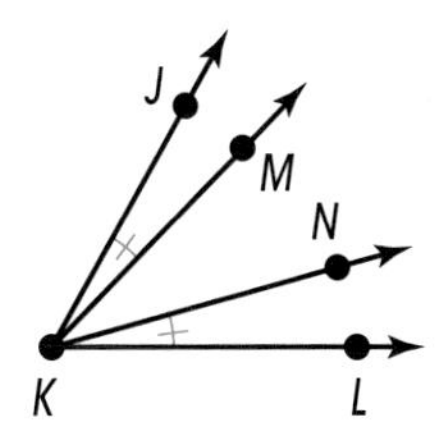

2. You are given:
$\angle PQT \cong \angle SQR$

Prove: $\angle PQS \cong \angle TQR$

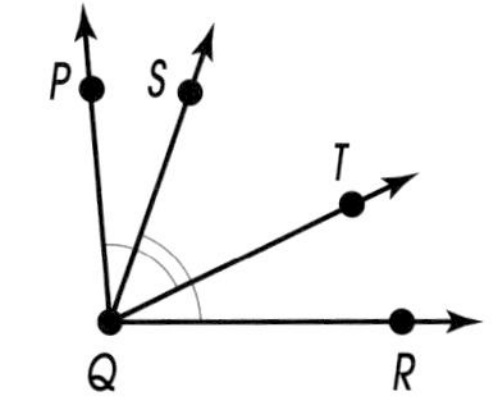

An alternate two-column proof lesson is provided on page 408 of the student book.

3·11 Proof: Angles and Line Segments

Getting Started Review supplementary and complementary angles.

> **THEOREM 6**
>
> If two angles are supplementary to the same angle, they are congruent.

Two angles can be supplementary to the same angle. Look at the diagrams below. $\angle A$ and $\angle C$ are both supplementary to $\angle B$.

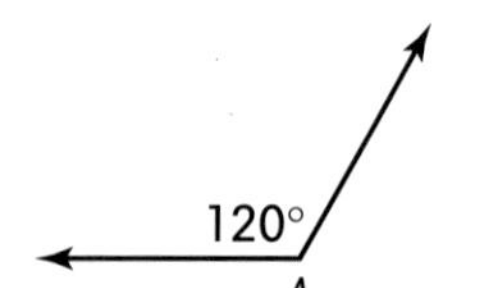

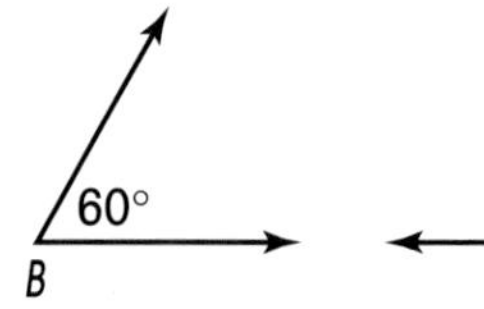

120° C

$$m\angle A + m\angle B = 180 \qquad m\angle B + m\angle C = 180$$

Notice that $\angle A$ and $\angle C$ are congruent. You can prove that if two angles are supplementary to the same angle, they are congruent.

EXAMPLE

You are given:

$\angle A$ and $\angle B$ are supplementary.
$\angle C$ and $\angle B$ are supplementary.

Prove: $\angle A \cong \angle C$

Begin with what you are given. Then, step by step use reasoning to reach the statement you need to prove.

You are given that $\angle A$ and $\angle B$ are supplementary. You are also given that $\angle C$ and $\angle B$ are supplementary. Supplementary angles are angles whose measures have a sum of 180°. So, $m\angle A + m\angle B = 180$ and $m\angle C + m\angle B = 180$.

Since both sums equal 180, they are equal to each other. $m\angle A + m\angle B = m\angle C + m\angle B$ You can subtract $m\angle B$ from both sides of this equation. Then, $m\angle A = m\angle C$. By definition of congruent angles, $\angle A \cong \angle C$.

So, you prove that $\angle A \cong \angle C$. ✓

Try This

You are given: $\overleftrightarrow{AB}$ intersects $\overleftrightarrow{CD}$.

Prove: $\angle 1 \cong \angle 3$

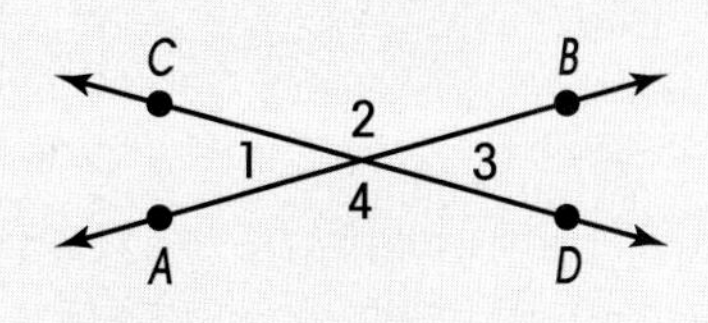

Look at the drawing. $\angle 1$ and $\angle 3$ are vertical angles. You need to prove that they are congruent.

You are given that $\overleftrightarrow{AB}$ ■ $\overleftrightarrow{CD}$. From the diagram you can see that $\angle 1$ and $\angle 2$ form a straight angle. Then, $m\angle 1 +$ ■ $= 180$. Notice, $\angle 2$ and $\angle 3$ also form a ■ angle. Then, $m\angle 2 + m\angle 3 =$ ■. intersects; $m\angle 2$; straight; 180

Since both sums equal 180, the angles are equal to each other. $m\angle 1 + m\angle 2 = m\angle 3 + m\angle 2$. You can subtract ■ from both sides of this equation. Then, $m\angle 1 =$ ■. By definition of ■ angles, $\angle 1 \cong \angle 3$. $m\angle 2$; $m\angle 3$; congruent

So, you prove that ■. ✓ $\angle 1 \cong \angle 3$

Practice

Write a proof for each of the following. See Additional Answers in the back of this book.

1. You are given:
$\angle D$ and $\angle E$ are complementary.
$\angle F$ and $\angle E$ are complementary.

Prove: $\angle D \cong \angle F$

2. You are given: $\overline{AB} \cong \overline{PQ}$
$\overline{BC} \cong \overline{QR}$

Prove: $\overline{AC} \cong \overline{PR}$

3. You are given: $\overline{JL} \cong \overline{EF}$
Point L is the midpoint of $\overline{JK}$.

Prove: $\overline{LK} \cong \overline{EF}$

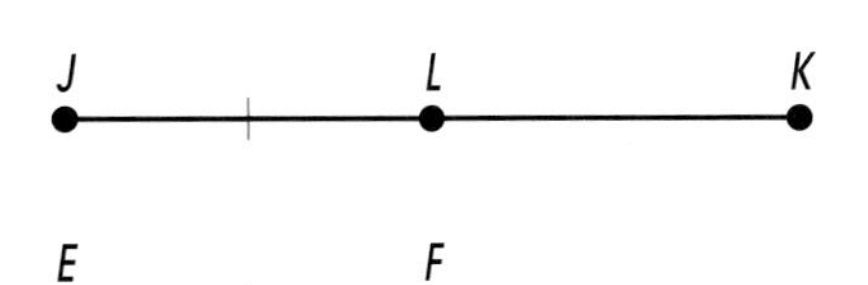

More Practice is provided in the Classroom Resource Binder.

Chapter 3 Review

Summary

You can look at the change in each number or shape to find a pattern in a sequence.
You can use patterns you know to reach a conclusion.
A conditional statement uses the words "if" and "then."
You can use the properties of equality to rewrite an equation.
You can use the properties of congruence to write a proof about line segments and angles.
To prove a statement is true, begin with what is given. Then write true statements using facts you know. Follow a logical order until you reach the statement you need to prove.
A paragraph proof, a two-column proof, and a flow proof are different forms of proof.
A counterexample can be used to prove that a conclusion is false.
An indirect proof is used when there are only two possibilities. If one possibility is false, the other must be true.

More vocabulary review is provided in the Classroom Resource Binder, page 23.

inductive reasoning
deductive reasoning
properties of congruence
conditional statement
indirect proof
hypothesis
counterexample
conclusion

Vocabulary Review

Complete the sentences with words from the box.

1. A ____ is the "then" part of a conditional statement. conclusion
2. A ____ shows the reasoning is false. counterexample
3. A proof that can be used when there are only two possibilities is an ____. indirect proof
4. A way to reach a conclusion based on a pattern is ____. inductive reasoning
5. A ____ uses the words "if" and "then." conditional statement
6. A ____ is the "if" part of a conditional statement. hypothesis
7. A way to reach a conclusion based on know facts is ____. deductive reasoning
8. ____ are used to prove statements about geometric figures, such as line segments and angles. properties of congruence

Chapter Quiz

More assessment is provided in the Classroom Resource Binder.

Continue each pattern.

1. 3, 6, 5, 8, 7, ____, ____ 10, 9

2. △ □ △ □ △ □

Write a conclusion for each conditional statement.

3. If $\angle A \cong \angle B$, then ____. $m\angle B \cong m\angle A$

4. If $CD = HI$, then ____. $HI = CD$

5. If C is between A and B, then ____. $AB = AC + CB$

6. If $\overrightarrow{QS}$ bisects $\angle PQR$, then ____. $\angle PQS \cong \angle SQR$

Match the statement on the left with the property on the right.

7. $\angle A \cong \angle A$ D

8. If $AB = CD$, then $CD = AB$. C

9. If $\angle 1 \cong \angle 2$ and $\angle 2 \cong \angle 3$, then $\angle 1 \cong \angle 3$. A

10. If $x = y$, then $x + 2 = y + 2$. B

A. Transitive property

B. Addition property

C. Symmetric property

D. Reflexive property

Problem Solving

11. Write an indirect proof.
You are given: n is a prime number.
Prove: n^2 is not a prime number.
See Additional Answers in the back of this book.

12. Write a proof. Choose the form you like best.
You are given: $\overrightarrow{KM}$ bisects $\angle JKL$
Prove: $m\angle JKM = \frac{1}{2} m\angle JKL$
Check students' work. Proofs will vary. The proof should show understanding of a bisector.

Additional Practice for this chapter is provided on page 395 of the student book.

A postulate is a statement that is accepted as true. A theorem is a statement you can prove to be true.

Write About Math

What is the difference between a postulate and a theorem?

Chapter 4 Perpendicular and Parallel Lines

The lanes in this pool help the racers swim as fast as they can in a straight line. What would happen if the pool lanes were not parallel? How would that affect the race?

Learning Objectives

- Identify parallel and perpendicular lines.
- Identify a perpendicular bisector.
- Identify angle pairs formed when parallel lines are cut by a transversal.
- Find the measure of alternate interior angles, same-side interior angles, and corresponding angles.
- Find the measure of angle pairs using a calculator.
- Draw a one-point perspective of parallel lines.
- Use parallel and perpendicular lines to determine the number of possible routes a taxicab can take.
- Write proofs about parallel lines.

ESL/ELL Note Have students add words to their vocabulary list. Have them draw a pair of parallel lines with a transversal and number the angles from 1 to 8. Then, list all pairs of alternate interior, same-side interior, and corresponding angles.

Words to Know

perpendicular lines	two lines that intersect to form a right angle
perpendicular bisector	a line that bisects a line segment and is perpendicular to the line segment
parallel lines	lines that lie in the same plane and do not intersect
transversal	a line that intersects two different lines at two different points
alternate interior angles	interior angles that are on opposite sides of a transversal
same-side interior angles	interior angles that are on the same side of the transversal
corresponding angles	angles on the same side of a transversal; one is an interior angle and one is an exterior angle
one-point perspective	a way to draw real objects on a flat surface as they appear in real life
horizon line	the level of the viewer's eyes as the viewer looks across a distance in a drawing
vanishing point	a point on the horizon where parallel lines appear to meet

Modern Art Project

The Dutch artist Piet Mondrian used parallel and perpendicular lines in some of his paintings. You can do something similar with paper and pencil or on the computer. Make a drawing using parallel and perpendicular lines. Use color to fill in the spaces between the lines. Trace a part of your drawing. Label the lines as parallel or perpendicular.

Project Students can begin work on the project after completing Lesson 4.3. See the Classroom Resource Binder for a scoring rubric to assess this project.

More practice is provided in the Workbook, page 19, and in the Classroom Resource Binder, page 34.

4·1 Perpendicular Lines

Getting Started
Review the Vertical Angle Theorem and the Angle Addition Postulate.

> **THEOREM 7**
> Perpendicular lines form four congruent right angles.

Perpendicular lines are two lines that intersect to form a right angle. Line segments and rays can also be perpendicular.

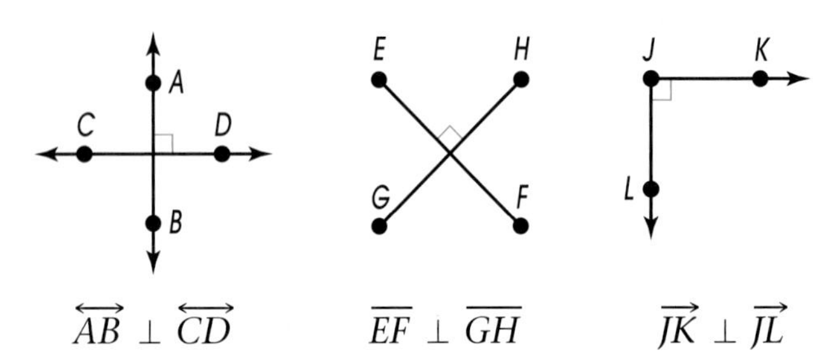

$\overleftrightarrow{AB} \perp \overleftrightarrow{CD}$ $\overline{EF} \perp \overline{GH}$ $\overrightarrow{JK} \perp \overrightarrow{JL}$

In the first diagram, $\overleftrightarrow{AB}$ is perpendicular to $\overleftrightarrow{CD}$.
The symbol $\perp$ means "is perpendicular to."
You can write $\overleftrightarrow{AB} \perp \overleftrightarrow{CD}$.

EXAMPLE 1

Use symbols to describe the diagram on the left.

$\angle RQT$ is a right angle.	$m\angle RQT = 90$
So, the line segment and the line are perpendicular.	$\overline{RS} \perp \overleftrightarrow{TU}$

You can write $\overline{RS} \perp \overleftrightarrow{TU}$

EXAMPLE 2

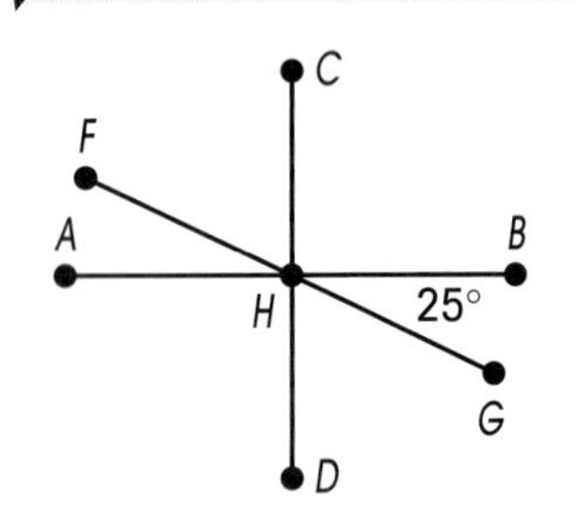

In the diagram, $\overline{AB} \perp \overline{CD}$. Find the measure of $\angle FHD$.

Write an equation.	$m\angle FHD = m\angle FHA + m\angle AHD$
$\angle FHA$ and $\angle BHG$ are vertical angles.	$m\angle FHA = m\angle BHG = 25$
$\angle AHD$ is a right angle.	$m\angle AHD = 90$
Substitute these values in the first equation.	$m\angle FHD = 25 + 90 = 115$

So, $\angle FHD$ is 115°.

Try This

In the diagram, $\overline{HL} \perp \overline{GJ}$.
Find the measure of $\angle HKM$.

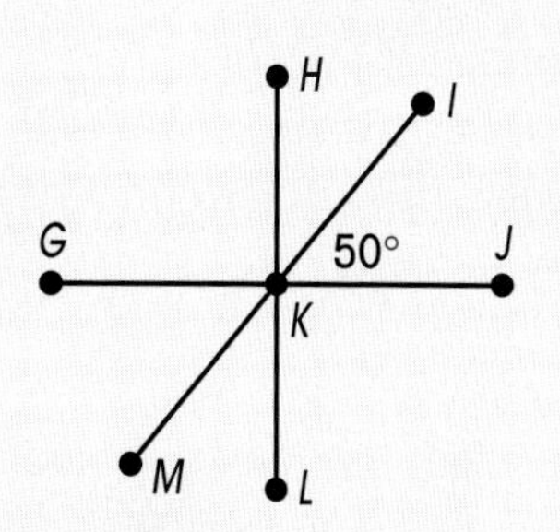

Write an equation.	
Use the Angle Addition Postulate.	$m\angle HKM = m\angle HKG + ■$ m∠GKM
∠GKM and ∠IKJ are ■ angles. vertical	$m\angle GKM = m\angle IKJ = ■$ 50
Because $\overline{HL} \perp \overline{GJ}$, ∠HKG is a ■ angle. right	$m\angle HKG = ■$ 90
Substitute 90 and 50 in the first equation.	$m\angle HKM = 90 + 50 = ■$ 140

So, $\angle HKM$ is ■. 140°

Practice

Decide if the lines or line segments are perpendicular. Write *yes* or *no*. If they are perpendicular, use the ⊥ symbol to write a statement.

1.

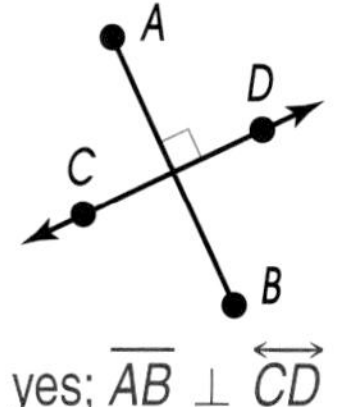

yes; $\overline{AB} \perp \overleftrightarrow{CD}$

2. E H G F

no

3.

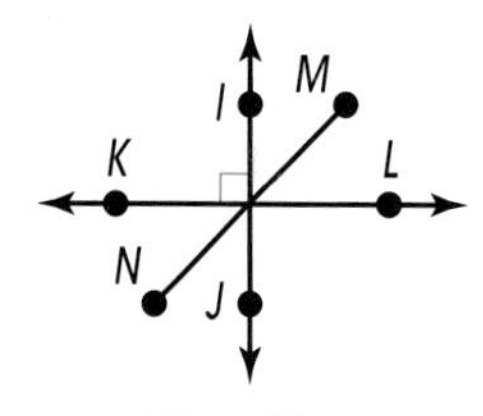

yes; $\overleftrightarrow{KL} \perp \overleftrightarrow{IJ}$

In the diagram, $\overrightarrow{TS} \perp \overleftrightarrow{EG}$. Find the measure of each angle.

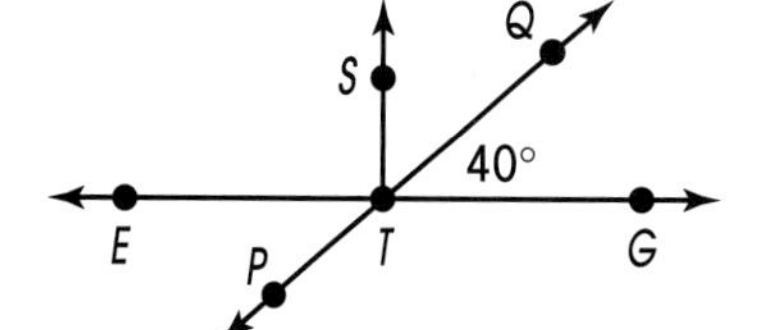

4. $\angle ETS$ 90°

5. $\angle QTG$ 40°

6. $\angle STQ$ 50°

7. $\angle PTG$ 140°

8. $\angle ETP$ 40°

9. $\angle PTS$ 130°

Try This

In the diagram, $\overleftrightarrow{JK} \perp \overline{LC}$.
$m\angle KQR = 2x$ and $m\angle RQC = x + 15$.
Find the measure of $\angle KQR$.

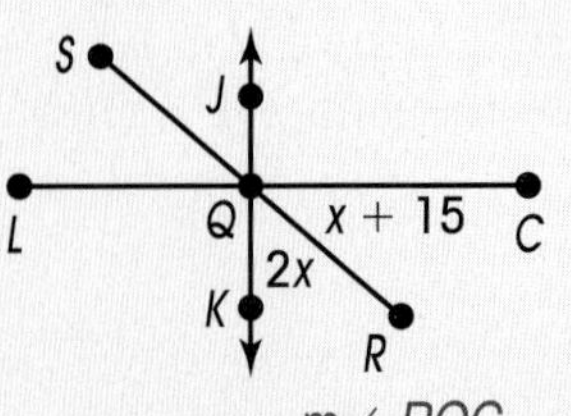

Write an equation. Use the Angle Addition Postulate. $m\angle KQR + \blacksquare = m\angle KQC$ m∠ RQC

Because $\overleftrightarrow{JK} \perp \blacksquare$, $\angle KQC$ is a ■ angle. $\overline{LC}$ right

Substitute the information you know in the equation. $2x + (x + 15) = \blacksquare$ 90

Solve for x.

$3x + 15 = \blacksquare$ 90

$3x = 75$

$x = \blacksquare$ 25

To find m∠ *KQR*, substitute ■ for x. 25

$m\angle KQR = 2(25)$

$m\angle KQR = \blacksquare$ 50

So, $\angle KQR$ is ■. 50°

Practice

In the diagram, $\overline{AC} \perp \overleftrightarrow{BD}$. Find the measure of each angle.

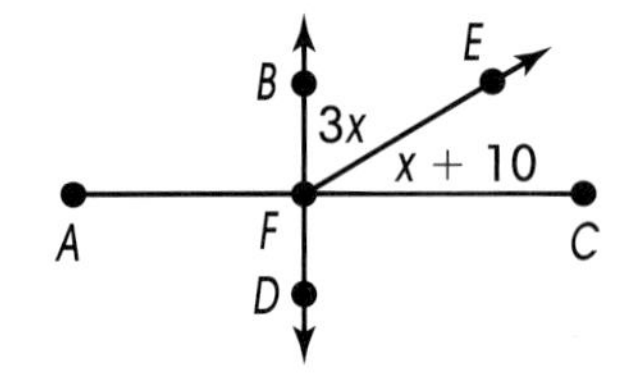

10. $\angle AFB$ 90°

11. $\angle BFE$ 60°

12. $\angle EFC$ 30°

13. $\angle EFD$ 120°

14. $\angle AFE$ 150°

15. $\angle AFC$ 180°

Share Your Understanding

16. Explain to a partner how you know when two lines are perpendicular. Have your partner make a sketch. Check students' work.

17. **CRITICAL THINKING** $\overline{KL}$ intersects $\overline{MN}$ at point *P*. Give a reason to justify the following statement. If $m\angle KPM$ is 90, then $\overline{KL} \perp \overline{MN}$. (Hint: Draw a diagram.) Because m∠ *KPM* is 90, the two line segments intersect to form a right angle. The two line segments are perpendicular.

CONSTRUCTION
A Perpendicular Line Through a Point on a Line

Follow the steps below to construct a line that is perpendicular to any point on the line.

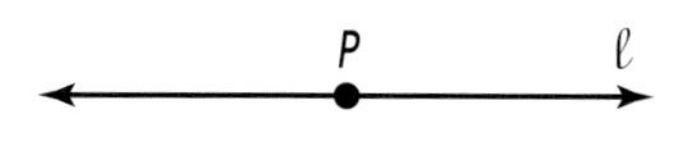

On a sheet of paper, copy line ℓ with point P on the line.

STEP 1 Place the compass on point P. Draw an arc that intersects line ℓ. Label the point X.

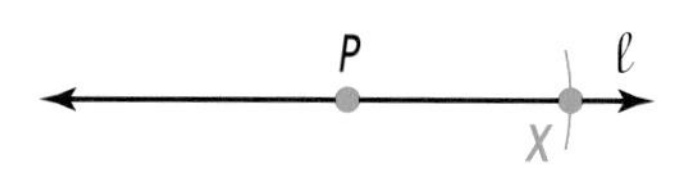

STEP 2 Do not change the compass setting. Draw an arc that intersects line ℓ on the other side of point P. Label the point Y.

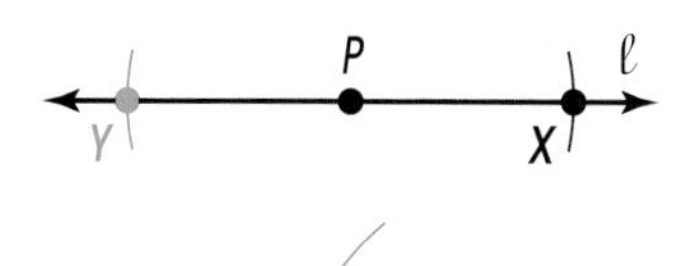

STEP 3 Increase the setting on the compass. Place the compass on point X. Draw an arc above or below point P.

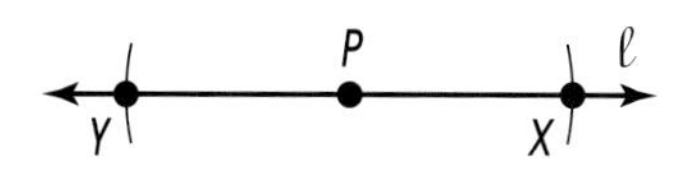

STEP 4 Do not change the compass setting. Place the compass on point Y. Draw an arc that intersects the arc you just drew above or below point P. Label the intersection Z.

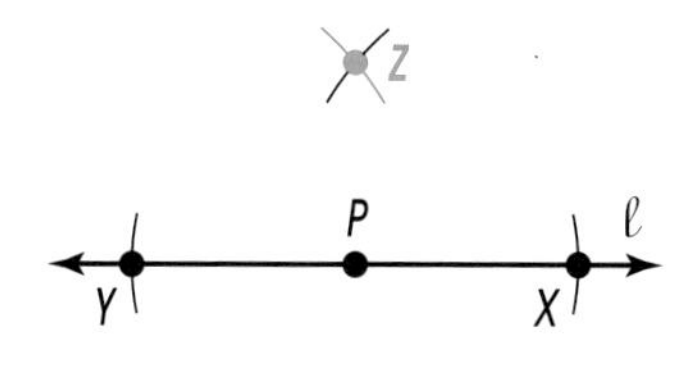

STEP 5 Draw a line that connects point P and point Z.

$\overleftrightarrow{PZ} \perp \overleftrightarrow{XY}$.

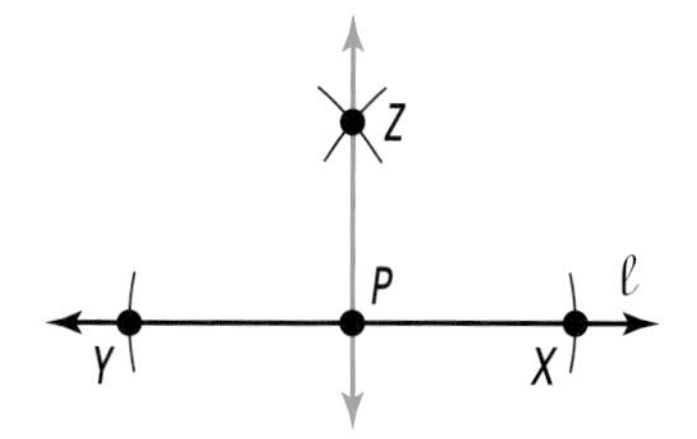

More practice is provided in the Workbook, page 19, and in the Classroom Resource Binder, page 34.

Perpendicular Bisector

Getting Started Review bisector of a line segment.

> **THEOREM 8**
>
> A point on the perpendicular bisector of a segment is equidistant from the endpoints of the segment.

You know that a bisector divides a line segment into two congruent line segments. A **perpendicular bisector** bisects a line segment and is perpendicular to the line segment. A perpendicular bisector can be a line, a line segment, or a ray.

EXAMPLE 1

Which diagram shows $\overline{AB}$ as a perpendicular bisector?

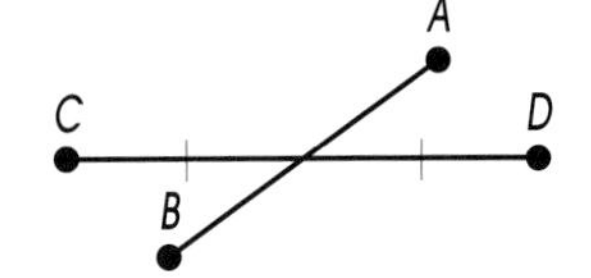

In the first diagram, $\overline{AB}$ bisects $\overline{CD}$. The intersecting lines do not form right angles. $\overline{AB}$ is not perpendicular to $\overline{CD}$. $\overline{AB}$ is NOT a perpendicular bisector.

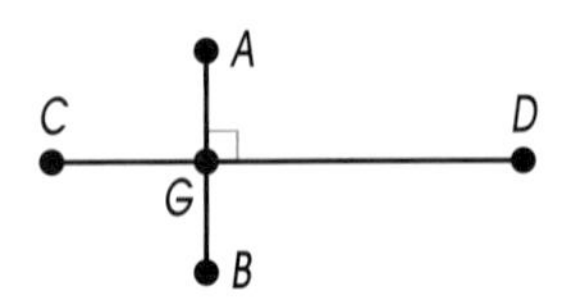

In the second diagram, $\overline{AB}$ is perpendicular to $\overline{CD}$. Because $\overline{CG}$ and $\overline{GD}$ are not congruent, $\overline{AB}$ does not bisect $\overline{CD}$. $\overline{AB}$ is NOT a perpendicular bisector.

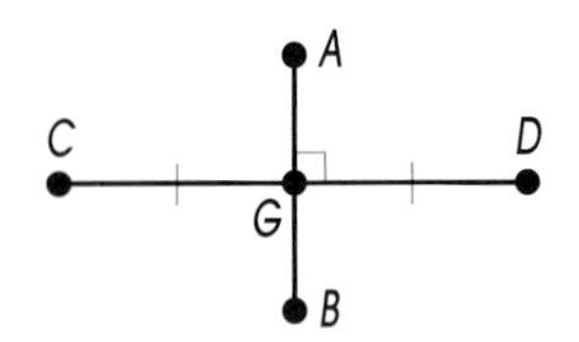

In the third diagram, $\overline{CG}$ and $\overline{GD}$ are congruent. $\angle AGD$ is 90°. $\overline{AB}$ is perpendicular to $\overline{CD}$ and $\overline{AB}$ bisects $\overline{CD}$.

So, $\overline{AB}$ is the perpendicular bisector of $\overline{CD}$.

The distance between any point on the perpendicular bisector and each endpoint is equal.

$\overrightarrow{JK}$ is the perpendicular bisector of $\overline{EF}$.
The distance between point E and point K is 12.
Find the distance between point K and point F.

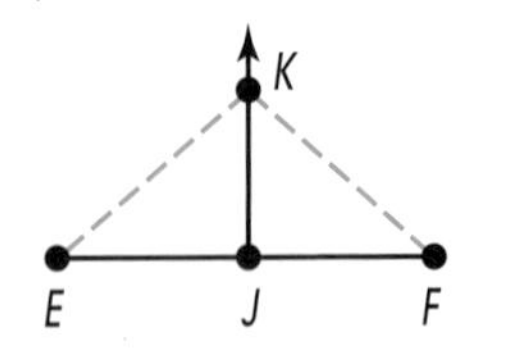

K is a point on the perpendicular bisector. Then, $EK = KF$.

$EK = 12$, so $KF = 12$

So, the distance between point K and point F is 12.

Try These

Line $\overleftrightarrow{RS}$ is the perpendicular bisector of $\overline{PQ}$.
Write *true* or *false* for each statement.
Explain your reasoning.

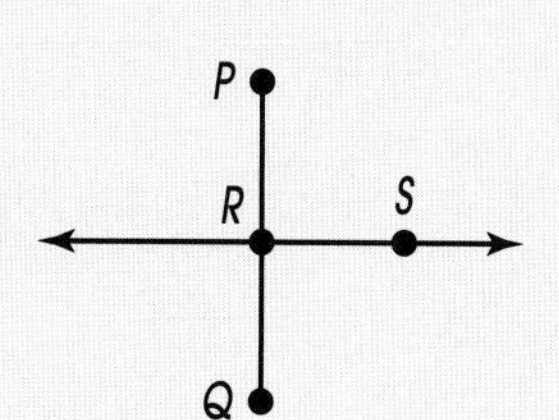

1. R is the midpoint of $\overline{PQ}$. True

   Because $\overleftrightarrow{RS}$ is the perpendicular ■, bisector
   it divides $\overline{PQ}$ into two congruent ■. line segments

2. $\overleftrightarrow{RS} \perp \overline{PQ}$ ■ True

   $\overleftrightarrow{RS}$ is the ■ bisector. perpendicular

3. $\angle PRS$ is an acute angle. False

   Perpendicular lines form ■ angles. right

Practice

Is $\overline{LM}$ the perpendicular bisector of $\overline{AB}$? Write *yes* or *no*. Explain.

1.

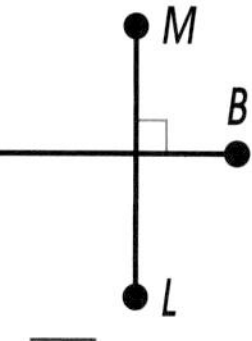

no; $\overline{LM}$ does not bisect $\overline{AB}$.

2.

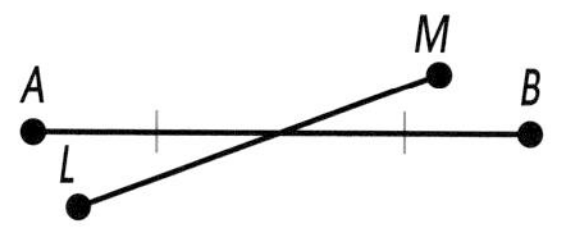

no; $\overline{LM}$ is not perpendicular to $\overline{AB}$.

3.

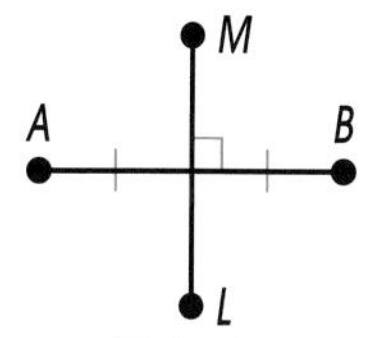

yes; $\overline{LM}$ is not perpendicular to $\overline{AB}$ and bisects $\overline{AB}$.

Look at the diagram below. $\overleftrightarrow{ST} \perp \overline{UV}$ at the midpoint.
Write *true* or *false* for each statement. Explain your reasoning.

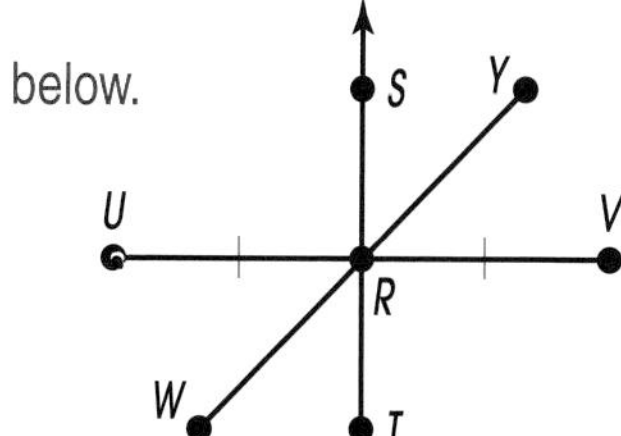

4. The distance from point U to point S is equal to the distance from point S to point V. See below.

5. $\overline{WY} \perp \overline{UV}$
   false; The intersecting lines do not form right angles.

6. $\angle URT$ is an acute angle.
   false; $\overleftrightarrow{ST} \perp \overline{UV}$ So, $\angle URT$ is a right angle.

7. R is the midpoint of $\overrightarrow{RS}$.
   false; R is the endpoint of a ray.

8. $\angle WRV$ is an obtuse angle.
   true; $\angle WRV$ is greater than 90°.

4. true; The distance between any point on the perpendicular bisector and each endpoint is equal.

Try This

$\overrightarrow{BD}$ is the perpendicular bisector of $\overline{AC}$.

Find the value of x.

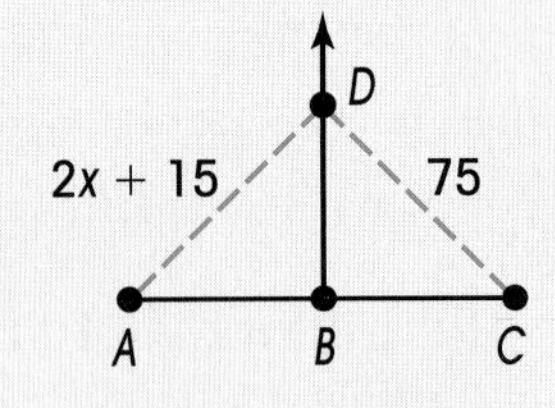

D is a point on the ■ bisector. So, DC = ■. perpendicular DA

DA = ■ and $DC = 75$. 2x + 15

Write an equation. $DA = DC$

Substitute the given values. $2x + 15 = 75$

Solve for x. $2x$ = ■ 60

x = ■ 30

The value of x is ■. 30

Practice

$\overrightarrow{HG}$ is the perpendicular bisector of $\overline{IK}$.

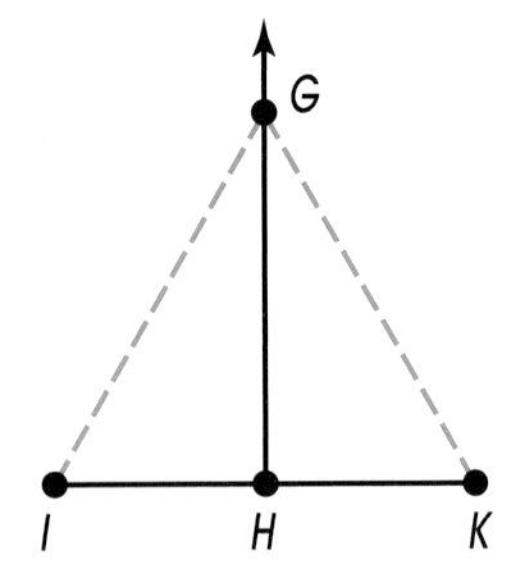

9. Find the value of x if $GI = x + 10$ and $GK = 29$. 19

10. Find the value of x if $GI = 54$ and $GK = x - 6$. 60

11. Find the distance between point G and point K if $GI = 2x - 23$ and $GK = x + 17$. 57

12. Find the distance between point G and point I if $GI = 3x - 20$ and $GK = x + 18$. 37

Share Your Understanding

13. Have a partner draw two intersecting line segments. Explain how you can find out whether one is the perpendicular bisector of the other. Be sure to use the words *bisect* and *perpendicular* in your explanation. Check students' work.

14. **CRITICAL THINKING** Explain to a partner why a line cannot have a perpendicular bisector. (Hint: Remember the definition of a line.)
A line extends endlessly both directions. So, you can never find the midpoint.

CONSTRUCTION
A Perpendicular Line Through a Point Not on a Line

Follow the steps below to construct a line that is perpendicular to line ℓ through point P.

On a sheet of paper, copy line ℓ and point P above line ℓ.

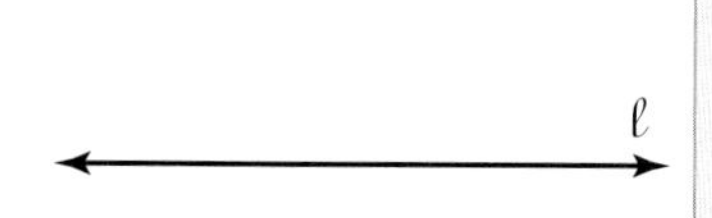

STEP 1 Place the compass on point P. Set the compass to a radius that intersects line ℓ. Draw an arc that intersects line ℓ. Label the point T.

Keep the compass center on point P. Draw an arc that intersects line ℓ on the opposite side of point T. Label the point U.

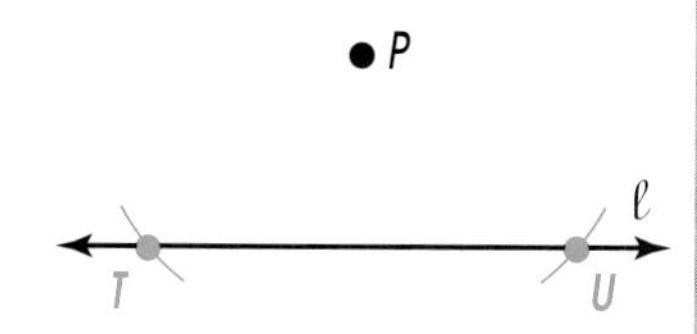

STEP 2 Put the compass on point T. Draw an arc above or below line ℓ that does not intersect point P.

Place the compass on point U. Draw an arc that intersects the arc you just drew. Label the intersection V.

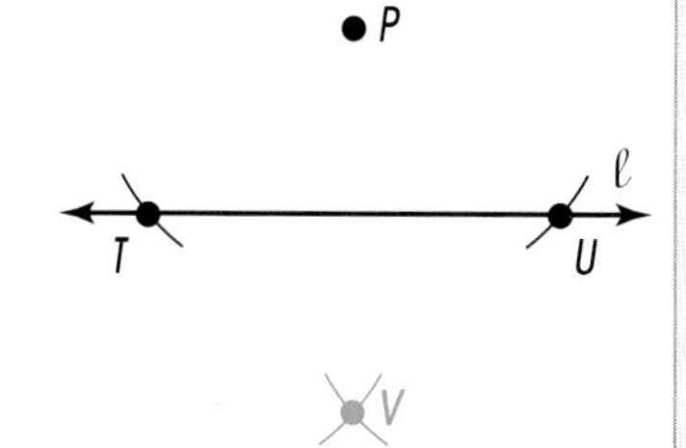

STEP 3 Draw a line that connects point P and point V.

$\overleftrightarrow{PV}$ is perpendicular to line ℓ.

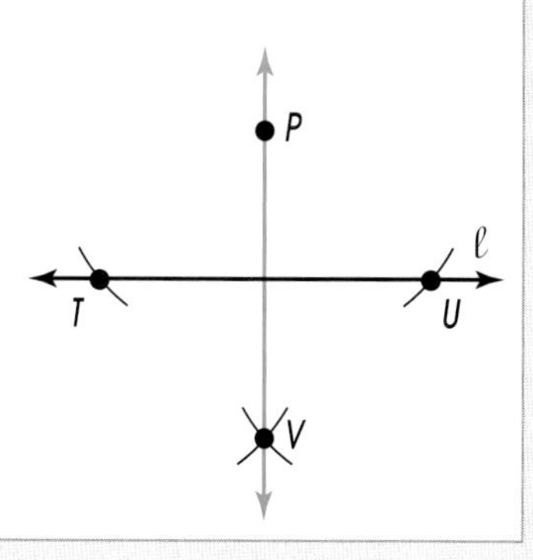

More practice is provided in the Workbook, page 20, and in the Classroom Resource Binder, page 35.

4·3 Parallel Lines

Getting Started Have students fold paper to illustrate parallel lines.

Remember
A plane is a flat surface with no thickness. Lines in the same plane are coplanar lines.

Parallel lines are lines that lie in the same plane and do not intersect. The distance between two parallel lines is always the same. Look at the diagram below.

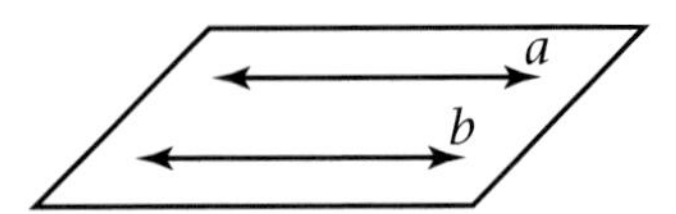

Line a is parallel to line b. The symbol $\parallel$ means "is parallel to." You can write $a \parallel b$. You can show that two lines in a diagram are parallel. Use small arrowheads.

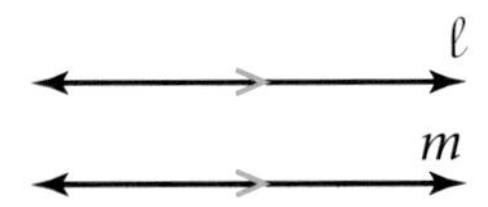

In the diagram above, $\ell \parallel m$.

EXAMPLE

Identify the parallel lines.

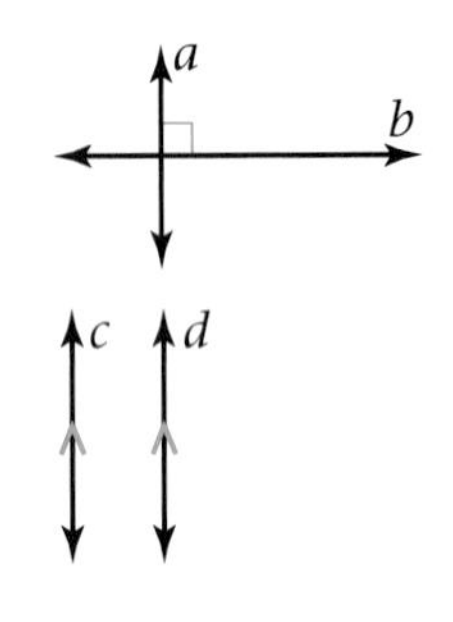

Lines a and b are perpendicular. They intersect, so they are not parallel.

Lines c and d are in the same plane. The arrowheads show that the lines are parallel. Lines c and d are parallel.

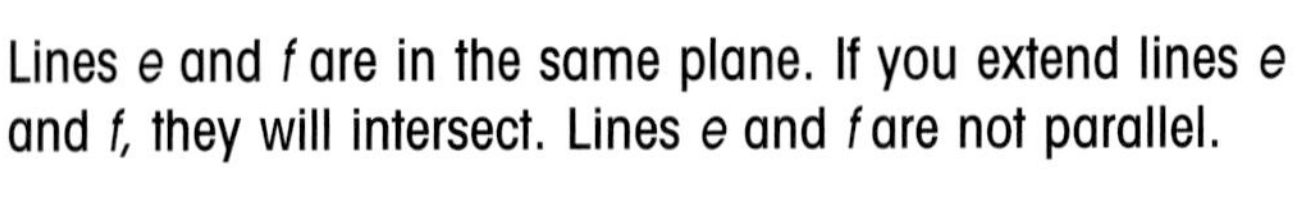

Lines e and f are in the same plane. If you extend lines e and f, they will intersect. Lines e and f are not parallel.

So, line c is parallel to line d. $c \parallel d$

Try This

Identify the parallel lines in the diagram.

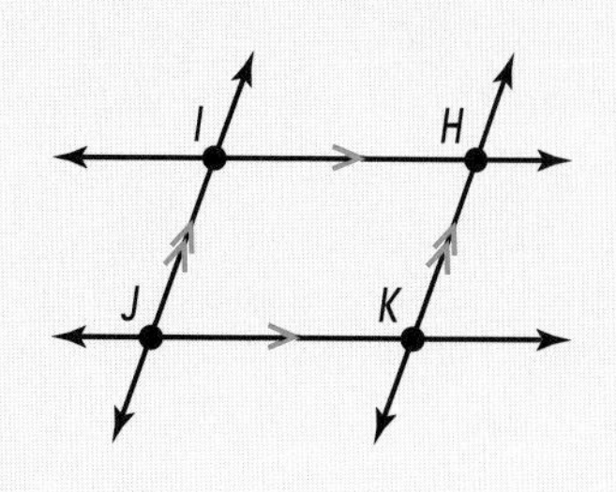

Look for the matching arrowheads in the diagram.
$\overleftrightarrow{IH}$ and $\overleftrightarrow{JK}$ each have one arrowhead.
$\overleftrightarrow{IJ}$ and ■ each have two arrowheads. $\overleftrightarrow{HK}$

$\overleftrightarrow{IH}$ is parallel to ■. $\overleftrightarrow{IJ}$ is parallel to ■.
$\overleftrightarrow{JK}$ $\overleftrightarrow{HK}$

Practice

Decide if the lines are parallel. Write *yes* or *no*. If they are parallel, use the || symbol to write a statement.

1. 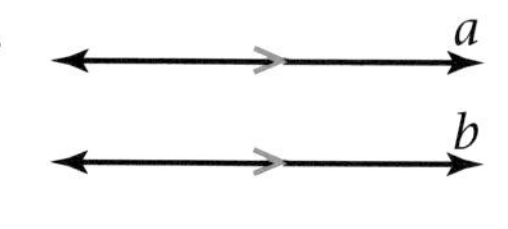

yes; $a \parallel b$

2. 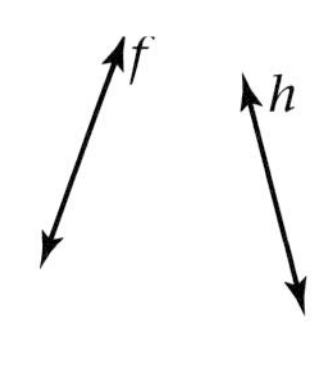

no

3. 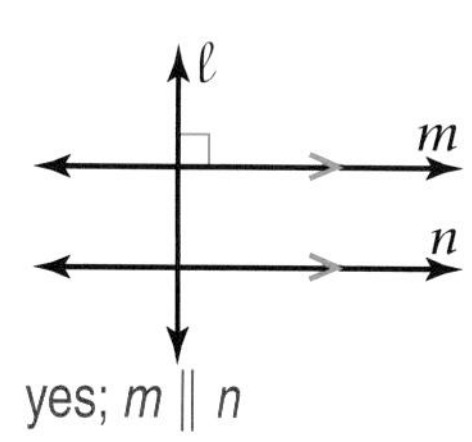

yes; $m \parallel n$

Identify the parallel lines in each diagram.

4. 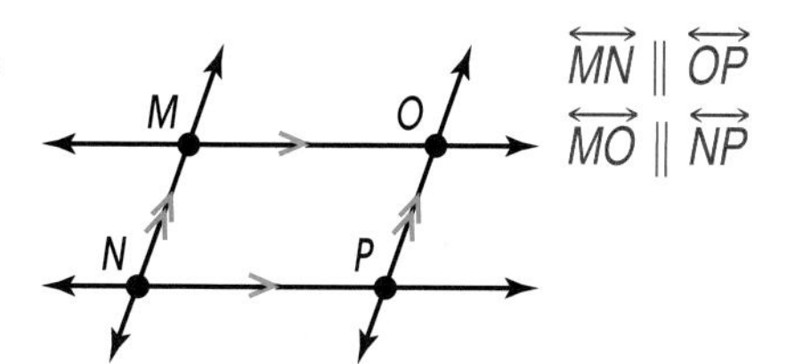

$\overleftrightarrow{MN} \parallel \overleftrightarrow{OP}$
$\overleftrightarrow{MO} \parallel \overleftrightarrow{NP}$

5. 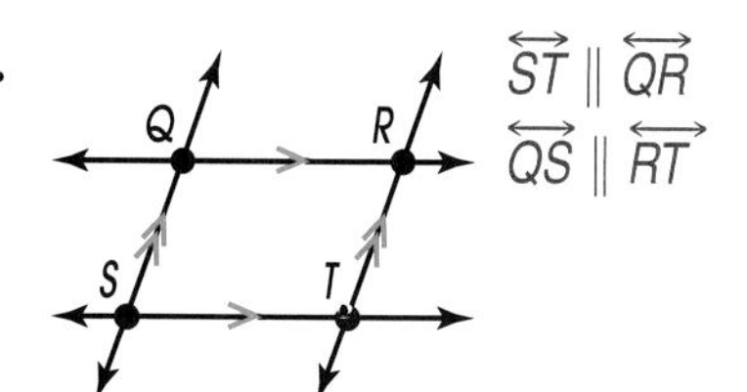

$\overleftrightarrow{ST} \parallel \overleftrightarrow{QR}$
$\overleftrightarrow{QS} \parallel \overleftrightarrow{RT}$

Share Your Understanding

6. Explain to a partner how parallel lines and perpendicular lines are different. Use drawings in your explanation.
Check students' work.

7. **CRITICAL THINKING** Write the symbol that completes each statement. ℓ ■ m ℓ ■ n m ■ n
Then, write your statement in words.
$\ell \perp m$; Line ℓ is perpendicular to line m.
$\ell \perp n$; Line ℓ is perpendicular to line n.
$m \parallel n$; Line m is parallel to line n.

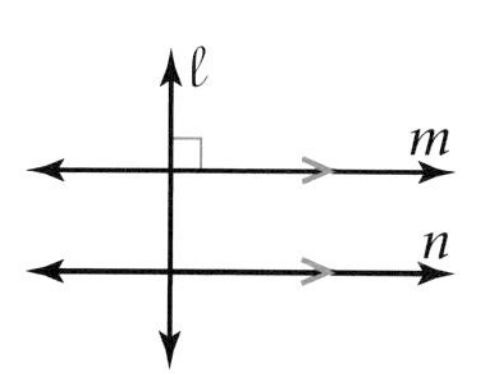

More practice is provided in the Workbook, page 20, and in the Classroom Resource Binder, page 35.

4·4 Parallel Lines with Transversals

Getting Started
Have students fold paper to create several pairs of parallel lines. Then draw the lines so that they can be seen. For each pair of parallel lines, have them draw a third line that intersects the parallel lines.

A line that intersects two parallel lines at two different points is a **transversal.**

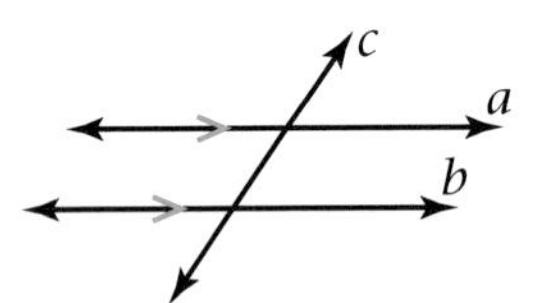

Line c crosses parallel lines a and b at two different points. Line c is a transversal.

EXAMPLE 1

Name the transversal.

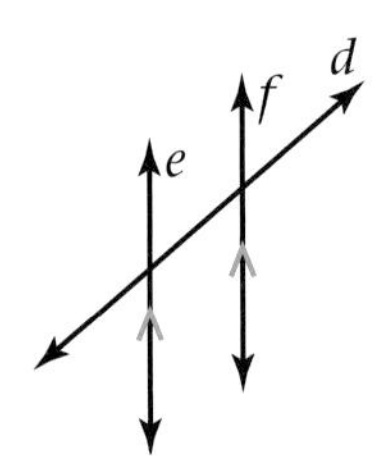

Line d crosses lines e and f at two different points.

So, line d is the transversal.

Angles form when a transversal intersects parallel lines. Angles inside parallel lines are interior angles. Interior angles on the same side of the transversal are **same-side interior angles.**

EXAMPLE 2

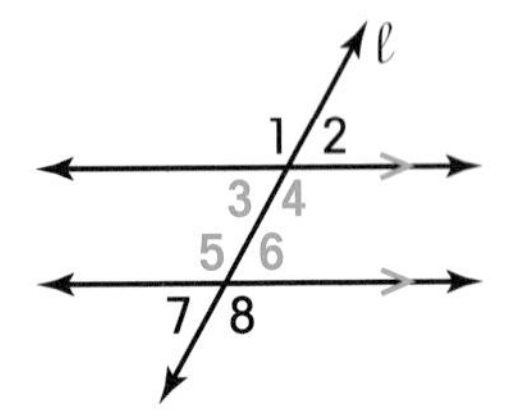

Name two pairs of interior angles on the same side of the transversal.

Look at the diagram on the left.

$\angle 3$, $\angle 4$, $\angle 5$, and $\angle 6$ are interior angles.
$\angle 3$ and $\angle 5$ are on the left side of the transversal.
They are both on the same side of the transversal.

So, $\angle 3$ and $\angle 5$ are same-side interior angles.
$\angle 4$ and $\angle 6$ are also same-side interior angles.

Interior angles on opposite sides of the transversal are **alternate interior angles.**

Try This

Name two pairs of alternate interior angles.

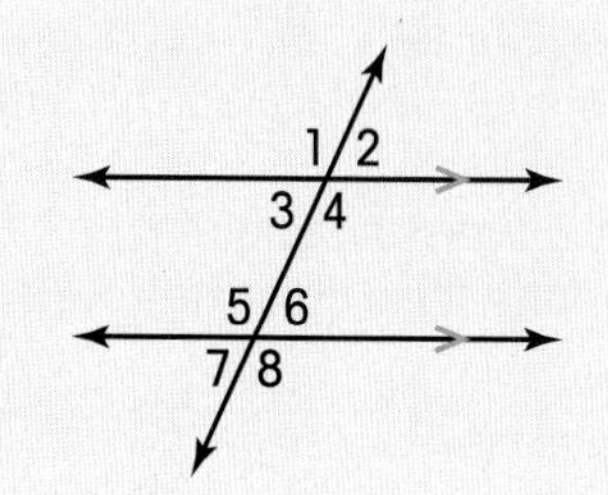

Alternate angles are on different sides of the transversal.

$\angle 3$ $\angle 5$

■, $\angle 4$, ■, and $\angle 6$ are interior angles.

$\angle 3$ and $\angle 6$ are alternate ■ angles. interior

$\angle 3$

So, ■ and $\angle 6$ are alternate interior angles.

$\angle 4$ and ■ are also alternate interior angles.

$\angle 5$

Practice

Name each transversal.

1. 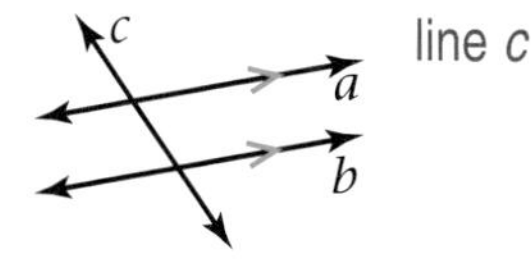

line c

2. 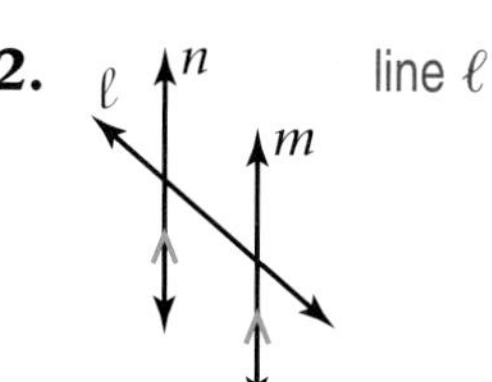

line ℓ

3. 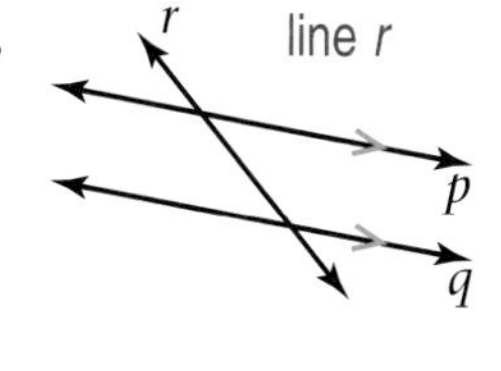

line r

Use the diagram on the right. Name all pairs of angles for each exercise.

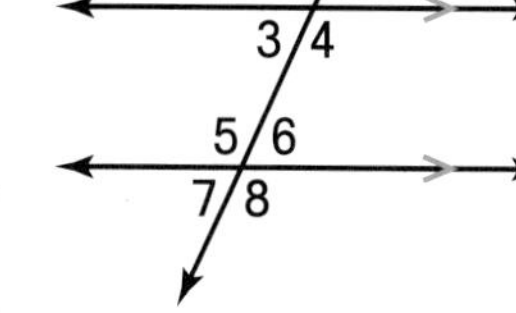

4. same-side interior angles
$\angle 3$ and $\angle 5$; $\angle 4$ and $\angle 6$

5. alternate interior angles
$\angle 3$ and $\angle 6$; $\angle 4$ and $\angle 5$

6. vertical angles
$\angle 1$ and $\angle 4$; $\angle 2$ and $\angle 3$; $\angle 5$ and $\angle 8$; $\angle 6$ and $\angle 7$

7. supplementary angles
Possible answers: $\angle 1$ and $\angle 2$; $\angle 1$ and $\angle 3$; $\angle 2$ and $\angle 4$; $\angle 3$ and $\angle 4$; $\angle 5$ and $\angle 6$; $\angle 5$ and $\angle 7$; $\angle 6$ and $\angle 8$; $\angle 8$ and $\angle 7$

Share Your Understanding

8. Have a partner draw two parallel lines and a transversal. Number the angles. Identify the same-side interior angles. Have your partner identify the alternate interior angles.
Check students' work.

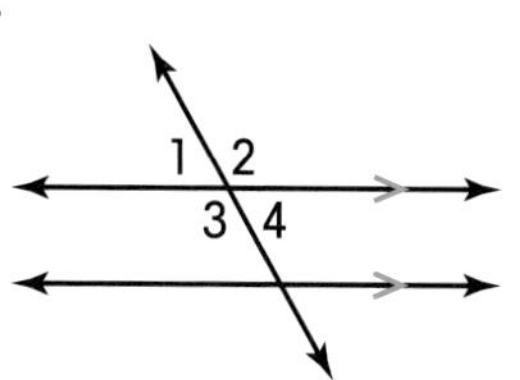

9. **CRITICAL THINKING** In the diagram on the right, $\angle 1$ is 65°. Find the measures of $\angle 2$, $\angle 3$, and $\angle 4$.
$\angle 2 = 115^\circ$; $\angle 3 = 115^\circ$; $\angle 4 = 65^\circ$

More practice is provided in the Workbook, page 21, and in the Classroom Resource Binder, page 36.

4•5 Alternate Interior Angles

Getting Started
Review congruent angles.

> POSTULATE 6 **Alternate Interior Angles Postulate**
>
> If parallel lines are cut by a transversal, then the alternate interior angles are congruent.

You learned that alternate interior angles are interior angles that are on opposite sides of the transversal. They are not adjacent. If the transversal intersects parallel lines, the alternate interior angles are congruent.

$\angle 3 \cong \angle 6$
$\angle 4 \cong \angle 5$

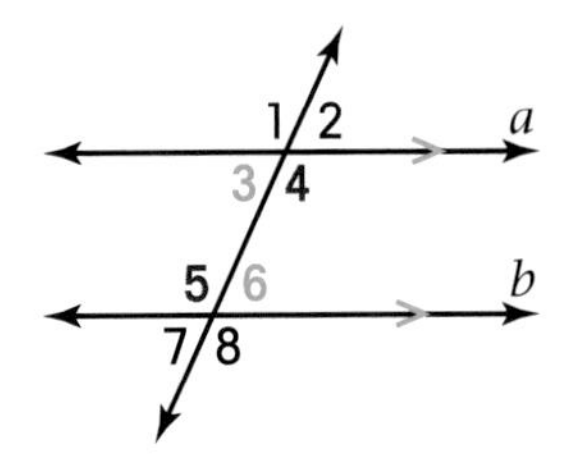

EXAMPLE 1

In the diagram, $c \parallel d$. Find the measure of $\angle 5$.

The 115° angle and $\angle 5$ are alternate interior angles.

Alternate interior angles are congruent.

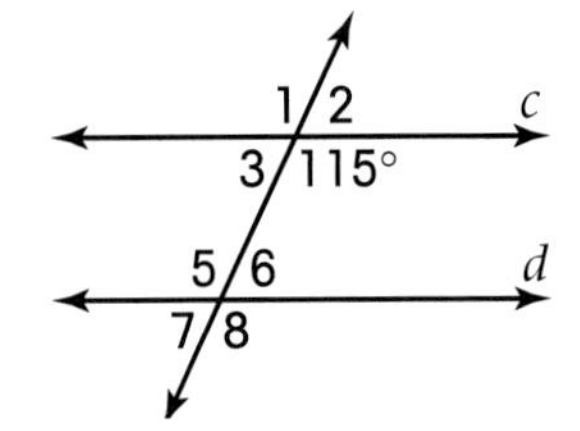

So, $\angle 5$ is 115°.

If you know the measure of one angle in the diagram above, you can find the measure of any other angle.

EXAMPLE 2

Use the diagram in Example 1. Find the measure of $\angle 6$.

$\angle 5$ and $\angle 6$ are supplementary angles.

Write an equation. $m\angle 5 + m\angle 6 = 180$

You know that $m\angle 5 = 115$. Substitute. $115 + m\angle 6 = 180$

Solve for $m\angle 6$. $m\angle 6 = 65$

So, $\angle 6$ is 65°.

Try This

In the diagram, $\ell \parallel m$.
Find the measure of $\angle 5$ and $\angle 6$.

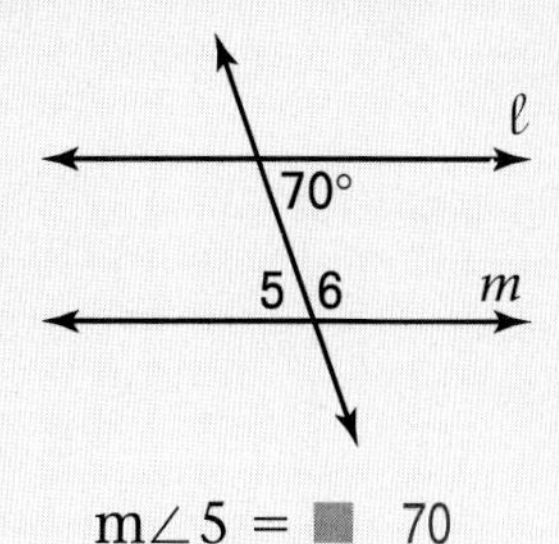

The 70° angle and ∠5 are ■ interior angles. alternate
Alternate interior angles are ■. congruent
$m\angle 5 =$ ■ 70

∠5 and ■ are supplementary angles. ∠6

Write an equation. $m\angle 5 + m\angle 6 =$ ■ 180
Substitute 70 for m∠5. 70 ■ $+ m\angle 6 = 180$
Solve for m∠6. $m\angle 6 =$ ■ 110

So, ∠5 is 70° and ∠6 is ■. 110°

Practice

In each diagram, $p \parallel q$. Find the measure of $\angle 5$ and $\angle 6$.

1.

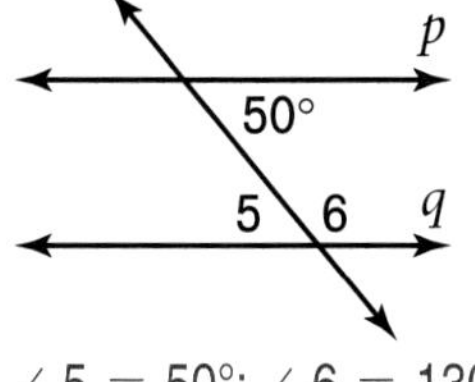

∠5 = 50°; ∠6 = 130°

2.

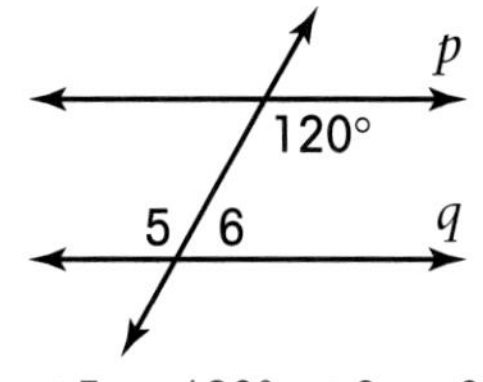

∠5 = 120°; ∠6 = 60°

3.

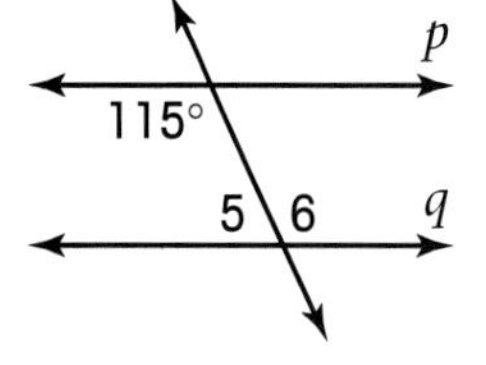

∠5 = 65°; ∠6 = 115°

Share Your Understanding

4. Have a partner draw two parallel lines and a transversal. Then, give the measure of one interior angle. Have your partner identify the alternate interior angle and its measure.
Check students' work.

5. **CRITICAL THINKING** In the diagram on the right, $k \parallel \ell$. If $m\angle 3 = 5x - 15$ and $m\angle 6 = 2x + 36$, find the measure of $\angle 5$. Show your work.
∠5 = 110°; Check students' work.

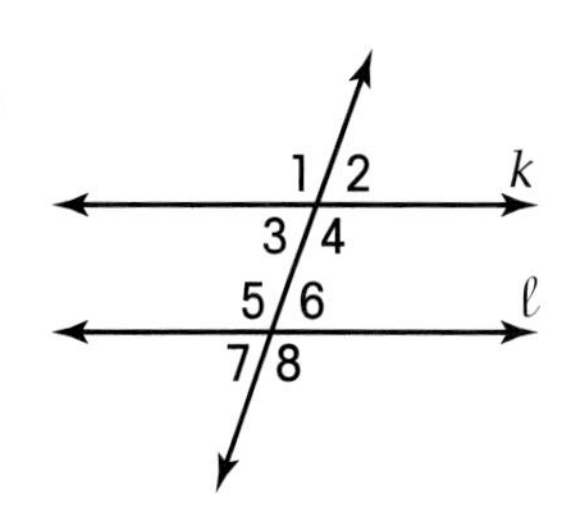

More practice is provided in the Workbook, page 21, and in the Classroom Resource Binder, page 36.

4-6 Same-Side Interior Angles

Getting Started Review supplementary angles.

THEOREM 9

If parallel lines are cut by a transversal, then the same-side interior angles are supplementary.

You learned that **same-side interior angles** are interior angles that are on the same side of the transversal. If the transversal intersects parallel lines, the same-side interior angles are supplementary.

$m\angle 3 + m\angle 5 = 180$
$m\angle 4 + m\angle 6 = 180$

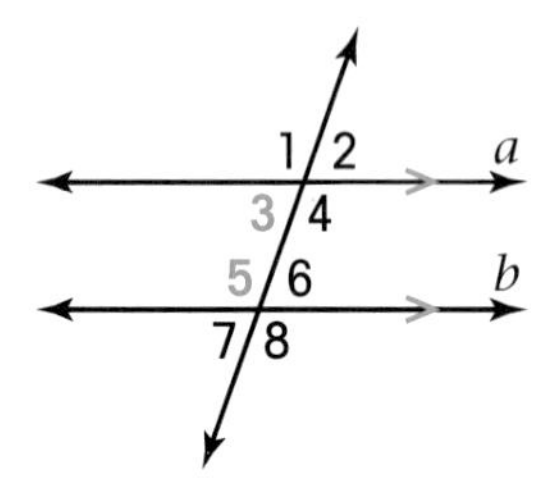

EXAMPLE 1

In the diagram, $c \parallel d$. Find the measure of $\angle 5$ and $\angle 8$.

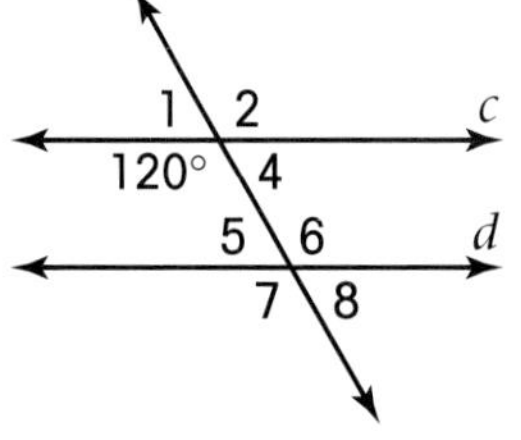

The 120° angle and $\angle 5$ are same-side interior angles. These angles are supplementary.

Write an equation. $120 + m\angle 5 = 180$
Solve for $m\angle 5$. $m\angle 5 = 60$

$\angle 5$ and $\angle 8$ are vertical angles.
$\angle 5$ and $\angle 8$ are congruent. $m\angle 8 = 60$

Remember
Vertical angles are formed by intersecting lines. Vertical angles are congruent.

So, $\angle 5$ is 60° and $\angle 8$ is 60°.

EXAMPLE 2

Use the diagram in Example 1.
Find the measure of $\angle 4$.

$\angle 5$ and $\angle 4$ are alternate interior angles.
Alternate interior angles are congruent. $\angle 4 \cong \angle 5$

Since $\angle 5$ is 60°, then $\angle 4$ is also 60°. $m\angle 4 = m\angle 5 = 60$

So, $\angle 4$ is 60°.

Try This

In the diagram, $s \parallel t$.
Find the measures of $\angle 5$ and $\angle 8$.

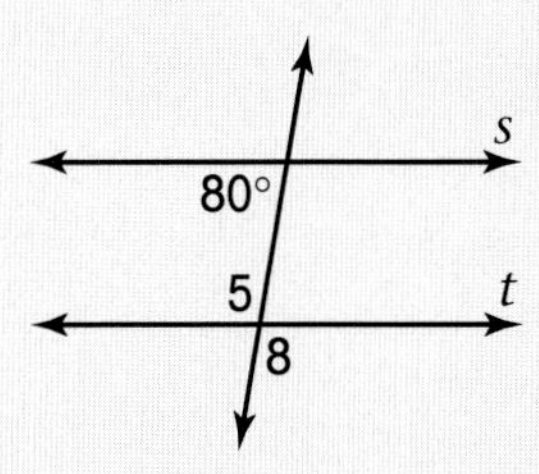

same-side
The 80° angle and $\angle 5$ are ■ interior angles.
They are supplementary. Write an equation. $80 + m\angle 5 =$ ■ 180
Solve for $m\angle 5$. $m\angle 5 =$ ■ 100

$\angle 8$
$\angle 5$ and ■ are vertical angles.
Vertical angles are ■. $m\angle 5 = m\angle 8$ $m\angle 8 =$ ■ 100
100° congruent
So, $\angle 5$ is ■ and $\angle 8$ is ■. 100°

Practice

In each diagram, $e \parallel f$. Find the measures of $\angle 5$ and $\angle 8$.

1.

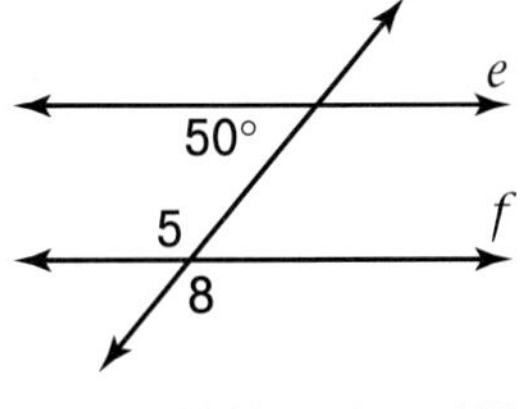

$\angle 5 = 130°$; $\angle 8 = 130°$

2.

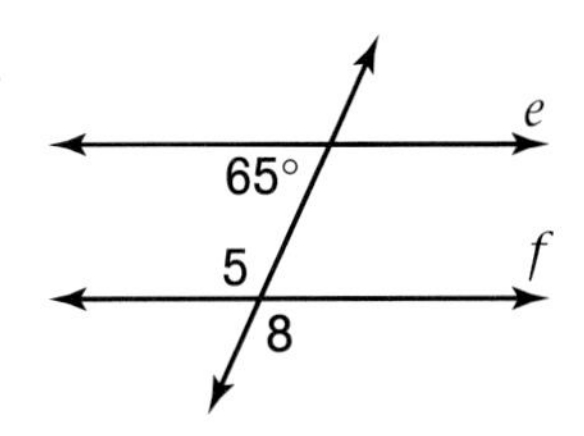

$\angle 5 = 115°$; $\angle 8 = 115°$

3.

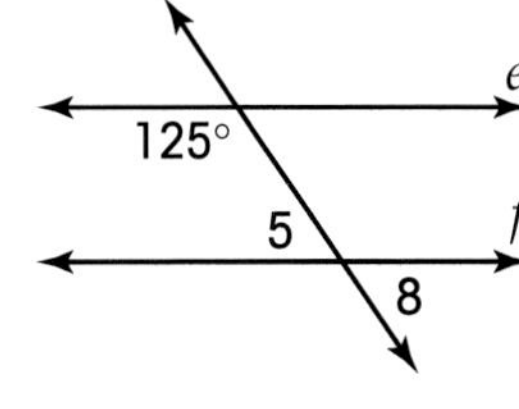

$\angle 5 = 55°$; $\angle 8 = 55°$

Share Your Understanding

4. In the diagram on the right, $k \parallel \ell$. Explain to a partner why $\angle 3$ and $\angle 5$ are supplementary. Have your partner explain why $\angle 1$ and $\angle 3$ are supplementary. See below.

5. **CRITICAL THINKING** In the diagram on the right, $k \parallel \ell$. If $\angle 4 = 3x - 20$ and $\angle 6 = 2x$, find the measure of $\angle 8$. Show your work. $\angle 8 = 100°$; Check students' work.

4. Line k and line ℓ are parallel lines cut by a transversal. $\angle 3$ and $\angle 5$ are supplementary because they are the same-side interior angles of these lines. $\angle 1$ and $\angle 3$ are supplementary because they form a straight angle.

More practice is provided in the Workbook, page 22, and in the Classroom Resource Binder, page 37.

Corresponding Angles

Getting Started Have students identify the exterior angles.

> **THEOREM 10**
> **Corresponding Angles**
> If parallel lines are cut by a transversal, then each pair of corresponding angles is congruent.

When a transversal intersects two lines it creates pairs of **corresponding angles**. Corresponding angles have the same position on each line. One angle is an interior angle. The other is an exterior angle. If the transversal intersects parallel lines, these angles are congruent.

$\angle 1 \cong \angle 5$
$\angle 3 \cong \angle 7$
$\angle 2 \cong \angle 6$
$\angle 4 \cong \angle 8$

EXAMPLE 1

In the diagram, $s \parallel t$. Find the measure of $\angle 6$.

The 70° angle and $\angle 6$ are corresponding angles.

Corresponding angles are congruent.

So, $\angle 6$ is 70°.

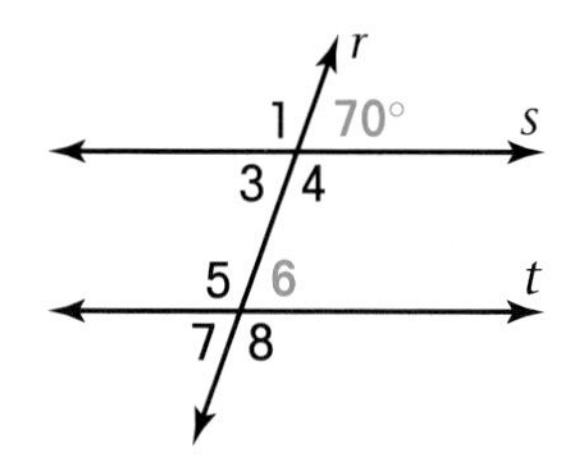

EXAMPLE 2

Use the diagram in Example 1. Find the measure of $\angle 4$.

$\angle 4$ and $\angle 6$ are same-side interior angles.
Same-side interior angles are supplementary.

Write an equation.	$m\angle 4 + m\angle 6 = 180$
Substitute 70 for $m\angle 6$.	$m\angle 4 + 70 = 180$
Solve.	$m\angle 4 = 110$

So, $\angle 4$ is 110°.

This is just one of several ways you can find $m\angle 4$.

Try This

In the diagram, $m \parallel n$ and $\angle 4$ is 115°. Find the measure of $\angle 7$ and $\angle 8$.

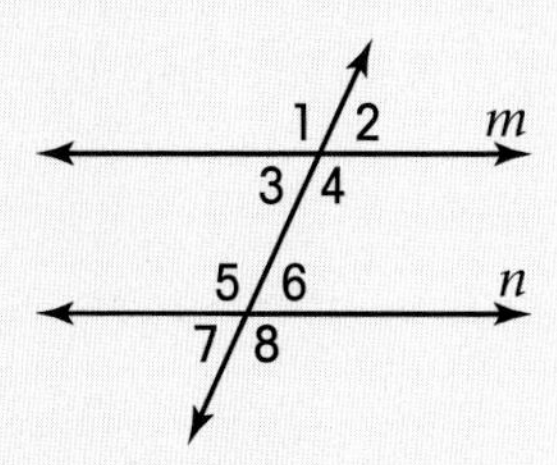

$\angle 4$ and $\angle 8$ are ■ angles. corresponding
$\angle 7$ and $\angle 8$ are ■ angles. Write an equation.
Substitute 115 for $m\angle 8$. supplementary

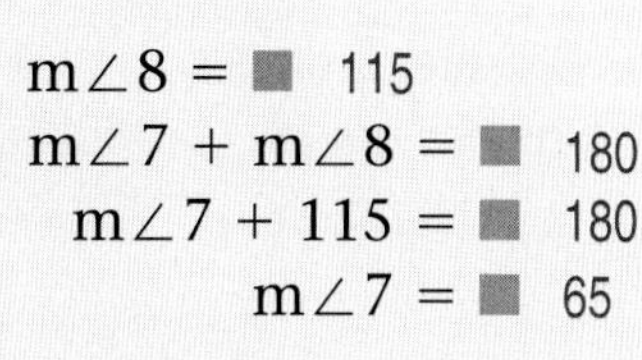

$m\angle 8 =$ ■ 115
$m\angle 7 + m\angle 8 =$ ■ 180
$m\angle 7 + 115 =$ ■ 180
$m\angle 7 =$ ■ 65

So, $\angle 7$ is ■ 65° and $\angle 8$ is ■ 115°.

Practice

In each diagram, $k \parallel \ell$. Find the measure of each angle named.

1.

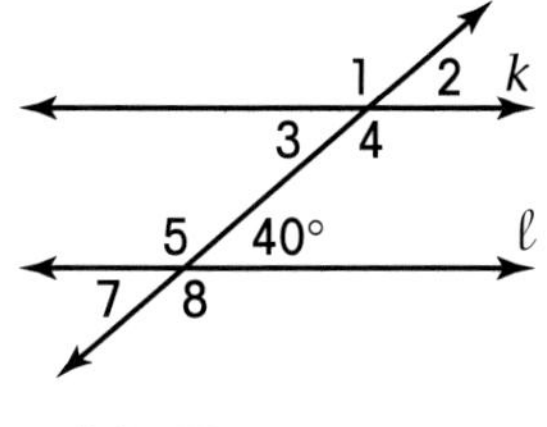

$\angle 2$ is ■. 40°

2.

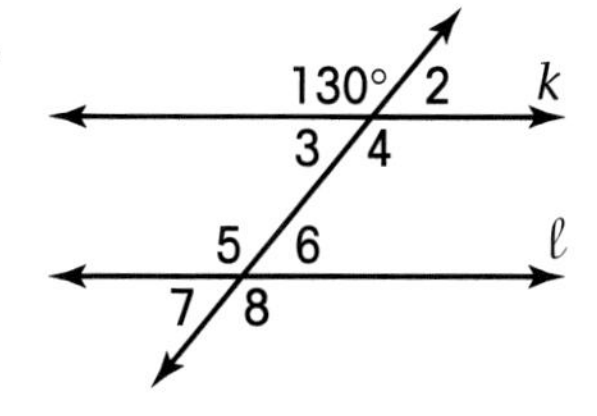

$\angle 5$ is ■. 130°

3.

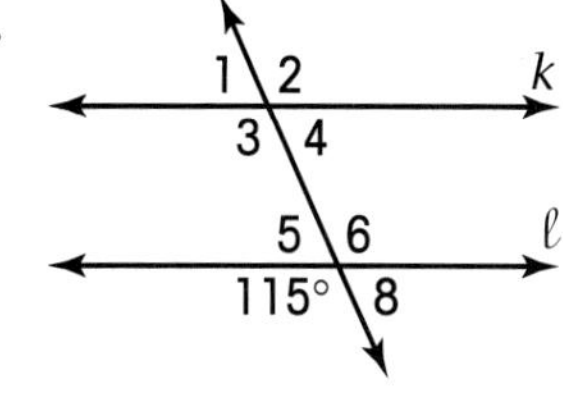

$\angle 3$ is ■. 115°

4.

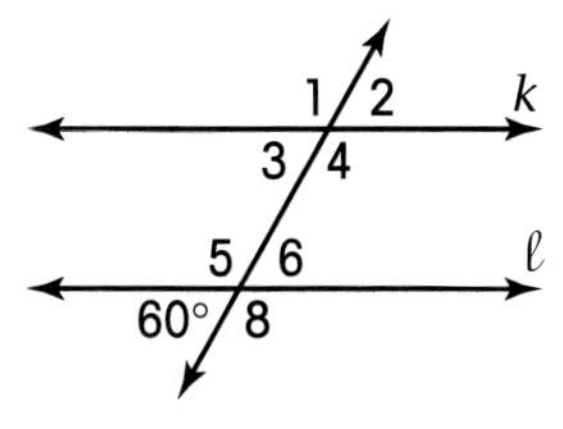

$\angle 2$ is ■. 60°

5.

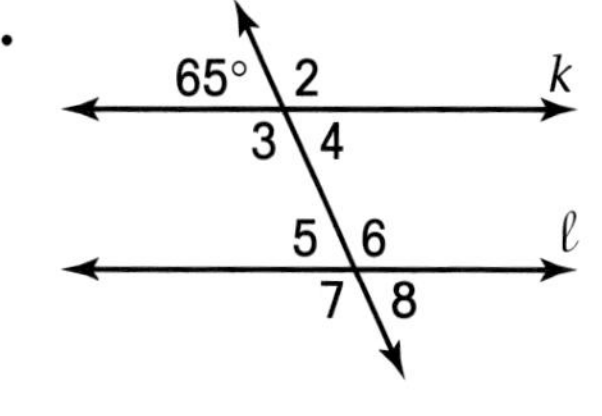

$\angle 6$ is ■. 115°

6.

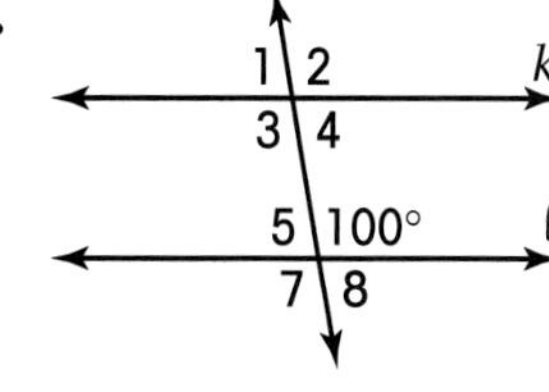

$\angle 4$ is ■. 80°

Try This

In the diagram, $\overleftrightarrow{AB} \parallel \overleftrightarrow{DE}$. Find the measure of $\angle ACG$.

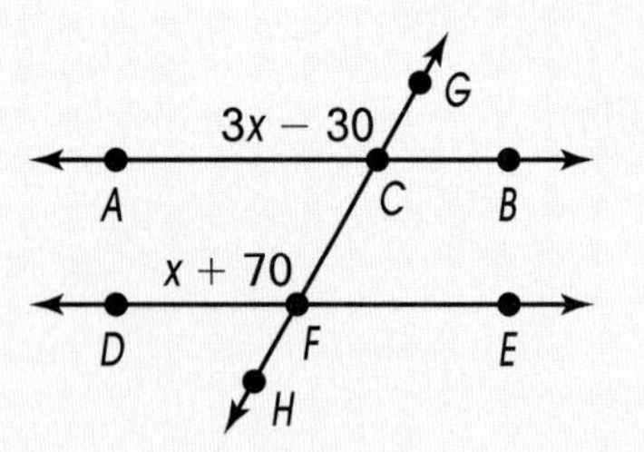

$\angle ACG$ and $\angle DFC$ are ■ angles. corresponding

Corresponding angles are ■. congruent

Write an equation. $3x - 30 = x + 70$

Solve for x.

$$3x = x + 100$$
$$2x = ■ \quad 100$$
$$x = ■ \quad 50$$

Substitute ■ (50) for x to find $m\angle ACG$. $3(■) - 30 = ■$ (50, 120)

So, $\angle ACG$ is ■. 120°

Practice

In each diagram, $c \parallel d$. Find the measure of $\angle 5$.

7.

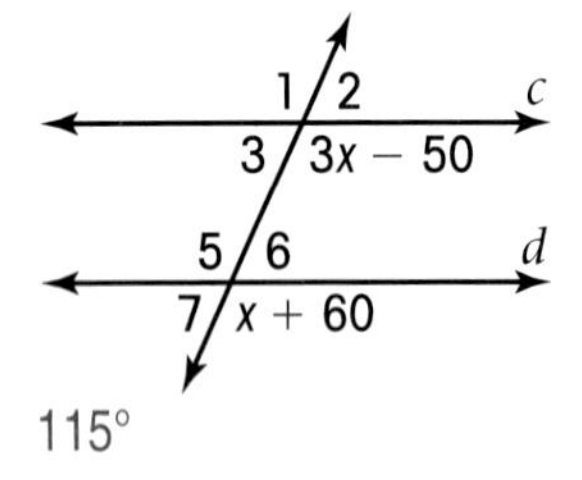

115°

8.

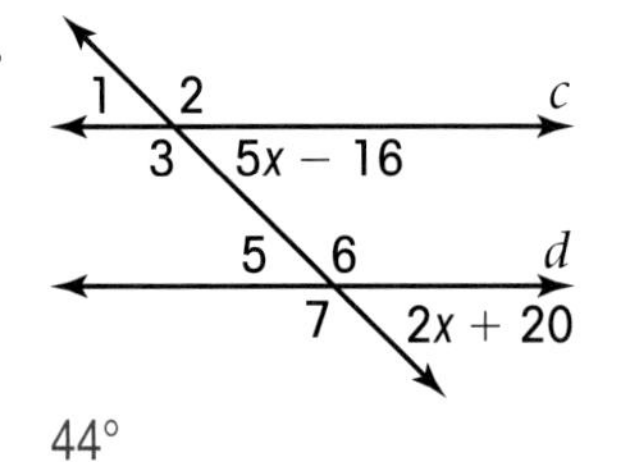

44°

9.

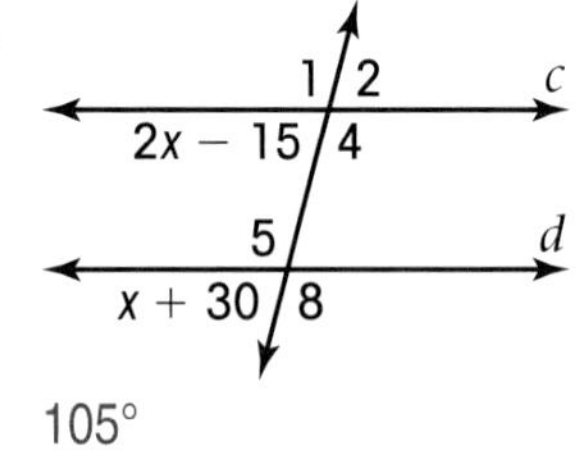

105°

Share Your Understanding

10. Work with a partner. Name each pair of corresponding angles in the diagram on the right. $\angle 1$ and $\angle 5$; $\angle 2$ and $\angle 6$; $\angle 3$ and $\angle 7$; $\angle 4$ and $\angle 8$

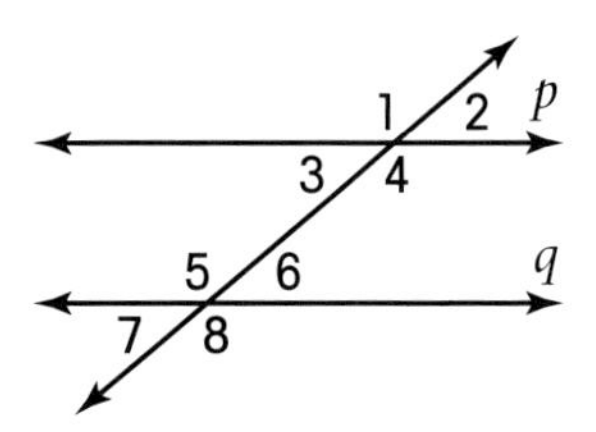

11. **CRITICAL THINKING** In the diagram on the right, $p \parallel q$. If $\angle 4$ is 135°, find the measure of $\angle 7$. 45°

CONSTRUCTION
A Line Parallel to a Given Line

POSTULATE 7
Parallel Postulate

Through a given point *P* not on a line, exactly one line may be drawn parallel to the line.

Follow the steps below to construct a line parallel to $\overleftrightarrow{AB}$ through point P. You will need to construct congruent corresponding angles.

On a sheet of paper, copy $\overleftrightarrow{AB}$ with point P above the line.

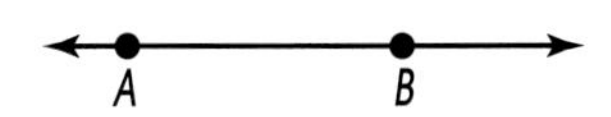

STEP 1 Draw a line to connect point A and point P. Use a straightedge. This will be the transversal.

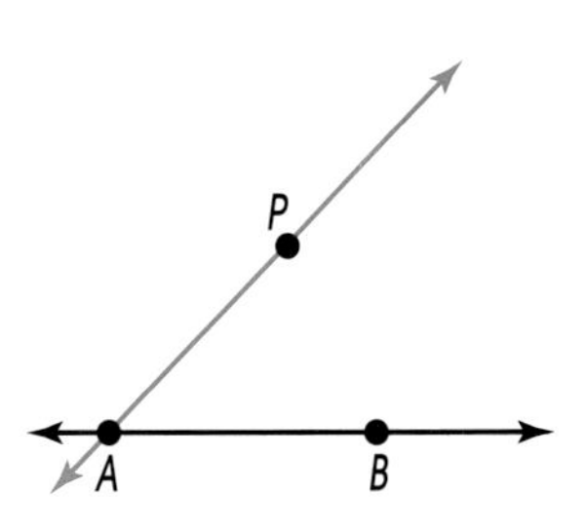

STEP 2 Place one end of the compass on point A. Draw an arc through $\overleftrightarrow{AP}$ so that it does not go past point P. Do not change the compass position. Now, place one end of the compass on point P. Draw another arc.

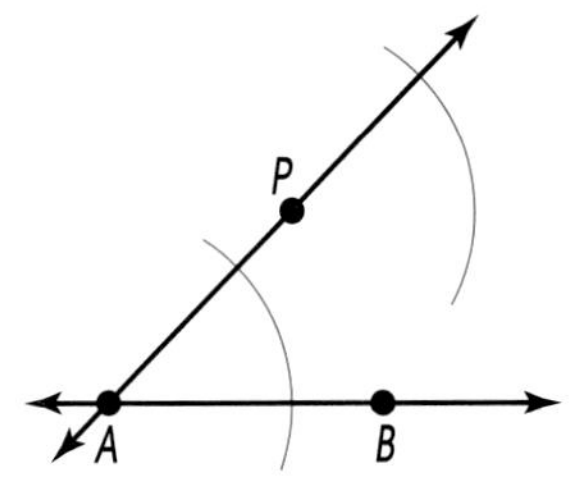

STEP 3 Place the compass end where the first arc intersects $\overleftrightarrow{AP}$. Open the compass so that the pencil end intersects $\overleftrightarrow{AB}$. Draw an arc. Do not change this compass position. Now, take the compass end and place it where the second arc intersects $\overleftrightarrow{AP}$. Draw an arc that intersects this second arc.

Label this point R. Draw $\overleftrightarrow{PR}$.

$\overleftrightarrow{PR}$ is parallel to $\overleftrightarrow{AB}$.

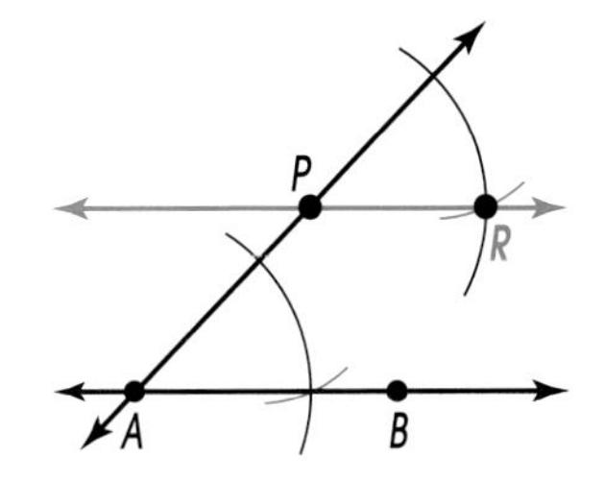

4·8 Calculator: Finding Angle Measures

Getting Started
Have students use color to identify the special angle pairs on a diagram on the board. Have them tell if the pairs are congruent or supplementary.

Sometimes the algebraic expressions for the angle measures include large numbers or decimals. You can use a calculator to help you find these measures.

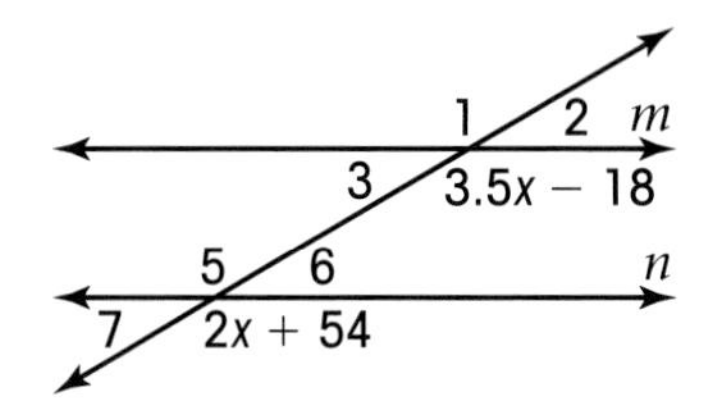

EXAMPLE

In the diagram above, $m \parallel n$. Find the measure of $\angle 1$.

Remember
Corresponding angles are congruent.

Write an equation. $3.5x - 18 = 2x + 54$
Use your calculator to solve for x.
To begin, add 18 to both sides.

Add 18 and 54 on the calculator. **Display**

PRESS: 5 4 + 1 8 = 72 $3.5x = 2x + 72$

Subtract $2x$ from both sides. **Display**

PRESS: 3 . 5 − 2 = 1.5 $1.5x = 72$

Divide both sides by 1.5. **Display**

PRESS: 7 2 ÷ 1 . 5 = 48 $x = 48$

Write the algebraic expression for one angle. $3.5x - 18$
Substitute 48 for x to find its measure. $3.5(48) - 18 = ?$

Do the calculations on the calculator. **Display**

PRESS: 3  . 5 × 4 8 − 1 8 = 150

The 150° angle and $\angle 1$ are vertical angles.

So, $\angle 1$ is 150°.

Practice

In each diagram, $k \parallel \ell$. Use your calculator to find the measure of $\angle 1$.

1. $x = 15$; $\angle 1$ is $110°$

$8.8x - 22$ / 2, k
3 / 4
$2.3x + 75.5$ / 6, ℓ
7 / 8

2. $x = 25$; $\angle 1$ is $120°$

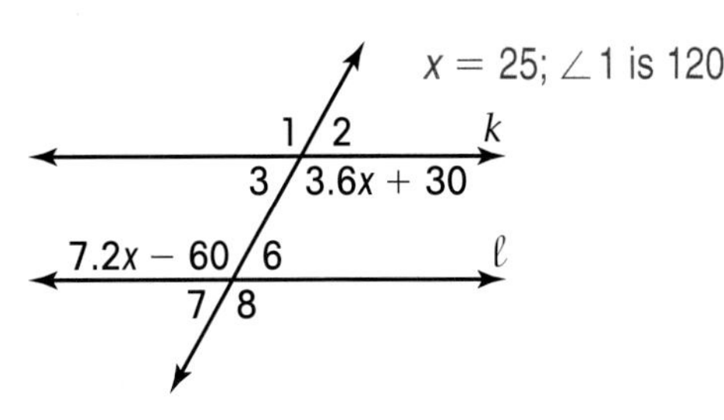

3. $x = 20$; $\angle 1$ is $115°$

1 / 2, k
$3.9x - 13$ / 4
5 / $1.7x + 31$, ℓ
7 / 8

4. $x = 10$; $\angle 1$ is $125°$

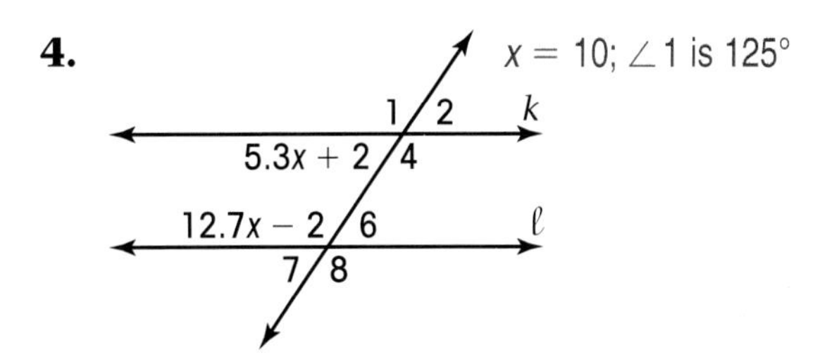

Math Connection

CONTOUR MAPS

A contour map uses lines called *isolines* to show the shape, or contour, of the land. Isolines are the same distance apart, but they are not in the same plane. If you follow an isoline, you stay at the same elevation, or height above sea level. Contour maps also use color bands within the isolines to indicate the elevation.

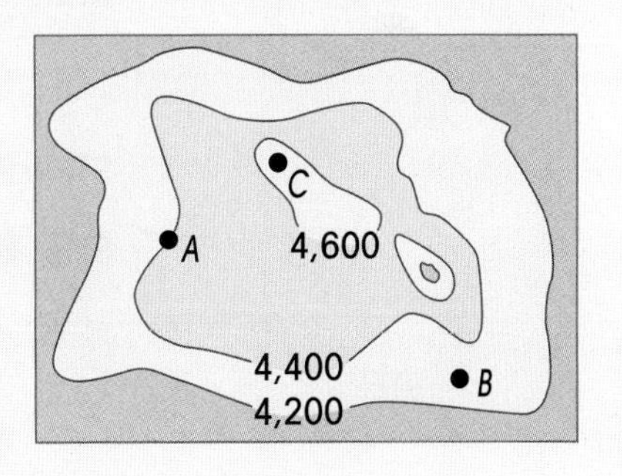

You may want to know the contour of the land if you go hiking. The closer the isolines are, the steeper the land. If you want an easy climb, look for isolines that are far apart.

More practice is provided in the Workbook, page 23, and in the Classroom Resource Binder, page 38.

Problem-Solving Skill: Draw a One-Point Perspective

Getting Started
Ask students how they might represent three-dimensional objects on a sheet of paper. Have them look at photos, paintings, or shop drawings.

Railroad tracks are parallel. They never meet. Yet, at a distance, they appear to meet on the horizon. The drawing below is in **one-point perspective**. It is a way to show the depth of real life on a flat surface.

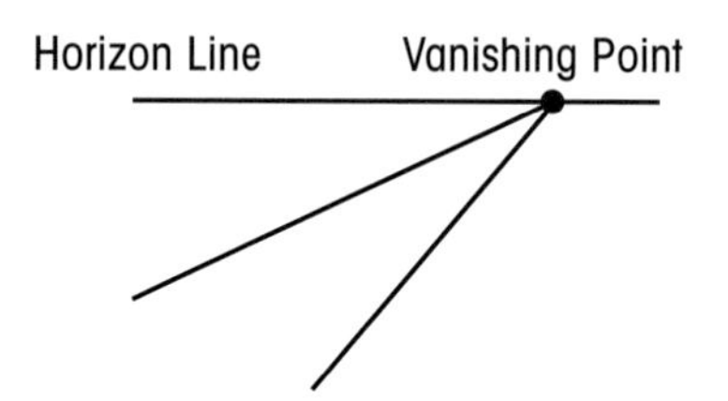

In the drawing, the **horizon line** is the level of the viewer's eyes as the viewer looks across a distance. The **vanishing point** is the point on the horizon where parallel lines appear to meet. You use a horizon line and a vanishing point to draw a three-dimensional object.

EXAMPLE

Draw a three-dimensional box.

STEP 1 Draw a horizon line and a vanishing point. Draw the front of the box. Connect each corner to the vanishing point.

STEP 2 Draw the back and sides of the box.

STEP 3 Erase the lines that are not part of the box.

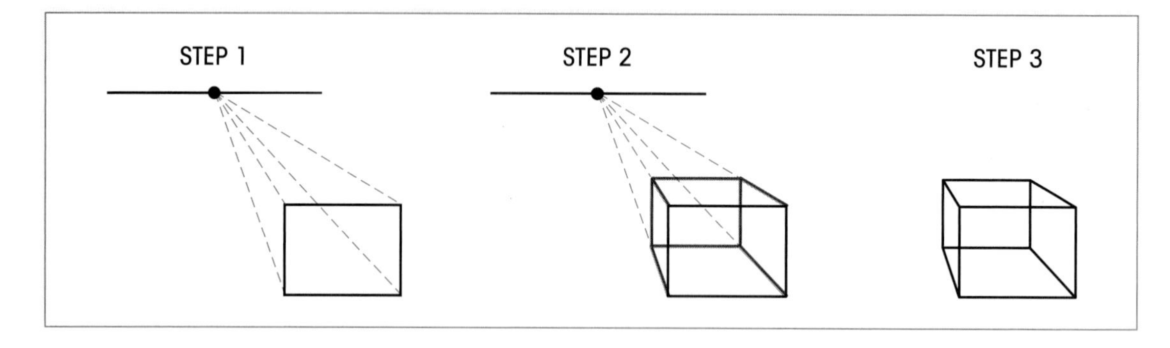

Try This

Locate the vanishing point in this drawing of a box.

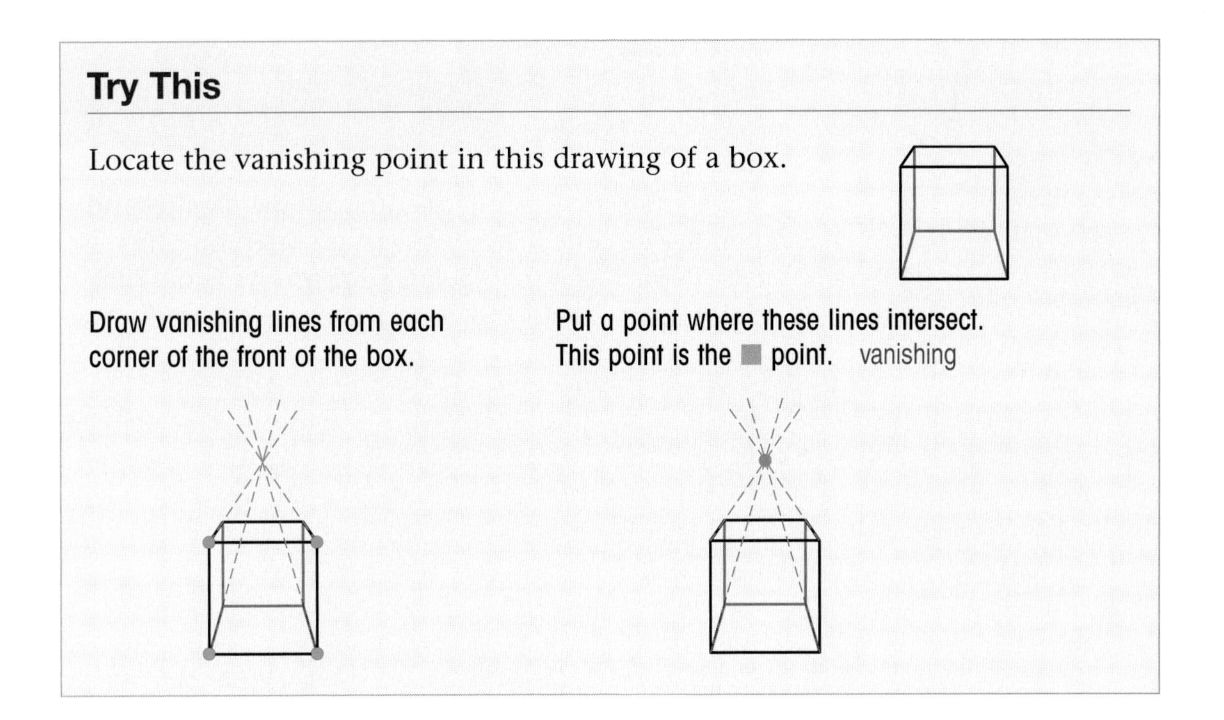

Draw vanishing lines from each corner of the front of the box.

Put a point where these lines intersect. This point is the ■ point. vanishing

Practice

Copy each problem. Locate and draw the vanishing point.

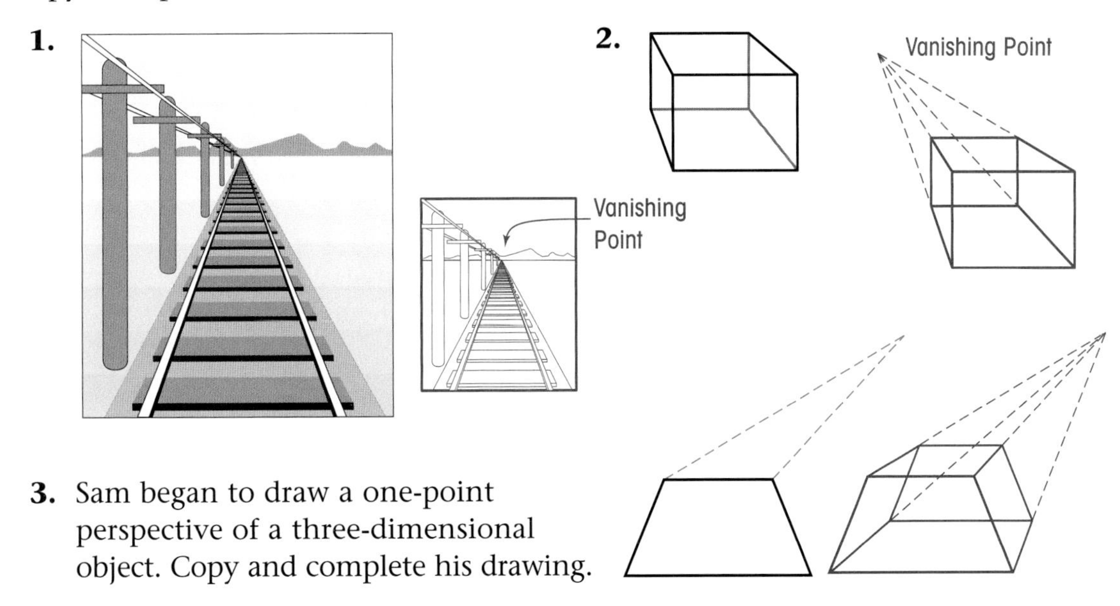

1.

2.

3. Sam began to draw a one-point perspective of a three-dimensional object. Copy and complete his drawing.

More practice is provided in the Workbook, page 24, and in the Classroom Resource Binder, page 39.

4-10 Problem-Solving Application: Taxicab Routes

Getting Started
Have students think of different routes they can take from home to school.

In some cities, streets and avenues follow parallel and perpendicular lines. These streets are often one-way streets. Look at the map below. The arrows show the direction cars should travel on the one-way streets.

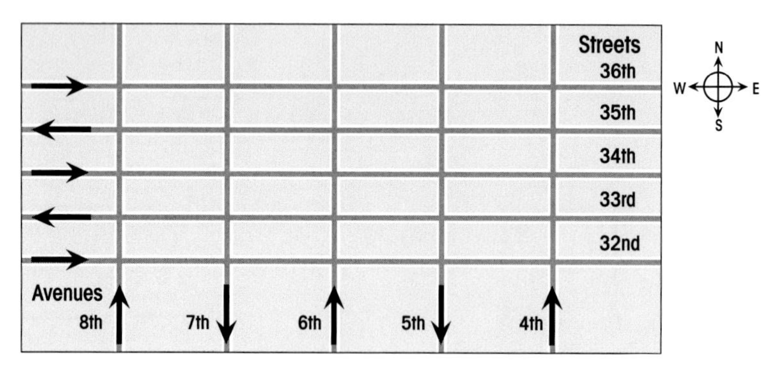

Because many streets and avenues are one way, a taxicab may not be able to travel directly from one place to another. There can be more than one route the taxicab can take.

EXAMPLE

Jake gets into a taxicab at 36th Street and 5th Avenue. He wants to go to 33rd Street and 8th Avenue. How many different routes can the taxicab take? The taxicab does not backtrack in one route.

First route:
The taxicab can travel south on 5th Avenue to 33rd Street.
Then, travel west on 33rd Street to 8th Avenue.

Second route:
The taxicab can travel south on 5th Avenue to 35th Street.
Then travel west on 35th Street to 7th Avenue.
Then travel south on 7th Avenue to 33rd Street.
Then travel west on 33rd Street to 8th Avenue.

The taxicab can take two different routes.

Try This

Use the map on page 114. A tourist gets into a taxicab at 35th Street and 5th Avenue. She wants to go to 32nd Street and 6th Avenue. Find the most direct route.

Travel west on ■ Street to 7th Avenue. 35th
(east)

Then, travel ■ on 7th Avenue to ■ Street. 32nd
(south)

Then, travel ■ on 32nd Street to ■ Avenue. 6th

This route has the fewest turns. This is the most direct route.

Practice

Use the map on page 114. Solve each problem.

1. Adela gets into a taxicab at 36th Street and 6th Avenue. She wants to go to 8th Avenue and 33rd Street. Find the most direct route. Travel east on 36th Street to 5th Avenue. Then, travel south on 5th Avenue to 33rd Street. Then, travel west on 33rd Street to 8th Avenue.
2. Bruce gets into a taxicab at 8th Avenue and 33rd Street. He wants to go to 36th Street and 6th Avenue. How many different routes can the taxicab take? There are two different routes.
3. Justine gets into a taxicab at 32nd Street and 4th Avenue. She wants to go to 36th Street and 8th Avenue. How many different routes can the taxicab take? There are three different routes.
4. Mari gets into a taxicab at 32nd Street and 6th Avenue. She wants to go to 36th Street and 4th Avenue. Find the most direct route. Travel north on 6th Avenue to 36th Street. Then, travel east on 36th Street to 4th Avenue.
5. Aaron gets into a taxicab at 34th Street and 8th Avenue. He wants to go to 32nd Street and 4th Avenue. Find the most direct route. Travel east on 34th Street to 5th Avenue. Then, travel south on 5th Avenue to 32nd Street. Then, travel east on 32nd Street to 4th Avenue. Or, travel east on 34th Street to 7th Avenue. Then, travel south on 7th Avenue to 32nd Street. Then, travel east on 32nd Street to 4th Avenue.

An alternate two-column proof lesson is provided on page 409 of the student book.

4·11 Proof: Proving Lines Are Parallel

POSTULATE 8

If two lines are cut by a transversal so that the alternate interior angles are congruent, the lines are parallel.

You learned that when a transversal cuts parallel lines, the alternate interior angles are congruent. The converse is also true. Look at the postulate on the left. You can use this postulate to prove lines are parallel.

EXAMPLE

Getting Started
Point out that the above postulate begins with the angles. Since the alternate interior angles are congruent, the lines are parallel.

You are given:
$\angle 3$ and $\angle 5$ are supplementary angles.

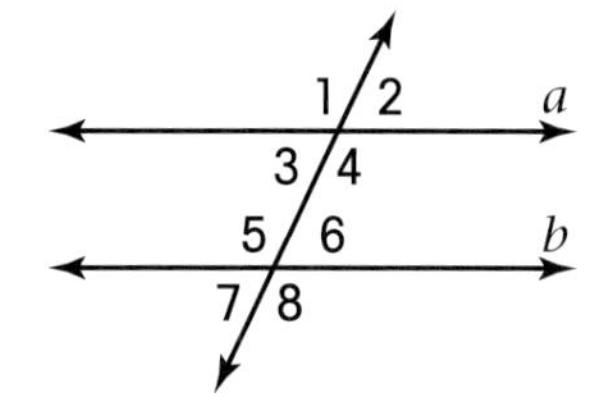

Prove: $a \parallel b$

You are given $\angle 3$ is supplementary to $\angle 5$. This means that $m\angle 3 + m\angle 5 = 180$ by definition of supplementary angles.

You also know that $m\angle 5 + m\angle 6 = 180$ because the angles form a straight angle.

Remember
If two angles are supplementary to the same angle, they are congruent.

Because $\angle 3$ and $\angle 6$ are supplementary to the same angle, they are congruent. $\angle 3$ and $\angle 6$ are alternate interior angles. By the postulate above, $a \parallel b$.

So, you prove that $a \parallel b$. ✓

You proved that if the same-side interior angles are supplementary, the lines are parallel. You will use this in the proof for Try This.

Try This

Copy and complete the proof.

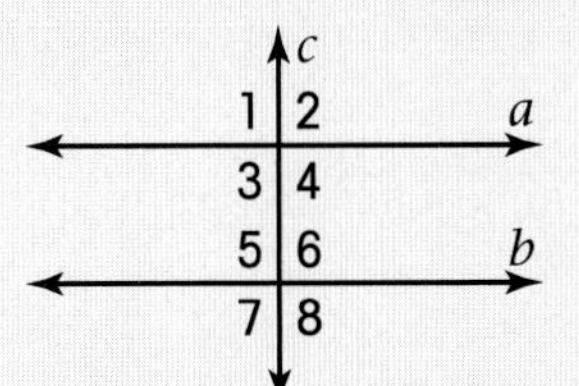

You are given: $a \perp c$ and $b \perp c$.

Prove: $a \parallel b$

You are given that a ■ c and b ■ c. $\perp$ $\perp$

Then, ∠4 is a ■ angle and ∠6 is a ■ angle. right right

m∠4 = ■ and m∠6 = ■. 90 90

∠4 and ∠6 are same-side ■ angles. interior

m∠4 + m∠6 = 90 + 90 = 180

Then, the same-side interior angles are ■. supplementary

So, you proved that ■. ✓ $a \parallel b$

Practice

Write a proof for each of the following. See Additional Answers in the back of this book.

1. You are given: $\angle 1 \cong \angle 5$
Prove: $m \parallel n$

2. You are given: $\angle 1 \cong \angle 8$
Prove: $\ell \parallel k$

More Practice is provided in the Classroom Resource Binder.

Chapter 4 Review

Summary

Perpendicular lines form four right angles.
A perpendicular bisector bisects a line segment and is perpendicular to the line segment.
Parallel lines lie in the same plane but do not intersect.
When parallel lines are cut by a transversal, the alternate interior angle pairs are congruent. The corresponding angle pairs are also congruent.
When parallel lines are cut by a transversal, the same-side interior angle pairs are supplementary.
To show objects in real life, you can draw a one-point perspective of parallel lines.
You can determine the number of possible routes a taxicab can take to get from one place to another by drawing the routes on a grid.
You can use theorems and postulates you know about parallel lines to prove a statement.

More vocabulary review is provided in the Classroom Resource Binder, page 33.

- perpendicular lines
- perpendicular bisector
- parallel lines
- transversal
- alternate interior angles
- same-side interior angles
- corresponding angles
- one-point perspective

Vocabulary Review

Complete the sentences with words from the box.

1. A ____ cuts a line segment into two equal segments and forms four right angles.
perpendicular bisector

2. ____ lie in the same plane, but do not intersect.
parallel lines

3. Line ℓ in the diagram is called a ____. transversal

4. In the diagram, $\angle 1$ and $\angle 5$ are ____. corresponding angles

5. In the diagram, $\angle 3$ and $\angle 6$ are ____.
alternate interior angles

6. In the diagram, $\angle 3$ and $\angle 5$ are ____.
same-side interior angles

7. A way to draw real objects on a flat surface is to use a ____. one-point perspective

8. Two lines that intersect to form right angles are____.
perpendicular lines

Chapter Quiz **More assessment** is provided in the Classroom Resource Binder.

Decide if the lines or line segments are perpendicular. Write *yes* or *no*. If they are perpendicular, use the ⊥ symbol to write a statement.

1.

no

2.

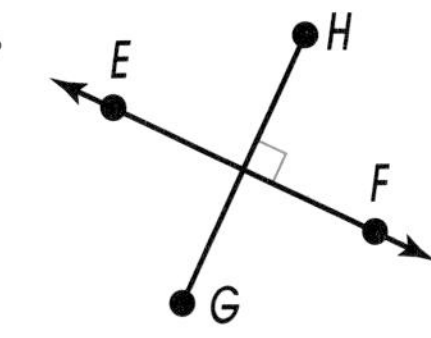

yes; $\overleftrightarrow{EF} \perp \overleftrightarrow{HG}$

3.

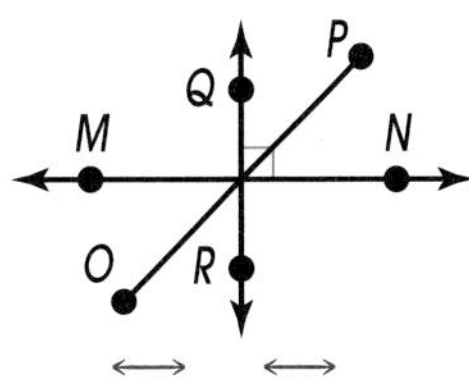

yes; $\overleftrightarrow{MN} \perp \overleftrightarrow{QR}$

In the diagram below, *m* || *n*. Use the diagram for exercises 4–10.

Name all pairs of angles for each exercise.

4. alternate interior angles
∠3 and ∠6; ∠4 and ∠5

5. same-side interior angles
∠3 and ∠5; ∠4 and ∠6

6. corresponding angles
∠1 and ∠5; ∠4 and ∠7; ∠2 and ∠6; ∠3 and ∠8

If ∠1 = 115°, find the measure of each angle.

7. ∠5 115° **8.** ∠4 115° **9.** ∠6 65° **10.** ∠8 65°

Problem Solving

11. Create a one-point perspective drawing of railroad tracks. Check students' work.

12. Look at the map on page 114. How many routes can a taxicab take from 33rd Street and 5th Avenue to 36th Street and 7th Avenue? There are 2 different routes.

Possible Answer: Alternate interior angles and corresponding angles are congruent. Same-side interior angles are supplementary.

Write About Math

If two parallel lines are cut by a transversal, which angles are congruent? Which angles are supplementary?

Additional Practice for this chapter is provided on p. 396 of the student book.

Unit 1 Review

Standardized Test Preparation This unit review follows the format of many standardized tests. A Scantron sheet is provided in the Classroom Resource Binder.

Write the letter of the correct answer.

1. $\overline{AB}$ is 18 cm long. Point M is the midpoint of $\overline{AB}$. Find AM.
A. 6 cm
(B.) 9 cm
C. 12 cm
D. 36 cm

2. $\angle A$ and $\angle B$ are complementary angles. $\angle A$ is 35°. Find m$\angle B$.
A. 20°
B. 35°
(C.) 55°
D. 145°

3. $\angle 1$ and $\angle 3$ are vertical angles. $\angle 1$ is 80°. Find m$\angle 3$.
A. 10°
B. 40°
(C.) 80°
D. 100°

4. Which of the following is the Reflexive Property of Equality?
A. If $a = b$, then $b = a$.
B. If $a = b$, then $a + 1 = b + 1$.
C. If $a = b$, then $3a = 3b$.
(D.) $a = a$

Use the diagram to answer Items 5 and 6. In this diagram, $p \parallel q$.

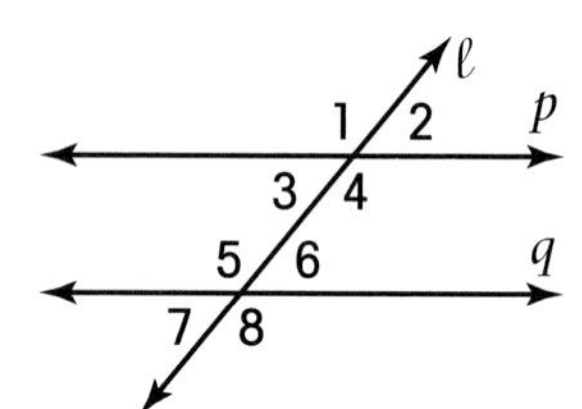

5. Which of the following is NOT true?
A. $\angle 1 \cong \angle 4$
(B.) $\angle 4 \cong \angle 6$
C. $\angle 3 \cong \angle 6$
D. $\angle 4 \cong \angle 8$

6. $\angle 3$ is 50°. Find m$\angle 5$.
A. 25°
B. 40°
C. 50°
(D.) 130°

Critical Thinking

In the diagram below, $\overleftrightarrow{AB} \parallel \overleftrightarrow{DE}$. Find m$\angle FCG$. 40°

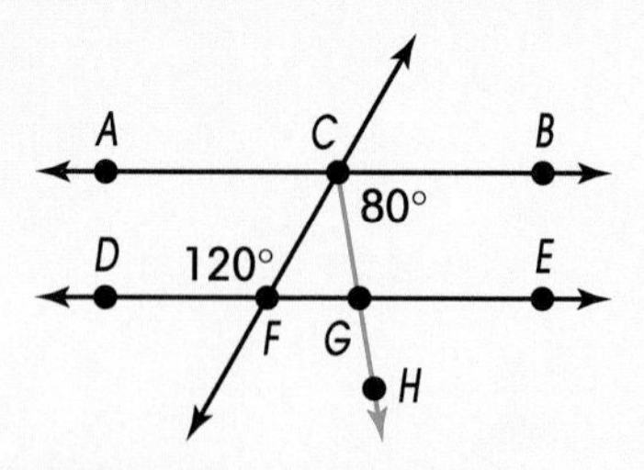

CHALLENGE Which angle is congruent to $\angle GCB$? Explain.

Possible answer: $\angle HGE$ is 80°. $\angle GCB$ and $\angle HGE$ are corresponding angles.

Unit Two

Chapter 5
Triangles

Chapter 6
Right Triangles

Chapter 7
Quadrilaterals and Polygons

Chapter 5 Triangles

The artist who made this sculpture used several different shapes. Which shape catches your eye? How does it interact with the other shapes?

Learning Objectives

- Classify triangles by angles and sides.
- Find the measure of interior and exterior angles.
- Use the properties of triangles.
- Find congruent triangles.
- Find the midsegment, median, altitude, and angle bisector of a triangle.
- Use a calculator to find missing triangle measures.
- Find the centroid of a triangle.
- Use properties of triangles to solve real-life problems.
- Write proofs using triangle properties.

ESL/ELL Note Explain in English that the prefix *tri-* means "three." Ask students to write words with the prefix *tri-* and to tell what the words mean. Some examples are *triangle*, *triple*, and *triplets*. You may want to illustrate your examples as well.

Words to Know

acute triangle a triangle with three acute angles

right triangle a triangle with one right angle

obtuse triangle a triangle with one obtuse angle

equiangular triangle a triangle with all angles equal

exterior angle the angle formed by extending one side of a triangle

remote interior angle an interior angle of a triangle not adjacent to the given exterior angle

scalene triangle a triangle with no congruent sides

isosceles triangle a triangle with two sides congruent

equilateral triangle a triangle with all sides congruent

median a line segment of a triangle that joins a vertex to the midpoint of the opposite side

altitude a line segment of a triangle that joins a vertex to the line containing the opposite side and that is perpendicular to that side

angle bisector a line segment that joins a vertex of a triangle to the opposite side and that bisects the angle

midsegment a line segment that joins the midpoints of two sides of a triangle

centroid the point where the three medians of a triangle meet

Triangle Truss Project

Because the triangle is rigid, it is used to support buildings and bridges. This support is called a truss.

Look for structures that use triangles for support. Draw or take a photo of each structure and write about it. Interview a builder in your town. Ask about the kinds of structural support used in buildings.

Project See the Classroom Resource Binder for a scoring rubric to assess this project.

More practice is provided in the Workbook, page 25, and in the Classroom Resource Binder, page 44.

5·1 Classifying Triangles by Angles

Getting Started
Review classifying angles. Be sure students understand the terms *acute*, *obtuse*, *right*, and *vertex*.

Math Fact
Use the vertices of a triangle to name it. Write $\triangle ABC$.

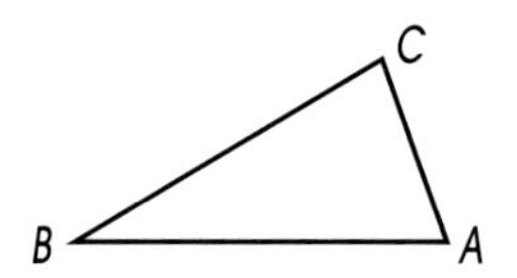

Other names for this triangle are $\triangle ACB$, $\triangle BCA$, $\triangle BAC$, $\triangle CBA$, and $\triangle CAB$.

The chart below shows how to use the angles of a triangle to classify a triangle.

Type of Triangle	Description	Example
Acute Triangle	All angles are acute.	70° 75° 35°
Right Triangle	One angle is a right angle.	
Obtuse Triangle	One angle is an obtuse angle.	110°
Equiangular Triangle	All angles are congruent.	

To classify a triangle, look at the angles. Then find the correct description in the chart above.

EXAMPLE 1

Classify $\triangle DRT$ as acute, right, obtuse, or equiangular.

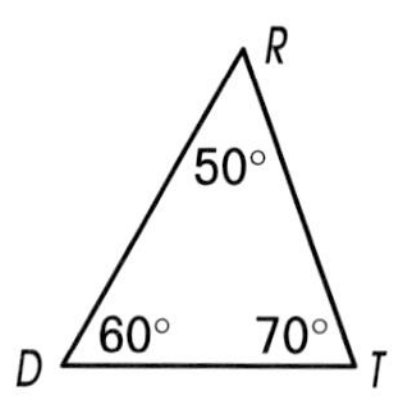

Look at each angle.
All the angles are less than 90°.
All the angles are acute.
The angles are not all equal.

So, $\triangle DRT$ is an acute triangle.

EXAMPLE 2

Classify $\triangle PSQ$ as acute, right, obtuse, or equiangular.

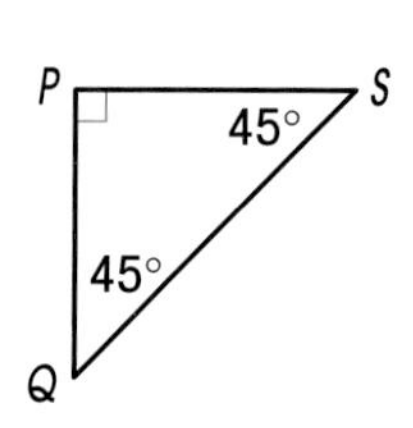

Look at each angle.
One angle is 90°.
That angle is a right angle.

So, $\triangle PSQ$ is a right triangle.

Try This

Draw obtuse triangle ABC. Make $\angle B$ a 130° angle.

Use your protractor to draw $\angle B$.
Complete $\triangle ABC$.
Look at each angle.
One angle is greater than 90° and less than ■. 180°

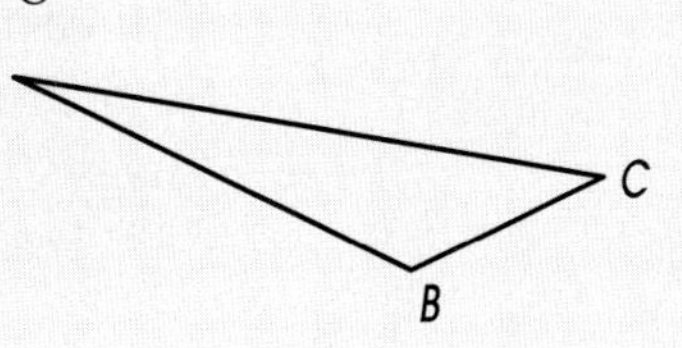

So, $\triangle ABC$ is an ■ triangle. obtuse

Practice

Classify each triangle. Write *acute, right, obtuse,* or *equiangular.*

1. 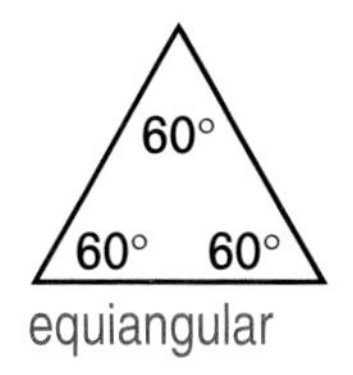

equiangular

2. 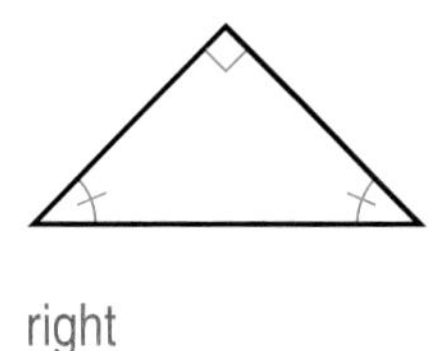
right

3. 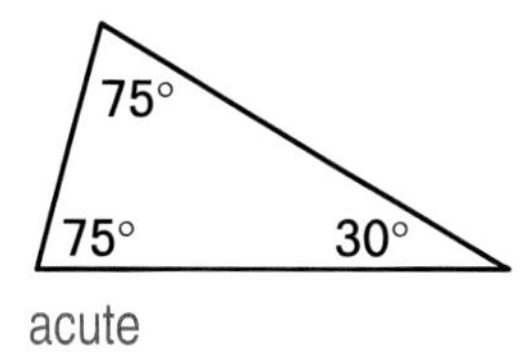

acute

4. 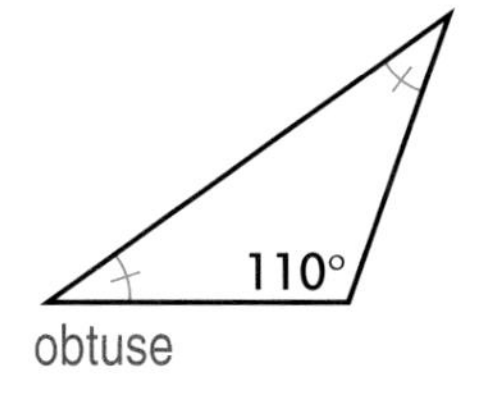

obtuse

5. 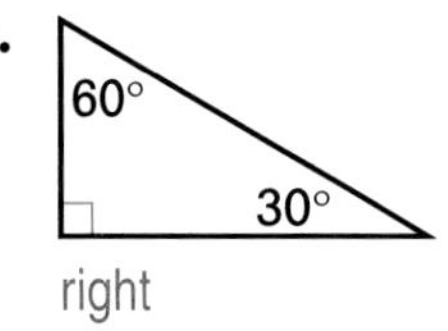

right

6. 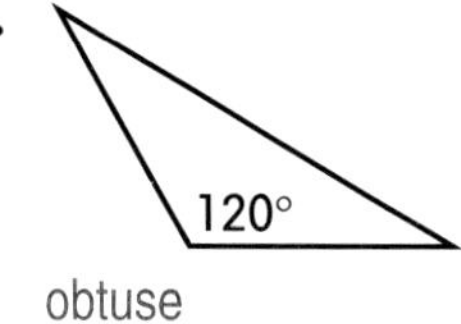

obtuse

Draw an example of each triangle. Check students' work.

7. acute **8.** obtuse **9.** right

10. In an acute triangle, all three angles are acute and in an obtuse triangle, one angle is obtuse.

Share Your Understanding

10. Explain to a partner the difference between an acute triangle and an obtuse triangle. Have your partner draw an example of each triangle. See above.

11. **CRITICAL THINKING** Copy this diagram. Make two right triangles by adding one line segment.

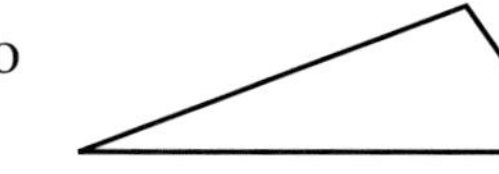

More practice is provided in the Workbook, page 25, and in the Classroom Resource Binder, page 44.

5·2 Angle Sum Theorem

THEOREM 11 **Angle Sum Theorem for Triangles**

The sum of the angles of a triangle is 180°.

You can measure each angle of a triangle. The Angle Sum Theorem says that the sum of the angles will always be 180°.

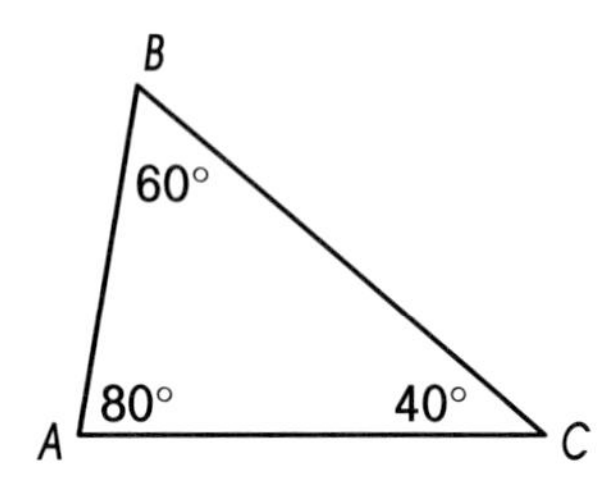

$$\angle A + \angle B + \angle C = 180°$$
$$80° + 60° + 40° = 180°$$

Getting Started
Have students draw a large triangle on a sheet of paper. Then cut out the angles and place them along a straight line to show that the sum is 180°.

If you know the measures of two angles of a triangle, you can find the measure of the third angle.

EXAMPLE 1

Find the measure of $\angle D$.

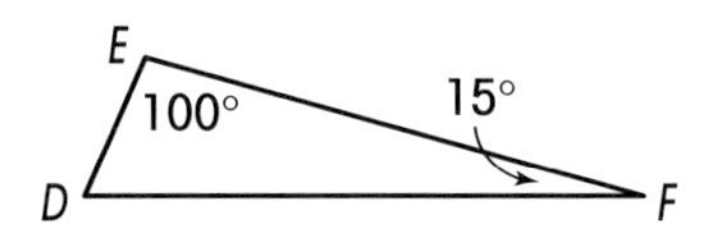

Write an equation. Use the Angle Sum Theorem for Triangles. Then, solve for $m\angle D$.

$$m\angle D + 100 + 15 = 180$$
$$m\angle D = 65$$

So, $\angle D$ is 65°.

EXAMPLE 2

Find the measure of $\angle G$.

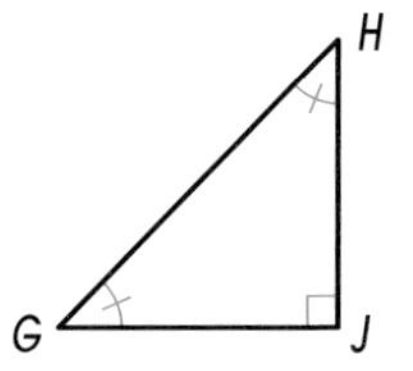

Write an equation. Use the Angle Sum Theorem for Triangles.

$$m\angle G + m\angle H + 90 = 180$$

Substitute $m\angle G$ for $m\angle H$.

$$m\angle G + m\angle G + 90 = 180$$

Solve for $m\angle G$.

$$2m\angle G + 90 = 180$$
$$2m\angle G = 90$$
$$m\angle G = 45$$

So, $\angle G$ is 45°.

Math Fact
The congruent marks in the triangle show that $\angle G \cong \angle H$. So, $m\angle G = m\angle H$.

Try This

Find the value of x and the value of y.

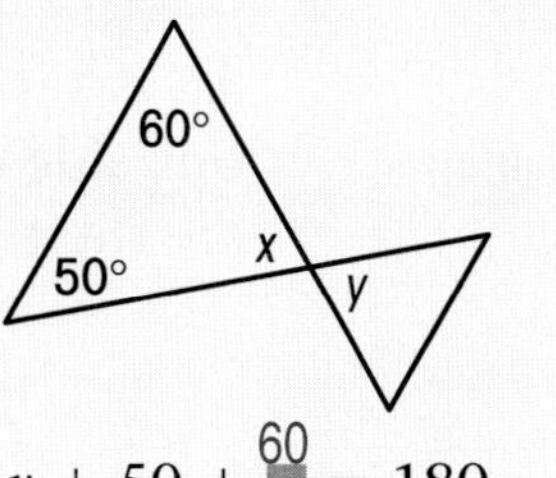

Write an equation. Use the Angle ■ Sum Theorem for Triangles.
Then, solve for x.

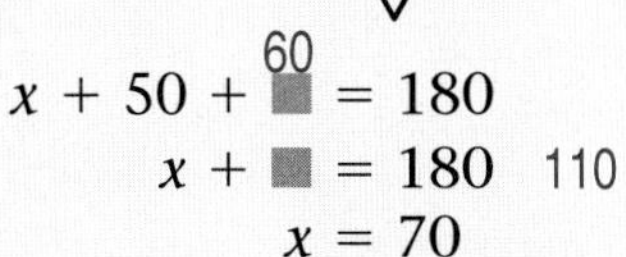

$$x + 50 + \blacksquare = 180 \quad 60$$
$$x + \blacksquare = 180 \quad 110$$
$$x = 70$$

Now, find the value of y. Notice that x and y are vertical angles. Vertical angles are ■. congruent

$$y = \blacksquare \quad 70$$

So, x is ■ 70° and y is ■. 70°

Practice

Find the value of x and the value of y.

1.

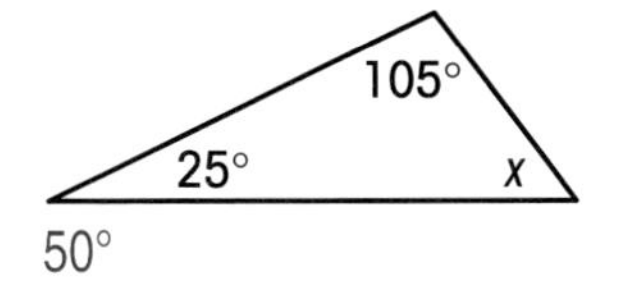

50°

2.

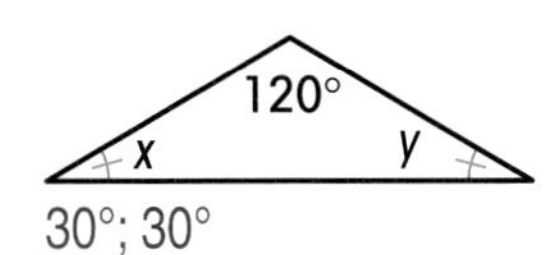

30°; 30°

3.

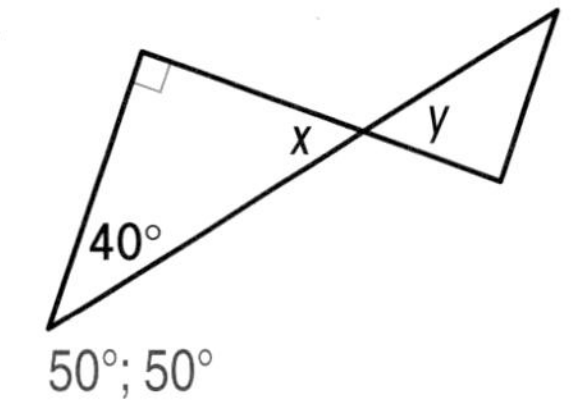

50°; 50°

Share Your Understanding

4. Explain the Angle Sum Theorem for Triangles to a partner. Have your partner draw an obtuse triangle and find the sum of the angles to check the theorem. Check students' work.

5. **CRITICAL THINKING** The measures of the angles of $\triangle ABC$ are x, $2x$, and $3x$. Classify the triangle as acute, right, or obtuse. Explain your thinking. The sum of the angles of a triangle is 180°. Then, $x + 2x + 3x = 180$. So, $x = 30$, $2x = 60$, and $3x = 90$. $\triangle ABC$ is a right triangle.

More practice is provided in the Workbook, page 25, and in the Classroom Resource Binder, page 45.

5·3 Exterior Angles of a Triangle

Getting Started Review supplementary angles.

> THEOREM 12
> **Exterior Angle Theorem**
> The measure of an exterior angle of a triangle equals the sum of the measures of the two remote interior angles.

An **exterior angle** of a triangle is formed by extending one side of the triangle. The **remote interior angles** are the two opposite angles inside the triangle.

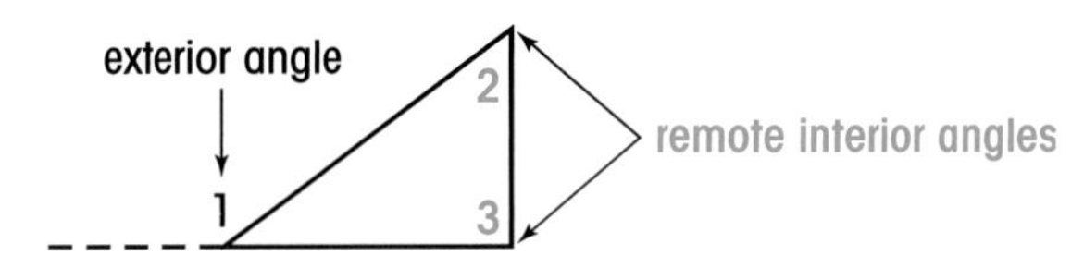

$$m\angle 1 = m\angle 2 + m\angle 3$$

You can find the measure of an exterior angle. Use the Exterior Angle Theorem.

EXAMPLE 1

Find the value of x.

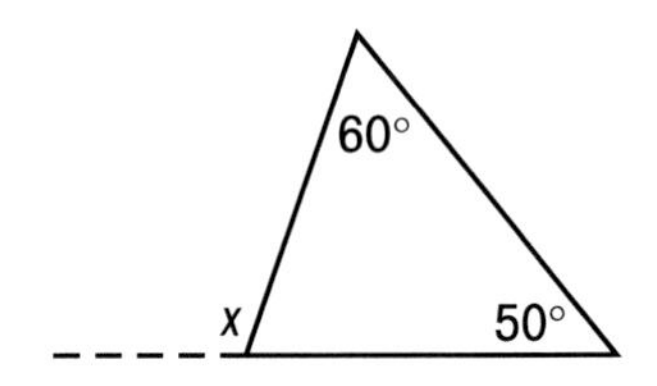

Write an equation. Use the Exterior Angle Theorem. $\quad x = 60 + 50$

Simplify. $\quad x = 110$

The exterior angle x is 110°.

You can find the measure of a remote interior angle. You need to know the measures of the exterior angle and the other remote interior angle.

EXAMPLE 2

Find the value of y.

Write an equation. Use the Exterior Angle Theorem $\quad 160 = y + 20$

Then, solve for y. $\quad 160 - 20 = y + 20 - 20$

$\quad 140 = y$

The remote interior angle y is 140°.

Try This

Find the value of y.

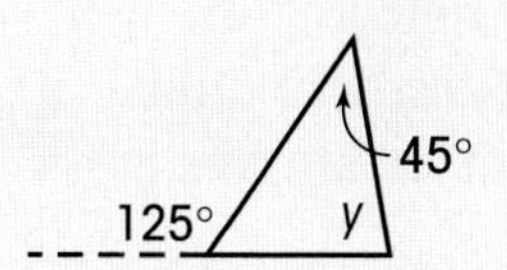

Write an equation. Use the Exterior Angle Theorem.

$125 = \blacksquare + y$ 45

Then solve for $\blacksquare$. y

$125 - \blacksquare = 45 + y - \blacksquare$ 45 45

80 $\blacksquare = y$

The remote interior angle y is $\blacksquare$. 80°

Practice

Find each value of x and y.

1.

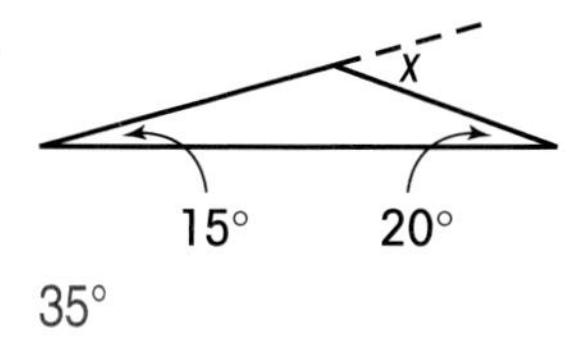

35°

2.

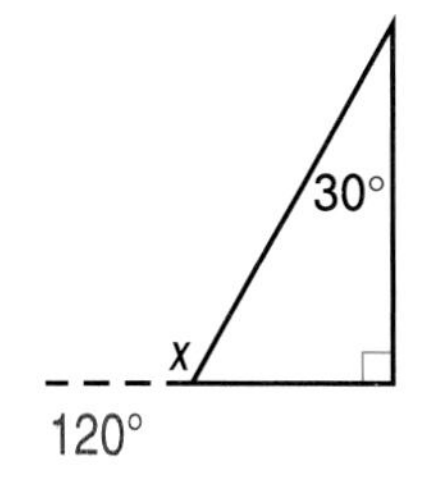

120°

3.

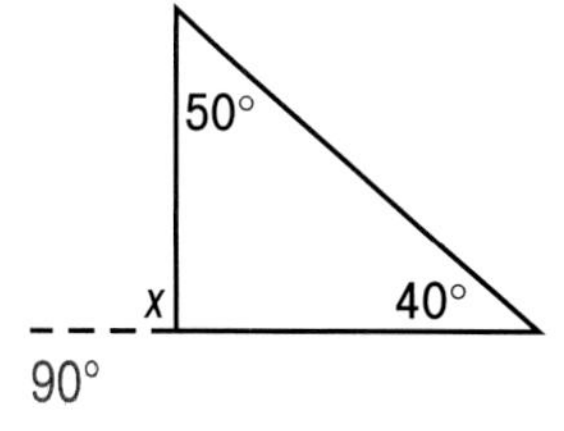

90°

4.

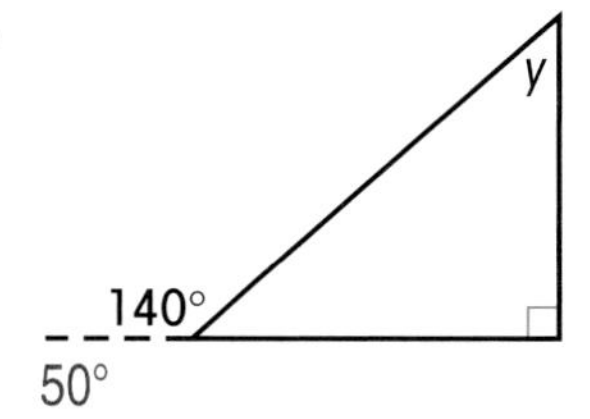

50°

5.

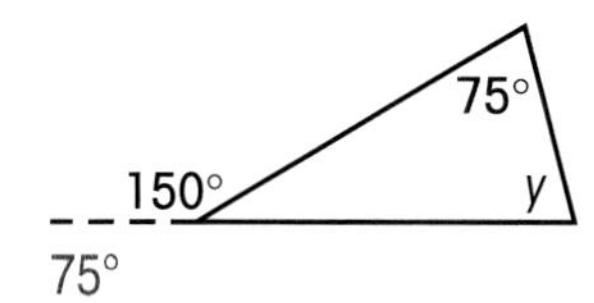

75°

6.

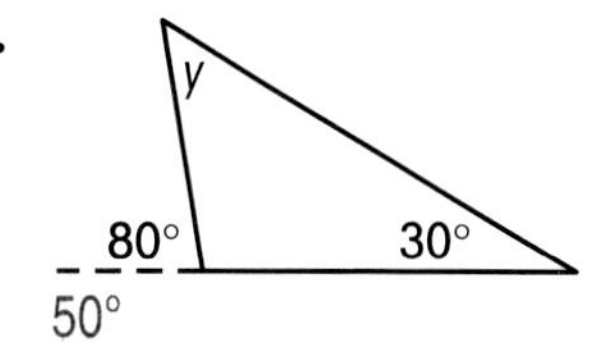

50°

Share Your Understanding

7. Explain to a partner how to find the measure of an exterior angle. Use the words *remote interior angles* in your explanation. Have your partner draw a diagram to illustrate.
Add the measures of the two remote interior angles. Check students' work.

8. **CRITICAL THINKING** In a triangle, the measure of an exterior angle is 103°. One of the remote interior angles is 56°. What are the measures of the other two angles in the triangle? (Hint: Draw a diagram.)
47° and 77°

More practice is provided in the Workbook, page 26, and in the Classroom Resource Binder, page 45.

5-4 Classifying Triangles by Sides

Getting Started
Review classifying triangles by angles. Be sure students can define *congruent.*

You learned to classify triangles by their angles. You can also classify triangles by their sides. The chart below shows the three ways to classify triangles by their sides.

Remember
The marks on the sides of triangles show congruence.

Type of Triangle	Description	Example
Scalene Triangle	No sides are congruent.	
Isosceles Triangle	Two sides are congruent.	
Equilateral Triangle	All sides are congruent.	

To classify a triangle, look at the sides. Then, find the correct description in the chart above.

EXAMPLE 1

Classify $\triangle ABC$ as scalene, isosceles, or equilateral.

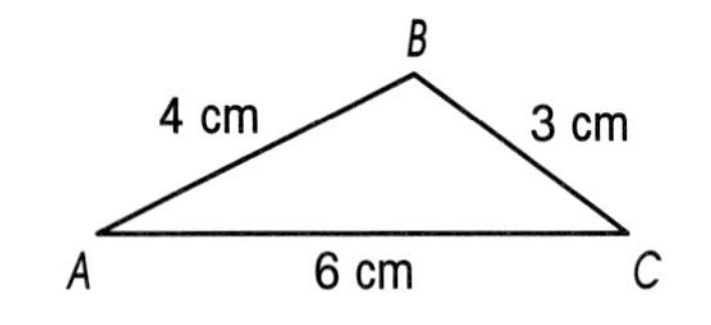

Look at the lengths of the sides. No two sides of the triangle are congruent.

So, $\triangle ABC$ is a scalene triangle.

EXAMPLE 2

Classify $\triangle DEF$ as scalene, isosceles, or equilateral.

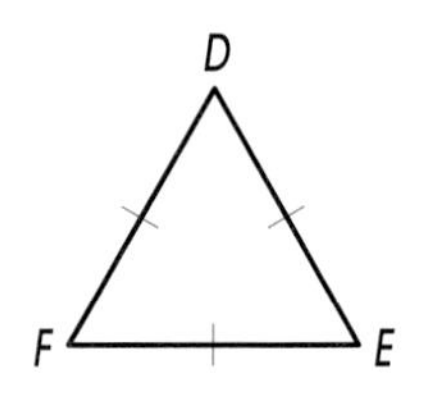

Look at the lengths of the sides. All three sides of the triangle are congruent.

So, $\triangle DEF$ is an equilateral triangle.

Try This

Classify $\triangle ABC$ by both angles and sides.

Look at the angles. $\angle C$ is a ■ angle. right

Look at the length of each ■. Two sides are ■. side; congruent

$\triangle ABC$ is a right triangle and an isosceles triangle.

So, $\triangle ABC$ is a right ■ triangle. isosceles

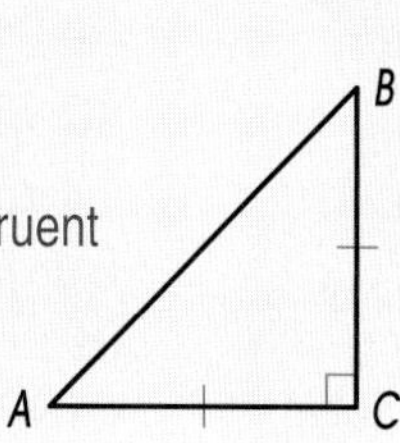

Practice

Classify each triangle by its sides. Write *scalene, isosceles,* or *equilateral.*

1.

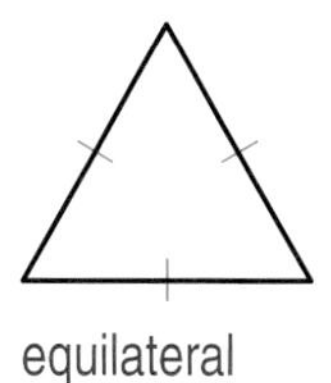

equilateral

2.

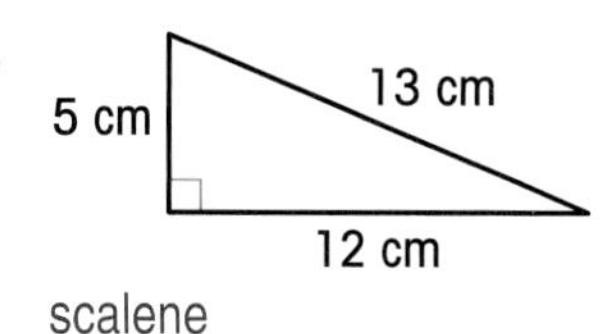

scalene

3.

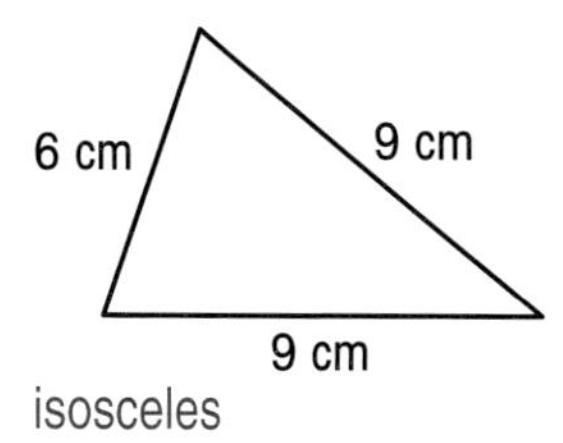

isosceles

Classify each triangle by its angles and sides.

4.

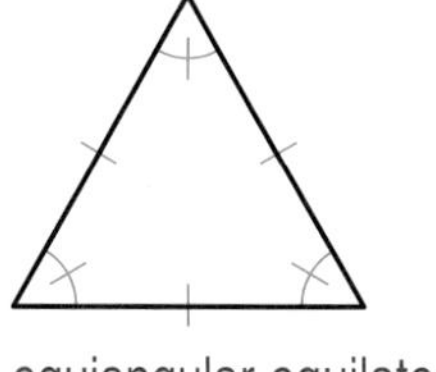

equiangular equilateral

5.

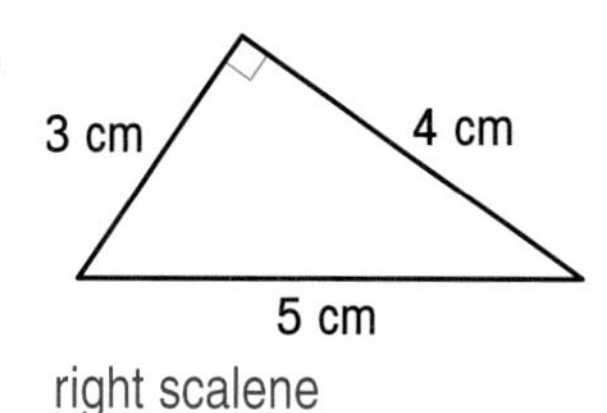

right scalene

6.

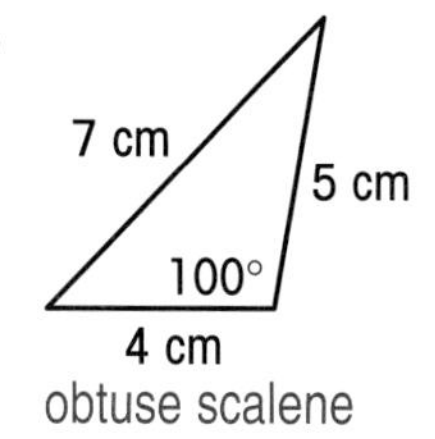

obtuse scalene

Share Your Understanding

7. Explain to a partner how you can identify an isosceles triangle. Have your partner explain how to identify a scalene triangle. Use the words *sides* and *congruent* in your explanations.
An isosceles triangle has 2 congruent sides. A scalene triangle has no congruent sides.

8. **CRITICAL THINKING** Use a ruler and a protractor to draw an equiangular equilateral triangle.
Check students' work.

More practice is provided in the Workbook, page 26, and in the Classroom Resource Binder, page 46.

5·5 Triangle Inequality Theorem

Getting Stated Review the inequality symbols.

THEOREM 13 Triangle Inequality Theorem

The sum of the lengths of any two sides of a triangle must be greater than the length of the third side.

Suppose you have three line segments 7 m, 10 m, and 13 m long. Can you make a triangle? Use the Triangle Inequality Theorem to find out. The sum of the lengths of any two sides must be greater than the length of the third side.

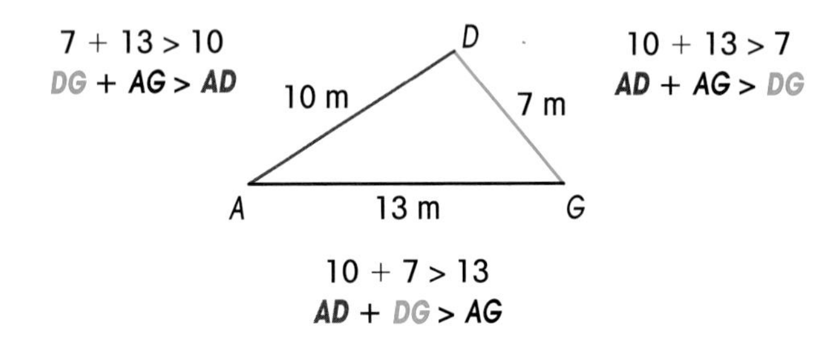

You can use the Triangle Inequality Theorem to decide if three given line segments can form a triangle.

EXAMPLE 1

Can segments 5 m, 8 m, and 12 m long form a triangle?

Find the sum of two lengths. $5 + 8 = 13$
Then, compare the sum to the third length. $13 > 12$

If the sum of any two lengths is greater than the third length, the segments can form a triangle.

$8 + 12 = 20$
$20 > 5$
$12 + 5 = 17$
$17 > 8$

Segments 5 m, 8 m, and 12 m long can form a triangle.

EXAMPLE 2

Can segments 3 ft, 9 ft, and 13 ft long form a triangle?

Find the sum of two lengths. $3 + 9 = 12$
Then, compare the sum to the third length. $12 \ngtr 13$

The sum is not greater than the third length.

Segments 3 ft, 9 ft, and 13 ft long cannot form a triangle.

Math Fact
The symbol $\ngtr$ means "is not greater than."

Try This

Can segments 16 yd, 35 yd, and 51 yd long form a triangle?

Find the sum of ■ lengths. two — $16 + 35 = ■$ 51

Then, compare the sum to the ■ length. third — $51 ■ \not> 51$

Segments 16 yd, 35 yd, and 51 yd long ■ form a triangle. cannot

Practice

Can segments with the given lengths form a triangle? Write *yes* or *no*. If your answer is *no*, tell why.

1. 4 m, 5 m, 16 m no; $4 + 5 \not> 16$

2. 4 in., 6 in., 10 in. no; $4 + 6 \not> 10$

3. 3 cm, 4 cm, 5 cm yes

4. 6 mm, 9 mm, 20 mm no; $6 + 9 \not> 20$

5. 5 dm, 12 dm, 15 dm yes

6. 7 m, 16 m, 21 m yes

7. 8 yd, 15 yd, 17 yd yes

8. 8 mi, 8 mi, 16 mi no; $8 + 8 \not> 16$

Can you form a triangle with the given segment lengths? If so, classify each triangle. Write *scalene*, *isosceles*, or *equilateral*.

9. 7 ft, 7 ft, 10 ft
yes; isosceles

10. 10 cm, 24 cm, 26 cm
yes; scalene

11. 15 m, 15 m, 15 m
yes; equilateral

Share Your Understanding

12. Take turns with a partner. Choose three numbers for the lengths of the sides of a triangle. Decide if you can form a triangle, using those lengths. Explain your thinking.
Check students' work.

13. **CRITICAL THINKING** The lengths of the sides of the triangle on the right are positive integers. What is the largest integer x can be? 15

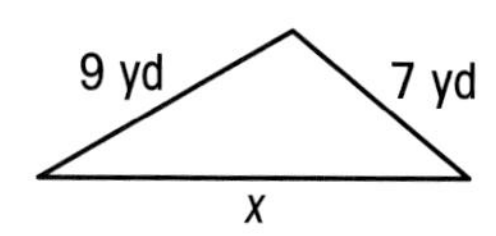

More practice is provided in the Workbook, page 27, and in the Classroom Resource Binder, page 46.

5·6 Isosceles Triangles

Getting Started Review the definition of an isosceles triangle.

THEOREM 14 Isosceles Triangle Theorem

If two sides of a triangle are congruent, then the angles opposite those sides are also congruent.

You learned that an isosceles triangle has two congruent sides. The theorem on the left tells you that the angles opposite those sides are also congruent.

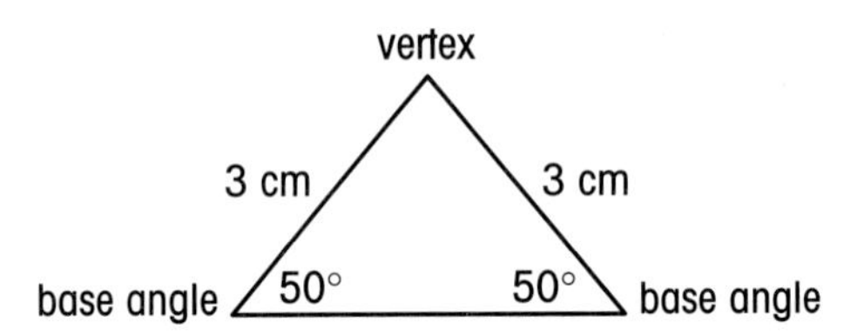

If you know the measure of a base angle, you can find the measure of the vertex angle.

EXAMPLE

$\triangle ABC$ is isosceles. Find the measure of $\angle B$.

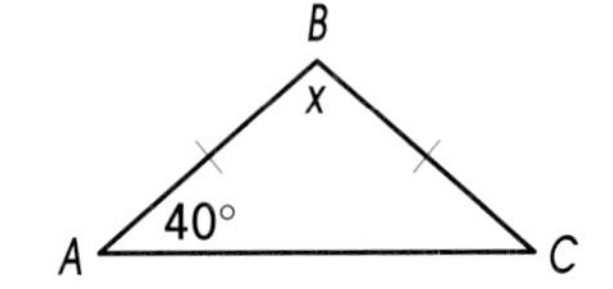

The Isosceles Triangle Theorem tells you that $m\angle A = m\angle C$. $\quad m\angle A = 40 \quad m\angle C = 40$

Write an equation. Use the Angle Sum Theorem. $\quad m\angle A + m\angle B + m\angle C = 180$

Substitute the values you know. $\quad 40 + x + 40 = 180$

Then, solve for x. $\quad x + 80 = 180$

$x = 100$

Remember
The sum of the angles of a triangle is 180°.

The vertex angle, $\angle B$, is 100°.

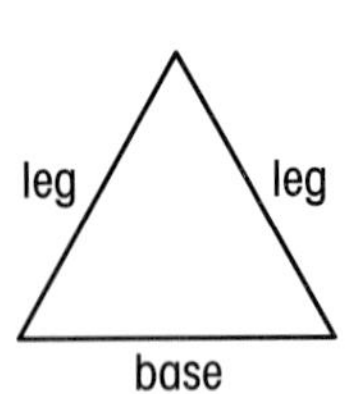

The converse of the Isosceles Triangle Theorem is also true. If two angles of a triangle are congruent, then the sides opposite those angles are also congruent.

The sides opposite the congruent angles are called legs. The side opposite the vertex is the base.

Try This

Find the length of side $\overline{DE}$.

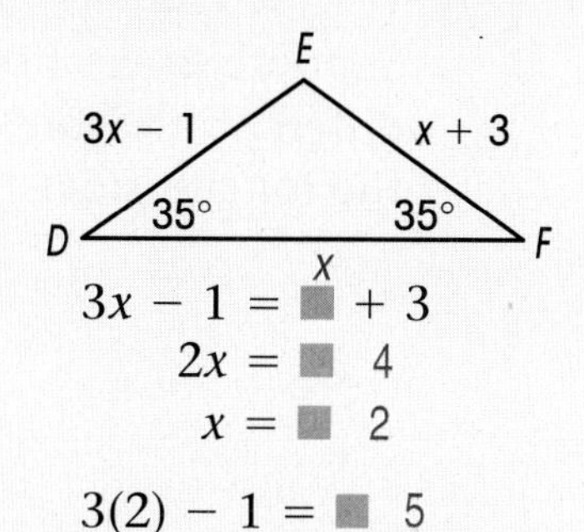

Because the base angles are congruent, $DE = ■$. FE

Write an equation. Solve for x.

$3x - 1 = ■ + 3$ (x)

$2x = ■$ 4

$x = ■$ 2

Substitute ■ (2) for x to find DE.

$3(2) - 1 = ■$ 5

The length of side $\overline{DE}$ is ■. 5

Practice

Find the value of x.

1. 50°

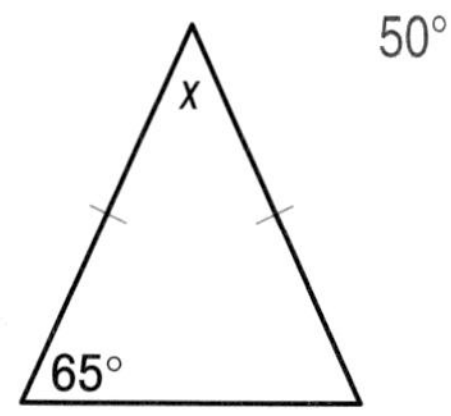

2. 90°

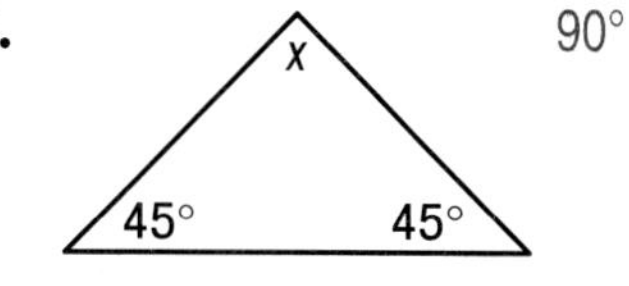

3. 10 in.

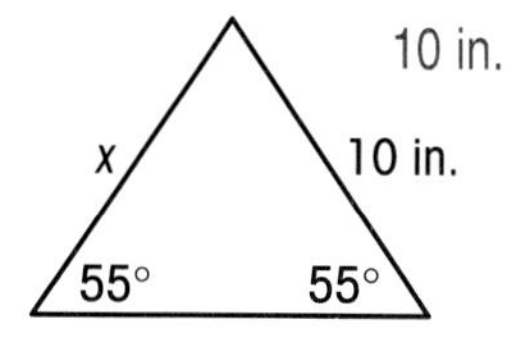

Find the length of each side of $\triangle HPJ$.

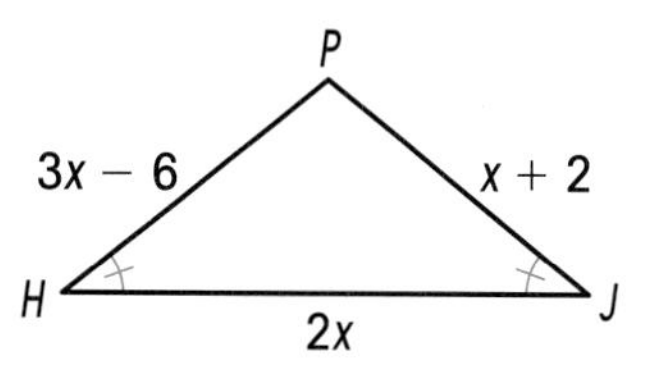

4. $\overline{HP}$ 6 **5.** $\overline{JP}$ 6 **6.** $\overline{HJ}$ 8

7. In an isosceles triangle, base angles are congruent. So, the sum of the base angles is 20°. Subtract 20° from the total angle sum, 180°. The vertex angle is the difference, 160°.

Share Your Understanding

7. Explain to a partner how to find the measure of the vertex angle in an isosceles triangle with a base angle that measures 10°.

8. **CRITICAL THINKING** The base of an isosceles triangle is 15 in. What is the smallest possible integer length for each leg?
8 in.

More practice is provided in the Workbook, page 27, and in the Classroom Resource Binder, page 47.

5·7 Side-Angle Relationship

THEOREM 15 Opposite Side-Angle Theorem

The longest side of a triangle is opposite the largest angle.

Getting Started
Have students draw a triangle. Have them find the longest side and the largest angle.

Suppose each of the three angles in a triangle has a different measure. The side opposite the largest angle is the longest side.

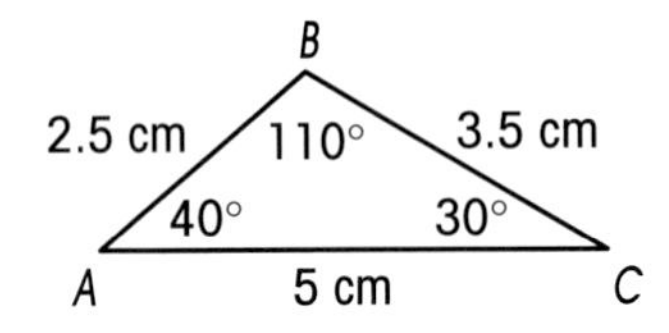

$\angle B$ is the largest angle in $\triangle ABC$.
$\overline{AC}$ is the longest side.

If you know the measures of the angles, you can find the longest side of a triangle.

EXAMPLE 1

Name the longest side of $\triangle DEF$.

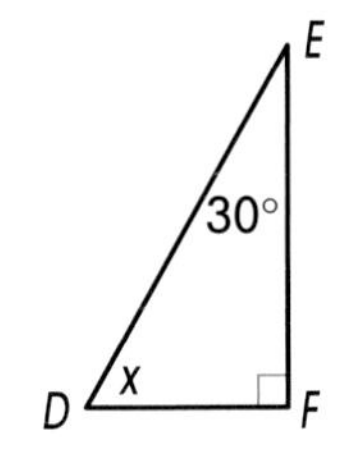

First, you need to find the largest angle. Use the Angle Sum Theorem.

Write an equation to find the unknown angle. Solve for x.

$$x + 30 + 90 = 180$$
$$x + 120 = 180$$
$$x = 60$$

$\angle D$ is 60°, $\angle E$ is 30°, and $\angle F$ is 90°.
$\angle F$ is the largest angle. The side opposite $\angle F$ is $\overline{DE}$.

The longest side of $\triangle DEF$ is $\overline{DE}$.

You can find the largest angle in a triangle if you know the measures of the sides.

EXAMPLE 2

Name the largest angle of $\triangle GHT$.

H
8 in.
4 in.
G
10 in.
T

Look at the length of each side.
The longest side is $\overline{GT}$.

The angle opposite $\overline{GT}$ is $\angle H$.

The largest angle of $\triangle GHT$ is $\angle H$.

Try This

Name the longest side of $\triangle JKL$.

First find the measure of $\angle K$.

Use the Angle ■ Theorem. Sum

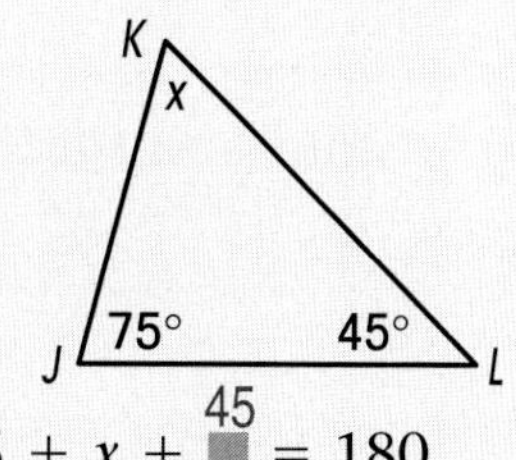

Write an equation. $75 + x + ■ = 180$ 45

Solve for x. $x + ■ = 180$ 120

$x = ■$ 60

■ is the largest angle. The side opposite $\angle J$ is ■. $\angle J$ $\overline{KL}$

The longest side of $\triangle JKL$ is ■. $\overline{KL}$

Practice

Name the longest side.

1.

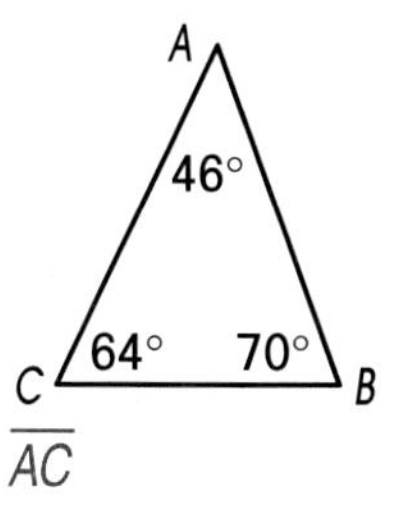

$\overline{AC}$

2.

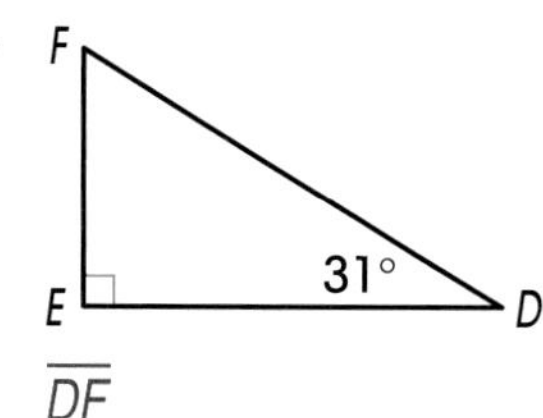

$\overline{DF}$

3.

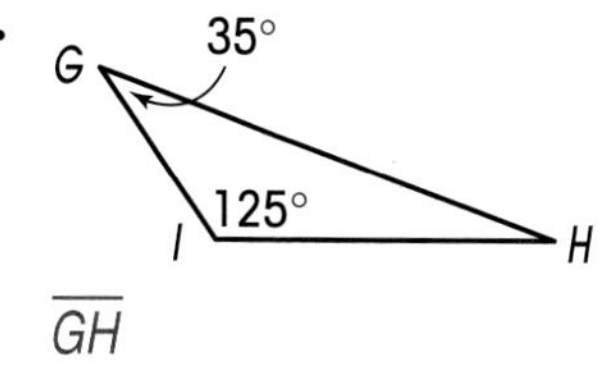

$\overline{GH}$

Name the largest angle.

4.

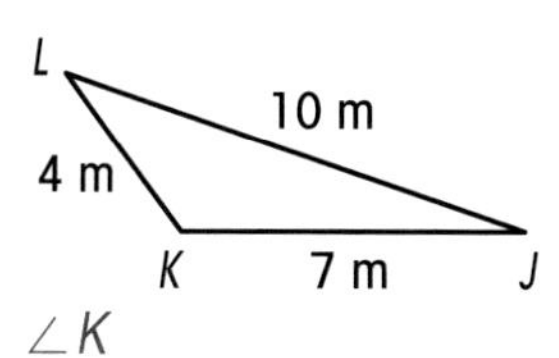

$\angle K$

5.

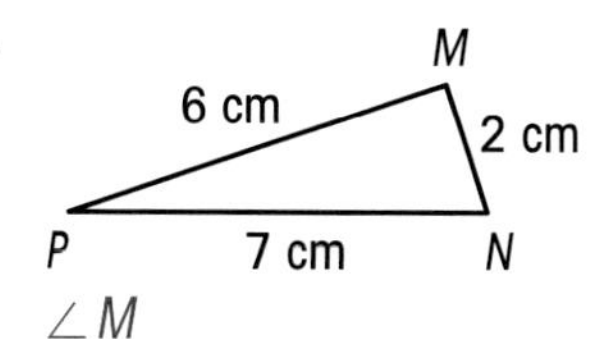

$\angle M$

6.

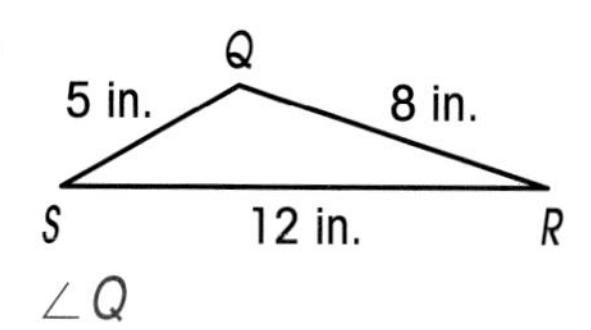

$\angle Q$

Share Your Understanding

7. Explain to a partner how to find the largest angle in a triangle. Use the words *angle*, *side*, *longest*, and *largest* in your explanation.
The largest angle of a triangle is opposite the longest side.

8. **CRITICAL THINKING** The vertex angle of an isosceles triangle is 50°. Which is longer, the base or a leg? Explain.
Each base angle is 65°. The leg is longer because it is opposite the larger angle.

More practice is provided in the Workbook, page 28, and the Classroom Resource Binder, page 48.

5-8 Congruent Triangles

Getting Started Review congruent line segments and angles.

You learned that line segments and angles are congruent if they have the same measure. Triangles can also be congruent. Two triangles are congruent if their corresponding sides and angles are congruent.

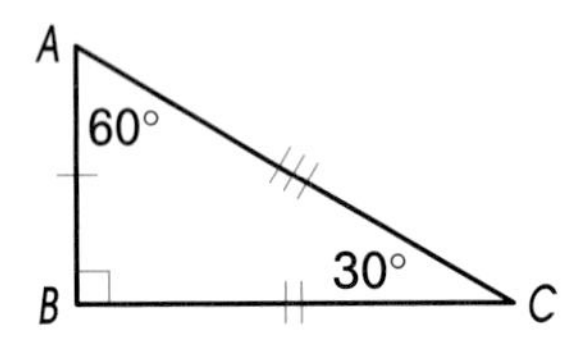

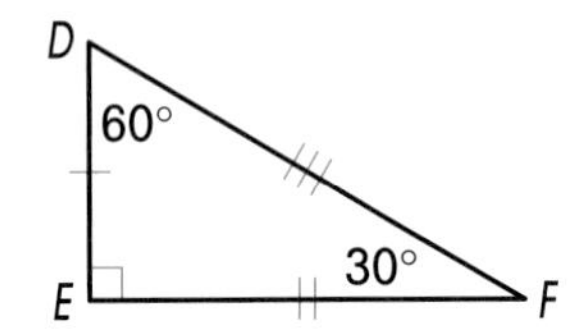

Math Fact
Matching parts are called corresponding parts.

$\triangle ABC$ is congruent to $\triangle DEF$. Write $\triangle ABC \cong \triangle DEF$.

When you write a congruence statement, the corresponding angles must be named in the same order for both triangles.

EXAMPLE

Write a congruence statement for these two congruent triangles.

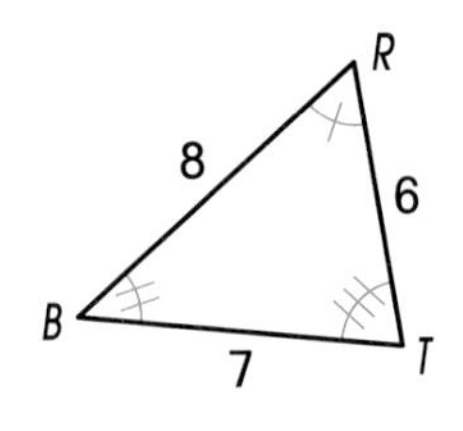

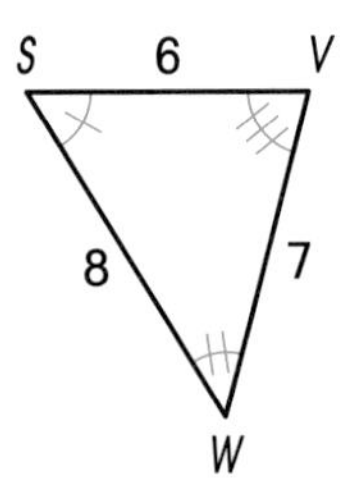

Match the congruent angles to write the congruence statement.

$\angle R \cong \angle S$
$\angle B \cong \angle W$
$\angle T \cong \angle V$
$\triangle RBT \cong \triangle SWV$

So, $\triangle RBT \cong \triangle SWV$.

Congruent triangles have the same size and shape.

Try This

In the diagram, $\triangle DLM \cong \triangle RKV$. Identify the corresponding parts. Find each unknown measure.

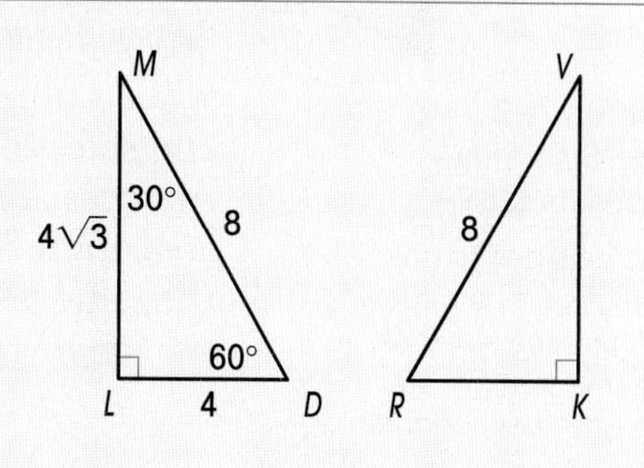

$\angle D \cong$ ■, so $\angle R =$ ■. $\angle R$; 60°

$\angle M \cong$ ■, so $\angle V =$ ■. $\angle V$; 30°

$\overline{DL} \cong$ ■, so $RK =$ ■. $\overline{RK}$; 4

$\overline{LM} \cong$ ■, so $KV =$ ■. $\overline{KV}$; $4\sqrt{3}$

Practice

In the diagram, $\triangle RTB \cong \triangle PSL$. Complete each statement.

1. $\angle R \cong$ ■ $\angle P$

2. $\angle T \cong$ ■ $\angle S$

3. $\angle B \cong$ ■ $\angle L$

4. $\overline{RT} \cong$ ■ $\overline{PS}$

5. $\overline{TB} \cong$ ■ $\overline{SL}$

6. $\overline{BR} \cong$ ■ $\overline{LP}$

In the diagram, $\triangle DSQ \cong \triangle JNM$. Complete each statement.

7. $m\angle D =$ ■ 45

8. $m\angle S =$ ■ 45

9. $m\angle M =$ ■ 90

10. $JN =$ ■ $6\sqrt{2}$

11. $NM =$ ■ 6

12. $DQ =$ ■ 6

13. Use the order in the congruence statement to identify the congruent parts. $\angle A \cong \angle J$. $\angle Y \cong \angle P$ and $\angle Z \cong \angle L$. When you name congruent sides, also go in order: $\overline{AY} \cong \overline{JP}$, $\overline{YZ} \cong \overline{PL}$, and $\overline{AZ} \cong \overline{JL}$.

Share Your Understanding

13. $\triangle AYZ$ is congruent to $\triangle JPL$. Explain to a partner how to find the congruent angles in the two triangles. Your partner should then explain how to find the congruent sides.

14. **CRITICAL THINKING** $\triangle PBX$ is congruent to $\triangle RSG$. Does this mean that $\triangle BPX$ must be congruent to $\triangle RGS$? If yes, tell why. If no, write a true congruence statement for $\triangle BPX$.

No; $\triangle BPX$ is not congruent to $\triangle RGS$. The letters for the angles need to match as in the first congruence statement. In that statement, B matches S. So, $\triangle BPX \cong \triangle SRG$.

More practice is provided in the Workbook, page 28, and in the Classroom Resource Binder, page 48.

5·9 Congruent Triangles: SSS and SAS

Getting Started Help students to identify an included angle.

> POSTULATE 9
> **Side-Side-Side Postulate (SSS)**
>
> If three sides of one triangle are congruent to the three sides of another triangle, the triangles are congruent.

You do not need to know the measures of all angles and sides to decide if two triangles are congruent. Two triangles are congruent if all three corresponding sides are congruent. Corresponding sides are sides that have the same length. $\overline{AD}$ corresponds to $\overline{QS}$. The order of the letters for naming the sides is important.

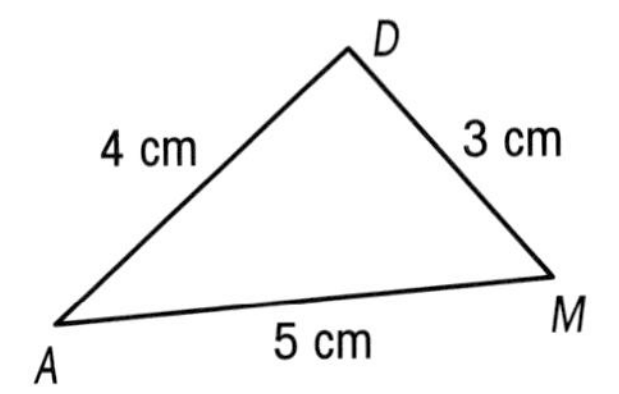

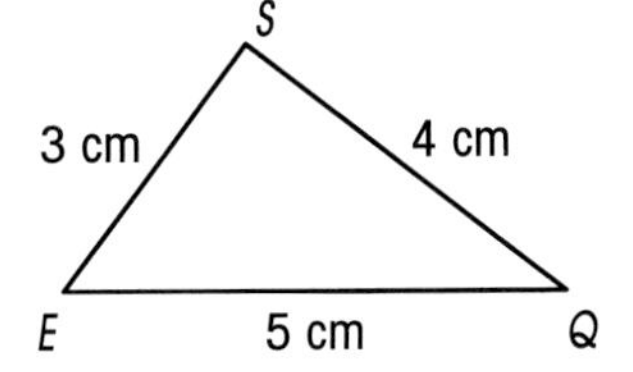

$\triangle ADM \cong \triangle QSE$ by the SSS postulate.

EXAMPLE

Write a congruence statement for these triangles. Name the postulate you used.

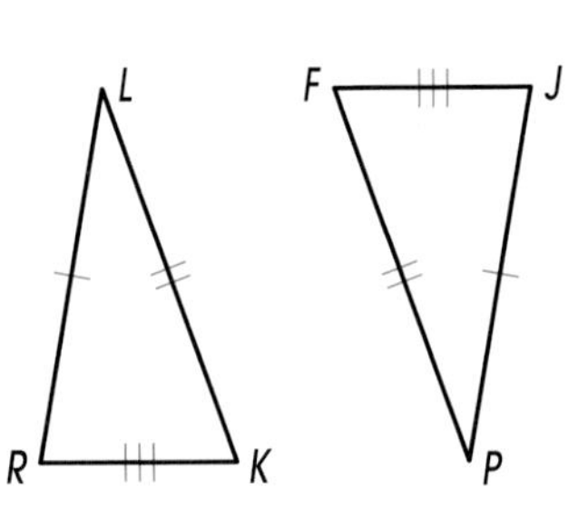

These triangles are congruent by SSS.
Now write a congruence statement.

Look for the corresponding angles. They are opposite the corresponding sides. Use the angles to write the congruence statement.

$\angle K \cong \angle F$
$\angle R \cong \angle J$
$\angle L \cong \angle P$
$\triangle KRL \cong \triangle FJP$

So, $\triangle KRL \cong \triangle FJP$, using the SSS postulate.

> POSTULATE 10
> **Side-Angle-Side Postulate (SAS)**
>
> If two sides and the included angle of one triangle are congruent to the corresponding parts of another triangle, the triangles are congruent.

Two triangles are also congruent if two corresponding sides and the included angle are congruent.

The included angle is the angle between two sides of a triangle.

Try This

Write a congruence statement for these triangles. Name the postulate that you used.

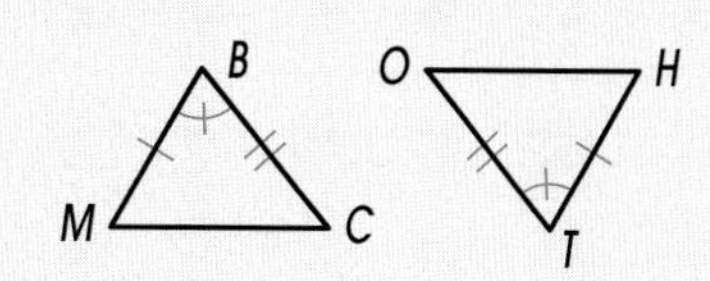

These triangles are congruent by SAS.

Look for the corresponding angles.
Begin with a pair of corresponding sides.

$\angle C$ is opposite $\overline{MB}$. ■ ($\angle O$) is opposite $\overline{HT}$. — $\angle C \cong$ ■ $\angle O$

Find the congruent angles in the diagram. — $\angle B \cong$ ■ $\angle T$

Look for the other pair of corresponding sides.
$\angle M$ is opposite $\overline{BC}$. ■ ($\angle H$) is opposite $\overline{TO}$. — $\angle M \cong$ ■ $\angle H$

Use the angles to write the congruence statement. — $\triangle CBM \cong \triangle OTH$

So, $\triangle CBM \cong \triangle OTH$, using the ■ (SAS) postulate.

Practice

Write a congruence statement for each pair of triangles. Name the postulate that you used.

1.

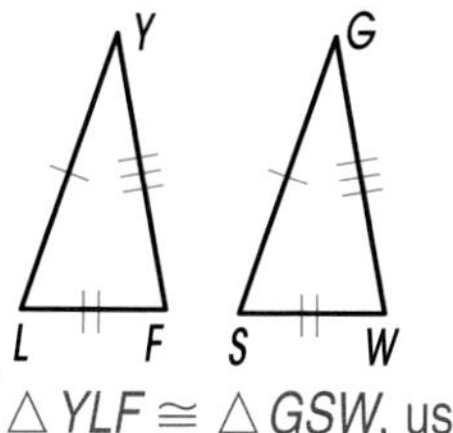

$\triangle YLF \cong \triangle GSW$, using the SSS postulate.

2.

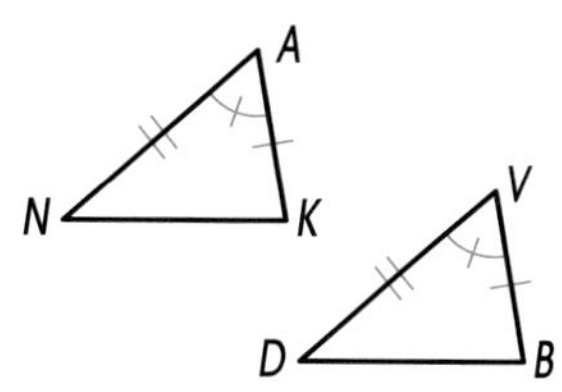

$\triangle KAN \cong \triangle BVD$, using the SAS postulate.

3.

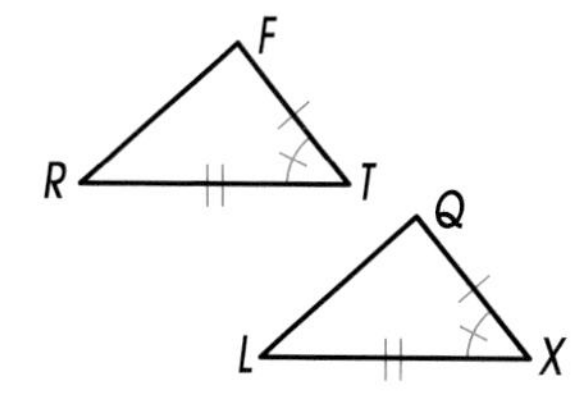

$\triangle FTR \cong \triangle QXL$, using the SAS postulate.

Share Your Understanding

4. Work with a partner. Explain how to use SAS to show that two triangles are congruent. Use the words *included angle* in your explanation. Show that two of the corresponding sides and the included angle of the two triangles are congruent.

5. **CRITICAL THINKING** True or false? $\triangle RLK \cong \triangle JLK$
Explain your thinking. True; SAS $\overline{LK} \cong \overline{LK}$ and $\overline{RK} \cong \overline{JK}$. Those are the sides. $\angle LKJ$ and $\angle LKR$ are supplementary. So $\angle LKJ$ is 90°. $\angle LKJ$ and $\angle LKR$ are the congruent included angles. So, $\triangle RLK \cong \triangle JLK$ by SAS.

L
R K J

More practice is provided in the Workbook, page 28, and the Classroom Resource Binder, page 48.

5·10 Congruent Triangles: ASA and AAS

Getting Started Have students identify an included side.

> POSTULATE 11
> **Angle-Side-Angle Postulate (ASA)**
>
> If two angles and the included side of one triangle are congruent to the corresponding parts of another triangle, the triangles are congruent.

You can use the SSS or SAS postulates to decide if two triangles are congruent. Here is another postulate you can use to decide if two triangles are congruent. Two triangles are congruent if two corresponding angles and the included side are congruent.

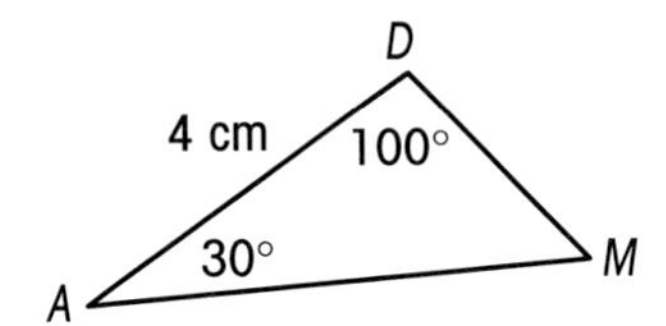

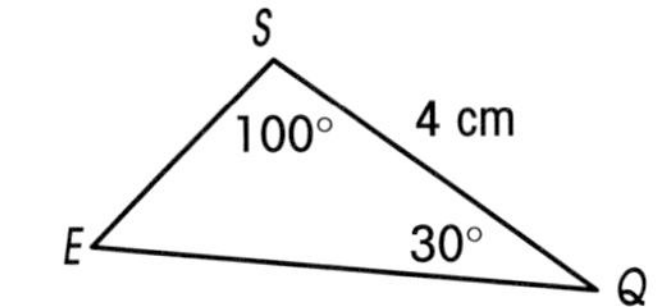

$\triangle ADM \cong \triangle QSE$ by the ASA postulate.

EXAMPLE

Write a congruence statement for these triangles. Name the postulate you used.

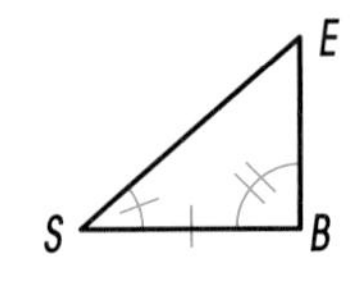

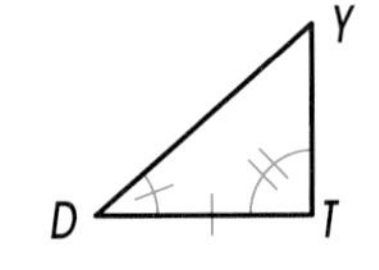

These triangles are congruent by ASA.
Now write a congruence statement.

Look for a pair of corresponding angles.	$\angle S \cong \angle D$
$\angle E$ and $\angle Y$ are opposite the corresponding sides.	$\angle E \cong \angle Y$
Look for the other pair of corresponding angles.	$\angle B \cong \angle T$

So, $\triangle SEB \cong \triangle DYT$, using the *ASA* postulate.

> THEOREM 16
> **Angle-Angle-Side Theorem (AAS)**
>
> If two angles and a non-included side of one triangle are congruent to the corresponding parts of another triangle, the two triangles are congruent.

Two triangles are also congruent if two corresponding angles and a corresponding side are congruent. The side does not have to be between the two angles. This is the Angle-Angle-Side Theorem. AAS is a theorem because it can be proved.

Try This

Write a congruence statement for these triangles. Name the postulate or theorem you used.

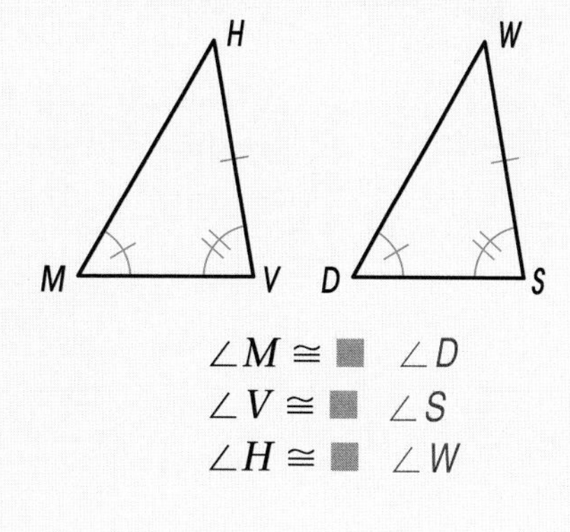

These triangles are congruent by AAS. Now write the congruence statement.

Find the corresponding congruent angles.

$\angle M \cong$ ■ $\angle D$
$\angle V \cong$ ■ $\angle S$
$\angle H \cong$ ■ $\angle W$

So, $\triangle MVH \cong$ ■ $\triangle DSW$, using the ■ AAS theorem.

Practice

Write a congruence statement for each pair of triangles. Name the postulate or theorem that you used. Write *AAS, SSS, ASA,* or *SAS.*

1.

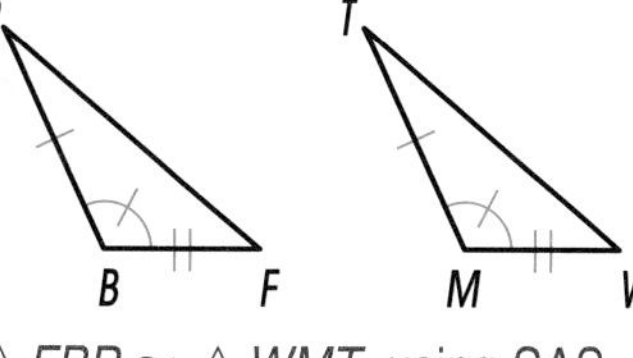

$\triangle FBP \cong \triangle WMT$, using SAS.

2.

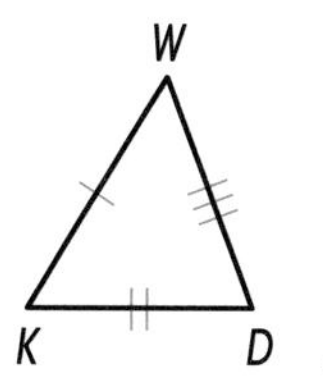

$\triangle WKD \cong \triangle LXV$, using SSS.

3.

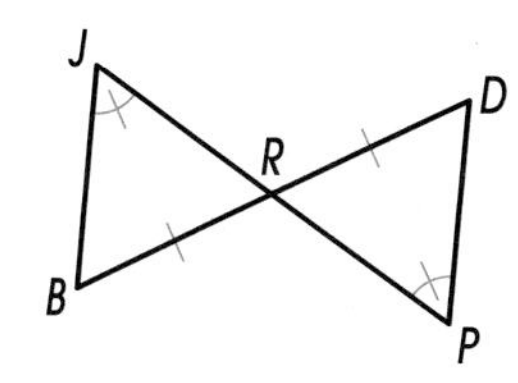

$\triangle BJR \cong \triangle DPR$, using AAS.

4. Two corresponding angles must be congruent in the two triangles. The included sides between these congruent angles must also be congruent. Then, you can use ASA to prove the triangles are congruent.

Share Your Understanding

4. Work with a partner. Explain how to use ASA to show that two triangles are congruent. Use the words *included side* in your explanation.
See above.

5. **CRITICAL THINKING** $\triangle RLK$ and $\triangle JLK$ are isosceles right triangles. Explain how you know that $\triangle RLK \cong \triangle JLK$. $\overline{RK} \cong \overline{LK}$ and $\overline{LK} \cong \overline{JK}$ because the triangles are isosceles. $\angle LKR \cong \angle LKJ$ because they are right angles. So, $\triangle RLK \cong \triangle JLK$ by SAS.

L

R K J

More practice is provided in the Workbook, page 29, and in the Classroom Resource Binder, page 49.

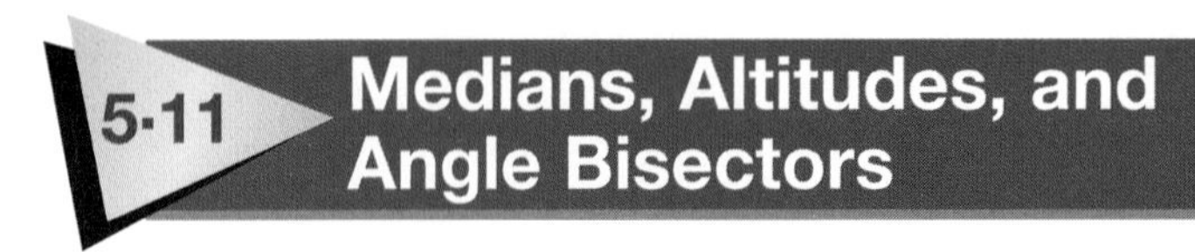

Medians, Altitudes, and Angle Bisectors

Getting Started
Review the words *midpoint* and *bisector*.

A triangle has three special line segments. Each line segment joins a vertex to the opposite side or the line containing the opposite side.

Line Segment	Description	Example
Median	Joins a vertex to the midpoint of the opposite side	
Altitude	Joins a vertex to the line containing the opposite side and is perpendicular to that side	
Angle Bisector	Joins a vertex to the opposite side and bisects the angle	

Math Fact
An altitude may be outside the triangle.

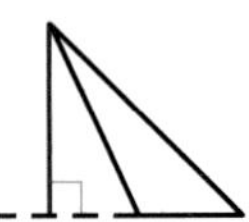

You can use the chart above to classify lines in a triangle.

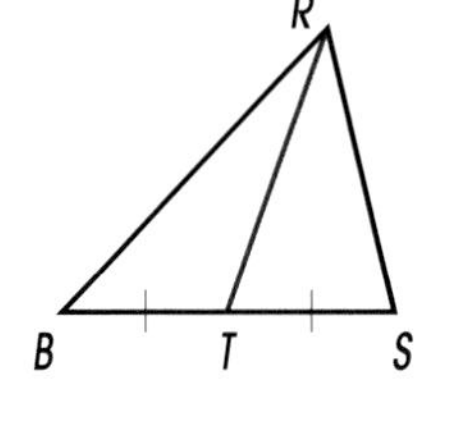

EXAMPLE 1

Classify $\overline{RT}$ in as many ways as you can.

$\overline{RT}$ joins vertex R to the opposite side, $\overline{BS}$.
$\overline{RT}$ is not perpendicular to $\overline{BS}$.
So, $\overline{RT}$ is not an altitude.

$\overline{RT}$ does not bisect $\angle R$.
So, $\overline{RT}$ is not an angle bisector.

$\overline{RT}$ bisects $\overline{BS}$.

So, $\overline{RT}$ is a median.

EXAMPLE 2

Classify $\overline{JK}$.

$\overline{JK}$ joins vertex J to the line containing $\overline{LB}$.
$\overline{LB}$ is the side opposite vertex J.
$\overline{JK}$ is perpendicular to $\overleftrightarrow{LB}$.

So, $\overline{JK}$ is an altitude.

Try This

$\triangle ABC$ is an isosceles triangle. Classify $\overline{BJ}$ in as many ways as you can.

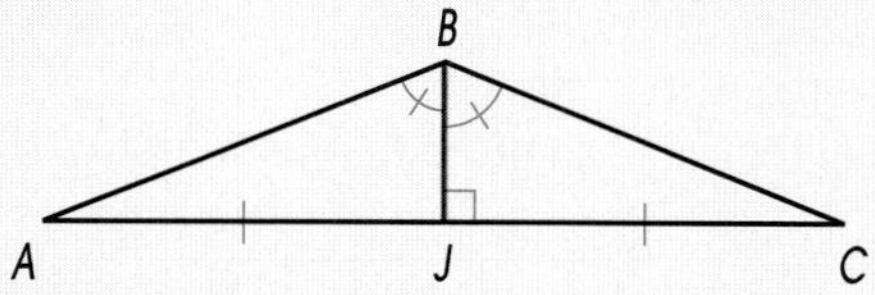

$\overline{BJ}$ joins vertex ■ (B) to the opposite side ■ ($\overline{AC}$).

■ ($\overline{BJ}$) is perpendicular to ■ ($\overline{AC}$).

$\overline{BJ}$ bisects ■ ($\overline{AC}$) and $\angle B$.

So, BJ is an ■ (altitude), a ■ (median), and an angle bisector of $\triangle ABC$.

B

A J C

Practice

Classify $\overline{GH}$ in as many ways as you can. Write *median, altitude,* or *angle bisector.*

1. angle bisector

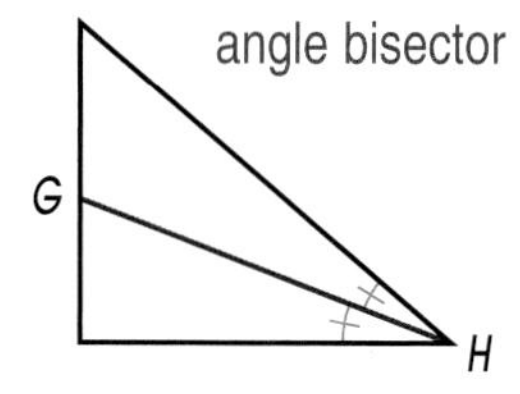

2. altitude

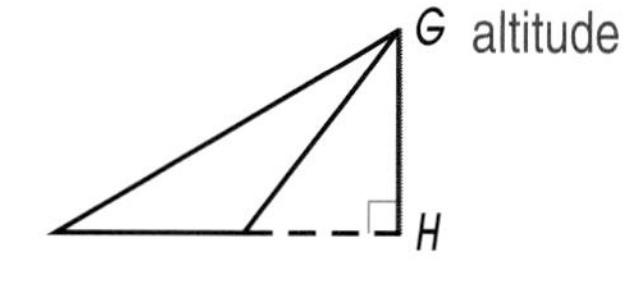

3. altitude, median, and angle bisector

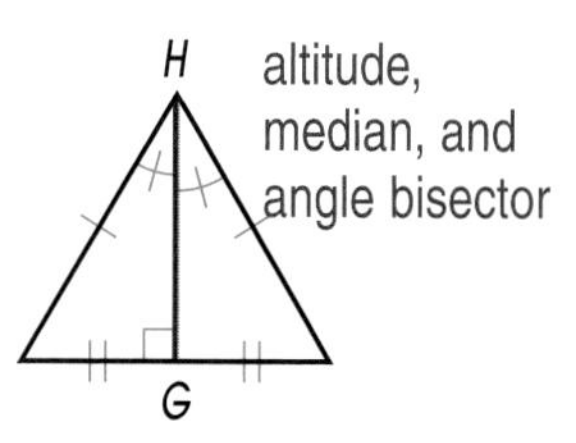

Use $\triangle JKL$ to name a line segment for each term.

4. median $\overline{JG}$

5. altitude $\overline{KF}$ or $\overline{JG}$

6. angle bisector $\overline{JG}$

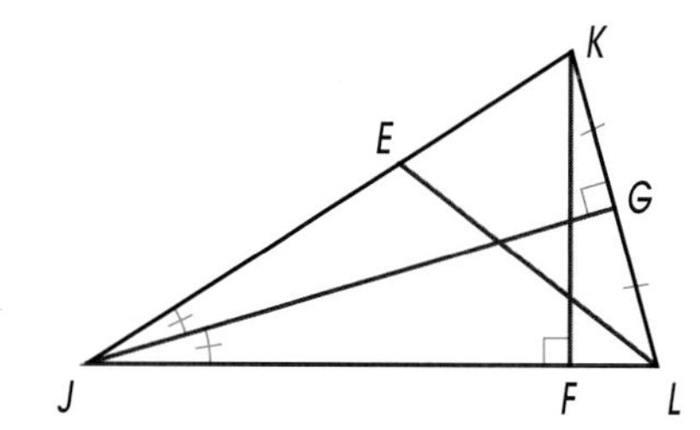

Share Your Understanding

7. Explain to a partner how a median, and an altitude from the same vertex will be different segments in a scalene triangle. Use a diagram and the words *segment, perpendicular,* and *congruent* in your explanation.
Check students' work.

8. **CRITICAL THINKING** Are the legs of a right triangle altitudes, medians, or angle bisector? Explain your thinking.
The legs of a right triangle are altitudes because each leg is perpendicular to the other leg.

5-12 Triangle Midsegment Theorem

Getting Started Review parallel lines and corresponding angles.

> THEOREM 17 **Triangle Midsegment Theorem**
>
> If a segment joins the midpoints of two sides of a triangle, then the segment is parallel to the third side and half its length.

A **midsegment** of a triangle joins the midpoints of any two sides. The midsegment is parallel to the third side and half its length.

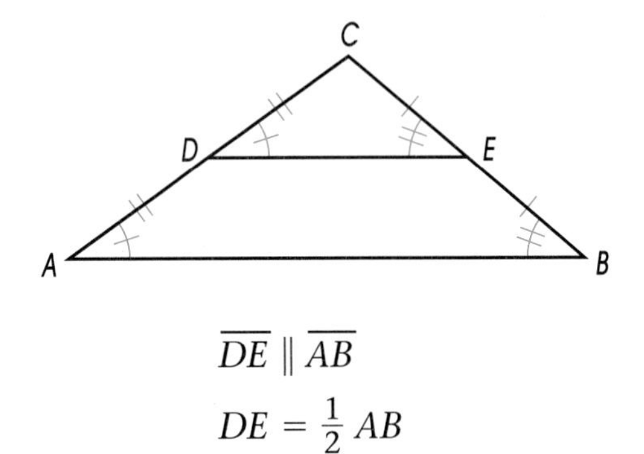

$$\overline{DE} \parallel \overline{AB}$$

$$DE = \frac{1}{2}AB$$

You can find the length of a side of a triangle if you know the length of the midsegment.

EXAMPLE

Find the length of $\overline{FG}$.

$FI = IH$ and $HJ = JG$. $\overline{IJ}$ is the midsegment of $\triangle FHG$. Write an equation. Use the Triangle Midsegment Theorem.	$IJ = \frac{1}{2}FG$
Substitute 25 for IJ.	$25 = \frac{1}{2}FG$
Solve for FG.	$50 = FG$

The length of $\overline{FG}$ is 50 feet.

You learned that if two lines are parallel, the corresponding angles are congruent. The midsegment and the third side of the triangle form corresponding angles. In $\triangle HFG$, $\angle HIJ$ and $\angle HFG$ are corresponding angles. Because $\angle HFG$ is 30°, $\angle HIJ$ is also 30°.

Try This

Find the length of $\overline{PQ}$. Find $m\angle L$.

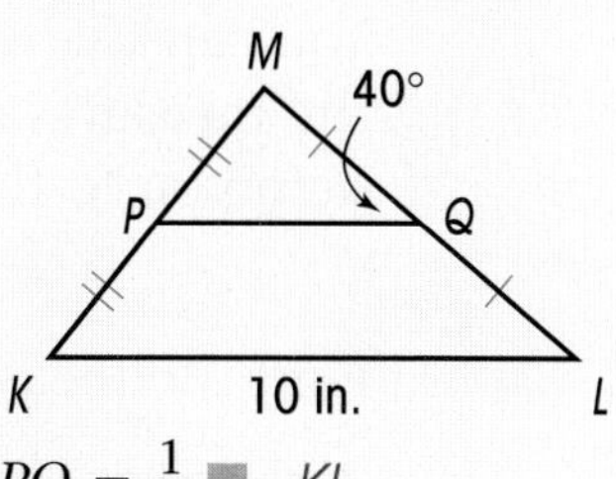

$\overline{PQ}$ is a ■ of $\triangle KLM$. midsegment

Write an equation. Use the Triangle Midsegment Theorem. $PQ = \frac{1}{2}$ ■ KL

Substitute 10 for ■. KL $PQ = \frac{1}{2} \cdot 10$

Solve for $\overline{PQ}$. $PQ =$ ■ 5

$\angle KLM$ and $\angle PQM$ are ■ angles. They are congruent. corresponding

The length of $\overline{PQ}$ is ■. 5 in. The measure of $\angle L$ is ■. 40°

Practice

Find the value of x or y in each triangle.

1. 9 cm

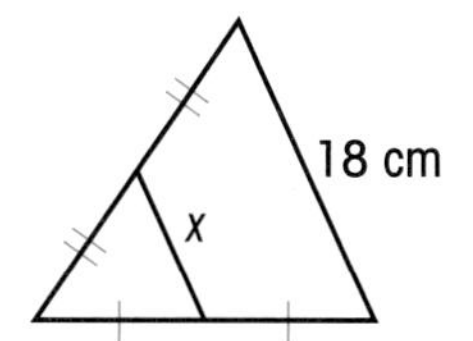

2. 5 m

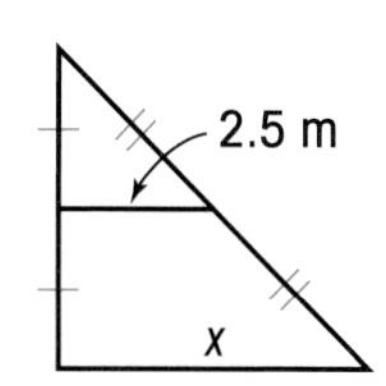

3. 11 in.

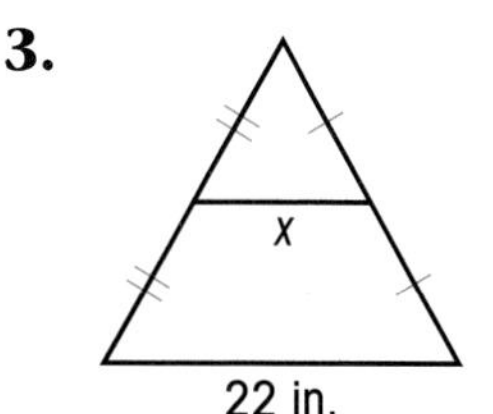

4. 35°

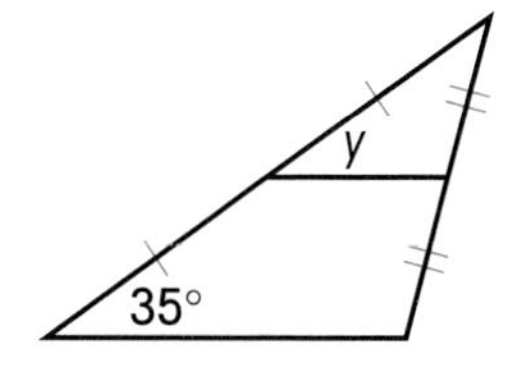

5. 90°

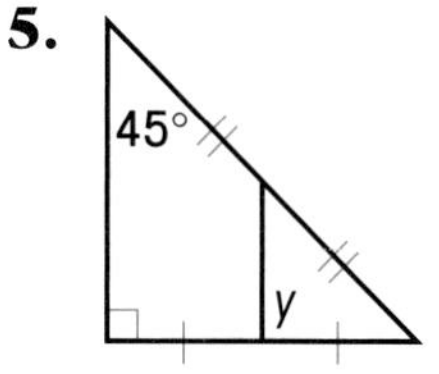

6. 50°

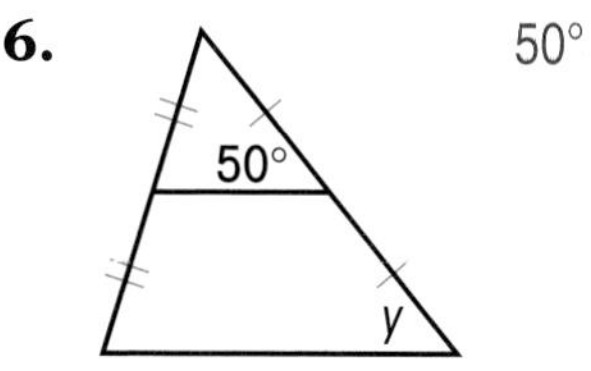

Share Your Understanding

7. Explain to a partner how to find the value of x in any exercise above. Use the words *midpoint, segment,* and *congruent* in your explanation. Answers will vary.

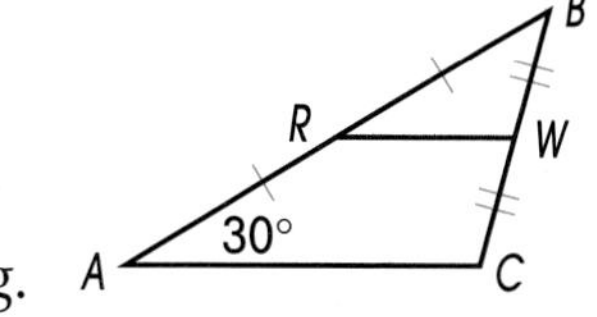

8. **CRITICAL THINKING** Find $m\angle WRA$. Explain your thinking.
150°; $\angle WRA$ and $\angle WRB$ are supplementary angles. So, $m\angle WRA + m\angle WRB = 180$. Since $180 - 30 = 150$, $m\angle WRA = 150°$.

5·13 Calculator: Measures of a Triangle

Getting Started
Review the Angle Sum Theorem, Isosceles Triangle Theorem, and Triangle Midsegment Theorem.

You can use what you learned about triangles to find the measures of the parts of a triangle. A calculator can help you to find some of these measures.

EXAMPLE 1

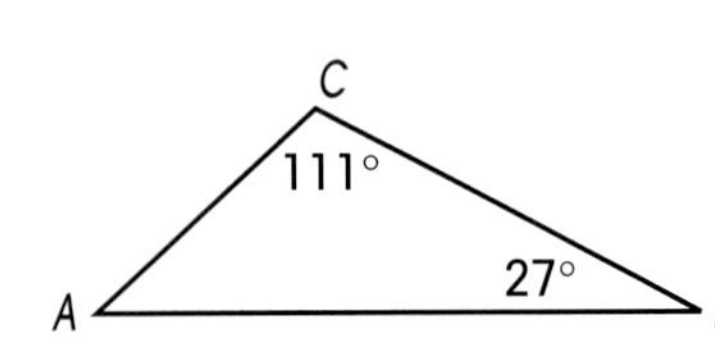

Find the measure of $\angle A$.

Write an equation.

Use the Angle Sum Theorem. $m\angle A + m\angle B + m\angle C = 180$

Substitute the values you know. $m\angle A + 27 + 111 = 180$

Add 27 and 111 on your calculator. **Display**

PRESS: [2] [7] [+] [1] [1] [1] [=] 138

$m\angle A + 138 = 180$

Subtract 138 from both sides of the equation. **Display**

PRESS: [1] [8] [0] [−] [1] [3] [8] [=] 42

$m\angle A = 42$

So, $\angle A$ is 42°.

EXAMPLE 2

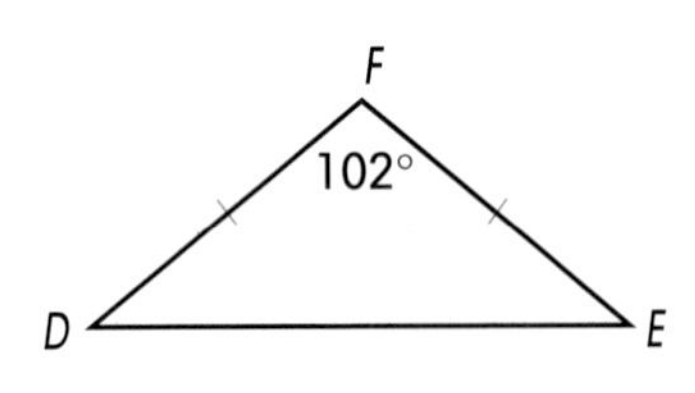

Find the measure of $\angle D$.

Write an equation.

Because $\triangle DEF$ is isosceles, $m\angle E = m\angle D$.

$m\angle D + m\angle E + m\angle F = 180$

$m\angle D + m\angle D + m\angle F = 180$

Substitute the values you know. $2m\angle D + 102 = 180$

Subtract 102 from both sides of the equation. **Display**

PRESS: [1] [8] [0] [−] [1] [0] [2] [=] 78

$2m\angle D = 78$

Divide both sides of the equation by 2. **Display**

PRESS: [7] [8] [÷] [2] [=] 39

$m\angle D = 39$

So, $\angle D$ is 39°.

Practice

Use a calculator to find the value of x.

1. 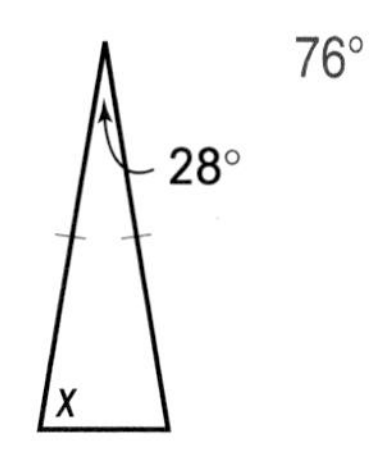

76°

2. 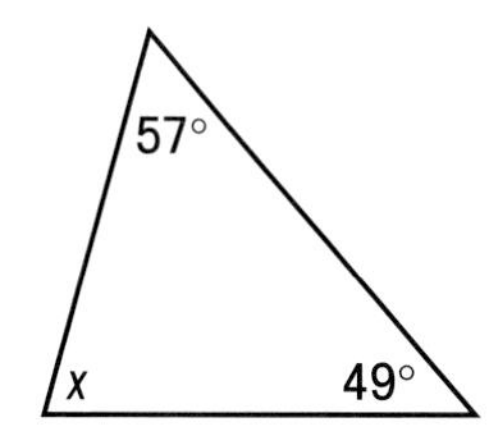

74°

3. 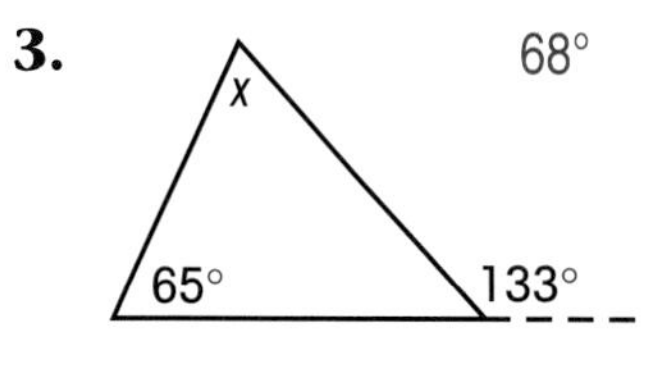

68°

4. 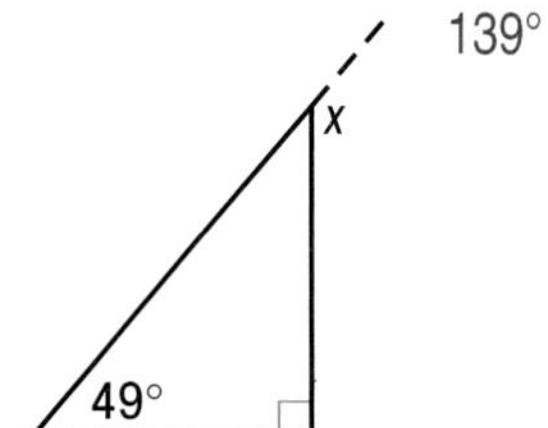

139°

5. 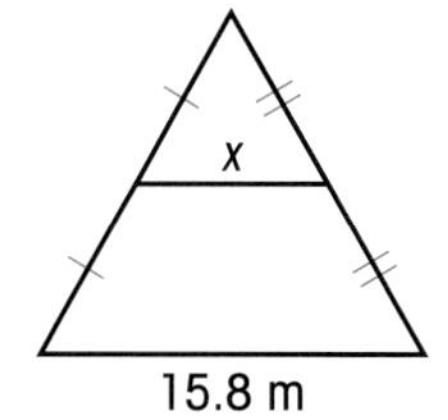

7.9 m

6. 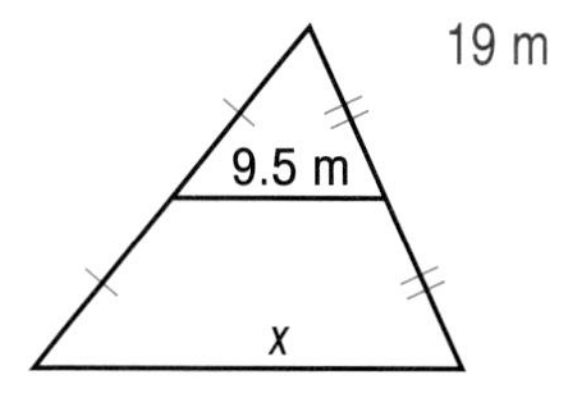

19 m

ON-THE-JOB MATH

TESSELLATIONS

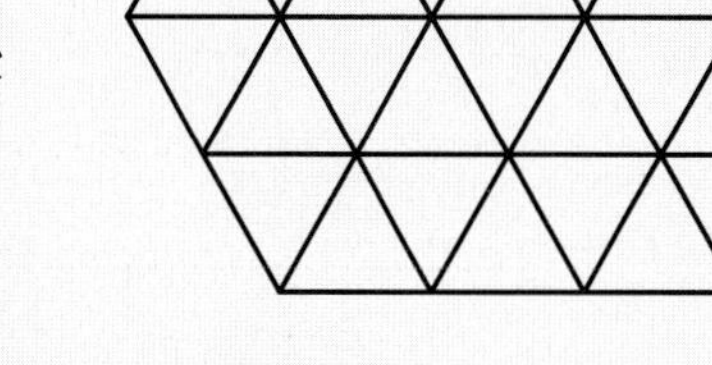

A mason works with tiles. The mason sets tiles on a floor so that they cover the floor without gaps or overlap. Such a pattern is called a tessellation. Only certain geometric shapes tessellate.

Make 32 congruent equilateral triangles on paper. Cut out the triangles. Then tessellate the triangles to create a pattern like the one on the right.

Notice that six triangles are arranged around a point. Because each vertex angle is 60°, the total number of degrees around a point is 6 • 60, or 360°. This is always true in a tessellation.

Create your own tessellation with triangles. Use congruent triangles. They do not have to be equilateral. Do all triangles tessellate?
Yes, all triangles tessellate. Check students' work.

More practice is provided in the Workbook, page 30, and the Classroom Resource Binder, page 50.

5·14 Problem-Solving Skill: Find the Centroid

Getting Started
Have volunteers predict what they think *centroid* might mean. [center]

Remember
A median is a segment from a vertex to the midpoint of the opposite side of a triangle.

The three medians of a triangle meet at a point inside the triangle. This is a special point called the **centroid.** Suppose you made a triangle out of cardboard. You could balance the triangle at the centroid.

You can use what you learned about medians to find the centroid of any triangle.

EXAMPLE

Find the centroid of an equilateral triangle.

Draw a large equilateral triangle on a piece of cardboard. Find the midpoint of each side. Then draw the median from each vertex to the midpoint of the opposite side.

Math Fact
The centroid is the center of gravity.

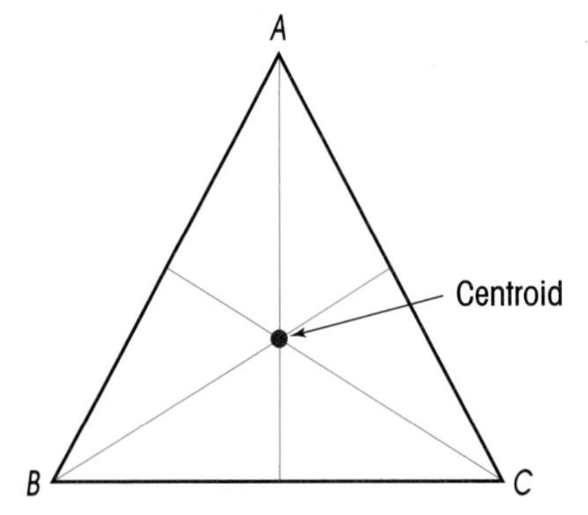

Cut out your triangle. Balance the triangle on a pencil point at the point where the three medians meet. If the triangle balances, you have found the centroid. ✓

The centroid of $\triangle ABC$ is the point where the three medians meet.

Try This

Find the centroid of an obtuse triangle.

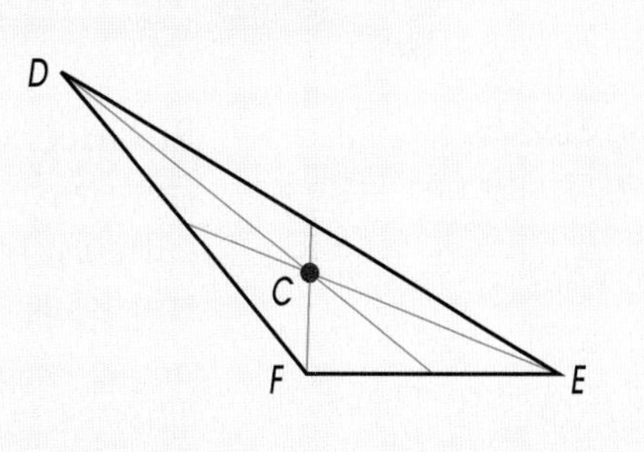

Copy $\triangle DEF$. Make the copy as large as possible. Do not include the medians.

To find the median of each side, use a ruler. First, find the midpoint of each side. Then, draw the median from a ■ to the ■ of the opposite side. vertex midpoint

Draw a median from each vertex. The point where the lines intersect is the ■. centroid

The centroid of $\triangle DEF$ in the diagram above is point ■. C

Practice

Make a cardboard model of this right triangle. Find the centroid. Then, balance the triangle on a pencil point at the centroid.

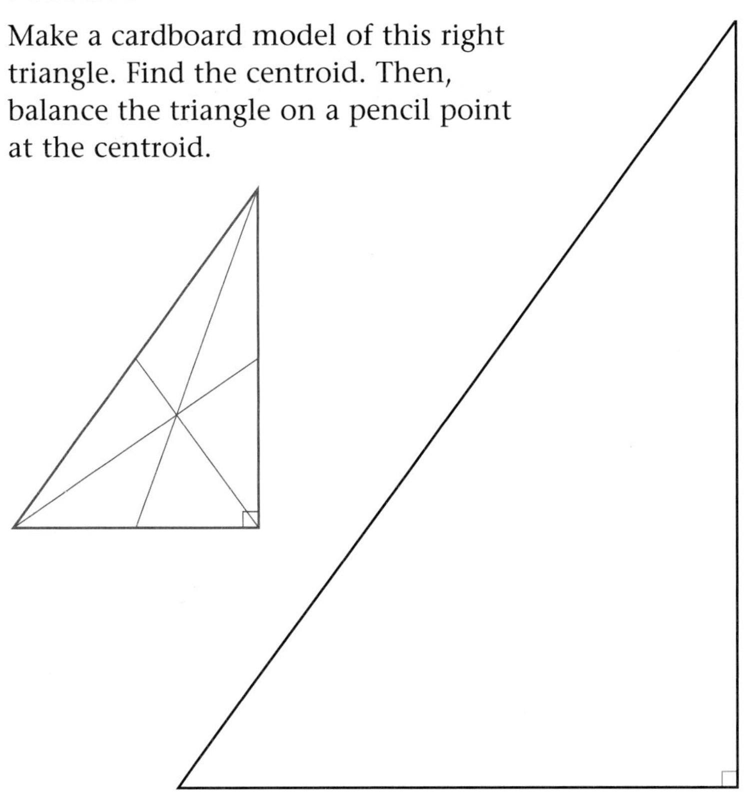

More practice is provided in the Workbook, page 31, and in the Classroom Resource Binder, page 51.

5·15 Problem-Solving Application: Engineering

Getting Started
Review the facts on triangles covered in the chapter. Have students illustrate each fact.

Engineers and carpenters use triangles in the framework of bridges and buildings. This framework is called a truss. You can use what you learned about triangles to solve problems in construction.

EXAMPLE 1

This king post truss is used for a roof. The king post is the median of the triangle formed by the rafters. The span of the roof is 24 feet. What is the length of $\overline{LR}$ and $\overline{RQ}$?

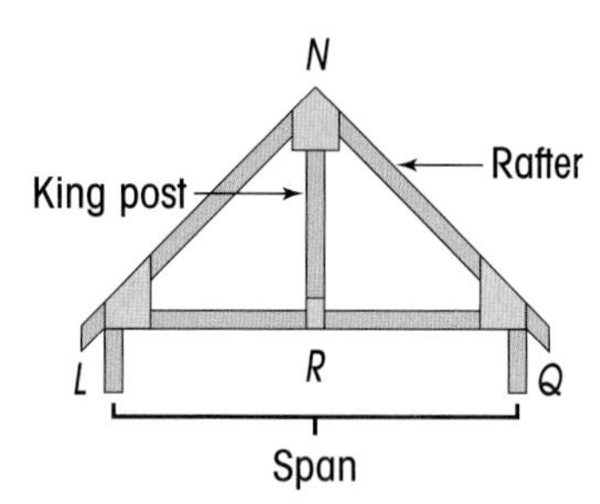

The median of a triangle joins a vertex to the midpoint of the opposite side.

Write an equation. Use the Midpoint Theorem. $LR = \frac{1}{2}LQ$

Simplify. $LR = \frac{1}{2} \cdot 24$

$LR = 12$

Both, $\overline{LR}$ and $\overline{RQ}$ are each 12 feet long.

EXAMPLE 2

This truss supports a roof. The distance from point M to point P is 18 feet. Point M and point P are midpoints of $\overline{LN}$ and $\overline{NQ}$. What is the span, $\overline{LQ}$, of this roof?

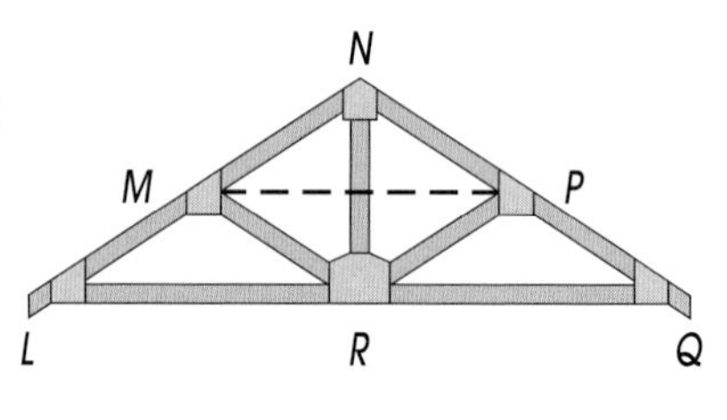

Write an equation. Use the Triangle Midsegment Theorem. $MP = \frac{1}{2}LQ$

Then solve for LQ. $18 = \frac{1}{2} \cdot LQ$

$36 = LQ$

The span of this roof is 36 feet.

Try This

The rafters that form this roof gable are congruent. At what angle were the rafters set to create this gable?

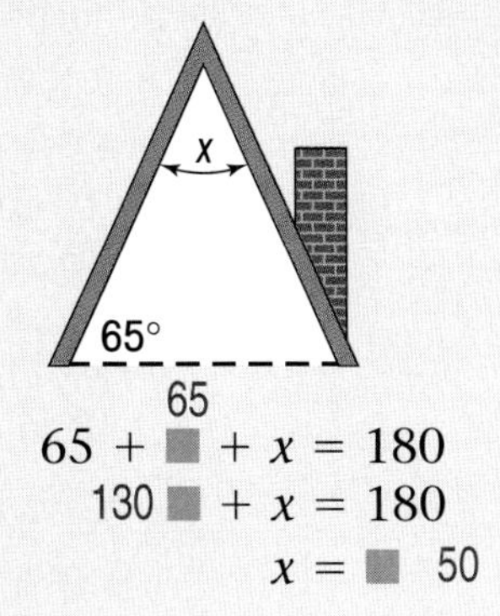

Because the rafters are congruent, they form part of an ■ triangle. Use this fact and the Angle ■ Theorem for Triangles to write an equation. Then, solve for the unknown angle.

isosceles

Sum

65

$65 + ■ + x = 180$

130 ■ $+ x = 180$

$x = ■$ 50

50°

The rafters are set at a ■ angle to form this gable.

Practice

Solve each problem. Show your work.

1. Two wire cables help to steady an antenna. What is the length of the cable on the left? How do you know?
   20 ft; Because the base angles are congruent, the opposite sides are congruent.

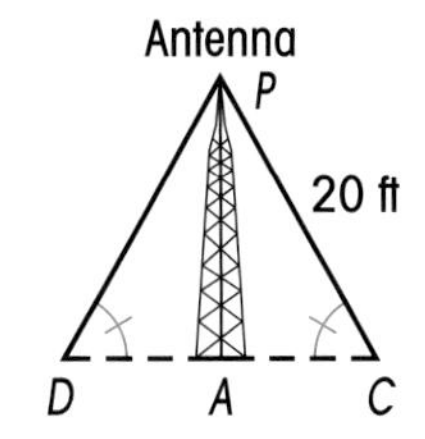

2. This truss is part of a bridge. It is made up of six congruent isosceles right triangles. How long is the distance from *A* to *B*? 168 ft

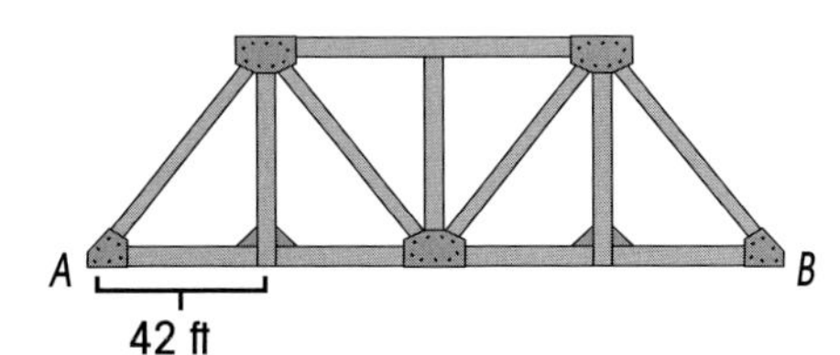

3. A machine shop is welding a triangular part for a bridge repair. One side is 5 ft long and the other side is 4 ft long. Is a 2-foot steel bar long enough for the third side? Explain.
   Yes; All three inequalities in the Triangle Inequality Theorem are true.

4. The rafters of a roof meet at a 70° angle. The rafters form part of an isosceles triangle. What is the measure of one of the base angles? (Hint: Draw a diagram.) 55°

An alternate two-column proof lesson is provided on page 410 of the student text.

5·16 Proof: CPCTC

Getting Started Review the corresponding parts of two congruent triangles.

Math Fact
Corresponding parts of congruent triangles are congruent. This statement is known as CPCTC.

Sometimes you need to prove that corresponding parts of two triangles are congruent. If you prove that the two triangles are congruent, you know that the corresponding parts are also congruent.

EXAMPLE 1

You are given: $\overline{AB} \parallel \overline{CD}$
E is the midpoint of $\overline{AD}$.

Prove: $\overline{AB} \cong \overline{DC}$

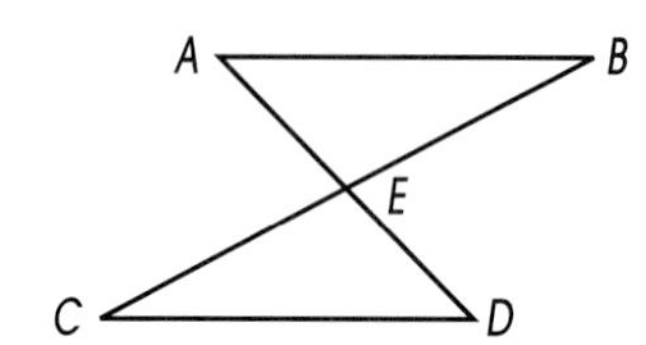

Copy the diagram. As you read the proof, mark off the congruent parts you find. Then, choose the postulate you can use to show that the triangles are congruent.

You are given that $\overline{AB} \parallel \overline{CD}$. This means that $\angle BAE \cong \angle CDE$. They are alternate interior angles. Because E is the midpoint of $\overline{AD}$, you know that $\overline{AE} \cong \overline{DE}$. You also know that $\angle AEB \cong \angle DEC$. They are vertical angles.

Now, $\triangle AEB \cong \triangle DEC$ by ASA. Then, $\overline{AB} \cong \overline{DC}$ because the segments are corresponding parts of congruent triangles, or CPCTC.

So, you prove that $\overline{AB} \cong \overline{DC}$. ✓

EXAMPLE 2

You are given: $\overline{FI} \cong \overline{HI}$
$\angle FIG \cong \angle HIG$

Prove: $\angle F \cong \angle H$

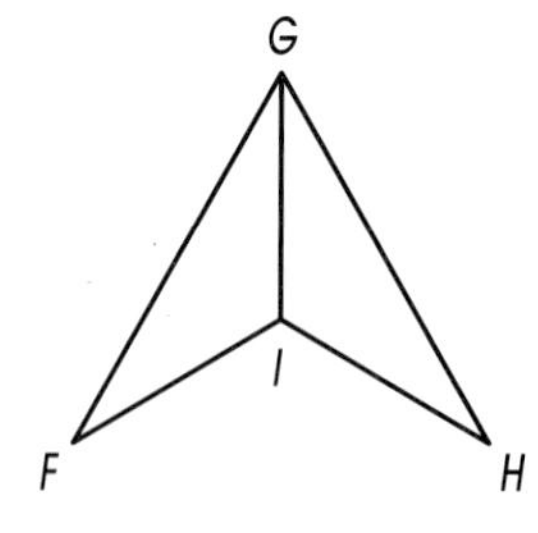

You are given that $\overline{FI} \cong \overline{HI}$ and $\angle FIG \cong \angle HIG$. You know that $\overline{IG} \cong \overline{IG}$. Now, $\triangle FIG \cong \triangle HIG$ by SAS. Then, $\angle F \cong \angle H$ by CPCTC.

So, you prove that $\angle F \cong \angle H$. ✓

Try This

Copy and complete the proof.

You are given: $\triangle JKL$ is isosceles.

$\overline{JK} \cong \overline{LK}$

$\overline{KM}$ bisects $\angle K$.

Prove: $\overline{JM} \cong \overline{LM}$

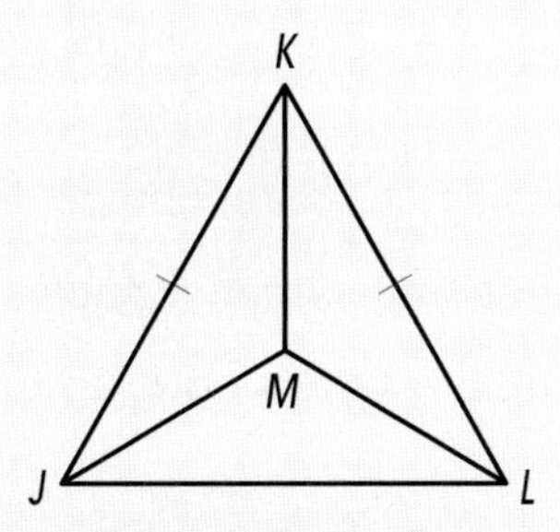

Copy the diagram. As you read the proof, mark off the congruent parts you find. Then, choose the postulate you can use to show that the triangles *JKM* and *LKM* are congruent.

You are given that $\triangle JKL$ is ■ isosceles *and that $\overline{JK} \cong$ ■* $\overline{LK}$*.*
You are also given that $\overline{KM}$ ■ bisects *$\angle K$. This means that $\angle JKM \cong$ ■* $\angle LKM$*. You know that $\overline{KM} \cong$ ■* $\overline{KM}$*. Now, $\triangle JKM \cong \triangle LKM$ by ■* SAS*. Then, $\overline{JM} \cong \overline{LM}$ by ■.* CPCTC

So, you prove that ■. ✓ $\overline{JM} \cong \overline{LM}$

Practice

Write a proof for each of the following. See Additional Answers in the back of this book.

1. You are given: $\angle Q \cong \angle S$
$\overline{QS}$ bisects $\overline{PR}$.

Prove: $\angle P \cong \angle R$

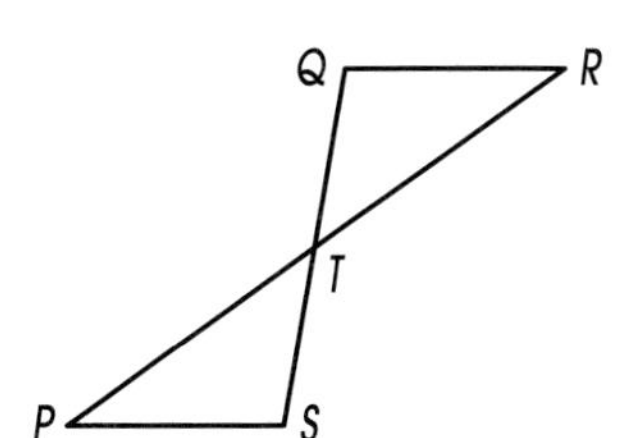

2. You are given: $\overline{DE} \cong \overline{FE}$
$\overline{EG}$ bisects $\angle E$.

Prove: $\overline{DG} \cong \overline{FG}$

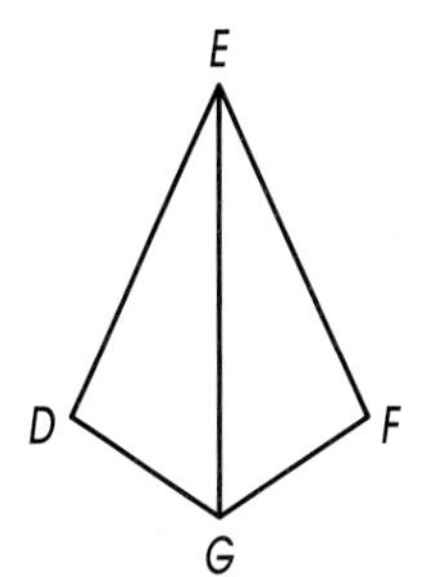

More Practice is provided in the Classroom Resource Binder.

Chapter 5 Review

Summary

A triangle can be classified as acute, obtuse, equiangular, or right.
The sum of the measures of the interior angles of a triangle is 180°.
The measure of an exterior angle is equal to the sum of the measures of the two remote interior angles.
A triangle can be classified as scalene, isosceles, or equilateral.
You can use the Triangle Inequality Theorem to decide whether any three given line segments can form a triangle.
The angles opposite congruent sides of a triangle are congruent.
The angle opposite the longest side of a triangle is the largest angle.
Two triangles are congruent by SSS, SAS, ASA, or AAS.
The median, altitude, and angle bisector are line segments in a triangle.
The midsegment of a triangle joins the midpoints of any two sides. It is parallel to the third side and half its length.
The centroid is the point where the three medians of a triangle meet.
You can use what you know about triangles to solve real-life problems.
You can use facts about triangles to prove a statement.

More vocabulary review is provided in the Classroom Resource Binder, page 43.

equilateral triangle
scalene triangle
midsegment
isosceles triangle
equiangular triangle
angle bisector

Vocabulary Review

Complete the sentences with words from the box.

1. A ____ is a line segment that joins the midpoints of two sides of the triangle. midsegment
2. A triangle with no congruent sides is called a ____. scalene triangle
3. An ____ is a triangle with all sides congruent. equilateral triangle
4. A triangle with all angles equal is called an ____. equiangular triangle
5. A line segment that joins a vertex to the opposite side and that bisects the angle is called an ____. angle bisector
6. An ____ is a triangle with two congruent sides. isosceles triangle

Chapter Quiz

More assessment is provided in the Classroom Resource Binder.

Classify each triangle by its angles and sides.

1.

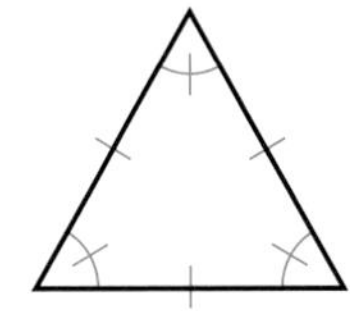

equiangular equilateral

2.

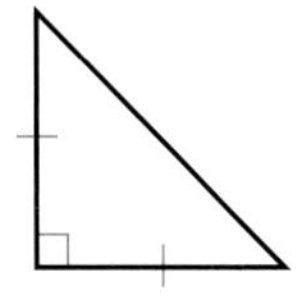

right isosceles

3.

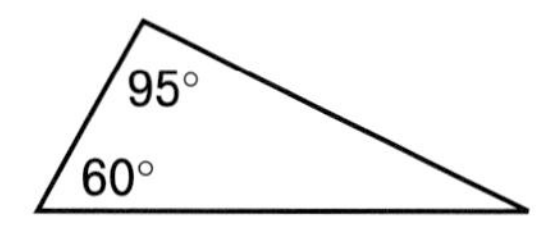

obtuse scalene

Find the value of *x*.

4.

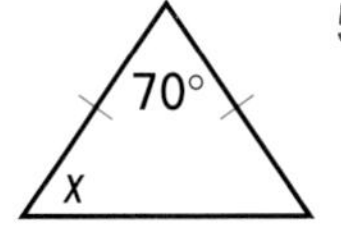

55°

5.

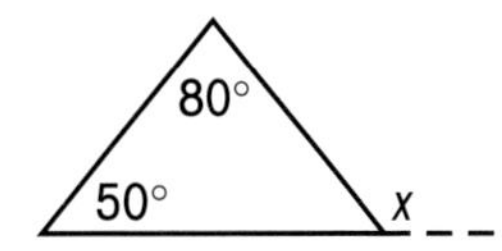

130°

6.

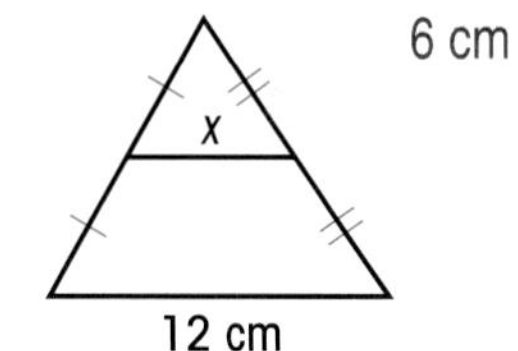

6 cm

7. Can segments 8 m, 12 m, and 25 m long form a triangle? Write *yes* or *no*. If your answer is *no*, tell why. No; 8 + 12 is ≯ 25.

Write a congruence statement for each pair of triangles. Name the postulate or theorem that you used.

8.

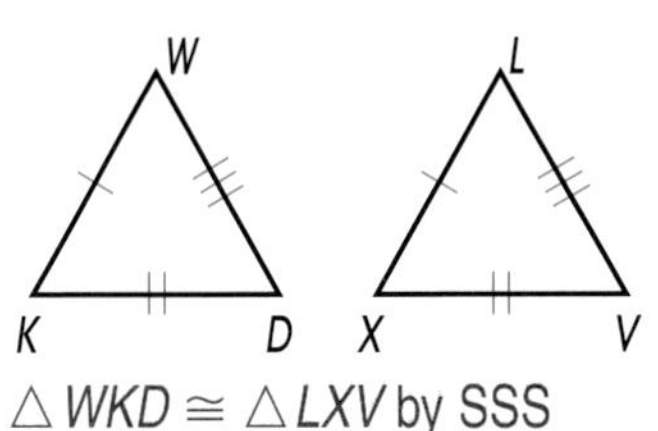

$\triangle WKD \cong \triangle LXV$ by SSS

9.

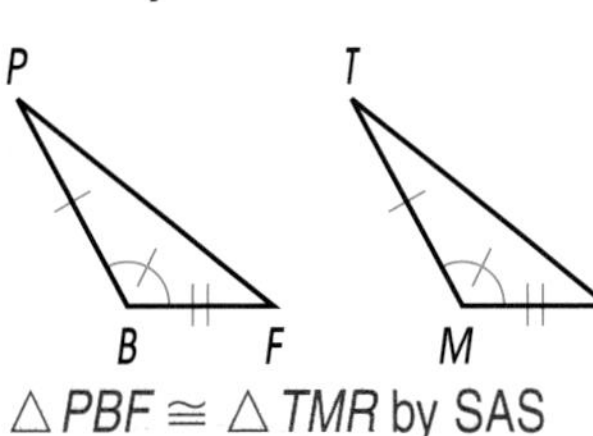

$\triangle PBF \cong \triangle TMR$ by SAS

10.

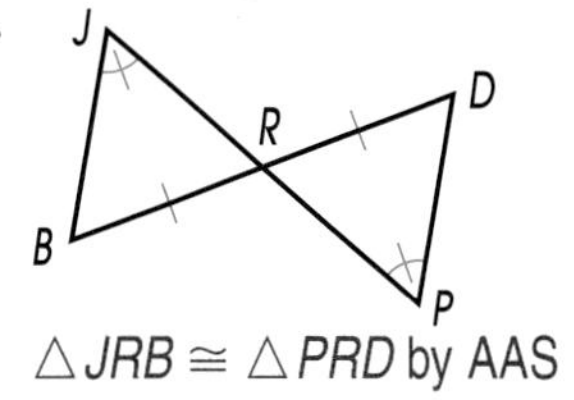

$\triangle JRB \cong \triangle PRD$ by AAS

Write About Math The altitude can be outside an obtuse triangle.

Problem Solving

Solve the problem. Show your work.

11. The span of a roof $\overline{PR}$ is 20 ft long. The rise $\overline{SQ}$ is a median of the triangle *PSR*. What is the length of $\overline{PQ}$? 10 ft

Write About Math

When can the altitude be a line segment outside a triangle? Use a diagram to explain. See above.

Additional Practice for this chapter is provided on p. 397 of the student book.

Chapter 6 Right Triangles

Right triangles can be found in many places. Which sail looks like a right triangle? Measure the sides of the right triangle. Which side is the longest side?

Learning Objectives

- Review squares and square roots.
- Identify parts of a right triangle.
- Use the Pythagorean Theorem.
- Use the special right triangle relationships.
- Use a calculator to identify right triangles.
- Draw a diagram to solve distance problems.
- Apply the Pythagorean Theorem to find indirect measurement.
- Write proofs using the Hypotenuse–Leg Theorem.

ESL/ELL Note Explain that *radical* comes from a Latin word that means "root." In math, *radical* means "the root of a number." In the context of everyday life, radical takes on another meaning (for example, a radical politician). Discuss how the meanings may be related. Use *radical* in a sentence.

Words to Know

square	the product of multiplying a number by itself
square root	the number that was squared
radical	a number with a square root symbol
right triangle	a triangle with one right angle
leg	each of the sides that form the right angle of a right triangle
hypotenuse	the side opposite the right angle of a right triangle
Pythagorean Theorem	a formula for finding the length of a side of a right triangle when you know the lengths of the other sides; $\text{leg}^2 + \text{leg}^2 = \text{hypotenuse}^2$
45°-45°-90° triangle	a right triangle whose acute angles both measure 45°
30°-60°-90° triangle	a right triangle whose acute angles measure 30° and 60°

Knotted Ropes Project

Ancient Egyptians used knotted ropes to form a right triangle to measure land. You can do the same. Use string about 18 inches long. Tie 13 knots equally spaced apart on your string. Then tie knots 1 and 13 together. Use this string to make a right triangle. With c as the longest side, show that $a^2 + b^2 = c^2$ for your right triangle.

Write a report about how the ancient Egyptians used their knotted ropes to measure land.

Project Students can begin work on the project after completing Lesson 6.5. Students having difficulty tying the knots might use a marker to denote 1" spaces on the string. See the Classroom Resource Binder for a scoring rubric to assess this project.

More practice is provided in the Classroom Resource Binder, page 56.

6·1 Algebra Review: Squares

Getting Started
Ask students to evaluate 4^2 and $4 \cdot 2$. Help them recognize that $4^2 = 4 \cdot 4 = 16$ and $4 \cdot 2 = 8$.

When you multiply a number by itself, the product is the **square** of the number. 7 squared is written 7^2.

$$7^2 = 7 \cdot 7 = 49$$

EXAMPLE

Find 9^2.

Multiply 9 by itself. $9^2 = 9 \cdot 9 = 81$

9^2 is 81.

The square of a whole number is a perfect square. 49 and 81 are perfect squares.

Try These

1. Find 50^2.
   Multiply ■ by itself. 50
   $50^2 = ■ \cdot ■ = ■$ 50 50 2,500
   The square of 50 is 50^2 or ■. 2,500

2. Find 13^2.
   Multiply ■ by itself. 13
   $13^2 = ■ \cdot ■ = ■$ 13 13 169
   The square of 13 is 13^2 or ■. 169

Practice

Find each square.

1. 8^2 64
2. 5^2 25
3. 6^2 36
4. 3^2 9
5. 4^2 16
6. 10^2 100
7. 12^2 144
8. 15^2 225

Share Your Understanding

9. Explain to a partner how to find the square of 14. Have your partner find 14^2. Multiply 14 by 14. $14 \times 14 = 196$

10. **CRITICAL THINKING** Ask a partner to find $3^2 + 4^2$. The sum is the square of what number? 25; 5

More practice is provided in the Classroom Resource Binder, page 56.

6·2 Algebra Review: Square Roots

Getting Started
Have students find the following squares:
6^2 [36]
10^2 [100]
12^2 [144]

A number that is squared is called the **square root**. The symbol for the square root is $\sqrt{}$.

$\sqrt{49}$ is 7 because $7 \cdot 7 = 49$.

EXAMPLE

Find $\sqrt{25}$.

Write 25 as the product of two equal factors. $25 = 5 \cdot 5$

Write the square root of 25. $\sqrt{25} = 5$

The square root of 25 is 5.

Try These

1. Find $\sqrt{16}$.

Write 16 as the product of two equal factors. $16 = 4 \cdot ■$ 4

Write the square root. $\sqrt{16} = ■$ 4

The square root of 16 is ■. 4

2. Find $\sqrt{15^2}$.

Write 15^2 as a product of two equal factors. $15^2 = ■ \cdot ■$ 15 15

Write the square root. $\sqrt{15^2} = 15$

The square root of 15^2 is ■. 15

Practice

Find the square root of each number.

1. $\sqrt{64}$ 8 **2.** $\sqrt{9}$ 3 **3.** $\sqrt{36}$ 6 **4.** $\sqrt{121}$ 11

5. $\sqrt{81}$ 9 **6.** $\sqrt{100}$ 10 **7.** $\sqrt{13^2}$ 13 **8.** $\sqrt{24^2}$ 24

Share Your Understanding

9. Think of a number from 5 to 12. Write its square. Have a partner find the square root of your square.
Answers will vary. Possible answer: Think 10. $10^2 = 10 \times 10 = 100$. $\sqrt{100} = 10$

10. **CRITICAL THINKING** Find $\sqrt{3^2 + 4^2}$. Now find $\sqrt{3^2} + \sqrt{4^2}$. Are the two expressions equal? Explain why or why not.
$\sqrt{3^2 + 4^2} = \sqrt{9 + 16} = \sqrt{25} = 5$
$\sqrt{3^2} + \sqrt{4^2} = 3 + 4 = 7$. The expressions are not equal.

More practice is provided in the Classroom Resource Binder, page 57.

6-3 Algebra Review: Simplifying Radicals

Getting Started
Have students find the perfect square in each of the following:

$4 \bullet 5 = 20$ [4]
$25 \bullet 3 = 75$ [25]
$16 \bullet 2 = 32$ [16]

Math Fact

$20 = \overbrace{4 \bullet 5}^{\text{factors}}$, where 4 is the perfect square

You have found the square root of a perfect square. A number with a square root symbol is called a **radical**. $\sqrt{20}$ is a radical. Sometimes you can simplify a radical that is not a perfect square. Look for a perfect square factor in the number under the square root symbol.

You can simplify $\sqrt{20}$. Write 20 as the product of a perfect square and another factor. Then, find the square root of the perfect square. Leave the remaining radical as is.

$$\sqrt{20} = \sqrt{4 \bullet 5} = \sqrt{4} \bullet \sqrt{5} = 2 \bullet \sqrt{5} = 2\sqrt{5}$$

EXAMPLE 1

Math Fact
Factors of 75
$75 = 1 \bullet 75$
$75 = 3 \bullet 25 \leftarrow$
$75 = 5 \bullet 15$

Simplify. $\sqrt{75}$

Write 75 as the product of a perfect square and another factor. Use $25 \bullet 3$.

$\sqrt{75} = \sqrt{25 \bullet 3}$
$= \sqrt{25} \bullet \sqrt{3}$

Find the square root of the perfect square.

$= 5 \bullet \sqrt{3}$
$= 5\sqrt{3}$

So, $\sqrt{75} = 5\sqrt{3}$.

The perfect square should be the largest one you can find.

EXAMPLE 2

Math Fact
Factors of 32
$32 = 1 \bullet 32$
$32 = 2 \bullet 16 \leftarrow$
$32 = 4 \bullet 8$

Simplify. $\sqrt{32}$

Write 32 as the product of a perfect square and another factor. Use $16 \bullet 2$.

$\sqrt{32} = \sqrt{16 \bullet 2}$
$= \sqrt{16} \bullet \sqrt{2}$

Find the square root of the perfect square.

$= 4 \bullet \sqrt{2}$
$= 4\sqrt{2}$

So, $\sqrt{32} = 4\sqrt{2}$.

Try These

1. Simplify. $\sqrt{8}$

Write 8 as a product of a perfect square and another factor.

$\sqrt{8} = \sqrt{4 \cdot 2}$
$= \sqrt{4} \cdot \sqrt{2}$

Find the square root of the perfect square.

$= \blacksquare \cdot \sqrt{2}$ (2)
$= \blacksquare \sqrt{\blacksquare}$ (2, 2)

So, $\sqrt{8} = 2\sqrt{2}$.

2. Simplify. $\sqrt{27}$

Write 27 as a product of a perfect square and another factor.

$\sqrt{27} = \sqrt{\blacksquare \cdot 3}$ (9)
$= \sqrt{\blacksquare} \cdot \sqrt{3}$ (9)

Find the square root of the perfect square.

$= \blacksquare \cdot \sqrt{3}$ (3)
$= \blacksquare \sqrt{\blacksquare}$ (3, 3)

So, $\sqrt{27} = 3\sqrt{3}$.

Practice

Simplify each radical.

1. $\sqrt{12}$ $2\sqrt{3}$ **2.** $\sqrt{24}$ $2\sqrt{6}$ **3.** $\sqrt{54}$ $3\sqrt{6}$

4. $\sqrt{18}$ $3\sqrt{2}$ **5.** $\sqrt{48}$ $4\sqrt{3}$ **6.** $\sqrt{28}$ $2\sqrt{7}$

7. $\sqrt{50}$ $5\sqrt{2}$ **8.** $\sqrt{80}$ $4\sqrt{5}$ **9.** $\sqrt{45}$ $3\sqrt{5}$

10. $\sqrt{75}$ $5\sqrt{3}$ **11.** $\sqrt{72}$ $6\sqrt{2}$ **12.** $\sqrt{63}$ $3\sqrt{7}$

13. $\sqrt{300}$ $10\sqrt{3}$ **14.** $\sqrt{125}$ $5\sqrt{5}$ **15.** $\sqrt{200}$ $10\sqrt{2}$

16. Write 500 as a product of a perfect square and another factor. Find the square root of the perfect square. $\sqrt{500} = \sqrt{100 \cdot 5} = \sqrt{100} \cdot \sqrt{5} = 10\sqrt{5}$

Share Your Understanding

16. Explain to a partner how to simplify $\sqrt{500}$. Use the words *product, perfect square,* and *square root* in your explanation.

17. **CRITICAL THINKING** Choose a perfect square and multiply it by 3. Use the product to write a radical. Ask a partner to simplify the radical. Check your partner's work.
Answers will vary. Possible answer: Choose 25; $25 \cdot 3 = 75$;
$\sqrt{75} = \sqrt{25 \cdot 3} = \sqrt{25} \cdot \sqrt{3} = 5\sqrt{3}$

More practice is provided in the Workbook, page 32, and in the Classroom Resource Binder, page 58.

6·4 Parts of a Right Triangle

Getting Started
Review that a right triangle has one right angle. Ask for volunteers to draw right triangles on the board using the corner of an index card to form the right angle.

A **right triangle** has one right angle. The sides of a right triangle have special names.

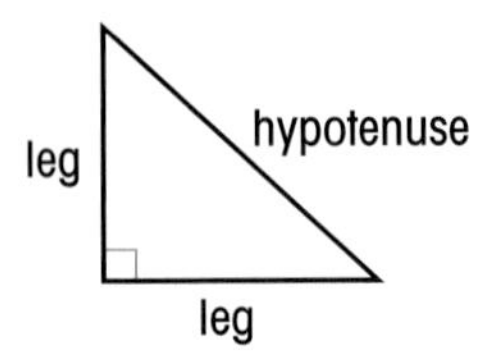

The side opposite the right angle is the **hypotenuse.** The hypotenuse is always the longest side. Each side that forms the right angle is a **leg.**

EXAMPLE 1

Name the hypotenuse and the legs.

Find the right angle. The hypotenuse is the side opposite the right angle. The other sides are the legs.

The hypotenuse is *c*. The legs are *a* and *b*.

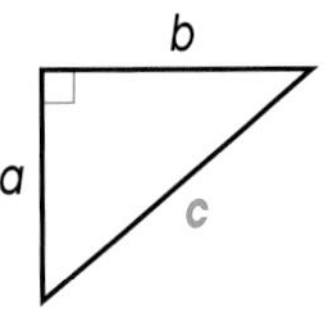

EXAMPLE 2

What is the length of the hypotenuse?

Find the right angle. The hypotenuse is the side opposite the right angle. Find the length of this side.

The length of the hypotenuse is 5 feet.

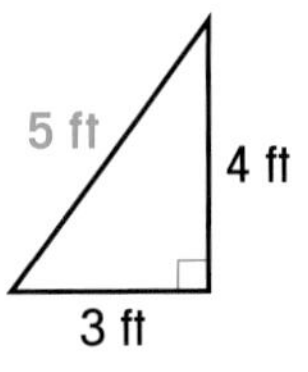

EXAMPLE 3

What is the length of each leg?

Find the right angle. The legs form the right angle.

Find the length of each leg.

The lengths of the legs are 5 inches and 12 inches.

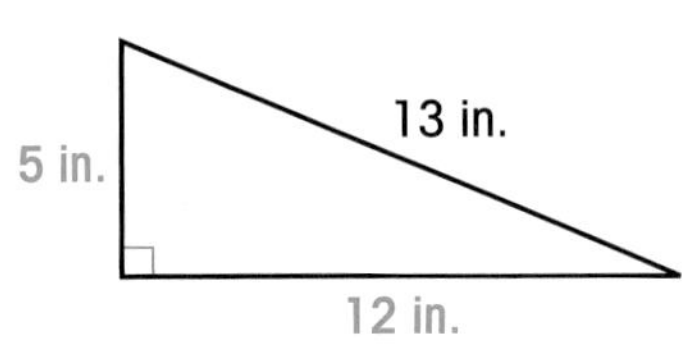

Try These

1. What is the length of the hypotenuse?

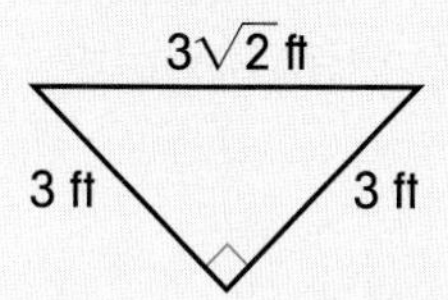

Find the right angle. The hypotenuse is the side opposite the right angle. Find the length of this side.

The length of the hypotenuse is ■. $3\sqrt{2}$ ft

2. What is the length of each leg?

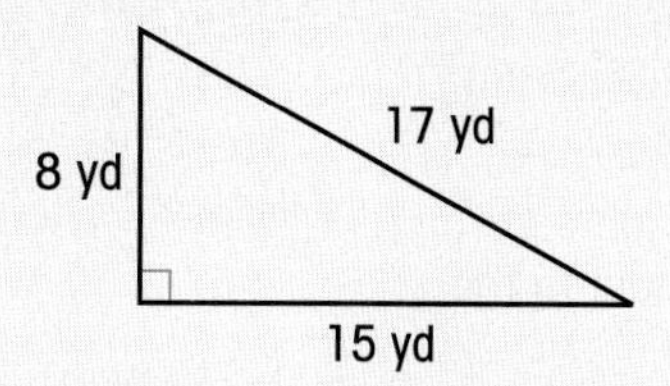

The legs form the right angle. Find the length of each leg.

The lengths of the legs are ■ 8 yd and ■. 15 yd

Practice

Write the length of the hypotenuse of each triangle.

1.

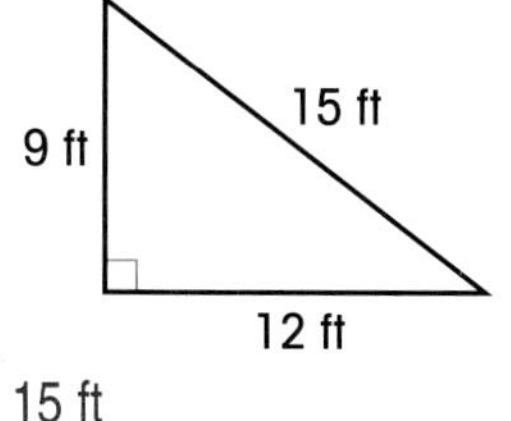

15 ft

2.

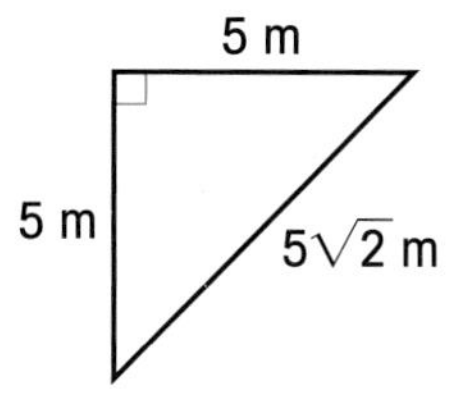

$5\sqrt{2}$ m

3.

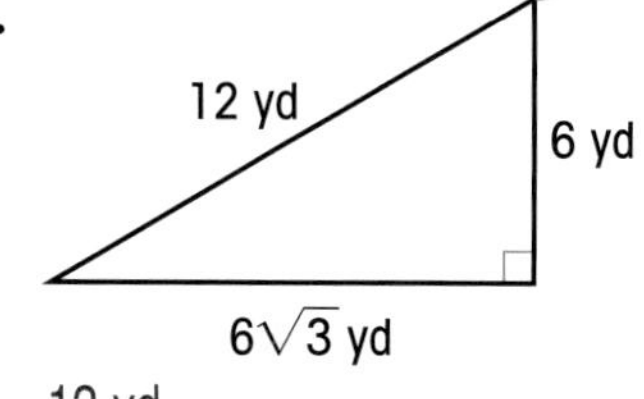

12 yd

4.

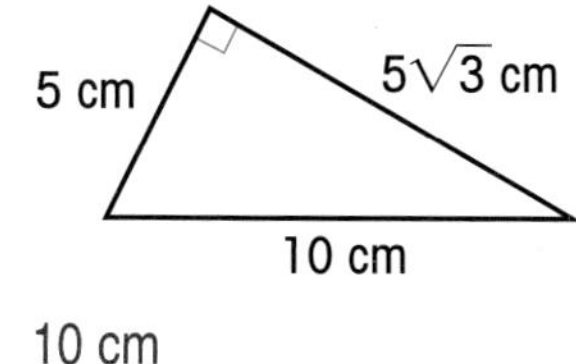

10 cm

5.

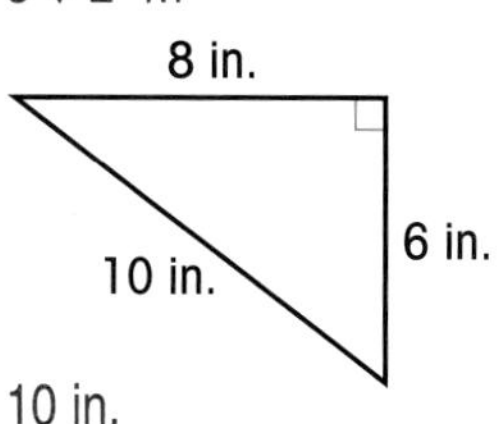

10 in.

6.

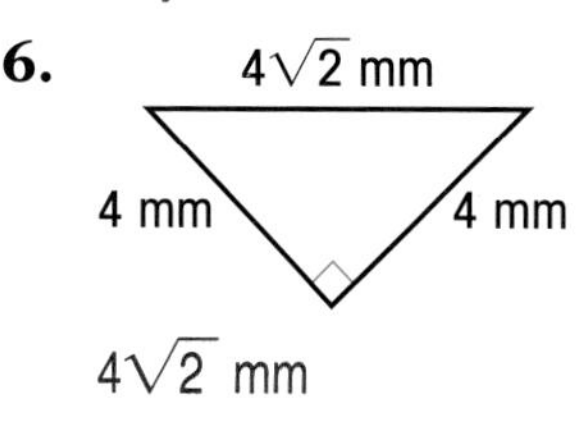

$4\sqrt{2}$ mm

Share Your Understanding

7. Explain to a partner how to locate the hypotenuse of a right triangle. Use the words *right angle* and *opposite*.
The hypotenuse is the side opposite the right angle.

8. **CRITICAL THINKING** The two legs of a certain right triangle are congruent. What are the measures of the opposite angles? Explain your thinking.
If the legs of a right triangle are congruent, then the angles opposite the legs must be congruent. Since the measure of a right angle is 90°, each of the other angles must be 90° ÷ 2 or 45°.

More practice is provided in the Workbook, page 32, and in the Classroom Resource Binder, page 59.

Getting Started
Review how to solve an addition equation.

6·5 Pythagorean Theorem

> THEOREM 18
> **Pythagorean Theorem**
> In a right triangle, the sum of the squares of the legs is equal to the square of the hypotenuse.
> $a^2 + b^2 = c^2$

You have identified the hypotenuse and legs of a right triangle. If you know the lengths of two sides, you can find the length of the third side. Use the formula called the **Pythagorean Theorem**.

$$\text{leg}^2 + \text{leg}^2 = \text{hypotenuse}^2$$

$$a^2 + b^2 = c^2$$

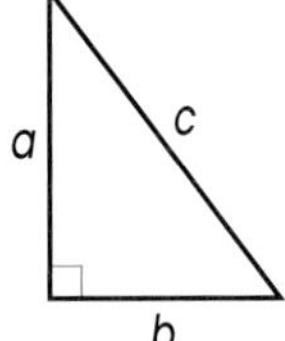

EXAMPLE 1

Find the length of c.

5 m
12 m
c

Write the Pythagorean Theorem. $a^2 + b^2 = c^2$

Substitute 5 for a and 12 for b. $5^2 + 12^2 = c^2$

Simplify. $25 + 144 = c^2$

$169 = c^2$

Find the square root of both sides. $\sqrt{169} = \sqrt{c^2}$

$13 = c$

The length of c is 13 meters.

EXAMPLE 2

Find the length of b.

4 in.
b
2 in.

Write the Pythagorean Theorem. $a^2 + b^2 = c^2$

Substitute 2 for a and 4 for c. $2^2 + b^2 = 4^2$

Solve for b. $4 + b^2 = 16$

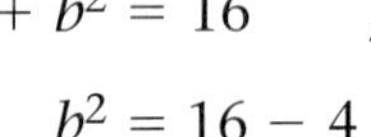

$b^2 = 16 - 4$

$b^2 = 12$

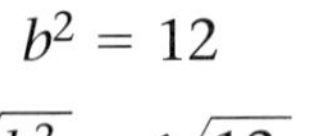

Find the square root of both sides. $\sqrt{b^2} = \sqrt{12}$

Simplify the radical. $b = \sqrt{4 \cdot 3} = 2\sqrt{3}$

So, b is $2\sqrt{3}$ inches long.

Try This

Find the length of c.

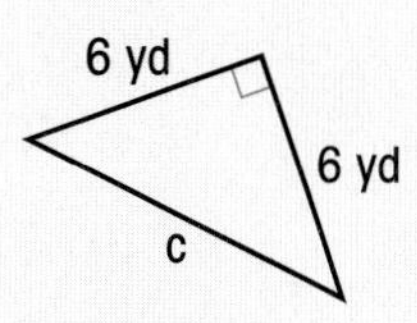

Write the Pythagorean Theorem. $a^2 + b^2 = c^2$

Substitute 6 for a and ■ for b. 6 $6^2 + ■^2 = c^2$

6

Simplify. 36 $36 + ■ = c^2$

$72 = c^2$

Find the square root of both sides. $\sqrt{72} = \sqrt{c^2}$

Simplify the radical. Use $72 = 36 \cdot 2$. $■\sqrt{2} = c$

6

So, c is ■ yards long.

$6\sqrt{2}$

Practice

Find the length of c.

1.

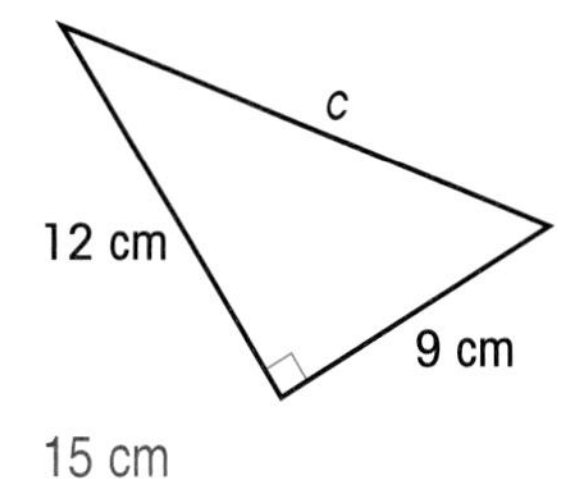

15 cm

2.

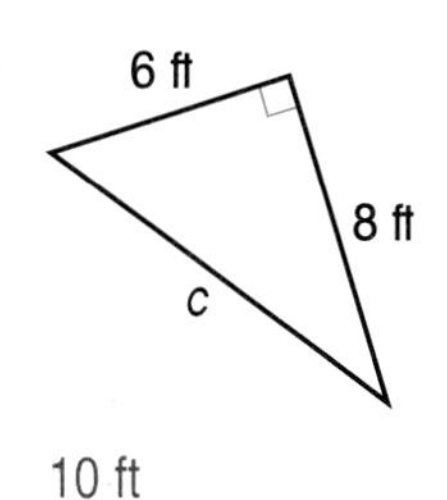

10 ft

3.

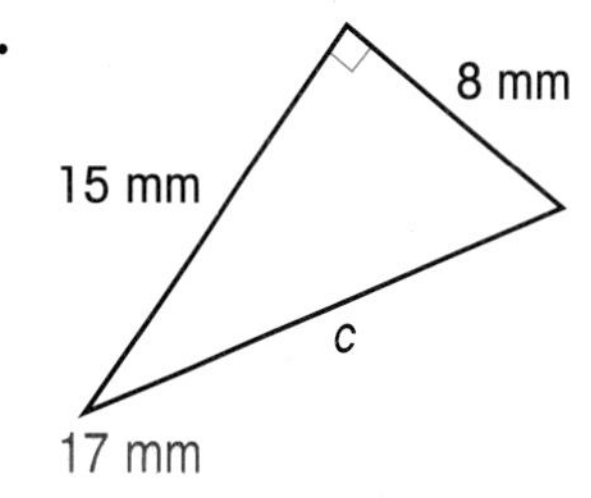

17 mm

4.

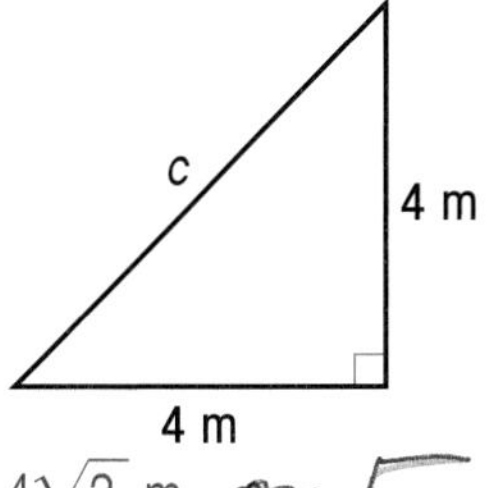

$4\sqrt{2}$ m or $\sqrt{32}$ or 5.66

5.

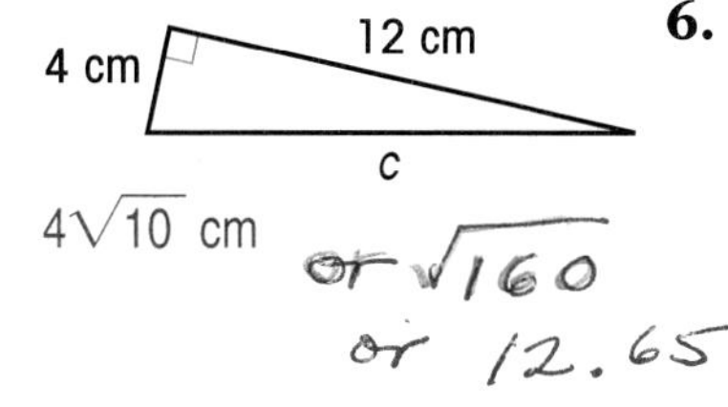

$4\sqrt{10}$ cm or $\sqrt{160}$ or 12.65

6.

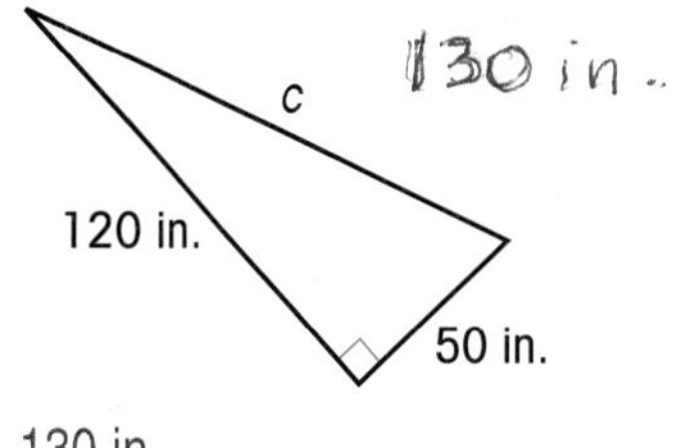

130 in.

130 in.

Try This

Find the length of b.

Write the Pythagorean Theorem.	$a^2 + b^2 = c^2$
Substitute ■ 8 for a and ■ 17 for c.	$8^2 + b^2 = 17^2$
Solve for b.	$64 + b^2 = 289$
	$b^2 = 289 - 64$
	$b^2 = 225$
Find the square root of both sides.	$\sqrt{b^2} = \sqrt{225}$
	$b = 15$

b

8 m

17 m

So, b is ■ 15 meters long.

Practice

Find the length of b.

7.

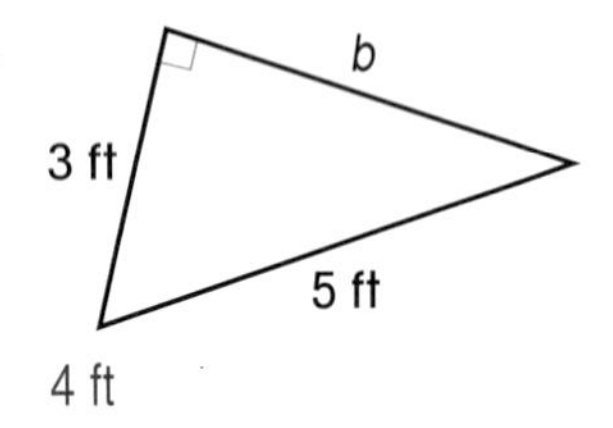

4 ft

8.

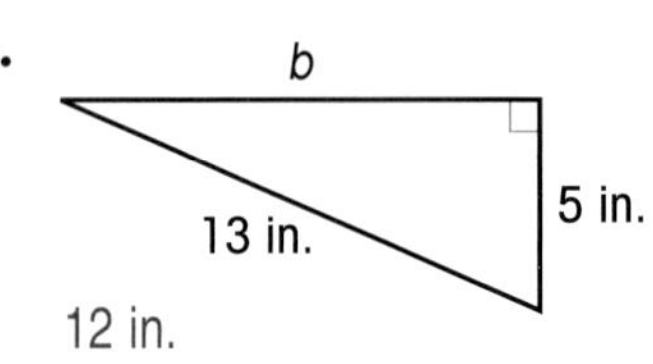

12 in.

9.

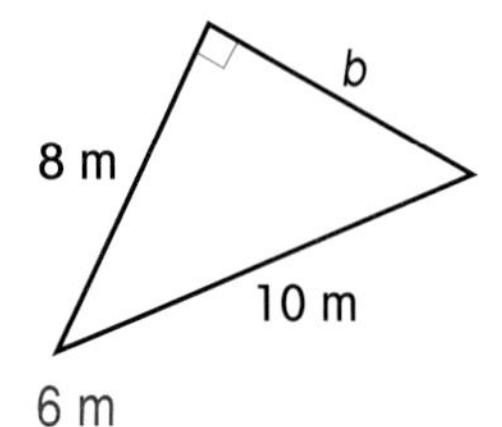

6 m

10.

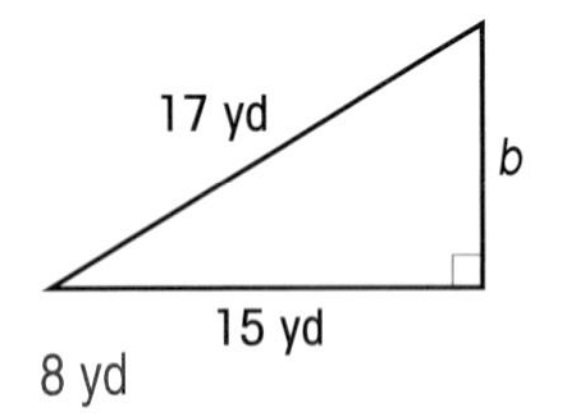

8 yd

11.

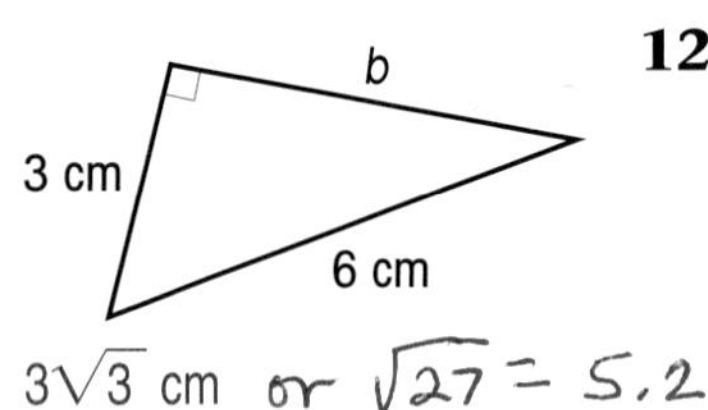

$3\sqrt{3}$ cm or $\sqrt{27} = 5.2$

12.

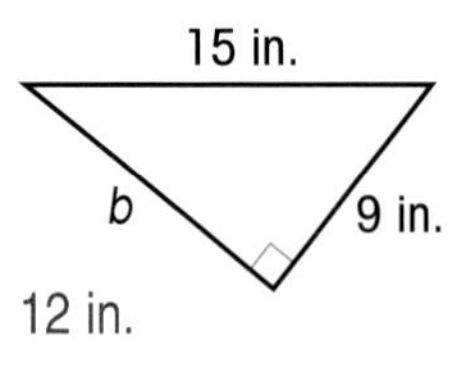

12 in.

13. Since the triangle is a right triangle, you can use the Pythagorean Theorem, $a^2 + b^2 = c^2$, to find the length of the missing side. Substitute 16 for a and 20 for c. Then solve for b. $b = 12$ ft

Share Your Understanding

13. Work with a partner to find b. Explain your thinking.

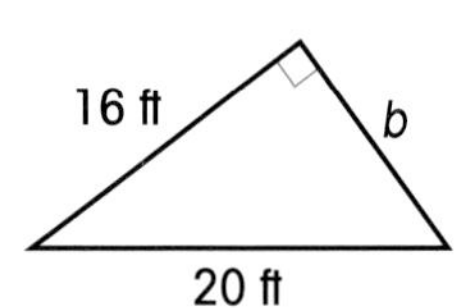

14. **CRITICAL THINKING** Draw different isosceles right triangles. Give the lengths of the legs. Find the length of the hypotenuse. What do you notice about the length of the hypotenuse and the length of a leg?

The hypotenuse is equal to the length of a leg times $\sqrt{2}$.

CONSTRUCTION
A Right Triangle

You learned how to construct a perpendicular to a line segment. You can use this skill to construct a right triangle. Follow the steps below.

STEP 1 Draw a line segment.

STEP 2 Place the compass on each endpoint. Draw arcs above and below the line segment. The arcs should intersect.

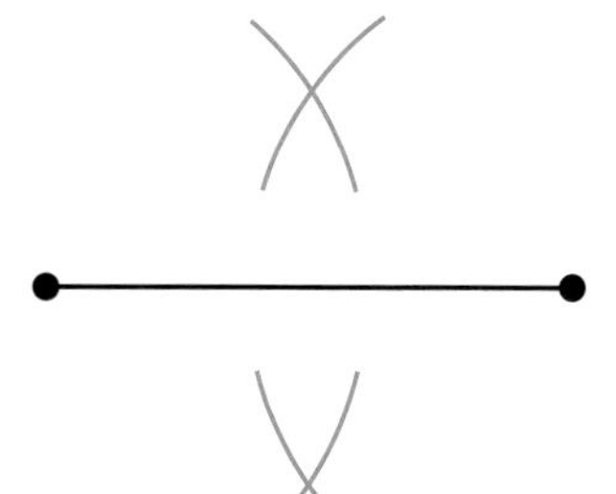

STEP 3 Connect the two points formed by the arcs. The line is perpendicular to the given line segment.

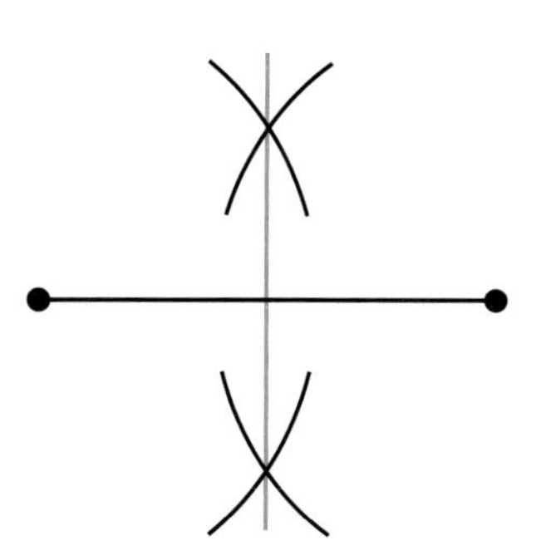

STEP 4 Draw a line from one endpoint to an intersection of the arcs. This is the hypotenuse. Check students' work.

$\triangle ABC$ is a right triangle.

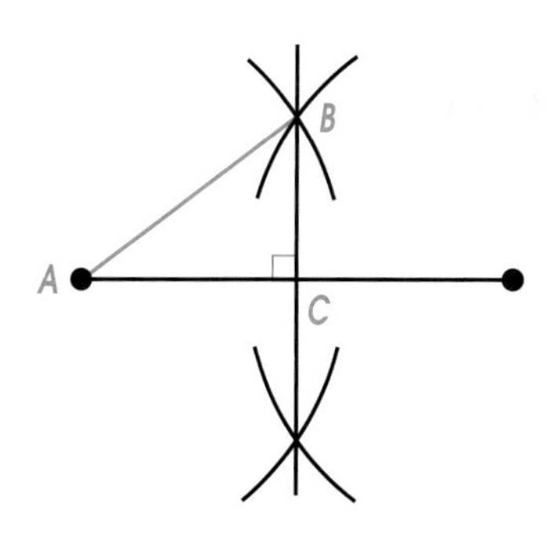

More practice is provided in the Workbook, page 33, and in the Classroom Resource Binder, page 60.

6-6 Special Right Triangle: 45°-45°-90°

Getting Started
Review the Pythagorean Theorem. Show that
$8^2 + 8^2 = (8\sqrt{2})^2$
$64 + 64 = 64 \cdot 2$
$128 = 128$

A **45°-45°-90° triangle** is an isosceles triangle. It has two congruent angles and two congruent sides. The hypotenuse of a 45°-45°-90° triangle is equal to the leg $\sqrt{2}$.

$$\text{leg} = \text{leg}$$

$$\text{hypotenuse} = \text{leg} \cdot \sqrt{2}$$

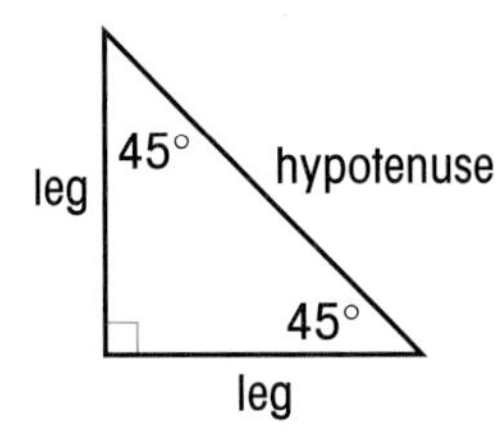

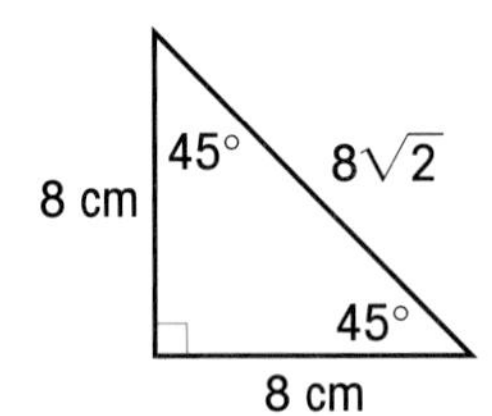

If you know the length of one side, you can find the lengths of the other two sides.

EXAMPLE 1

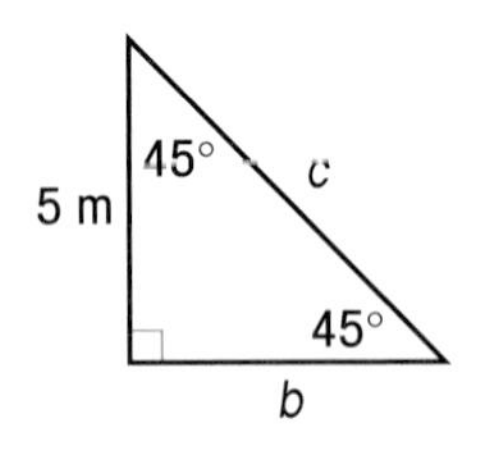

Find the unknown lengths of the sides.

Write the formula for the hypotenuse.	$\text{hypotenuse} = \text{leg} \cdot \sqrt{2}$
Substitute 5 for the leg.	$\text{hypotenuse} = 5 \cdot \sqrt{2}$ $c = 5\sqrt{2}$
Find the length of the other leg.	$\text{leg} = \text{leg}$ $b = 5$

So, c is $5\sqrt{2}$ meters and b is 5 meters.

EXAMPLE 2

a
45°
b
7√2 ft
45°

Find the unknown lengths of the sides.

Write the formula for the hypotenuse.	$\text{hypotenuse} = \text{leg} \cdot \sqrt{2}$
The hypotenuse is $7\sqrt{2}$.	$\text{hypotenuse} = 7 \cdot \sqrt{2}$
Find the length of the leg.	$\text{leg} = 7$

So, a is 7 feet and b is 7 feet.

Try This

Find the unknown lengths of the sides.

Write the formula for the hypotenuse.	hypotenuse = leg • $\sqrt{2}$
Substitute 6 for the leg.	hypotenuse = ■ • $\sqrt{2}$ 6
Find the length of the other leg.	leg = ■ 6

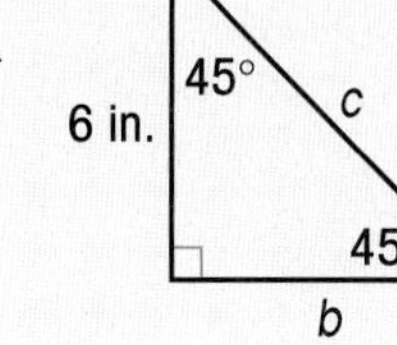

So, b is ■ 6 inches and c is ■ $6\sqrt{2}$ inches.

Practice

Find the unknown lengths of the sides of each triangle.

1.

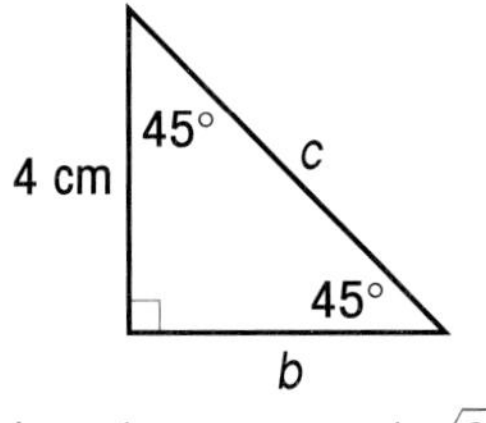

$b = 4$ cm; $c = 4\sqrt{2}$ cm

2.

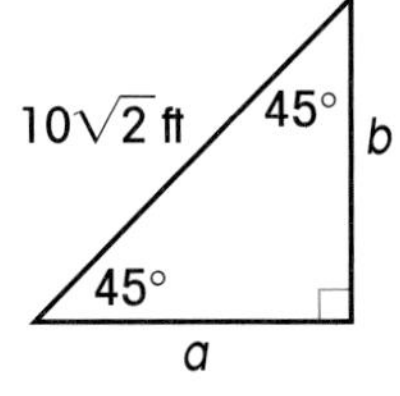

$a = 10$ ft; $b = 10$ ft

3.

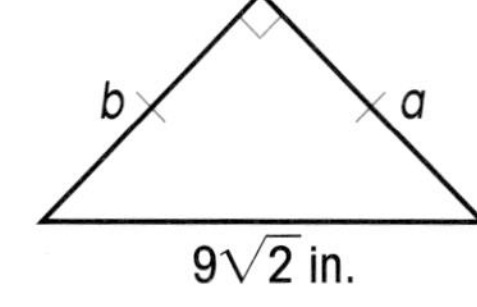

$a = 9$ in.; $b = 9$ in.

4.

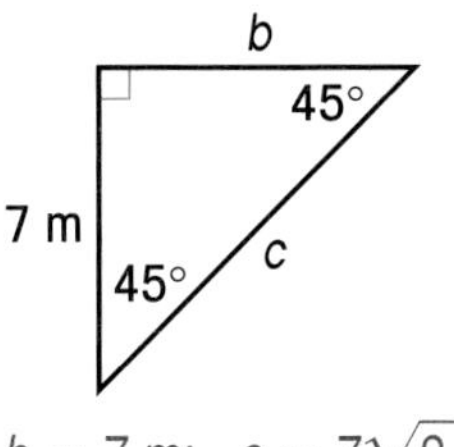

$b = 7$ m; $c = 7\sqrt{2}$ m

5.

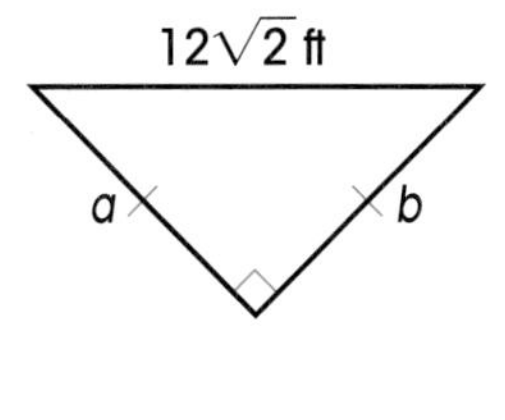

$a = 12$ ft; $b = 12$ ft

6.

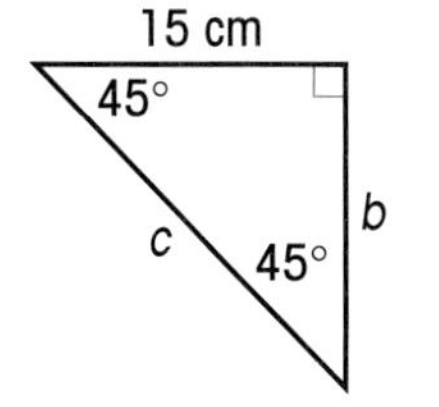

$b = 15$ cm; $c = 15\sqrt{2}$ cm

Share Your Understanding

7. Draw your own 45°-45°-90° triangle. Give the length of one leg. Ask a partner to explain how to find the lengths of the other two sides. Check your partner's work. See below.

8. **CRITICAL THINKING** Use the Pythagorean Theorem to show that the hypotenuse = leg • $\sqrt{2}$ for an isosceles right triangle.
Answers will vary.

7. Possible answer: In a 45°-45°-90° triangle, the legs are of equal length, and the length of the hypotenuse is equal to the length of a leg times $\sqrt{2}$.

More practice is provided in the Workbook, page 34, and in the Classroom Resource Binder, page 61.

6·7 Special Right Triangle: 30°-60°-90°

Getting Started
Draw a 30° -60° -90° triangle on the board. Use one color for the short side and for the label of the 30° angle. Then use different colors for the other two sides and angles.

You can use a shortcut to find the lengths of the unknown sides of a **30°-60°-90° triangle**. First, you need to identify the short leg. The short leg is always opposite the 30° angle.

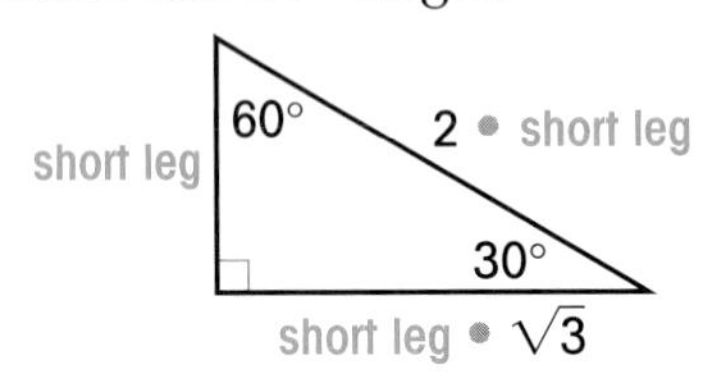

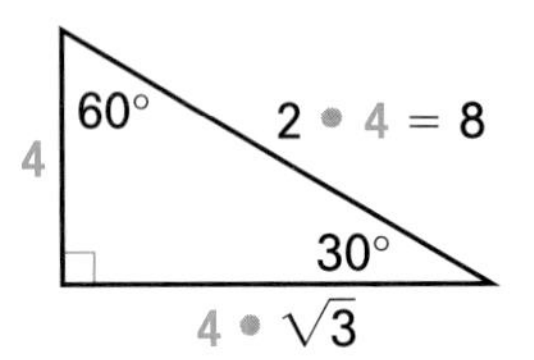

long leg = short leg • $\sqrt{3}$
hypotenuse = 2 • short leg

EXAMPLE 1

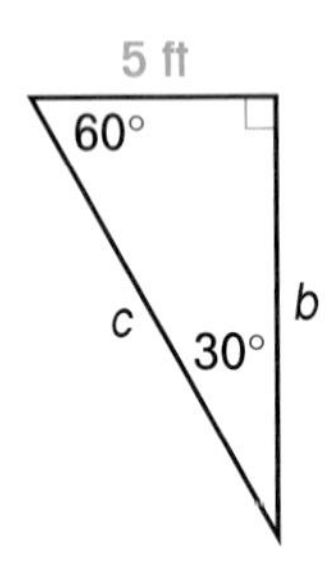

Find the unknown lengths of the sides.

Write the formula for the hypotenuse.	hypotenuse = 2 • short leg
Substitute 5 for the short leg.	hypotenuse = 2 • 5 = 10
Write the formula for the long leg.	long leg = short leg • $\sqrt{3}$
Substitute 5 for the short leg.	long leg = 5 • $\sqrt{3}$ = $5\sqrt{3}$

So, c is 10 feet and and b is 5 $\sqrt{3}$ feet.

EXAMPLE 2

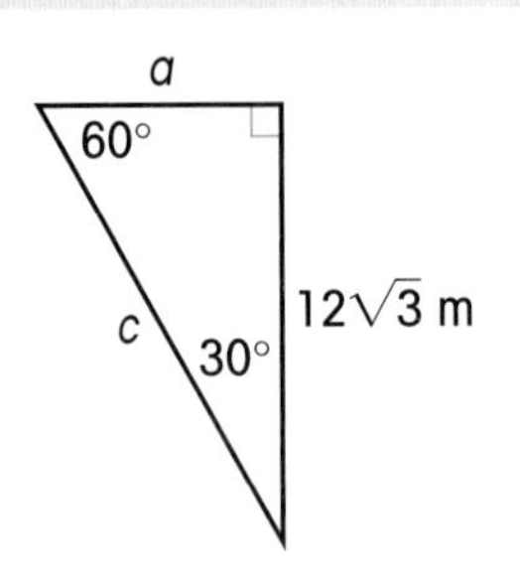

Find the unknown lengths of the sides.

Write the formula for the side you know. Substitute 12 for the short leg.	long leg = short leg • $\sqrt{3}$ = $12\sqrt{3}$
Find the short leg.	short leg = 12
Write the formula for the hypotenuse.	hypotenuse = 2 • short leg
Substitute 12 for the short leg.	hypotenuse = 2 • 12 = 24

So, a is 12 meters and c is 24 meters.

Try This

Find the unknown lengths of the sides.

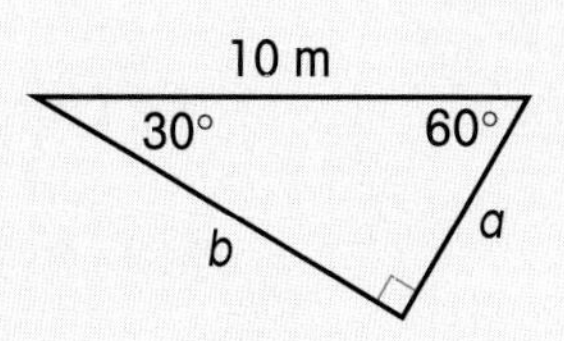

Write the formula for the side you know.	hypotenuse = 2 • short leg
Substitute 10 for the hypotenuse.	10 = 2 • short leg
Find the short leg.	10 = 2 • ■ 5
	short leg = ■ 5
Now find the length of the long leg.	long leg = short leg • $\sqrt{3}$ = ■$\sqrt{3}$ 5
Write a formula for the long leg.	

So, a is 5 meters and b is $5\sqrt{3}$ meters.

Practice

Find the unknown lengths of the sides of each triangle.

1. 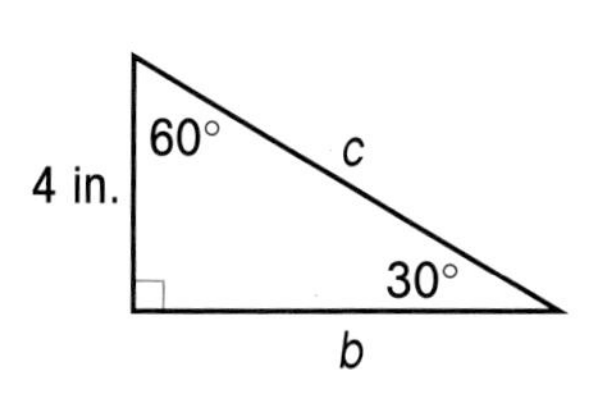

$b = 4\sqrt{3}$ in.; $c = 8$ in.

2. 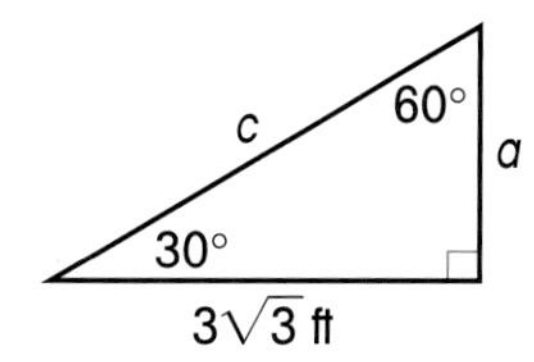

$a = 3$ ft; $c = 6$ ft

3. 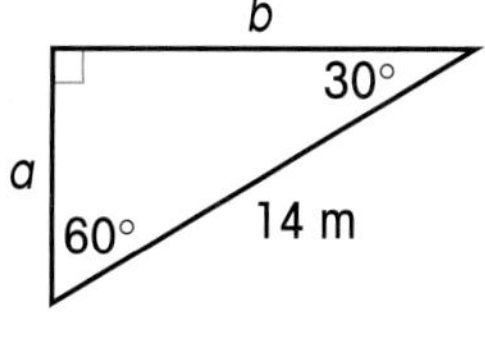

$a = 7$ m; $b = 7\sqrt{3}$ m

4. 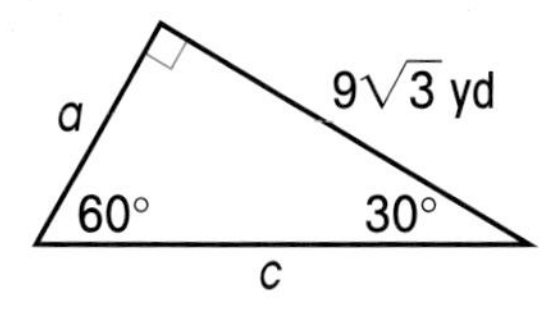

$a = 9$ yd; $c = 18$ yd

5. 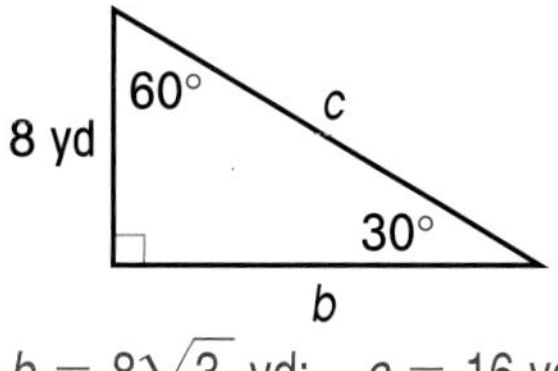

$b = 8\sqrt{3}$ yd; $c = 16$ yd

6. 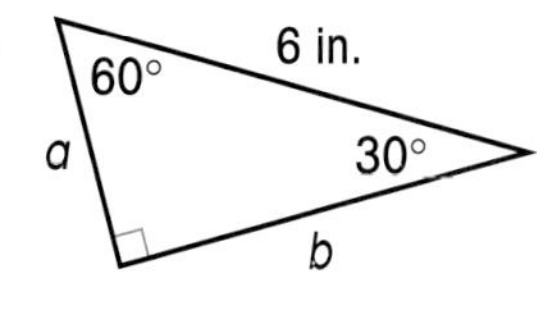

$a = 3$ in.; $b = 3\sqrt{3}$ in.

Share Your Understanding

7. Explain to a partner how to find the long leg in exercise **3.** See above.

7. The hypotenuse equals 2 times the short leg, so the short leg is 7 m long. The long leg is $\sqrt{3}$ times the short leg. The long leg = $7\sqrt{3}$ m.

8. CRITICAL THINKING The short leg of a 30°-60°-90° triangle is 1 m. The long leg is $\sqrt{3}$ m. Show that $a^2 + b^2 = c^2$.
$a^2 + b^2 = c^2$, $1^2 + (\sqrt{3})^2 = 2^2$, $1 + 3 = 4$.

6·8 Calculator: Pythagorean Triples

Getting Started Review that 5^2 is 5 • 5.

The Converse of the Pythagorean Theorem can help you tell whether a triangle is a right triangle. You just need to know the lengths of all of the sides.

Substitute all the lengths of the sides of the triangle in the Pythagorean Theorem. If the equation is true, the triangle is a right triangle. You can use a calculator to do the calculations.

THEOREM 19
Converse of the Pythagorean Theorem

If a, b, and c are the lengths of the sides of a triangle so that $a^2 + b^2 = c^2$, then the triangle is a right triangle.

EXAMPLE

A triangle has sides of 5 cm, 12 cm, and 13 cm. Is this a right triangle?

Write the Pythagorean Theorem. $a^2 + b^2 = c^2$

Substitute 5 for a, 12 for b, ? for =, and 13 for c. Use a ? for = because you do not know if both sides are equal. $5^2 + 12^2 \; ? \; 13^2$

Students not using a scientific calculator will need to use the memory key.

Use a scientific calculator. Find the value of the left side of the equation.

Display

PRESS: 5 × 5 + 1 2 × 1 2 = 169

Clear the display.

PRESS: 0 0

Find the value of the right side of the equation.

PRESS: 1 3 × 1 3 = 169

The products are the same. $169 = 169$

Both sides of the equation are equal. $5^2 + 12^2 = 13^2$

A triangle with sides of 5 cm, 12 cm, and 13 cm is a right triangle.

Practice

Tell whether the three lengths can be the sides of a right triangle. Write *yes* or *no*. Explain your answers.

1. 4 ft, 8 ft, 10 ft no
2. 6 ft, 8 ft, 10 ft yes
3. 8 m, 17 m, 15 m yes
4. 10 m, 12 m, 15 m no
5. 12 in., 16 in., 20 in. yes
6. 12 in., 24 in., 12 in. no
7. 10 cm, 20 cm, 35 cm no
8. 10 cm, 24 cm, 26 cm yes
9. 15 yd, 12 yd, 9 yd yes
10. 5 yd, 12 yd, 15 yd no

On-the-Job Math

CARPENTERS

Carpenters put together things that are made of wood. They work with their hands to build some of the things that we use every day. Chairs, desks, tables, and even houses are built by carpenters.

There are two types of carpenters, a rough carpenter and a finish carpenter. Rough carpenters assemble the framework of a building. Finish carpenters work on the inside of the building. Some finish carpenters are specialists, like cabinetmakers. Some carpenters learn their trade through on-the-job training. Some attend a technical or vocational school. Others are in an apprenticeship program.

Carpenters use the 30°-60°-90° triangle in their work.

More practice is provided in the Workbook, page 35, and in the Classroom Resource Binder, page 62.

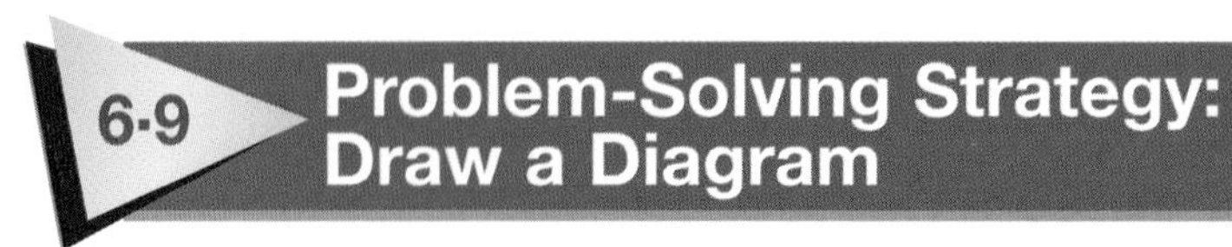

6·9 Problem-Solving Strategy: Draw a Diagram

Getting Started
On the board, show the directions for north, south, east, and west.

Sometimes you can draw a diagram to help you solve a word problem.

EXAMPLE

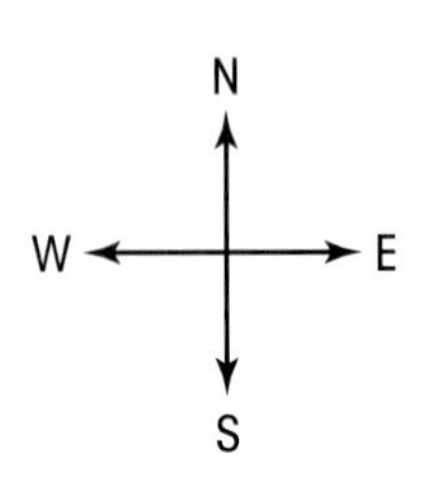

Frank walks from his house to school. He walks 12 blocks north. Then he walks 5 blocks east. A bird flies the shortest distance from Frank's house to school. What is the distance the bird flies?

READ **What do you need to find out?**
The shortest distance from the house to school.

PLAN **What do you need to do?**
Draw a diagram. Show the direction Frank walks and the direction the bird flies.

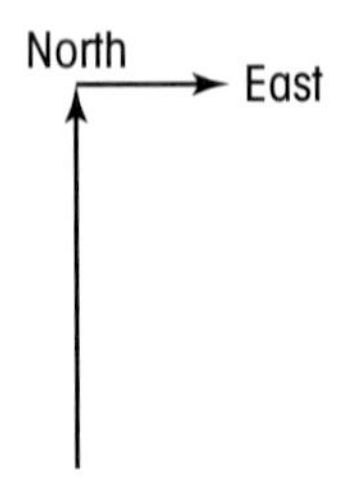

DO **Follow the plan.**
Draw the diagram.

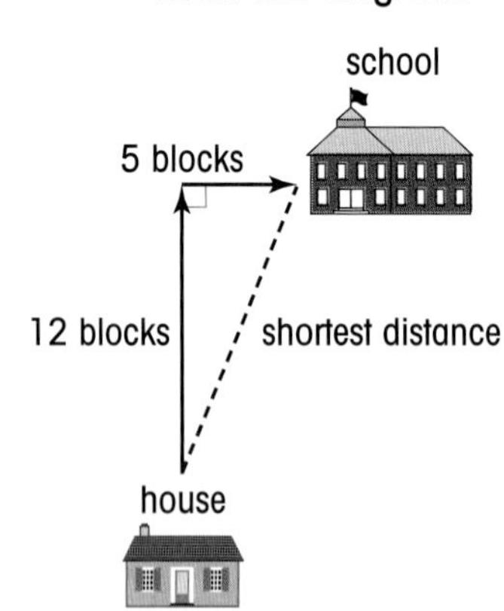

Use the Pythagorean Theorem.

$$a^2 + b^2 = c^2$$
$$12^2 + 5^2 = c^2$$
$$144 + 25 = c^2$$
$$169 = c^2$$
$$\sqrt{169} = \sqrt{c^2}$$
$$13 = c$$

CHECK **Does your answer make sense?**

12 blocks + 5 blocks = 17 blocks

Frank walks 17 blocks.

13 blocks < 17 blocks ✓

The bird flies 13 blocks.

Try This

To go to the library, Frank walks 6 blocks south of the school. Then, he walks 8 blocks west. A bird flies the shortest distance. How far does the bird fly?

Draw a diagram. First show the distance Frank walks to the library. Then draw a line to show the shortest distance.

Use the Pythagorean Theorem to find the distance the bird flies.

$6^2 + \blacksquare^2 = c^2$ 8

$\blacksquare + \blacksquare = c^2$ 36, 64

$100 = \blacksquare^2$ c

$\sqrt{100} = \sqrt{\blacksquare}$ c^2

$\blacksquare = c$ 10

The bird flies ■ blocks. 10

Practice

Solve each problem. Be sure to draw a diagram.

1. Green Mountain is 12 kilometers north of Deep Lake. Little Pond is 5 kilometers west of Green Mountain. How far is Little Pond from Deep Lake? 13 km

2. Lake City is 10 miles south of Red Hill. River Town is 24 miles east of Lake City. How far is Red Hill from River Town? 26 mi

3. A lighthouse is 70 meters east of High Point. A ship is 70 meters north of the lighthouse. How far is the ship from High Point? (Hint: Draw a 45°-45°-90° triangle.)
   $70\sqrt{2}$ m or about 98.98 m

More practice is provided in the Workbook, page 36, and in the Classroom Resource Binder, page 63.

Problem-Solving Application: Indirect Measurement

Getting Started
Review the Pythagorean Theorem. Ask for a volunteer to solve for b.
$5^2 + b^2 = 13^2$
$[b = 12]$

You can use right triangles to measure the lengths of objects. This is very helpful when the object is difficult to measure. For example, a ladder leaning against a building forms a right triangle.

EXAMPLE 1

A ladder is 26 feet long. The bottom is resting 10 feet from the building. How far up the side of the building does the ladder reach?

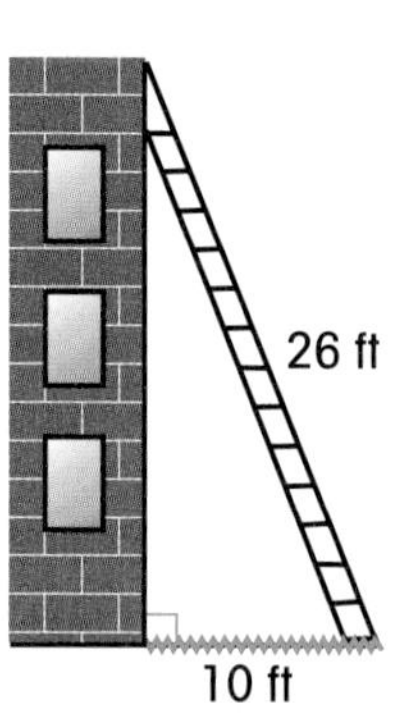

Look for a right triangle. Then use the Pythagorean Theorem to find the missing side.

$$a^2 + b^2 = c^2$$
$$a^2 + 10^2 = 26^2$$
$$a^2 + 100 = 676$$
$$a^2 = 676 - 100$$
$$a^2 = 576$$
$$\sqrt{a^2} = \sqrt{576}$$
$$a = 24$$

The ladder reaches 24 feet up the side of the building.

EXAMPLE 2

A ladder is leaning against a building. It makes a 60° angle with the ground. It is 8 feet from the building. How long is the ladder?

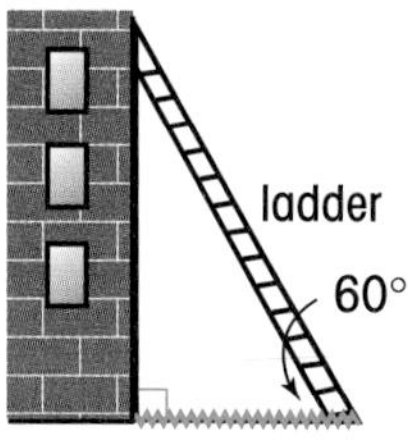

This is a 30°-60°-90° triangle. Find the hypotenuse.

hypotenuse = 2 • short leg
hypotenuse = 2 • 8 = 16

The ladder is 16 feet long.

Try This

A baseball diamond is in the shape of a square. Each side is 90 feet long. What is the distance from first base to third base?

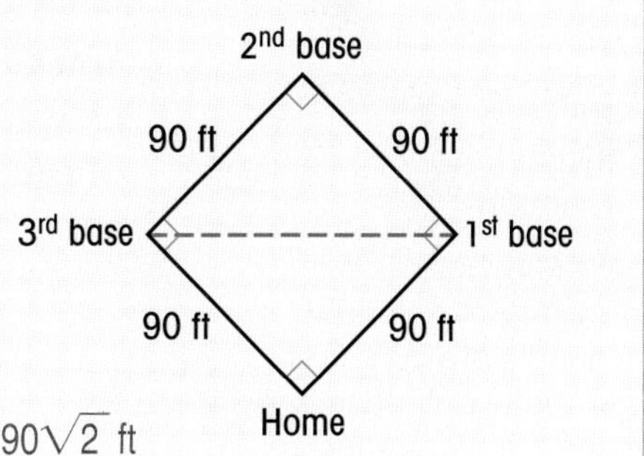

Draw a dotted line from first base to third base. This is the ■ of a 45°-■°-90° triangle. hypotenuse 45

Find the hypotenuse. hypotenuse = ■ $\sqrt{2}$ 90

The distance from first base to third base is ■. $90\sqrt{2}$ ft

Practice

Solve each problem. Show your work.

1. A wire is attached to the top of a pole and meets the ground 6 feet from the base of the pole. The wire makes a 60° angle with the ground. How long is the wire? 12 ft

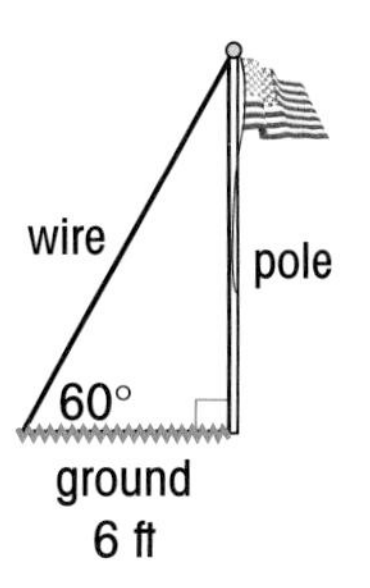

2. A surveyor measured the distance from each end of a pond to point *C*. He made a 90° angle. See the diagram at the right. What is the length across the pond from point *A* to point *B*? 150 m

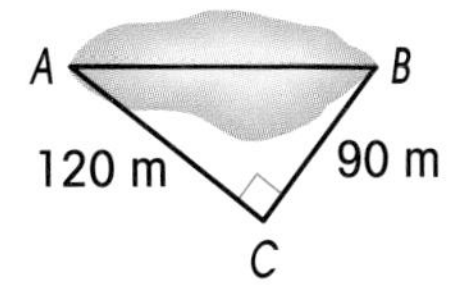

3. A garage is 8 feet tall. A ladder to the top of the garage is 10 feet long. How far away from the garage is the bottom of the ladder? 6 ft

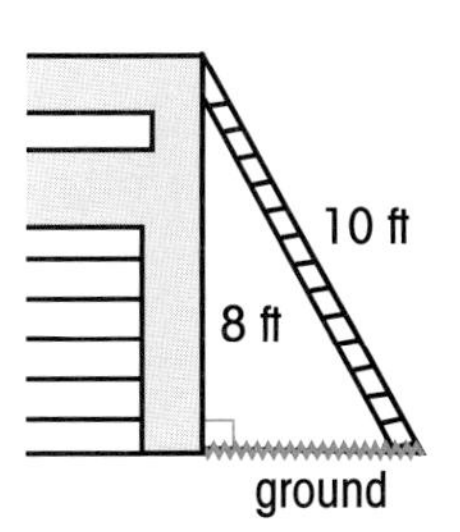

6·11 Proof: Hypotenuse–Leg Theorem

Getting Started Review congruent triangles.

THEOREM 20
Hypotenuse–Leg Theorem

If the hypotenuse and one leg of one right triangle are congruent to the hypotenuse and one leg of another right triangle, then the triangles are congruent.

You can prove that two right triangles are congruent by using the Hypotenuse–Leg Theorem. If the hypotenuse of each triangle and one pair of corresponding legs are congruent, the triangles are congruent.

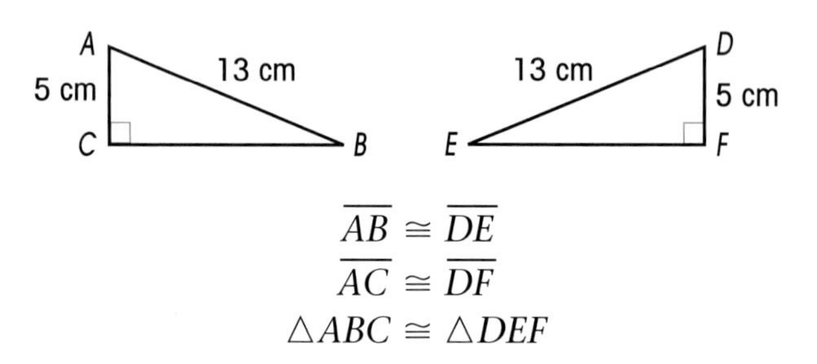

$$\overline{AB} \cong \overline{DE}$$
$$\overline{AC} \cong \overline{DF}$$
$$\triangle ABC \cong \triangle DEF$$

You can use this theorem in a proof.

EXAMPLE

You are given:

$\angle B$ and $\angle D$ are right angles.

$\overline{AB} \cong \overline{CD}$

Prove: $\triangle ABC \cong \triangle CDA$

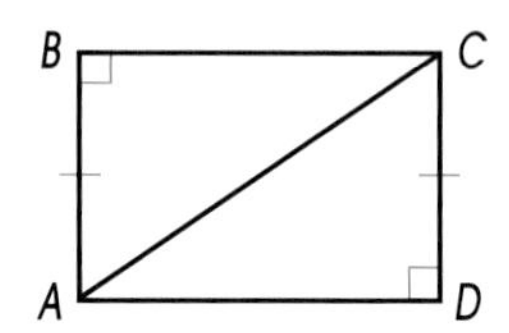

Math Fact

Information you are given is information you know.

You are given that $\angle B$ and $\angle D$ are right angles. This means that $\triangle ABC$ and $\triangle CDA$ are right triangles.

You are also given that the corresponding legs $\overline{AB}$ and $\overline{CD}$ are congruent. Because the hypotenuse of both triangles is the same line segment, $\overline{AC}$, the hypotenuses of both triangles are congruent. Then, by the Hypotenuse-Leg Theorem, the triangles are congruent.

So, you prove that $\triangle ABC \cong \triangle CDA$. ✓

Try This

Copy and complete the proof.

You are given:

$\angle H$ and $\angle K$ are right angles.

$\overline{GH} \cong \overline{JK}$

$\overline{HK}$ bisects $\overline{GJ}$

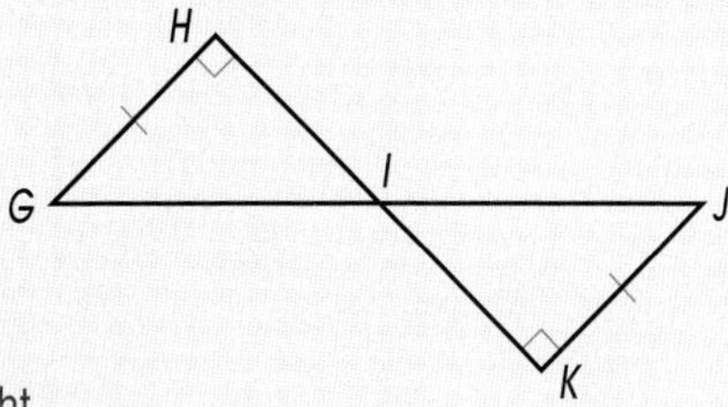

Prove: $\triangle GHI \cong \triangle JKI$

You know that $\angle H$ and $\angle K$ are ■ angles. This means that $\triangle GHI$ and $\triangle JKI$ are ■ triangles. right right

You know that corresponding legs are congruent because you are given that ■ $\cong$ ■. You are also given that $\overline{HK}$ bisects $\overline{GJ}$. This means that $\overline{GI} \cong$ ■. $\overline{GH}, \overline{JK}$ $\overline{JI}$

Thus, the hypotenuses of both triangles are ■. Then, by the ■ Theorem, the triangles are congruent. congruent Hypotenuse–Leg

So, you prove that ■. ✓$\triangle GHI \cong \triangle JKI$

Practice

Write a proof for each of the following. See Additional Answers in the back of this book.

1. You are given:
$\overline{MO} \perp \overline{LO}$
$\overline{MO} \perp \overline{MN}$
$\overline{LM} \cong \overline{NO}$
Prove: $\triangle LMO \cong \triangle NOM$

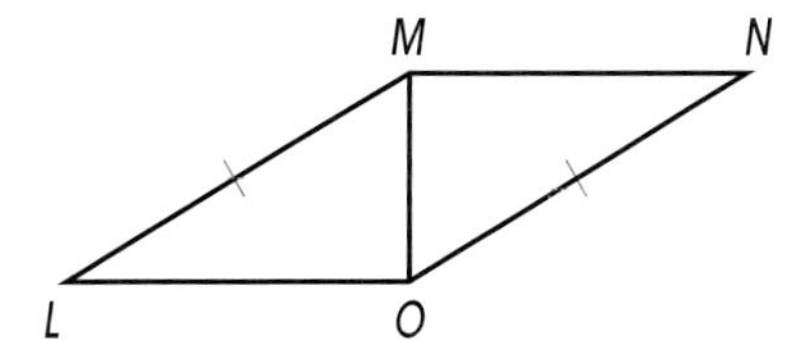

2. You are given:
$\overline{QP} \cong \overline{TS}$
$\overline{SP}$ is the perpendicular bisector of $\overline{QT}$

Prove: $\triangle PQR \cong \triangle STR$

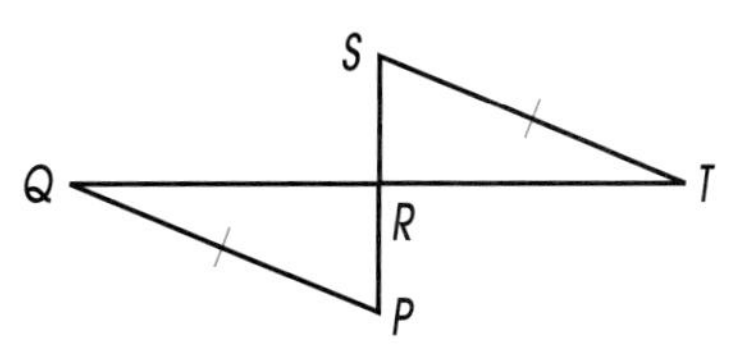

More Practice is provided in the Classroom Resource Binder.

Chapter 6 Review

Summary

To find the square of a number, multiply the number by itself.
The number that is squared is the square root.
To simplify a radical, look for the product of a perfect square and another factor. Then find the square root of the perfect square and leave the remaining radical as is.
Any right triangle has one right angle, two legs, and one hypotenuse.
The formula for the Pythagorean Theorem is $\text{leg}^2 + \text{leg}^2 = \text{hypotenuse}^2$.
For the 45°-45°-90° right triangle, hypotenuse = leg • $\sqrt{2}$
For the 30°-60°-90° right triangle, hypotenuse = 2 • short leg, and long leg = short leg • $\sqrt{3}$.
You can use the Pythagorean Theorem to determine whether a triangle is a right triangle.
You can draw a diagram to help you solve distance problems.
The Pythagorean Theorem can be used to find indirect measurement.

More vocabulary review is provided in the Classroom Resource Binder, page 55.

square
square root
radical
right triangle
leg
hypotenuse
Pythagorean Theorem
45°-45°-90° triangle
30°-60°-90° triangle

Vocabulary Review

Complete the sentences with the words from the box.

1. A right triangle with two equal angles is a ____. 45°-45°-90° triangle
2. The symbol $\sqrt{8}$ is a ____. radical
3. Use the ____ to find the length of a side of a right triangle, when you know the lengths of the other two sides. Pythagorean Theorem
4. Each side that forms the right angle of a right triangle is called a ____. leg
5. A ____ is the product of multiplying a number by itself. square
6. The hypotenuse of a ____ is 2 • short leg. 30°-60°-90° triangle
7. The number that is squared is the ____. square root
8. A triangle with one right angle is called a ____. right triangle
9. The ____ is the side of a right triangle opposite the right angle. hypotenuse

Chapter Quiz

More assessment is provided in the Classroom Resource Binder.

Simplify each expression.

1. 12^2 144 **2.** $\sqrt{81}$ 9 **3.** $\sqrt{48}$ $4\sqrt{3}$ **4.** $\sqrt{50}$ $5\sqrt{2}$

Find the unknown lengths of the sides.

5.

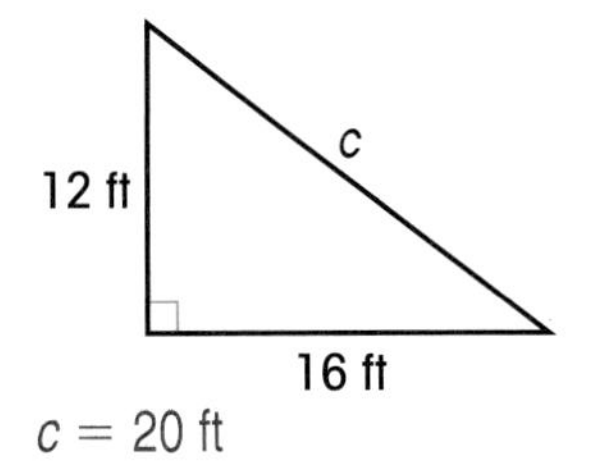

$c = 20$ ft

6.

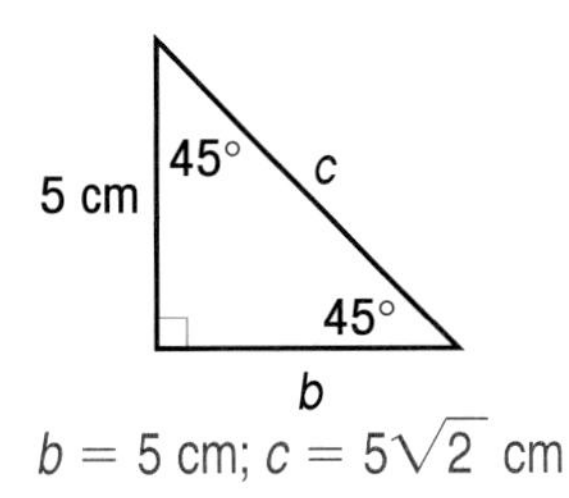

$b = 5$ cm; $c = 5\sqrt{2}$ cm

7.

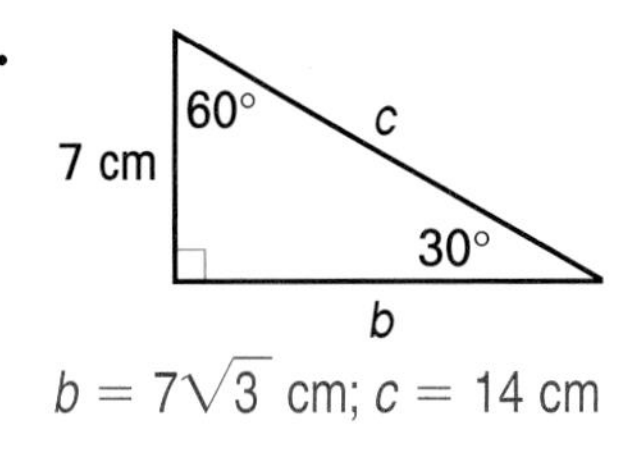

$b = 7\sqrt{3}$ cm; $c = 14$ cm

8.

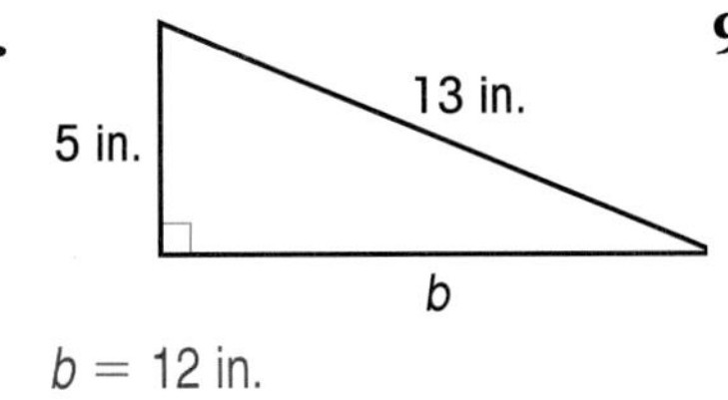

$b = 12$ in.

9.

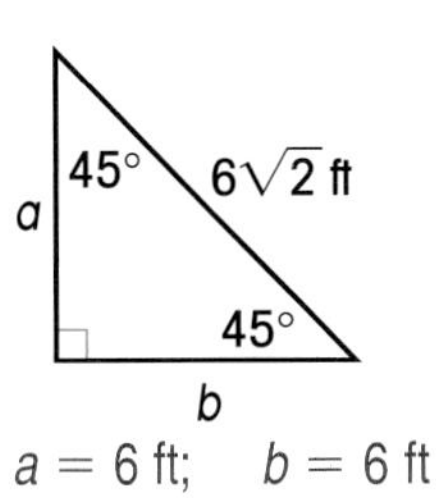

$a = 6$ ft; $b = 6$ ft

10.

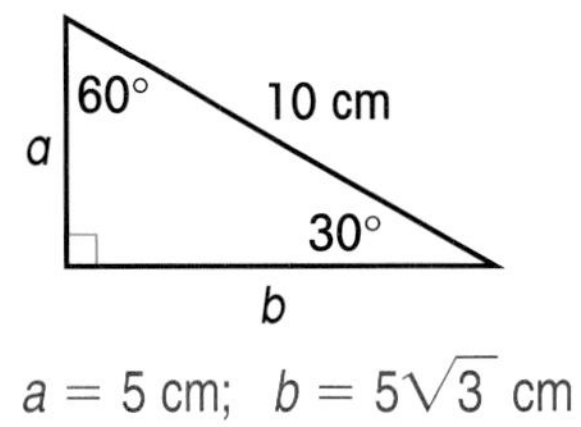

$a = 5$ cm; $b = 5\sqrt{3}$ cm

Tell whether the three lengths can form the sides of a right triangle. Write *yes* or *no*.

11. 12 ft, 15 ft, 20 ft No **12.** 8 cm, 15 cm, 17 cm Yes

Write About Math No. The hypotenuse must be longer than the legs. In an equilateral triangle, all the legs are equal.

Problem Solving

Solve each problem. Show your work.

13. A ladder is 17 feet long. It reaches 15 feet up the side of a building. How far from the base of the building is the bottom of the ladder? 8 ft

14. Sue walks from her house to school. She walks 6 blocks south, then she walks 8 blocks west. What is the shortest distance from Sue's house to school? 10 blocks

Write About Math

Can a right triangle be equilateral? Explain why or why not.

Additional Practice for this chapter is provided on page 398 of the student book.

Chapter 7

Quadrilaterals and Polygons

These kites are shapes called quadrilaterals. What do you notice about the kite frame? How do you think it affects the shape of the kite?

Learning Objectives

- Classify polygons by the number of sides.
- Recognize a parallelogram, a rectangle, a square, a rhombus, a trapezoid, and an isosceles trapezoid.
- Use the properties of parallelograms to find lengths of sides and angle measures.
- Use the properties of diagonals of parallelograms to find lengths and angle measures.
- Find the length of a midsegment of a trapezoid.
- Find the measures of interior angles and exterior angles of polygons.
- Write proofs about the properties of parallelograms.

ESL/ELL Note Explain that in English the prefixes of the names of polygons often indicate the number of sides. For example, *tri* for triangle, *quad* for quadrilateral, *penta* for pentagon, and so on. Have students write the name of each polygon and circle the prefix.

Words to Know

polygon	a closed geometric figure made up of at least three line segments, or sides
parallelogram	a special quadrilateral with both pairs of opposite sides parallel
rectangle	a parallelogram with four right angles
rhombus	a parallelogram with four congruent sides
square	a parallelogram with four right angles and four congruent sides
diagonal	a line segment that connects one vertex to an opposite vertex
trapezoid	a quadrilateral with only one pair of parallel sides
isosceles trapezoid	a special trapezoid with congruent legs; each pair of base angles is congruent
regular polygon	a polygon that has all congruent sides and all congruent angles

Triangle Project

Draw four congruent isosceles right triangles on a sheet of construction paper. Cut out each triangle. Put two or four triangles together to form a square, a rectangle, a parallelogram, and a trapezoid. Then, use three triangles to form a polygon that is not a rectangle, square, parallelogram, or trapezoid.

Project Students can begin work on the project after completing Lesson 7.5. See the Classroom Resource Binder for a scoring rubric to assess this project.

Possible answers:

hexagon

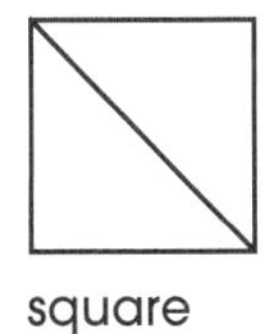

square

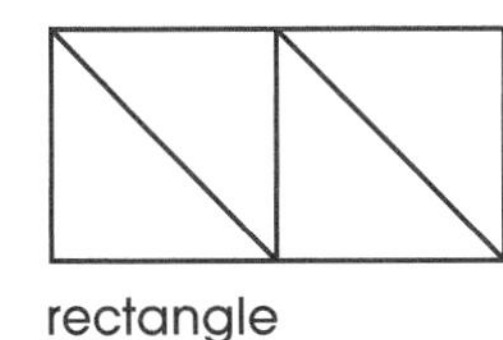

rectangle

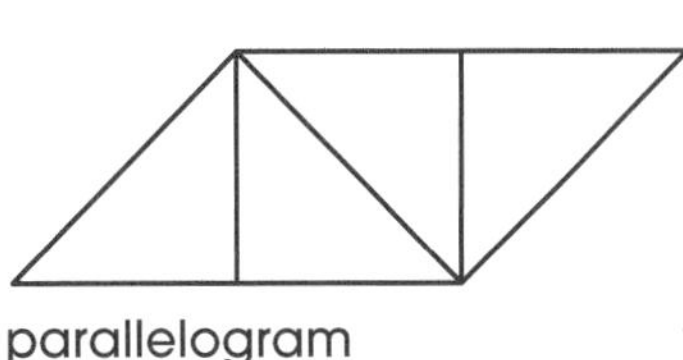

parallelogram

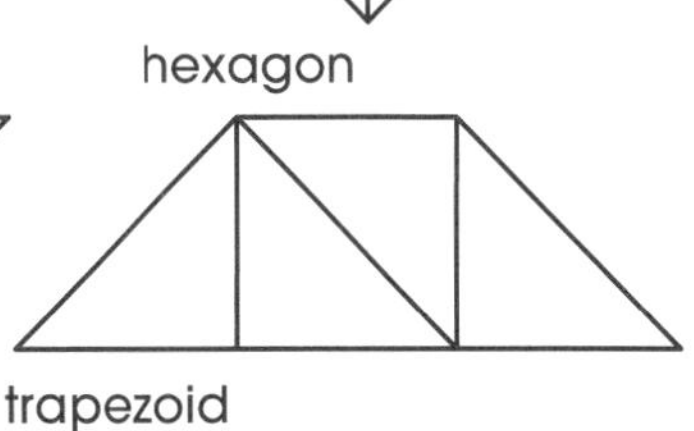

trapezoid

More practice is provided in the Workbook, page 37, and in the Classroom Resource Binder, page 68.

7·1 Polygons

Getting Started Have students list and draw all the types of polygons they know.

A **polygon** is a closed geometric figure made up of at least three line segments, or sides. Each side is connected to another side at an endpoint. These endpoints are called vertices.

Hexagon *KACEGI*

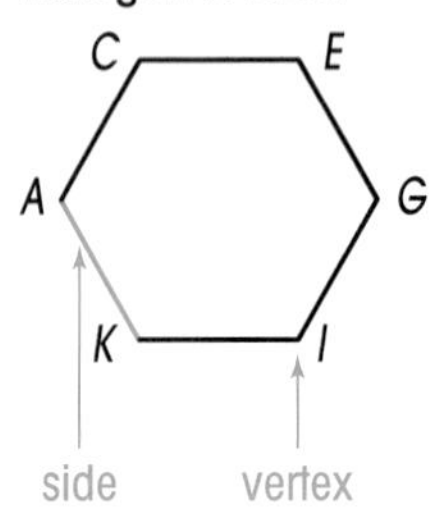

When naming a polygon, always list the vertices in consecutive order. Consecutive vertices are next to each other. The polygon on the left could be named polygon *KACEGI*, polygon *ECAKIG*, and so on.

The chart below shows the basic types of polygons.

Type of Polygon	Number of Sides
Triangle	3
Quadrilateral	4
Pentagon	5
Hexagon	6
Heptagon	7
Octagon	8
Nonagon	9
Decagon	10
n-gon	*n*

Math Fact
A polygon with *n* number of sides is an *n*-gon. For example, a 15-gon has 15 sides.

The chart above will help you classify polygons.

EXAMPLE

Classify the polygon on the right. Then, name it.

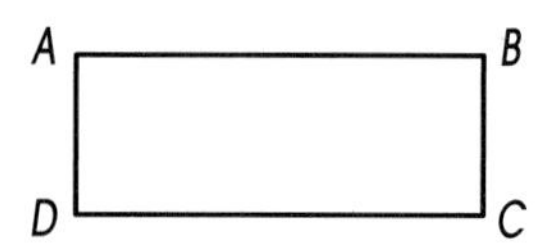

Count the number of sides. 4

Find 4 in the chart above. A quadrilateral has 4 sides.

Name the quadrilateral. quadrilateral *ABCD*

The polygon could be called quadrilateral *ABCD*.

Try This

Draw pentagon *OPQRS*.

Find the number of sides in a pentagon. ■ sides 5

Draw a ■-sided figure and label it. 5

A possible answer is given for the labeled drawing.

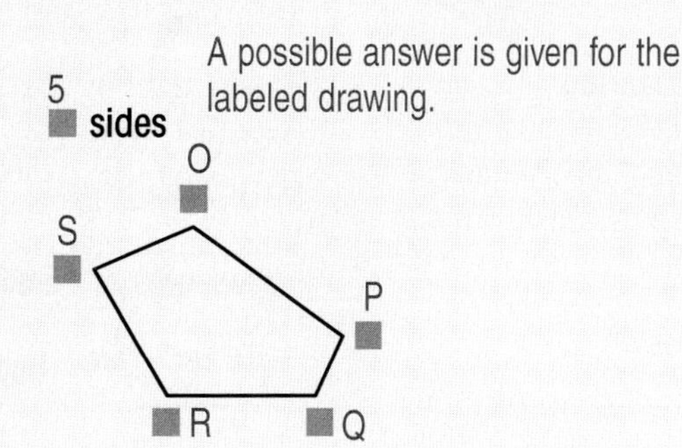

Practice

Classify each polygon. Then, name it.

1.

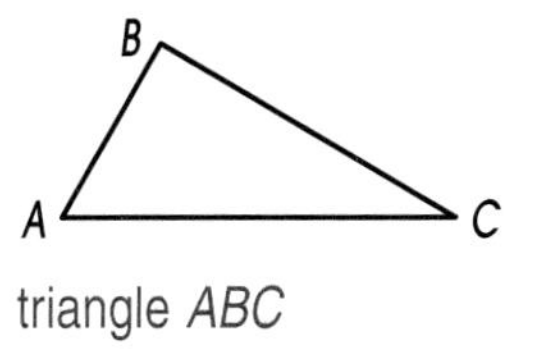

triangle *ABC*

2.

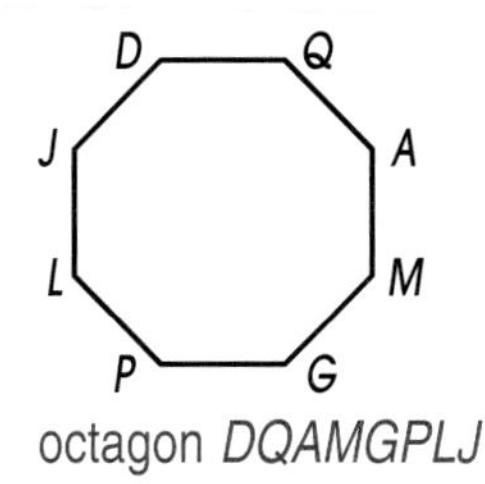

octagon *DQAMGPLJ*

3.

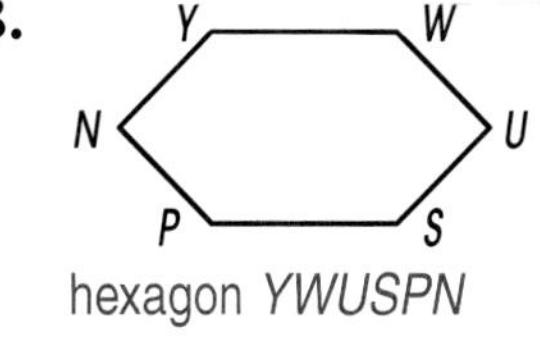

hexagon *YWUSPN*

Draw and label each polygon. Check students' work.

4. pentagon *ABCDE*

5. heptagon *FGHIJKL*

6. quadrilateral *MNOP*

7. octagon *QRSTUVWX*

8. triangle *XYZ*

9. hexagon *FEDCBA*

Share Your Understanding

10. Draw a polygon. Have a partner classify the polygon. Then, have your partner draw a polygon for you to classify. Check students' work.

11. **CRITICAL THINKING** Draw a pentagon. Draw as many lines from one vertex to its opposite vertices as possible. How many triangles did you make? 3 triangles

More practice is provided in the Workbook, page 37, and in the Classroom Resource Binder, page 68.

7·2 Parallelograms

Getting Started
Point out that a figure may have more than one name. For example a parallelogram is also a polygon.

Math Fact

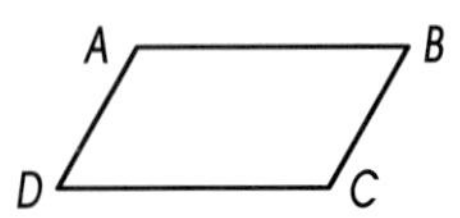

Parallelogram *ABCD* can also be written as ▱*ABCD*.

A **parallelogram** is a quadrilateral with opposite sides parallel. The chart below shows the properties of a parallelogram.

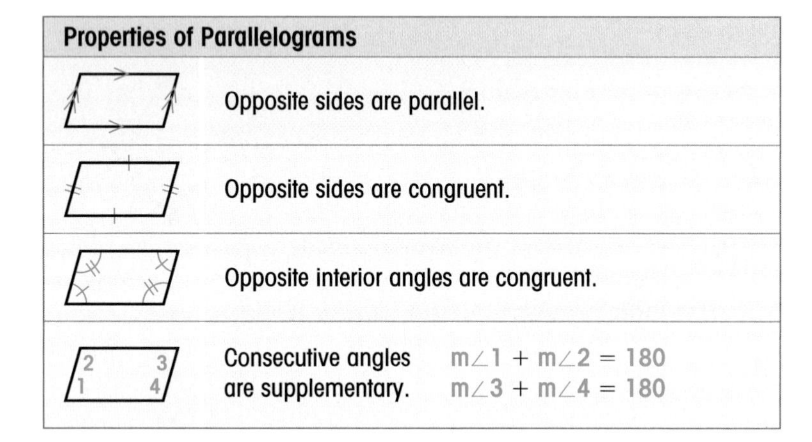

Properties of Parallelograms		
	Opposite sides are parallel.	
	Opposite sides are congruent.	
	Opposite interior angles are congruent.	
2 3 1 4	Consecutive angles are supplementary.	$m\angle 1 + m\angle 2 = 180$ $m\angle 3 + m\angle 4 = 180$

The chart above can help you find angle measures.

EXAMPLE

Find the measures of $\angle C$, $\angle D$, and $\angle B$ in ▱$ABCD$.

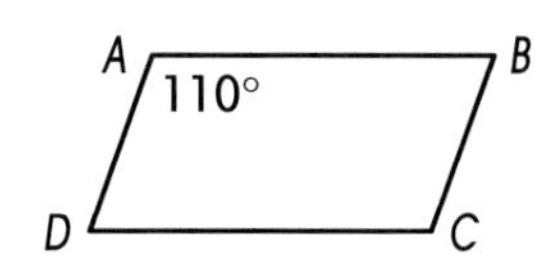

Opposite angles are congruent in parallelograms.

$$\angle A \cong \angle C$$
$$m\angle A = m\angle C$$
$$110 = m\angle C$$

Consecutive angles in a parallelogram are supplementary.

$$m\angle A + m\angle D = 180$$
$$110 + m\angle D = 180$$
$$m\angle D = 70$$

Opposite angles are congruent in parallelograms.

$$\angle D \cong \angle B$$
$$m\angle D = m\angle B$$
$$70 = m\angle B$$

So, $\angle C$ is 110°, $\angle D$ is 70°, and $\angle B$ is 70°.

Try This

Find the length of $\overline{EG}$ in $\square EGIK$.

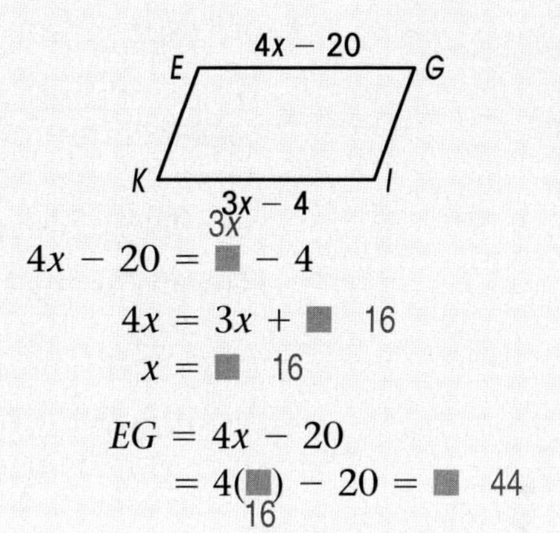

Opposite sides of a parallelogram are ■. So, $\overline{EG} \cong$ ■. Write an equation. congruent $\overline{KI}$

$4x - 20 =$ ■ $- 4$ 3x

Solve for x.

$4x = 3x +$ ■ 16

$x =$ ■ 16

Substitute 16 for x to find EG.

$EG = 4x - 20$

$= 4($■$) - 20 =$ ■ 44 (16)

The length of $\overline{EG}$ is ■. 44

Practice

Find the length of each side and the measure of each angle.

F, H, 42 m, 106°, M, 36 m, J

1. $\overline{FM}$ 42 m **2.** $\overline{FH}$ 36 m **3.** $\overline{HJ}$ 42 m

4. $\angle F$ 106° **5.** $\angle H$ 74° **6.** $\angle M$ 74°

Find the length of $\overline{JK}$ in each diagram.

7. K, L, 3x − 14, 58, J, M — 58

8.
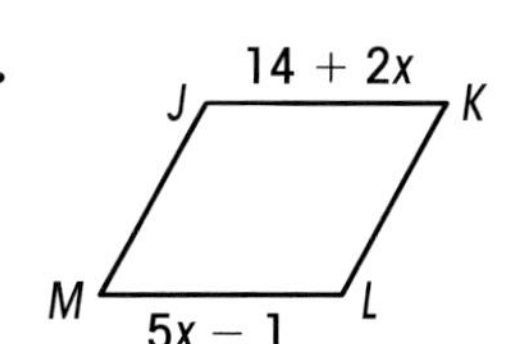

24

9.
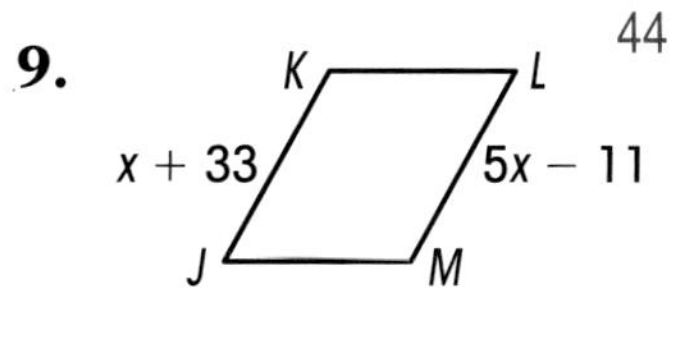

44

10. Opposite sides are parallel. Opposite sides are congruent. Opposite interior angles are congruent. Consecutive angles are supplementary.

Share Your Understanding

10. Write about the four properties of a parallelogram. Use the words *opposite, parallel, congruent,* and *supplementary* in your writing. See above.

11. **CRITICAL THINKING** Work with a partner to find the sum of the interior angles of a parallelogram. (Hint: Use the property that states two consecutive angles are supplementary.) 360°

More practice is provided in the Workbook, page 38, and in the Classroom Resource Binder, page 69.

7-3 Special Parallelograms: Rectangle, Square, and Rhombus

Getting Started
Students can make a parallelogram from thin strips of cardboard and paper fasteners. Have them move the sides to form 90° angles. Introduce the word *rectangle*.

Math Fact
The diagram below shows how special parallelograms are related.

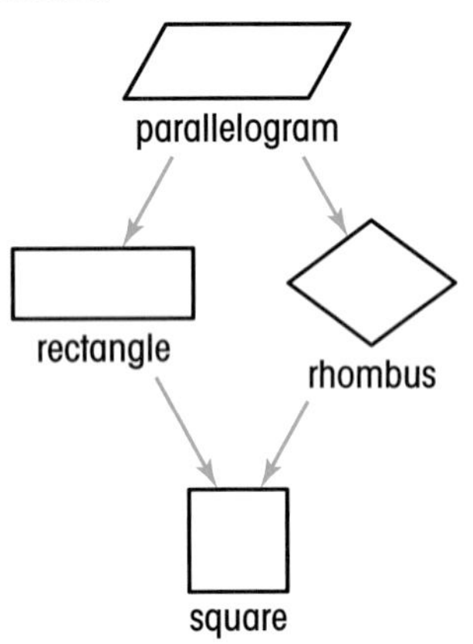

There are three special parallelograms. A **rectangle** is a parallelogram with four right angles. A **rhombus** is a parallelogram with four congruent sides. A **square** is a parallelogram with four right angles and four congruent sides.

The chart below lists the special properties of a rectangle, a rhombus, and a square. These figures also have the properties of a parallelogram.

Special Parallelograms		
Rectangle		All four angles are right angles.
Rhombus		All four sides are congruent.
Square		All four sides are congruent. All four angles are right angles.

Use the chart above to classify parallelograms.

EXAMPLE

▱*JKLM* is a rectangle.
Find the length of $\overline{JM}$.

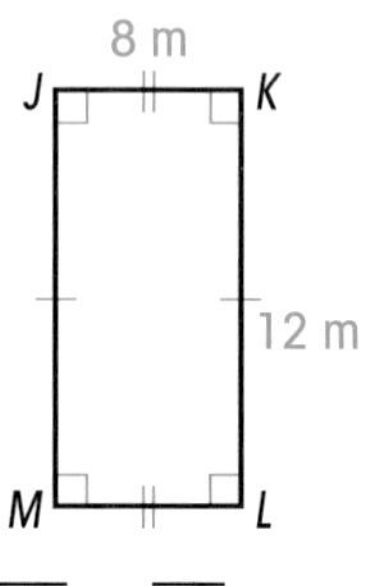

A rectangle is also a parallelogram. The opposite sides are congruent.
$\overline{JM} \cong \overline{KL}$
$JM = KL$

Substitute 12 for *KL*.
$JM = 12$

The length of $\overline{JM}$ is 12 meters.

Try This

▱$ABCD$ is a rhombus.
Find the measure of $\angle D$.

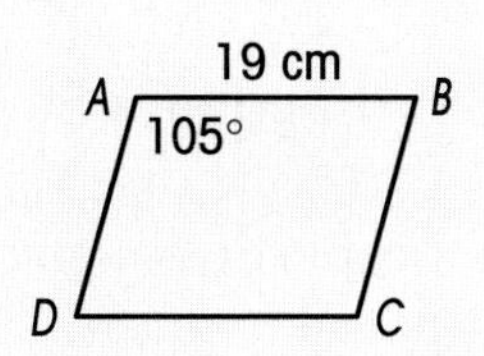

A rhombus is also a ■. parallelogram
Consecutive angles are supplementary.
Write an equation. $m\angle A + m\angle D =$ ■ 180
Substitute 105 for ■. $m\angle A$ — 105 ■ $+ m\angle D = 180$
Solve for $m\angle D$. $m\angle D =$ ■ 75

So, $\angle D$ is ■. 75°

Practice

▱$DFHJ$ is a rectangle. Find the length of each side and the measure of each angle.

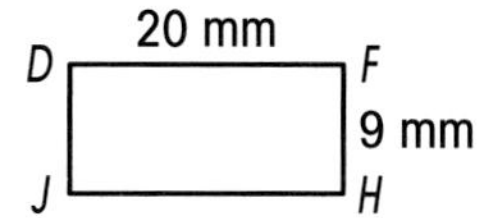

1. $\overline{JH}$ 20 mm **2.** $\overline{DJ}$ 9 mm **3.** $\angle H$ 90°

▱$LNPQ$ is a rhombus. Find the length of each side and the measure of each angle.

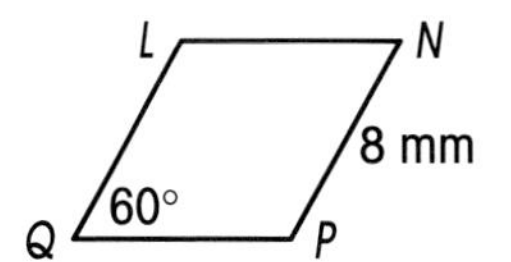

4. $\overline{LN}$ 8 mm **5.** $\overline{QL}$ 8 mm **6.** $\overline{QP}$ 8 mm

7. $\angle N$ 60° **8.** $\angle L$ 120° **9.** $\angle P$ 120°

10. Possible answer: A rectangle has four right angles. A parallelogram has consecutive angles that are supplementary. A rhombus has four congruent sides. A rectangle has four congruent angles.

Share Your Understanding

10. Explain to a partner the differences between a rectangle and a parallelogram. Have your partner explain the differences between a rectangle and a rhombus. See above.

11. **CRITICAL THINKING** Complete the hypothesis for the following conditional statement. If a parallelogram ________________, then it is a square. has four right angles and four congruent sides

More practice is provided in the Workbook, page 38, and in the Classroom Resource Binder, page 70.

Diagonals of Parallelograms

Getting Started
Have students review the similarities among quadrilaterals.

A **diagonal** of a parallelogram is a line segment that connects one vertex to an opposite vertex. The properties of the diagonals depend on the shape of the parallelogram.

Figure	Properties of Diagonals
Parallelogram	Diagonals bisect each other.
Rectangle	Diagonals bisect each other. Diagonals are congruent.
Rhombus	Diagonals bisect each other. Diagonals are perpendicular. Diagonals bisect each pair of opposite angles.
Square	Diagonals bisect each other. Diagonals are congruent. Diagonals are perpendicular. Diagonals bisect each pair of opposite angles.

You can use the properties of diagonals to solve problems.

EXAMPLE

In ▱$JCNR$, JN is 20 cm and RA is 11 cm.

Find the length of $\overline{AN}$ and $\overline{RC}$.

The diagonals in a parallelogram bisect each other.

Divide JN by 2 to find AN.

$$AN = JN \div 2 = 20 \div 2 = 10$$

Multiply RA by 2 to find RC.

$$RC = RA \cdot 2 = 11 \cdot 2 = 22$$

So, AN is 10 cm and RC is 22 cm.

Try This

▱$PQKN$ is a rhombus. $\angle KNP$ is 60°. Find the measure of $m\angle KNQ$.

In a rhombus, the diagonals ■ the angles. bisect

$m\angle KNQ = 60 \div$ ■ 2

$=$ ■ 30

So, $\angle KNQ$ is ■. 30°

Practice

▱$EFHG$ is a rectangle. EA is 8 in. Find the length of each line segment and the measure of each angle.

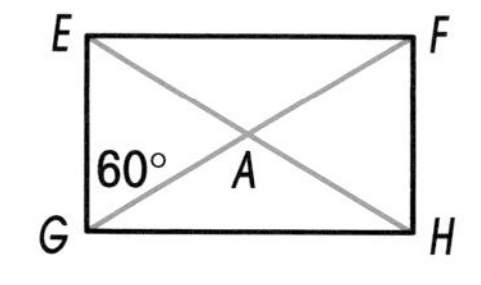

1. $\overline{EH}$ 16 in. **2.** $\overline{FG}$ 16 in. **3.** $\overline{GA}$ 8 in.

4. $\angle EGH$ 90° **5.** $\angle GFH$ 60° **6.** $\angle EFG$ 30°

▱$ACDG$ is a square. AD is 18 mm. Find the length of each line segment and the measure of each angle.

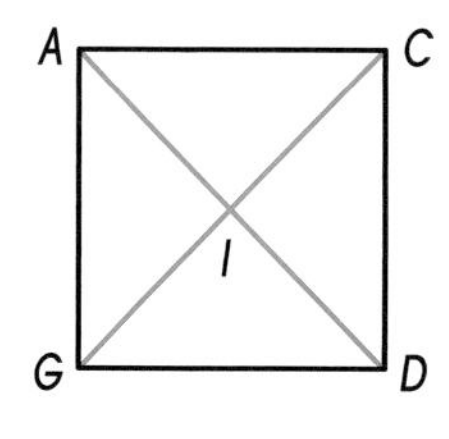

7. $\overline{ID}$ 9 mm **8.** $\overline{CG}$ 18 mm **9.** $\overline{CI}$ 9 mm

10. $\angle AIG$ 90° **11.** $\angle AGI$ 45° **12.** $\angle AGD$ 90°

Share Your Understanding

13. Draw a diagram of a parallelogram, a rectangle, a rhombus, and a square. Have a partner draw the diagonals for each figure. Tell how the figures are the same and how they are different. Check students' work.

14. **CRITICAL THINKING** For which kinds of parallelograms are the diagonals perpendicular? What do you think these parallelograms have in common?
rhombus and square; They both have four congruent sides.

More practice is provided in the Workbook, page 39, and in the Classroom Resource Binder, page 70.

7·5 Trapezoids

Getting Started Review the midsegment of a triangle.

THEOREM 21
Trapezoid Midsegment Theorem

The length of the midsegment of a trapezoid is half the sum of the lengths of both bases.

A **trapezoid** is a quadrilateral with only one pair of parallel sides. The two parallel sides are called the bases. The legs are the two sides that are not parallel.

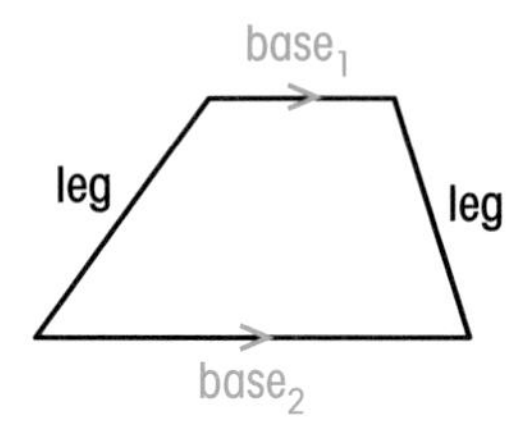

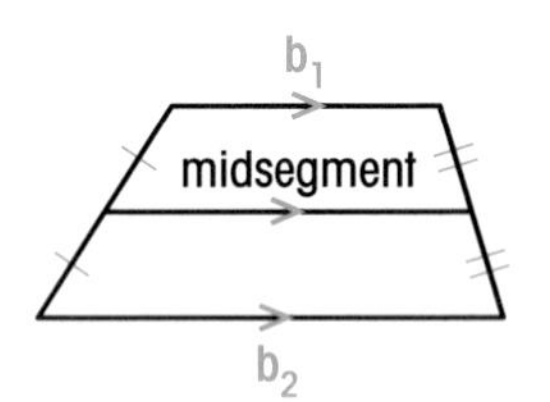

The midsegment of a trapezoid is a line segment that joins the midpoints of the legs. The midsegment is parallel to both bases. The length of the midsegment is half the sum of the lengths of the bases.

$$\text{midsegment} = \frac{1}{2}(b_1 + b_2)$$

You can find the length of the midsegment if you know the length of each base.

EXAMPLE

Quadrilateral *KMTU* is a trapezoid. Find the length of the midsegment $\overline{XP}$.

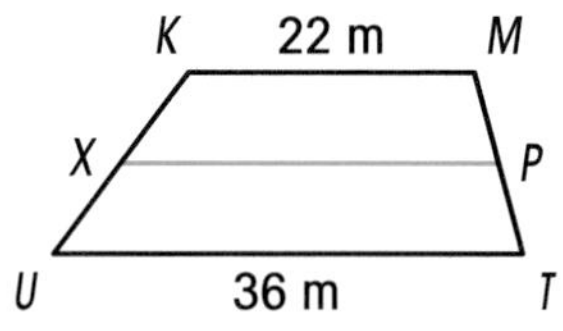

Write the equation for the length of the midsegment. $\text{midsegment} = \frac{1}{2}(b_1 + b_2)$

Substitute the given values. $XP = \frac{1}{2}(22 + 36)$

Simplify. $XP = \frac{1}{2}(58)$

$XP = 29$

The length of the midsegment $\overline{XP}$ is 29 meters.

You can find the length of a base if you know the length of the midsegment and the length of the other base.

Try This

Find the length of b_1 in the trapezoid on the right.

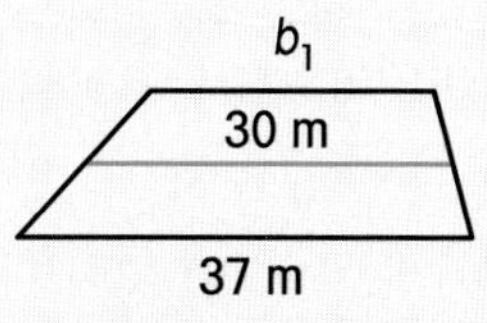

Write the equation for the length of the ■. midsegment

$$\text{midsegment} = \frac{1}{2}(b_1 + b_2)$$

Substitute the given values.

$$30 ■ = \frac{1}{2}(b_1 + 37)$$

Solve for b_1. First, multiply by 2.

$$30 \cdot 2 = \frac{1}{2} \cdot 2\ (b_1 + 37)$$

$$60 ■ = b_1 + 37$$

$$23 ■ = b_1$$

The length of b_1 is ■. 23 m

Practice

Find the length of the midsegment in each trapezoid.

1. 24 m

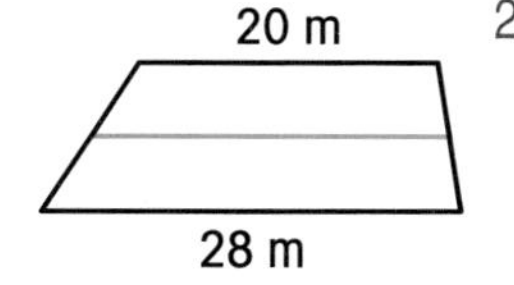

2. 42 cm

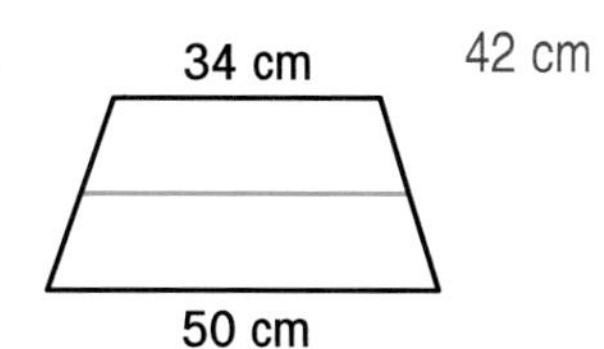

3. 23 mm

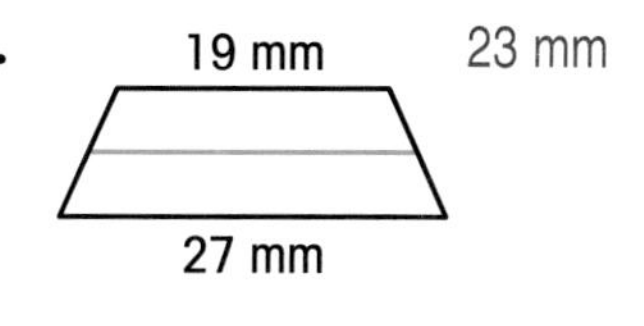

Find the unknown base of each trapezoid.

4. 15 in.

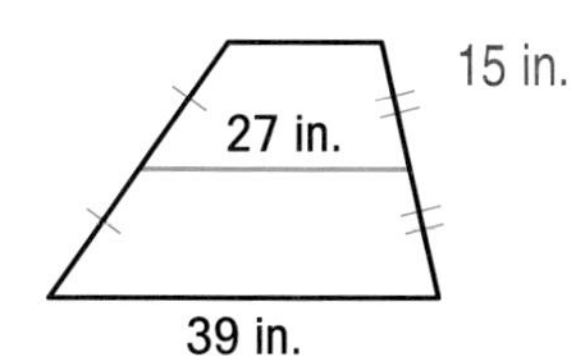

5. 40 m

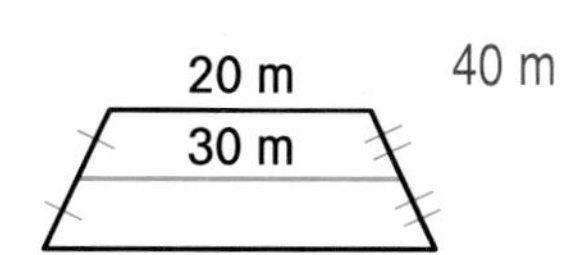

6. 32 ft

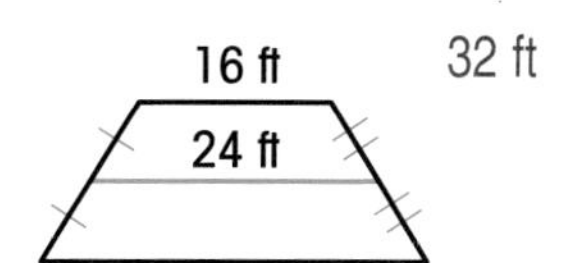

Share Your Understanding

7. Explain to a partner how to find the length of the midsegment in a trapezoid. Use the words *sum* and *lengths*. Find half of the sum of the lengths of the bases.

8. **CRITICAL THINKING** Compare finding the midsegment of a triangle to finding the midsegment of a trapezoid. A triangle has one base. The midsegment is half of this base. A trapezoid has two bases. The midsegment is half of the sum of these bases.

More practice is provided in the Workbook, page 39, and in the Classroom Resource Binder, page 71.

7·6 Isosceles Trapezoids

Getting Started
Review the properties of an isosceles triangle. Have students cut out a large isosceles triangle from a sheet of cardboard. Then, have them cut off a piece from the top of the triangle to make an isosceles trapezoid. Have them measure the legs and the angles and share their findings.

You learned that an isosceles triangle has two congruent sides and two congruent angles.

An **isosceles trapezoid** has congruent legs and each pair of base angles is congruent.

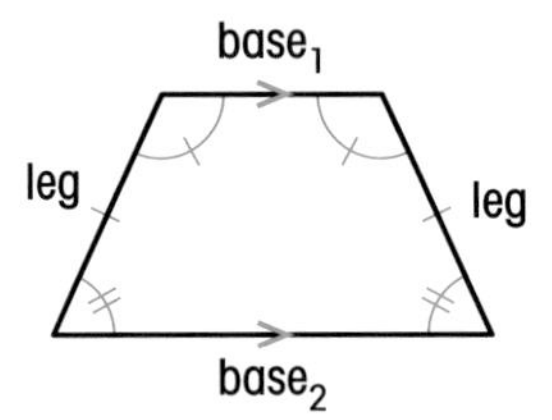

You can use what you know about an isosceles trapezoid to find the measure of a leg or an angle.

EXAMPLE 1

Trapezoid $ABCD$ is isosceles. Find BC.

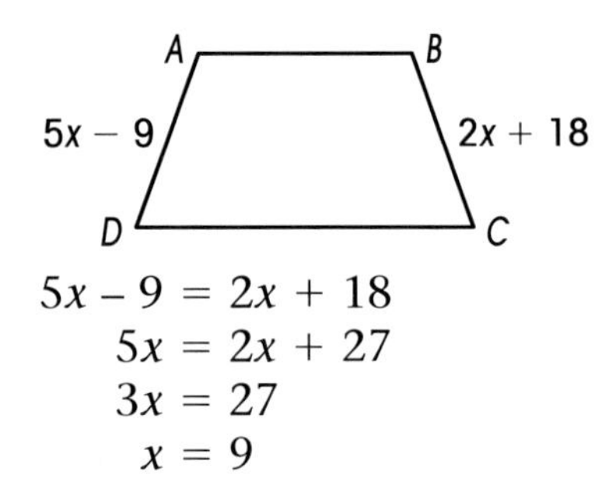

In an isosceles trapezoid, the legs are congruent. Write an equation. Solve for x.

$$5x - 9 = 2x + 18$$
$$5x = 2x + 27$$
$$3x = 27$$
$$x = 9$$

Substitute 9 for x to find BC.

$$BC = 2(9) + 18 = 36$$

So, BC is 36.

EXAMPLE 2

Trapezoid $EGID$ is isosceles. Find $m\angle G$.

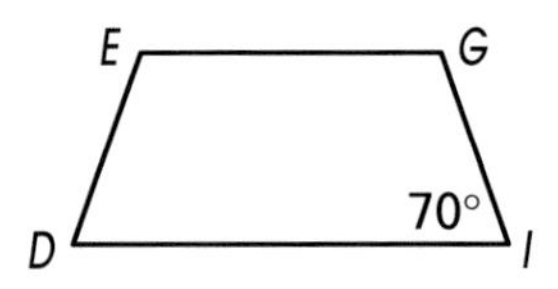

$\angle I$ and $\angle G$ are same-side interior angles. They are supplementary. Write an equation.

$$m\angle I + m\angle G = 180$$
$$70 + m\angle G = 180$$
$$m\angle G = 110$$

So, $\angle G$ is 110°.

Try This

Trapezoid *KTLP* is isosceles.
Find the measure of $\angle L$.

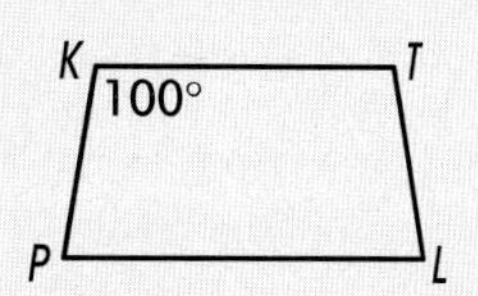

In an isosceles trapezoid, each pair of base angles is ■. congruent

$\angle K \cong \angle T$

$\angle L$ and $\angle T$ are ■. supplementary Write an equation.

Substitute ■ 100 for m$\angle T$. Then, solve for m$\angle L$.

$m\angle K = m\angle T =$ ■ 100

$m\angle L + m\angle T =$ ■ 180

$m\angle L + 100 =$ ■ 180

$m\angle L =$ ■ 80

So, $\angle L$ is ■. 80°

Practice

Trapezoid *SPLM* is isosceles. Complete each congruence statement.

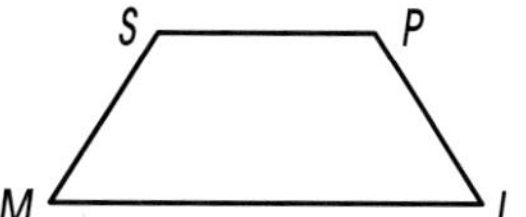

1. $\overline{PL} \cong$ ■ $\overline{SM}$

2. $\overline{SM} \cong$ ■ $\overline{PL}$

3. $\angle P \cong$ ■ $\angle S$

4. $\angle S \cong$ ■ $\angle P$

5. $\angle M \cong$ ■ $\angle L$

6. $\angle L \cong$ ■ $\angle M$

Trapezoid *HIJK* is isosceles. Find the length of each side and the measure of each angle.

H I 105° $5x - 3$ $2x + 6$ K J

7. $\overline{IJ}$ 12

8. $\overline{HK}$ 12

9. $\angle I$ 105°

10. $\angle H$ 105°

11. $\angle K$ 75°

12. $\angle J$ 75°

Share Your Understanding

13. Explain to a partner the difference between a trapezoid and an isosceles trapezoid. Use the word *congruent* in your explanation. A trapezoid has one pair of parallel sides. An isosceles trapezoid also has two congruent legs and each pair of base angles is congruent.

14. **CRITICAL THINKING** The height of a trapezoid is the perpendicular distance between the two bases. Find the height of this trapezoid. (Hint: Use the right triangle.) 4

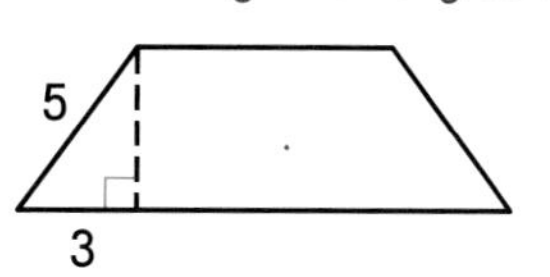

7·7 Calculator: Exterior Angles of Regular Polygons

Getting Started Have students draw the exterior angles of a triangle.

> THEOREM 22
> **Exterior-Angle Sum Theorem for Polygons**
>
> The sum of the measures of the exterior angles of a polygon is 360°.

An exterior angle of a polygon is formed by extending one side of the polygon. The sum of the measures of all of the exterior angles of any polygon is 360°.

A **regular polygon** has all congruent sides. All interior angles are congruent. All exterior angles are congruent.

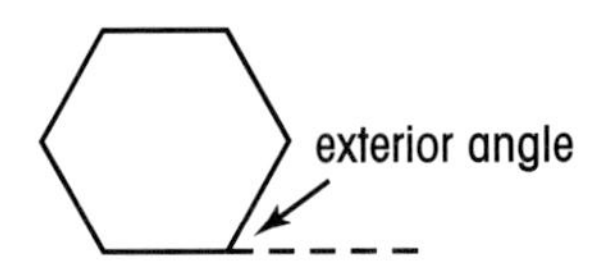

To find the measure of one exterior angle, divide 360 by the number of sides of the polygon.

EXAMPLE 1

Find the measure of an exterior angle of a regular octagon.

An octagon has 8 sides.
Divide 360 by 8. $360 \div 8$

Use your calculator to divide. **Display**

PRESS: 3 6 0 ÷ 8 = 45

An exterior angle of a regular octagon is 45°.

You can find the number of sides of a regular polygon if you know the measure of one exterior angle.

EXAMPLE 2

Name the regular polygon that has a 90° exterior angle.

Divide 360 by the measure of an exterior angle. $360 \div 90$

Use your calculator to divide. **Display**

PRESS: 3 6 0 ÷ 9 0 = 4

The regular polygon has 4 sides. It is a regular quadrilateral, or square.

Practice

Find the measure of an exterior angle for each regular polygon.

1. pentagon 72°

2. hexagon 60°

3. nonagon 40°

4. heptagon about 51°

5. decagon 36°

6. 15-gon 24°

Name the regular polygon that has each exterior angle.

7. 30° 12-gon

8. 36° decagon

9. 45° octagon

10. 60° hexagon

11. 40° nonagon

12. 20° 18-gon

Math Connection

KITES

This kite is a quadrilateral. The diagonals of the kite are perpendicular. The longer diagonal bisects the shorter diagonal and bisects the angles.

A kite can be made to fly in the wind. It needs to be made of lightweight material and include a tail. The tail helps to balance the kite.

In his famous experiment of 1752, Benjamin Franklin hung a metal key from a kite line. The key attracted electricity from the air during a storm. This experiment demonstrated the electrical nature of lightning.

Today, kites come in many different shapes. You can make a kite. Look for a book on kite making in your school or local library. The perfect wind to fly kites has a speed between 8 and 20 miles per hour.

More practice is provided in the Workbook, page 40, and in the Classroom Resource Binder, page 72.

7·8 Problem-Solving Skill: Interior-Angle Sum of a Polygon

Getting Started Have students review the sum of the interior angles of a triangle.

> THEOREM 23
> **Interior-Angle Sum Theorem for Polygons**
>
> The sum of the measures of the interior angles of a polygon with n sides is $(n - 2) \cdot 180$.

An interior angle of a polygon is formed by two consecutive sides. You can find the sum of the measures of interior angles of any polygon.

First, divide the polygon into triangles. Then, multiply the number of triangles by 180. You can do this for any polygon. Look at the chart below.

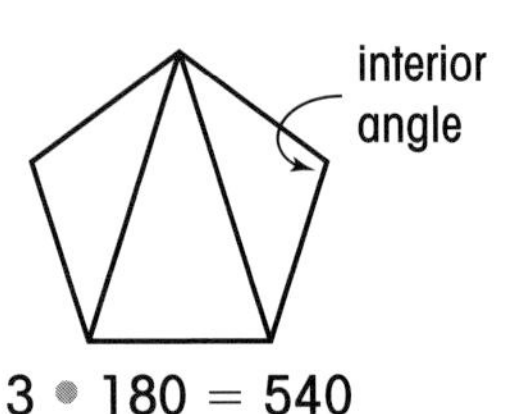

$3 \cdot 180 = 540$

	Pentagon	Hexagon	Heptagon
Number of Sides (n)	5	6	7
Number of Triangles ($n - 2$)	3	4	5
Interior-Angle Sum $(n - 2) \cdot 180$	540	720	900

You can also use a formula to find the interior-angle sum of a polygon.

$$\text{interior-angle sum} = (n - 2) \cdot 180$$

EXAMPLE 1

Remember
An octagon has 8 sides.

Find the interior-angle sum of an octagon.

Write the formula.	interior-angle sum $= (n - 2) \cdot 180$
Substitute 8 for n.	$= (8 - 2) \cdot 180$
Simplify.	$= 1{,}080$

The interior-angle sum of an octagon is 1,080°.

If you know that the polygon is a regular polygon, you can find the measure of an interior angle.

EXAMPLE 2

Find the measure of an interior angle of a regular octagon.

An octagon has 8 sides. Divide the interior-angle sum of an octagon by 8. $1{,}080 \div 8 = 135$

An interior angle of a regular octagon is 135°.

Try This

A polygon has an interior-angle sum of 540°.
Find the number of sides. Name the polygon.

Write the interior-angle sum formula.	interior-angle sum $= (n - 2) \cdot 180$
Substitute the given value.	540 ■ $= (n - 2) \cdot 180$
Solve for n. Divide both sides by 180.	3 ■ $= n - 2$
Add 2 to both sides.	5 ■ $= n$

The polygon has ■ 5 sides. It is a ■. pentagon

Practice

Find the interior-angle sum of each polygon.

1. hexagon 720°
2. octagon 1,080°
3. 11-gon 1,620°
4. quadrilateral 360°
5. nonagon 1,260°
6. heptagon 900°
7. decagon 1,440°
8. pentagon 540°
9. 12-gon 1,800°

Find the measure of an interior angle of each regular polygon.

10. hexagon 120°
11. decagon 144°
12. nonagon 140°

Name the polygon that has each interior-angle sum.

13. 720° hexagon
14. 1,440° decagon
15. 1,080° octagon

More practice is provided in the Workbook, page 41, and in the Classroom Resource Binder, page 73.

7-9 Problem-Solving Application: Tiling a Surface

Getting Started
Have students cut out several identical shapes in different color paper. Tell them to form patterns with their shapes.

In some homes, floors and walls are covered with tiles set in a pattern. The tiles are set without overlaps or gaps. This is a tessellation.

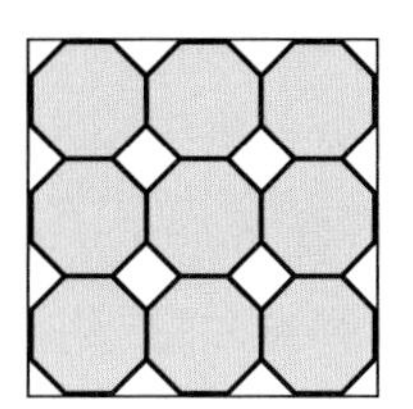

If you know the pattern, then you can solve problems about tiling.

EXAMPLE

Maria wants to tile her bathroom floor in the pattern shown. Each small tile has 3-in. sides. Each large tile has 6-in. sides. The bathroom is 6 ft by 8 ft. How many large tiles and how many small tiles will Maria need?

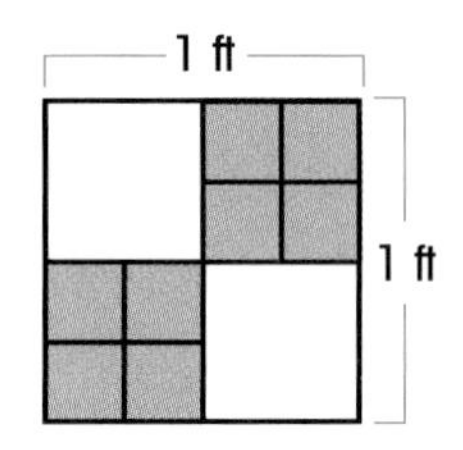

Look at the drawing. Count the number of small tiles and large tiles in 1 square foot.	2 large tiles 8 small tiles
The bathroom is 6 ft by 8 ft.	
Multiply to find the number of each type of tile in one 6-ft row of the pattern.	$6 \cdot 2 = 12$ large tiles $6 \cdot 8 = 48$ small tiles
Multiply to find the number of each type of tile in 8 rows.	$8 \cdot 12 = 96$ large tiles $8 \cdot 48 = 384$ small tiles

Maria needs 96 large tiles and 384 small tiles.

Try This

Steve has $600 to spend on bricks for his walkway. Each brick costs $1.20. They are 6 in. long and 3 in. wide. Steve's walkway is 22 ft by 3 ft. Does he have enough money to make this pattern?

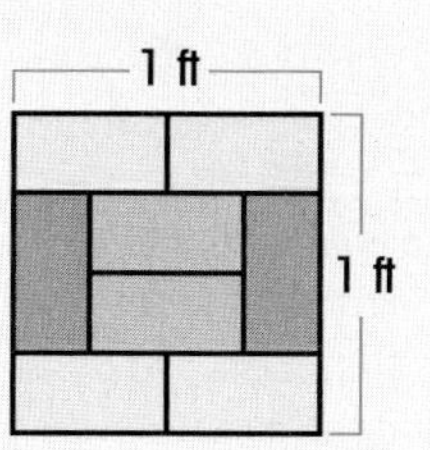

Find the number of bricks in 1 ■ foot. square — 8 ■ bricks

The walkway is 22 ft by ■. 3 ft

■ to find the number of bricks in one 22-ft row of the pattern. Multiply — $22 \cdot 8 =$ ■ bricks 176

Multiply to find the number of bricks in 3 rows. — $3 \cdot$ ■ $=$ ■ bricks 176 528

Multiply to find the cost. — $(1.20)(528) =$ ■ 633.60

Compare Steve's budget with the cost. — $600 < ■ $633.60

Steve ■ enough money to complete the walkway with this pattern. does not have

Practice

Use the drawing on the right to solve each problem.

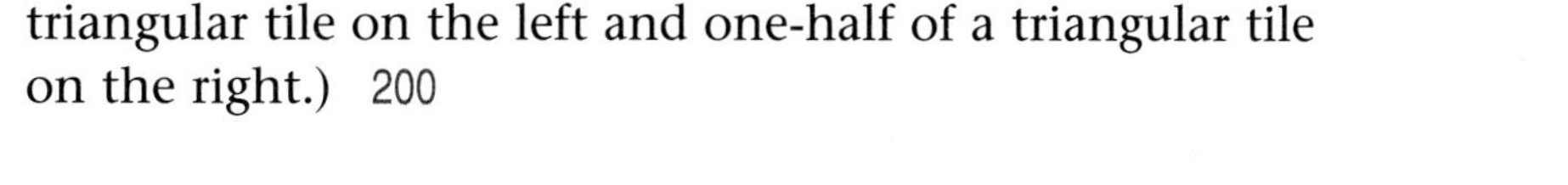

1. Chris used this pattern to make a 6-ft border. He used 100 square tiles in all. How many triangular tiles did he use? (Hint: There is one-half of a triangular tile on the left and one-half of a triangular tile on the right.) 200

2. Square tiles cost $0.55 each and triangular tiles cost $0.25 each. How much will it cost Chris to make the border? $105.00

3. Chris liked the border. So, he used 600 triangular tiles and 300 square tiles for his patio. How much did it cost to tile the patio? $315.00

An alternate two-column proof lesson is provided on page 412 of the student book.

Proof: Proving a Quadrilateral Is a Parallelogram

Getting Started Review the properties of a parallelogram.

You learned that if a quadrilateral is a parallelogram, the diagonals bisect each other. Suppose you have a quadrilateral and you know that the diagonals bisect each other. Is this quadrilateral a parallelogram? This is a theorem you can prove.

> **THEOREM 24**
>
> If a quadrilateral has diagonals that bisect each other, then the quadrilateral is a parallelogram.

EXAMPLE

You are given: quadrilateral *ABCD*
$\overline{AC}$ bisects $\overline{BD}$.
$\overline{BD}$ bisects $\overline{AC}$.

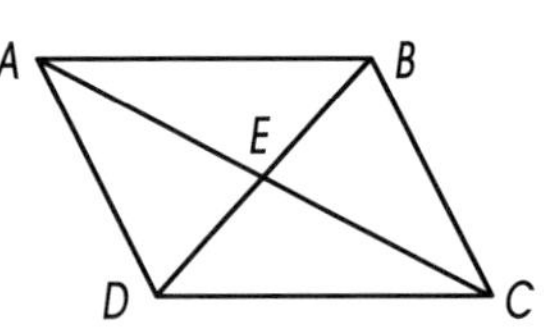

Prove: Quadrilateral *ABCD* is a parallelogram.

Copy the diagram. As you read the proof, mark off the congruent parts you find. Then, choose the postulate you can use to show that the triangles are congruent.

You are given that the diagonals bisect each other. This means that $\overline{AE} \cong \overline{CE}$ and that $\overline{BE} \cong \overline{DE}$. You also know that $\angle BEA \cong \angle DEC$ because the angles are vertical angles. Then, $\triangle BEA \cong \triangle DEC$ by SAS. Because these triangles are congruent, $\angle BAE \cong \angle DCE$. These are alternate interior angles for $\overline{AB}$ and $\overline{CD}$. So, $\overline{AB} \parallel \overline{CD}$.

You can also show that $\triangle BEC \cong \triangle DEA$ by SAS. Because these triangles are congruent, $\angle EBC \cong \angle EDA$. These are alternate interior angles for $\overline{BC}$ and $\overline{DA}$. So, $\overline{BC} \parallel \overline{DA}$.

Because the opposite sides of quadrilateral ABCD are parallel, the quadrilateral is a parallelogram.

So, you prove that quadrilateral *ABCD* is a parallelogram. ✓

You can prove that if a quadrilateral has the properties of a parallelogram, then it is a parallelogram.

Try This

Copy and complete the proof.

You are given: quadrilateral *HIJK*

$\angle H \cong \angle J$

$\angle K \cong \angle I$

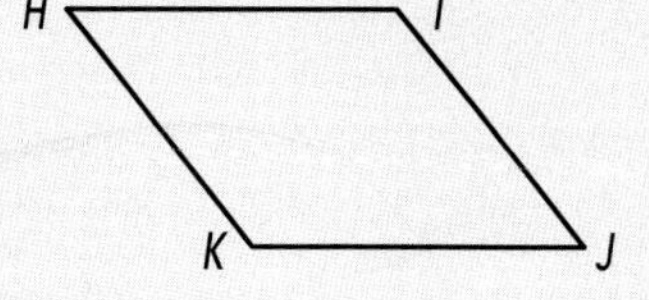

Prove: Quadrilateral *HIJK* is a parallelogram.

You are given $\angle H \cong$ ■ [$\angle J$] and $\angle K \cong \angle I$. This means that $m\angle H =$ ■ [$m\angle J$] and $m\angle K =$ ■ [$m\angle I$]. You know that the interior-angle sum of a quadrilateral is ■ [360°]. Then, $2m\angle H + 2m\angle K = 360$ or $m\angle H + m\angle K =$ ■ [180]. $\angle H$ and $\angle K$ are same-side ■ [interior] angles. Because they are supplementary, $\overline{HI} \parallel$ ■ [$\overline{KJ}$].

You can also show that $m\angle H + m\angle I =$ ■ [180]. $\angle H$ and $\angle I$ are ■ [same-side] interior angles. Because they are ■ [supplementary], $\overline{HK} \parallel$ ■ [$\overline{IJ}$].

Because the opposite sides of quadrilateral HIJK are ■ [parallel], the quadrilateral is a parallelogram.

So, you prove that quadrilateral *HIJK* is a parallelogram. ✓

Practice

Write a proof for each of the following. See Additional Answers in the back of this book.

1. You are given: quadrilateral *CDEF*

   $\overline{CD} \cong \overline{EF}$

   $\overline{CF} \cong \overline{ED}$

   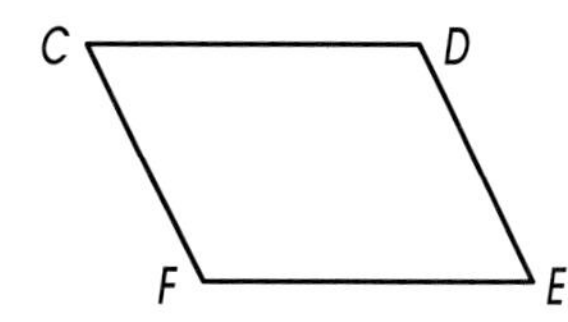

   Prove: Quadrilateral *CDEF* is a parallelogram. (Hint: Draw diagonal $\overline{FD}$. Prove the triangles are congruent.)

2. You are given: quadrilateral *PQRS*

   $\overline{PQ} \parallel \overline{SR}$

   $\overline{QP} \cong \overline{SR}$

   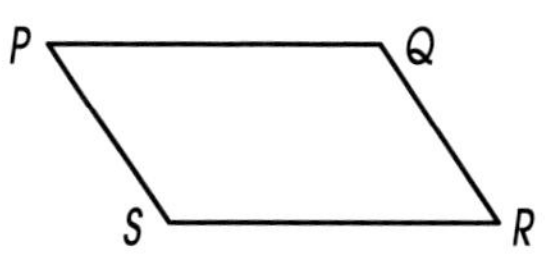

   Prove: Quadrilateral *PQRS* is a parallelogram. (Hint: Draw diagonal $\overline{SQ}$. Prove the triangles are congruent.)

More Practice is provided in the Classroom Resource Binder.

Chapter 7 Review

Summary

You can classify polygons by the number of sides.
The parallelogram, the rectangle, the square, the rhombus, and the trapezoid are special quadrilaterals.
The sides, angles, and diagonals of special quadrilaterals have certain properties.
Use $\frac{1}{2}(b_1 + b_2)$ to find the length of a midsegment of a trapezoid.
The sum of all exterior angles of any polygon is 360°.
Use $360 \div n$ to find the measure of one exterior angle of a regular polygon with n sides.
Use $(n - 2) \bullet 180$ to find the interior-angle sum of a regular polygon with n sides.
To find the measure of an interior angle of a regular polygon, divide the interior-angle sum by the number of sides.
You can tessellate certain polygons to cover a surface.

More vocabulary review is provided in the Classroom Resource Binder, page 67.

polygon
parallelogram
rectangle
isosceles trapezoid
square
rhombus
trapezoid
regular polygon

Vocabulary Review

Complete the sentences with the words from the box. Use each term only once.

1. A trapezoid with congruent legs is an ____. isosceles trapezoid
2. A quadrilateral with only one pair of parallel sides is a ____. trapezoid
3. A closed geometric figure made up of at least three line segments is a ____. polygon
4. A parallelogram with four right angles and four congruent sides is a ____. square
5. A parallelogram with four right angles is a ____. rectangle
6. A parallelogram with four congruent sides is a ____. rhombus
7. A special quadrilateral with both pairs of opposite sides parallel is a ____. parallelogram
8. A polygon with all congruent angles and congruent sides is a ____. regular polygon

Chapter Quiz

More assessment is provided in the Classroom Resource Binder.

1. Name a polygon with eight sides. octagon

Complete each statement.

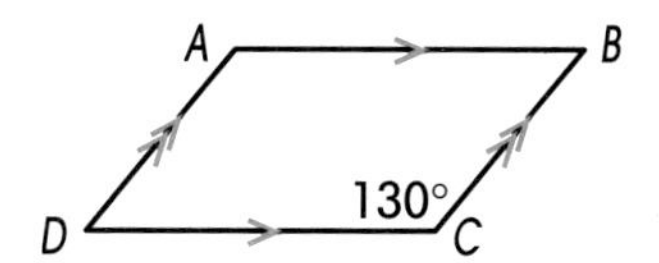

2. $\angle A \cong$ ■ $\angle C$

3. $\overline{AD} \cong$ ■ $\overline{BC}$

4. $\overline{AB} \parallel$ ■ $\overline{DC}$

5. $m\angle B =$ ■ $m\angle D$

6. A parallelogram with four congruent sides is a ____. rhombus or square

▱ *HJKP* is a rectangle. *HM* is 8 cm. Find the length of each line segment and the measure of each angle.

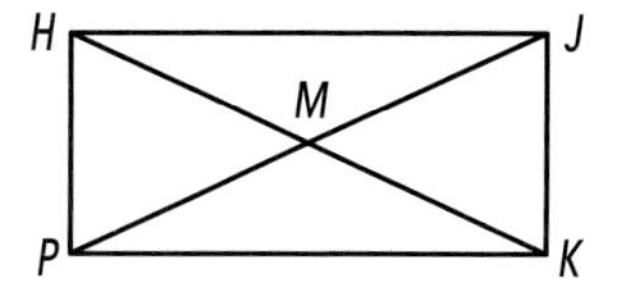

7. $\overline{HK}$ 16 cm

8. $\overline{PJ}$ 16 cm

9. $\angle HPK$ 90°

10. $\overline{MK}$ 8 cm

Trapezoid *PQRS* is isosceles. *PS* is 8 m. Find the length of each line segment and the measure of each angle.

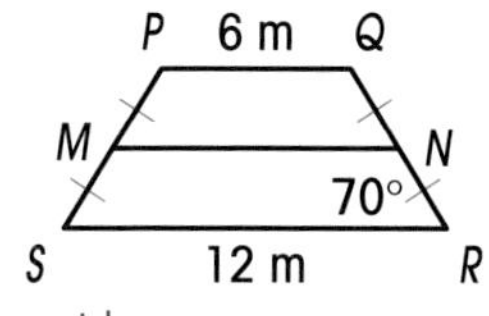

11. $\overline{MN}$ 9 m

12. $\angle S$ 70°

13. $\angle Q$ 110°

14. $\overline{QR}$ 8 m

Write About Math No, a 140° interior angle cannot have a 75° exterior angle. An exterior angle and an interior angle are supplementary. 140 + 75 = 215, not 180.

Problem Solving

Solve each problem. Show your work.

15. Find the measure of an exterior angle of a regular decagon. 36°

16. Find the measure of an interior angle of a regular pentagon. 108°

> **Write About Math**
> A regular nonagon has a 140° interior angle. Can it have an exterior angle of 75°? Explain your answer.
> See above.

Additional Practice for this chapter is provided on p. 399 of the student book.

Standardized Test Preparation This unit review follows the format of many standardized tests. A Scantron sheet is provided in the Classroom Resource Binder.

Unit 2 Review

Write the letter of the correct answer.

1. Choose the measures that can form the sides of a right triangle.
A. 5 ft, 5 ft, 12 ft
(B.) 7 m, 24 m, 25 m
C. 6 in., 6 in., 12 in.
D. none of the above

2. Find the measure of $\angle E$.

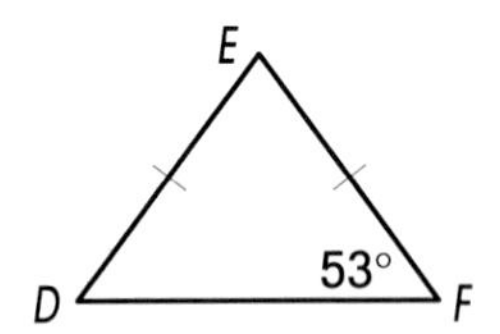

A. 53°
B. 90°
(C.) 74°
D. 180°

3. A triangle can be
I. isosceles and right
II. equiangular and scalene
III. equilateral and right
(A.) I only
B. II only
C. III only
D. II and III only

4. Find the length of the hypotenuse.
A. 7 in.
B. $7\sqrt{3}$ in.
(C.) $7\sqrt{2}$ in.
D. 14 in.

5. $\triangle ABC \cong \triangle DEF$. Which side corresponds to $\overline{AB}$?
A. $\overline{BC}$
(B.) $\overline{DE}$
C. $\overline{EF}$
D. $\overline{DF}$

6. Find the length of the midsegment.

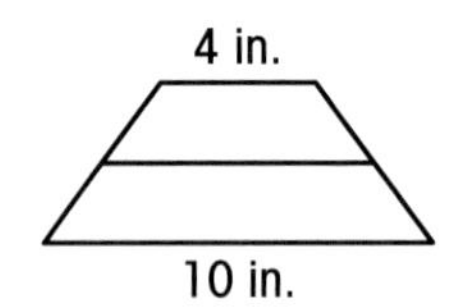

A. 5 in.
B. 6 in.
C. 6.5 in.
(D.) 7 in.

Possible answers: right isosceles triangle, quadrilateral, rectangle, trapezoid, and square

Critical Thinking

Name the shapes you see in the diagram below. See above.

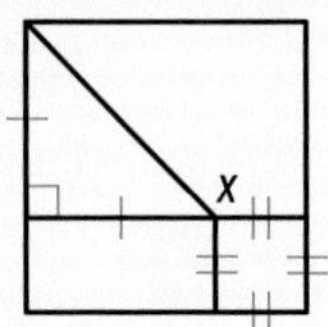

Find the measure of $\angle x$ inside the puzzle. 135°

CHALLENGE How many shapes are there in all? Possible answer: 11

Critical Thinking See the Classroom Resource Binder for a scoring rubric to assess the Critical Thinking Question.

Unit Three

Chapter 8 Perimeter and Area

In this farm field, the farmer planted crops in carefully marked areas. How do you think farmers decide on the amount of area for each crop?

Learning Objectives

- Find the perimeter of polygons.
- Find the area of rectangles, squares, parallelograms, triangles, and trapezoids.
- Use a calculator to find the area of a regular polygon.
- Simplify problems to make them easier to solve.
- Find the area of a shape that can be divided into polygons.
- Write a proof for equal areas.

ESL/ELL Note Relate *perimeter* to the words *rim* and *border*. Underline *rim* in *perimeter*. Explain that these three words mean the same thing. To convey the meaning of area, length, and width, draw and label a rectangle.

Words to Know

perimeter	the distance around a polygon
formula	a rule that uses letters to represent measures
area	the number of square units needed to cover a surface
base	a side of a polygon to which an altitude is drawn
height	the length of the altitude of a polygon
apothem	a perpendicular line segment from the center of a regular polygon to any base

Sizing Up Polygons Project

Guess which of the following polygons has the greatest perimeter. Then, guess which polygon has the greatest area.

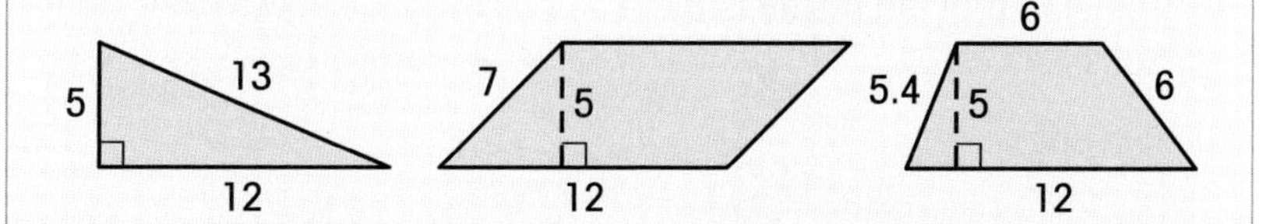

Find the actual perimeter and area of each polygon. Write a report about what you discover.

The parallelogram has the greatest area and the greatest perimeter.

Project Students can begin the project after Lesson 8.3 and complete it after Lesson 8.5. See the Classroom Resource Binder for a scoring rubric to assess this project.

More practice is provided in the Workbook, page 42, and in the Classroom Resource Binder, page 78.

8·1 Perimeter of Polygons

Getting Started
Measure the perimeter of an object such as a desk or a book by placing string on the edge of the object. Mark or cut the string. Lay the string on a ruler to find the perimeter.

Remember
All sides of a regular polygon are congruent.

The **perimeter** of a polygon is the distance around the polygon. To find the perimeter, add the length of each side. You can find the perimeter of this polygon.

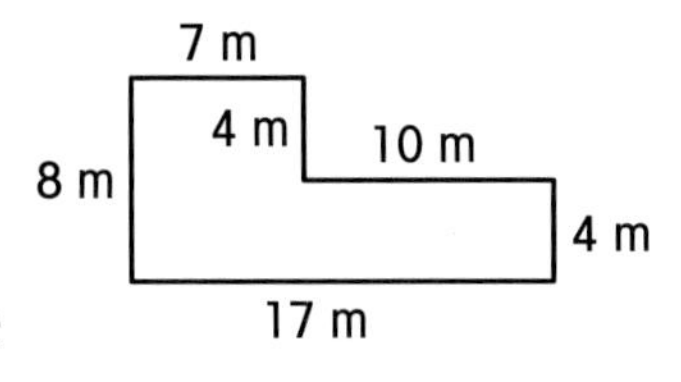

$$\text{Perimeter} = 7 + 4 + 10 + 4 + 17 + 8 = 50 \text{ m}$$

A **formula** is a rule that uses letters to represent measures. You can use a formula to find the perimeter of a regular polygon.

$$\text{Perimeter} = \text{number of sides} \cdot \text{length of one side}$$
$$P = ns$$

EXAMPLE 1

Find the perimeter of the regular hexagon.

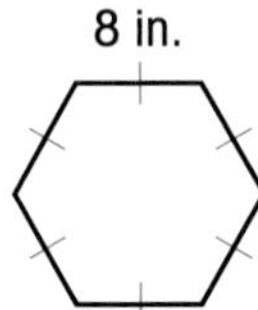

Write the formula. $P = ns$

There are 6 sides.
Substitute 6 for n. $= 6s$

The length of each side is 8.
Substitute 8 for s. $= 6 \cdot 8$

Simplify. $= 48$

The perimeter of the regular hexagon is 48 in.

There is also a formula for the perimeter of a rectangle.

$$\text{Perimeter} = 2 \cdot \text{length} + 2 \cdot \text{width}$$
$$P = 2l + 2w$$

EXAMPLE 2

Find the perimeter of the rectangle.

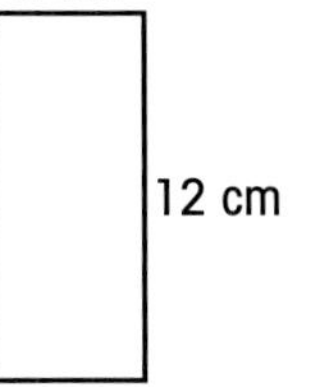

Write the formula. $P = 2l + 2w$

Substitute 12 for l and 5 for w. $= 2(12) + 2(5)$

Simplify. $= 24 + 10 = 34$

The perimeter of the rectangle is 34 cm.

Try This

The perimeter of a rectangle is 42 ft. The length is 12 ft. Find the width.

12 ft
w
P = 42 ft

Write the formula for the ■ of a rectangle. perimeter $P = 2l + 2w$

Substitute 42 for P and ■ for l. 12 $42 = 2(12) + 2w$

Solve for w. $42 = ■ + 2w$ 24

18 $■ = 2w$

9 $■ = w$

The width of the rectangle is ■. 9 ft

Practice

Find the perimeter of each polygon.

1. 80 ft

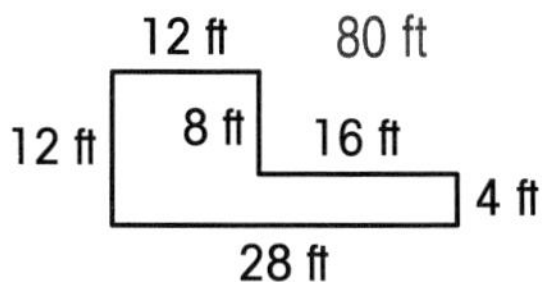

2. 200 yd

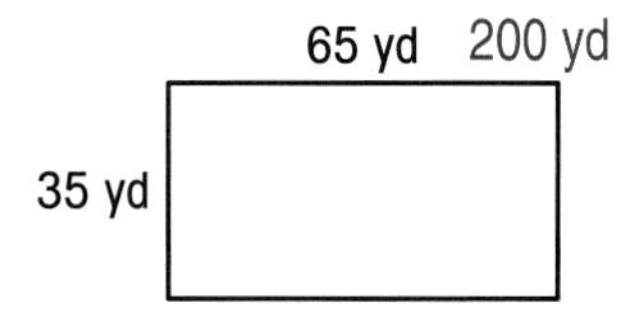

3. 90 cm

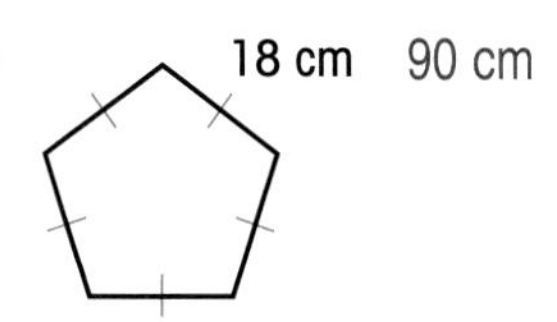

Find the value of x in each polygon.

4. 25 mm

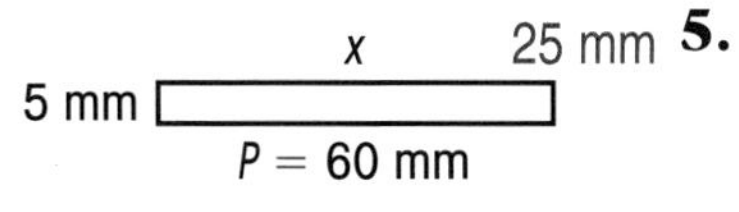

5. 7 ft

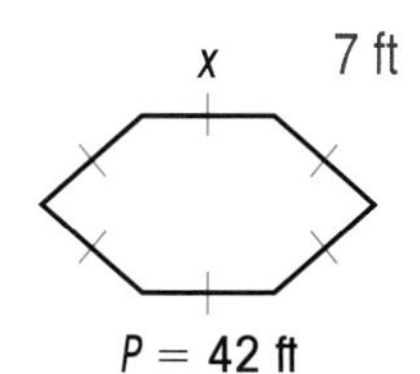

6. 5 m

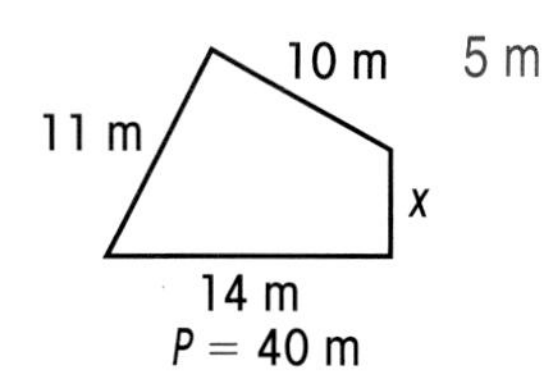

Share Your Understanding

7. Explain to a partner how to find the perimeter of a regular octagon. Use the words *sides, congruent,* and *length* in your explanation. Since all sides of a regular polygon are congruent, the perimeter of a regular octagon is the product of the number of sides times the length of one side.

8. **CRITICAL THINKING** The perimeter of a rectangle is 18 m. The length is twice the width. What is the length and width of the rectangle? The length is 6 m. The width is 3 m.

More practice is provided in the Workbook, page 43, and in the Classroom Resource Binder, page 79.

8·2 Area of Rectangles and Squares

Getting Started
Have students find the squares of the following numbers: 5, 8, 13, 15, and 25.

The **area** of a polygon is the number of square units needed to cover the surface.

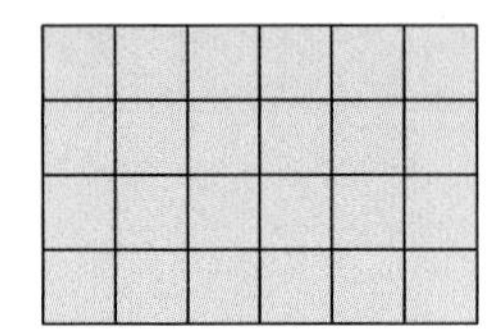

Math Fact
Area is always measured in square units. Some units of area are the square inch ($in.^2$) and the square meter (m^2).

The area of this rectangle is 24 square units.

You can use a formula to find the area of a rectangle or the area of a square.

Rectangle	**Square**
Area = length • width	Area = side • side = $(side)^2$
$A = lw$	$A = s^2$

EXAMPLE 1

12 in.

5 in.

Find the area of the rectangle on the left.

Write the formula for area of a rectangle. $A = lw$

Substitute the given values. $= 12 \cdot 5$

Simplify. $= 60$

The area of the rectangle is 60 square inches, or 60 $in.^2$

EXAMPLE 2

6 cm

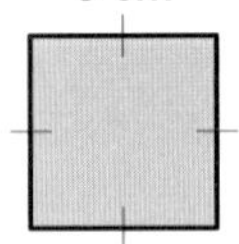

Find the area of the square on the left.

Write the formula for area of a square. $A = s^2$

Substitute the given value. $= 6^2$

Simplify. $= 36$

The area of the square is 36 square centimeters, or 36 cm^2.

Try This

The area of a rectangle is 78 m^2.
The width is 6 m. Find the length.

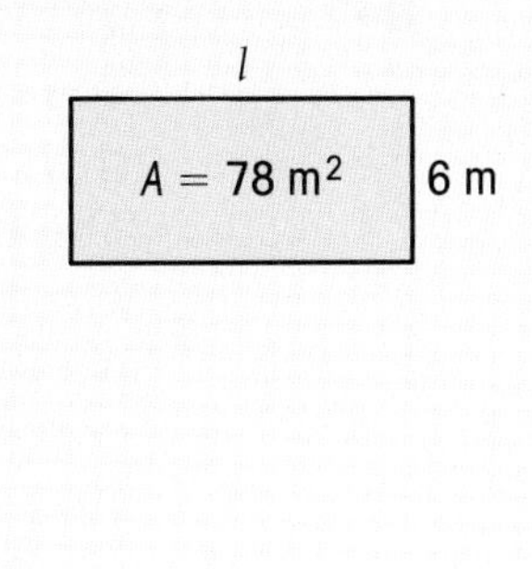

Write the formula for ■ of a rectangle. (area)	$A = lw$
Substitute the given values.	$78 = l \cdot 6$
Solve for l.	$\frac{78}{■} = l$ (6)
	$13\ ■ = l$

The length of the rectangle is ■. 13 m

Practice

Find the area of each polygon.

1. 78 m^2

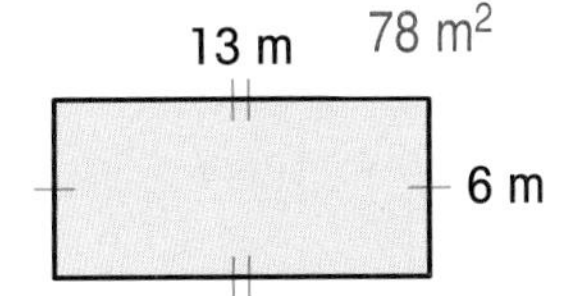

2. 49 $in.^2$

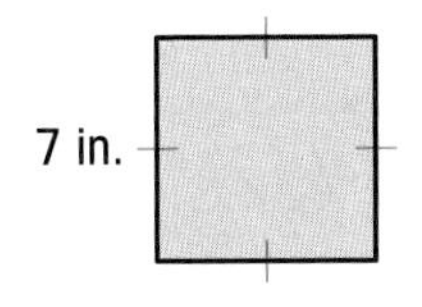

3. 36 ft^2

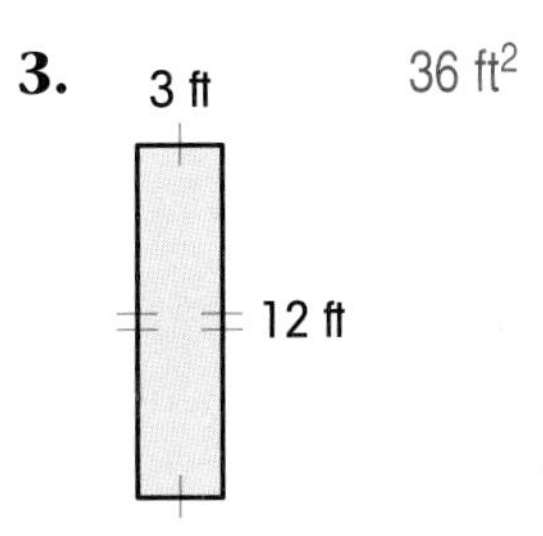

Find the value of x in each polygon.

4. 9 m

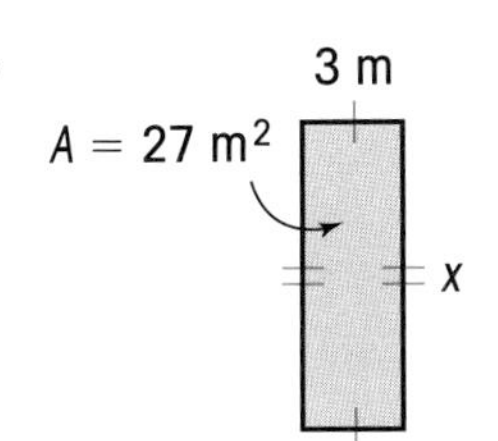

5. 2 yd

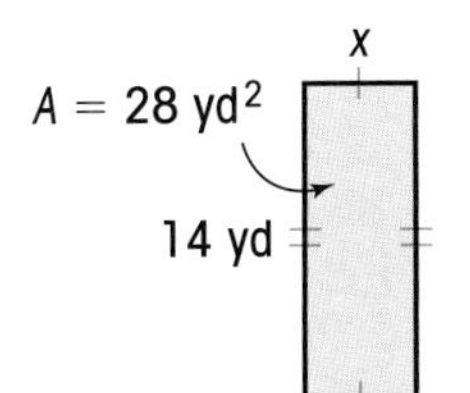

6. 15 mm

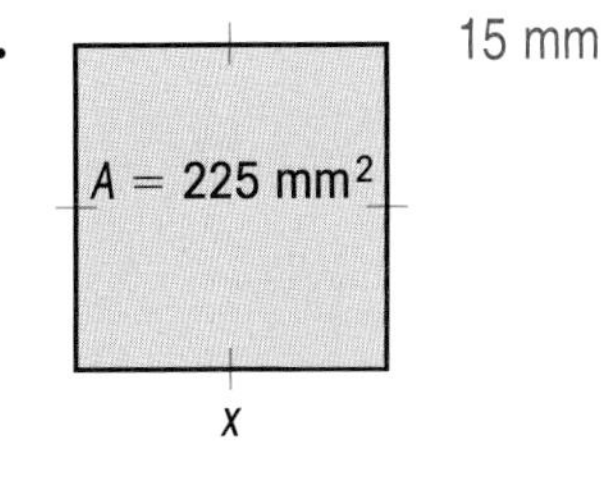

Share Your Understanding

7. Draw a rectangle. Label the length 8 ft and the width 2 ft. Have a partner draw a square. Label one side 4 ft. Exchange drawings and find the areas. Check each other's work.
Area of the rectangle is 16 ft^2. Area of the square is 16 ft^2.

8. **CRITICAL THINKING** Find the greatest possible area of a rectangle with a perimeter of 24 mm. 36 mm^2

More practice is provided in the Workbook, page 43, and in the Classroom Resource Binder, page 79.

8-3 Area of Parallelograms

Getting Started Review the altitude of a triangle.

You can turn a rectangle into a parallelogram. The area stays the same. The names of the parts are different. The **base** of a parallelogram can be any one of the sides. The **height** is the length of the altitude. The height of a parallelogram is always perpendicular to the base.

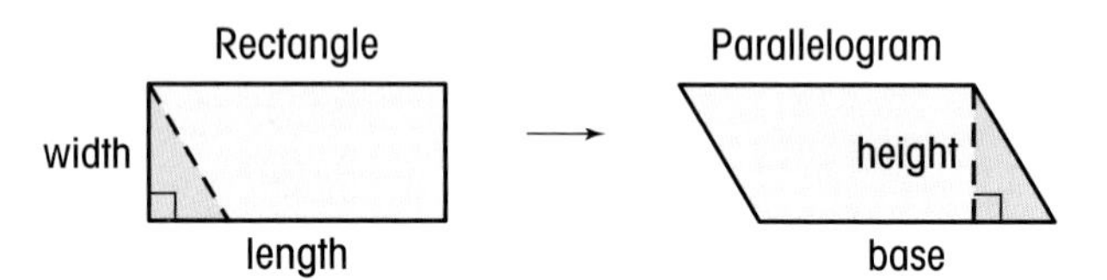

Area = length • width
$A = lw$

Area = base • height
$A = bh$

You can use the formula to find the area of a parallelogram.

EXAMPLE 1

Find the area of ▱*ABCD*.

Write the formula for area.	$A = bh$
Substitute the given values.	$= 6 \cdot 2$
Simplify.	$= 12$

The area of ▱*ABCD* is 12 cm^2.

If you know the area of a parallelogram and the height, you can find the length of the base.

EXAMPLE 2

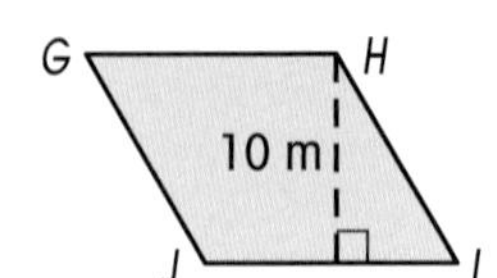

The area of ▱*GHIJ* is 125 m^2. Find the length of the base.

Write the formula for area.	$A = bh$
Substitute the given values.	$125 = b(10)$
Solve for *b*.	$12.5 = b$

The base of ▱*GHIJ* is 12.5 m.

Try This

The area of $\square KLMN$ is 112 in.^{2}
Find the height.

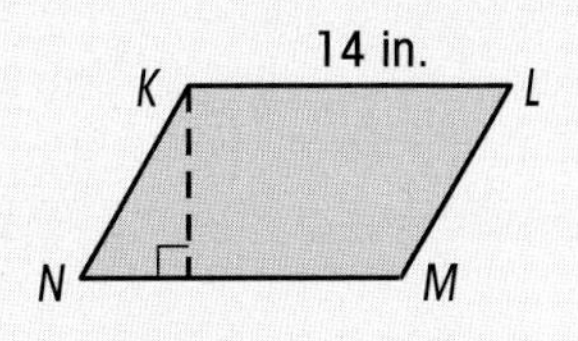

Write the formula for ■. area $A = bh$

Substitute the given values. 112 ■ $= 14h$

Solve for ■. h 8 ■ $= h$

The height of $\square KLMN$ is ■. 8 in.

Practice

Find the area of each parallelogram.

1.

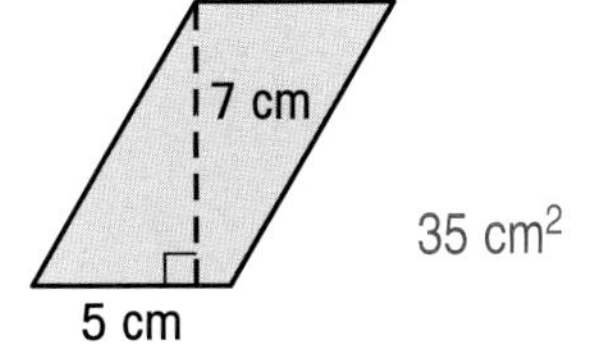

35 cm^2

2.

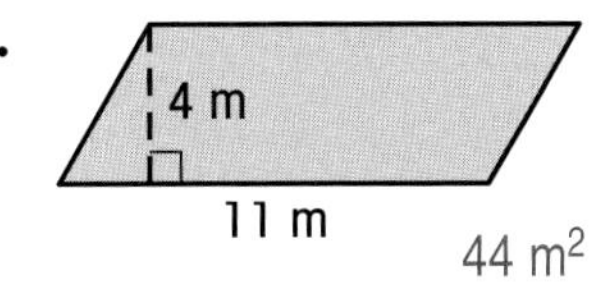

44 m^2

3.

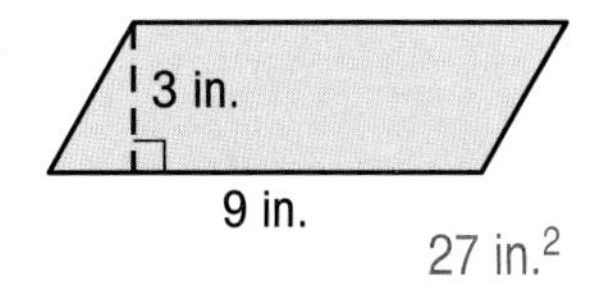

27 in.^{2}

The area of each parallelogram is 120 ft^2. Find the value of x in each parallelogram.

4.

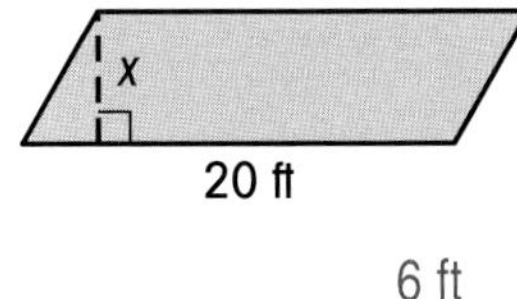

6 ft

5.

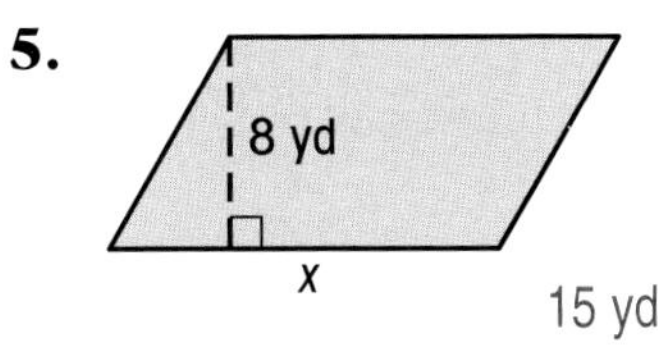

15 yd

6.

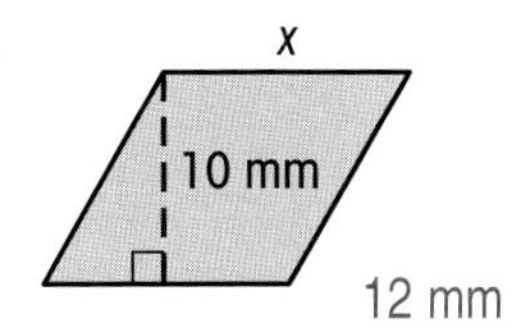

12 mm

Share Your Understanding

7. Explain to a partner how to find the area of a parallelogram. Use the words *multiply* and *base* in your explanation. Multiply base times height.

8. **CRITICAL THINKING** Find the area of $\square PQRS$. (Hint: Use the Pythagorean Theorem to find the height.) 24 m^2

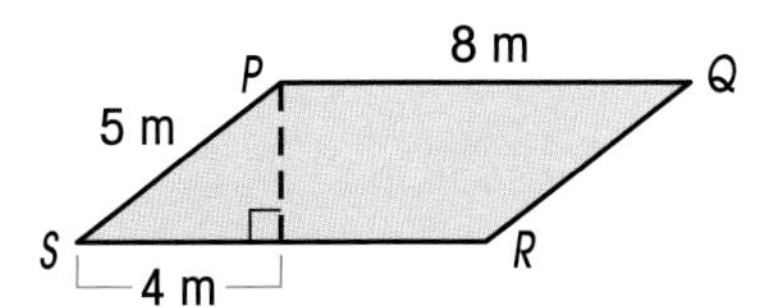

More practice is provided in the Workbook, page 44, and in the Classroom Resource Binder, page 80.

8·4 Area of Triangles

Getting Started
Give students copies of the following triangles: acute, right, obtuse, isosceles, and scalene. Have them turn each triangle into a parallelogram.

You can cut a parallelogram in half to make two triangles. The area of one triangle is one-half the area of the parallelogram.

Parallelogram

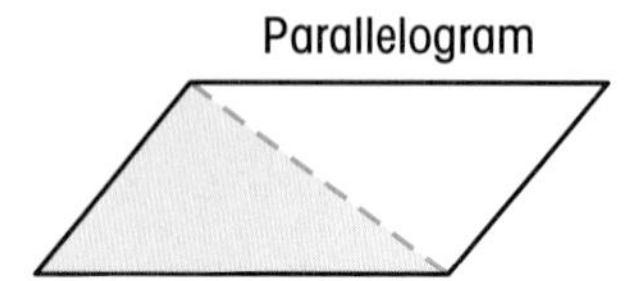

Triangle

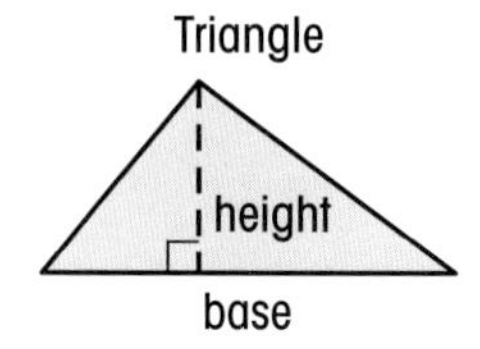

Area = base • height

$A = bh$

Area = $\frac{1}{2}$ • base • height

$A = \frac{1}{2}bh$

You can use the formula to find the area of a triangle.

EXAMPLE 1

Find the area of $\triangle ABC$.

The legs of this right triangle are the height and the base.

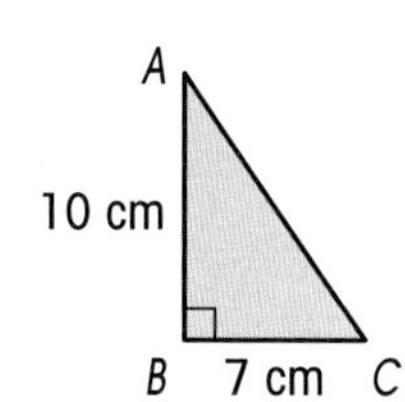

Write the formula for area of a triangle.	$A = \frac{1}{2}bh$
Substitute the given values.	$= \frac{1}{2}(7)(10)$
Simplify.	$= 35$

The area of $\triangle ABC$ is 35 cm^2.

Sometimes the height is outside the triangle.

EXAMPLE 2

Find the area of $\triangle DEG$.

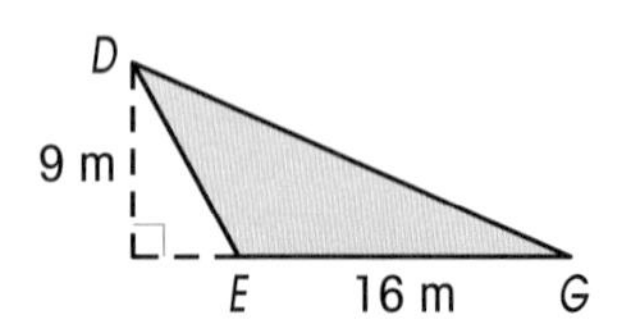

Write the formula for area of a triangle.	$A = \frac{1}{2}bh$
Substitute the given values.	$= \frac{1}{2}(16)(9)$
Simplify.	$= 72$

The area of $\triangle DEG$ is 72 m^2.

Try This

The area of $\triangle HKL$ is 192 ft^2.
Find the length of the base.

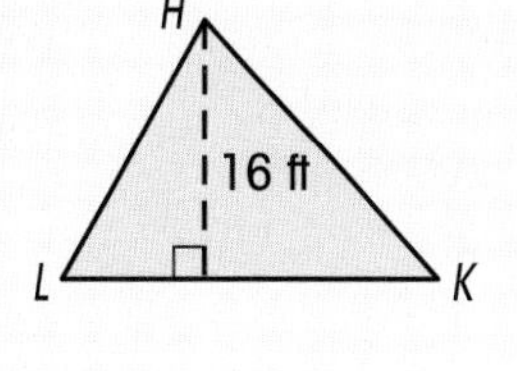

Write the formula for area. $A = \frac{1}{2}bh$

Substitute the given values. $192\ \blacksquare = \frac{1}{2}(b)(16)$

Solve for b. $192\ \blacksquare = 8b$

$24\ \blacksquare = b$

The base of $\triangle HKL$ is ■. 24 ft

Practice

Find the area of each triangle.

1. 72 cm^2

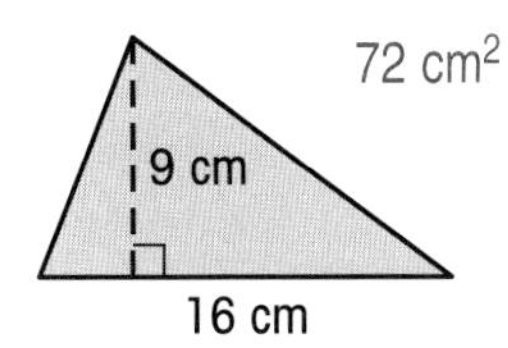

2. 27.5 ft^2

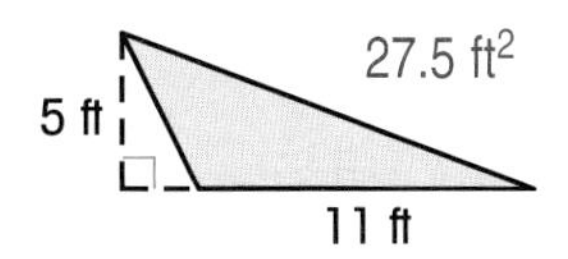

3. 70 in.^{2}

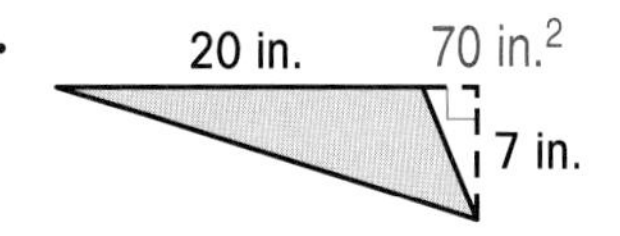

Find the length of each base. The area is given.

4. 14 ft

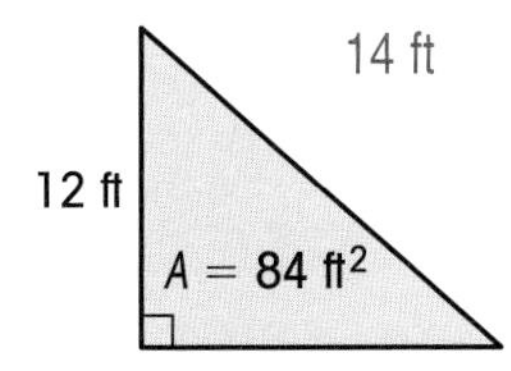

5. 5 in.

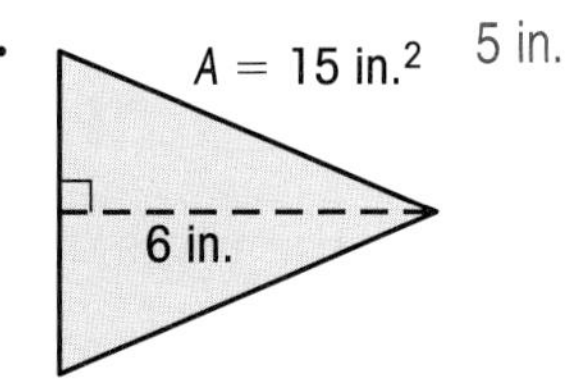

6. 45 m

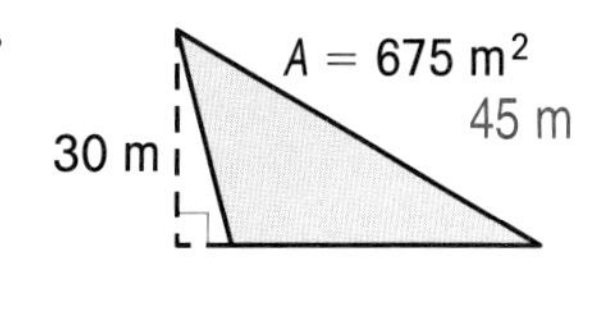

Share Your Understanding

7. Explain to a partner how to find the area of a right triangle. Use the words *multiply* and *half* in your explanation. Multiply one half times the length of the base times the height.

8. **CRITICAL THINKING** Find the perimeter of $\triangle KLM$. (Hint: First find LM. Then use the Pythagorean Theorem.) 24 yd

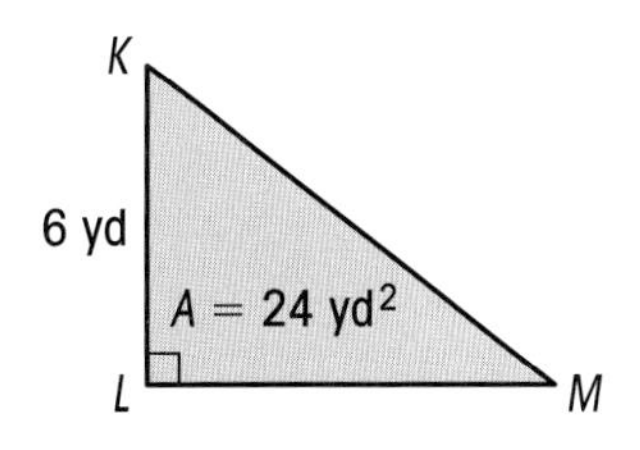

More practice is provided in the Workbook, page 45 and in the Classroom Resource Binder, page 81.

8·5 Area of Trapezoids

Getting Started
Show students a square divided into two congruent triangles. Give them the base and the height of the triangles. Have them figure out the area of the square, using the formula for area of a triangle.

You can cut a trapezoid to make two triangles. Each base of the trapezoid is a base of one triangle. The height of the trapezoid is the height of each triangle.

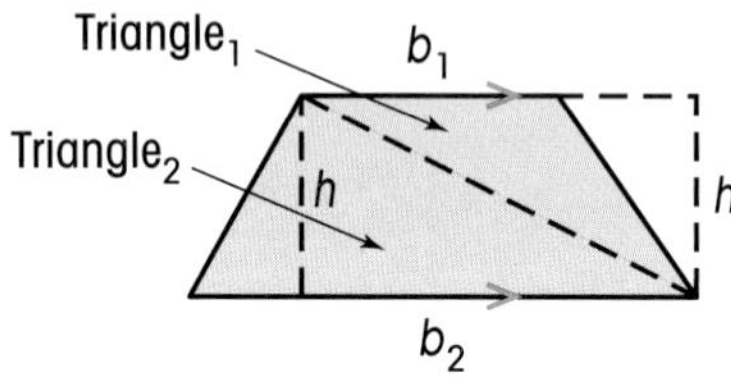

Triangle$_1$ $A = \frac{1}{2}b_1h$ Triangle$_2$ $A = \frac{1}{2}b_2h$

The area of the trapezoid is the sum of these areas.

Math Fact
Factor out the common factor $\frac{1}{2}h$.

$$\text{Area} = \frac{1}{2}b_1h + \frac{1}{2}b_2h$$

$$A = \frac{1}{2}h(b_1 + b_2)$$

You can use the formula to find the area or height.

EXAMPLE 1

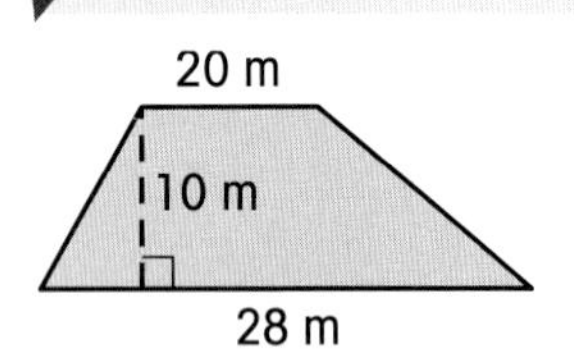

Find the area of the trapezoid on the left.

Write the formula for area of a trapezoid. $A = \frac{1}{2}h(b_1 + b_2)$

Substitute the given values. $= \frac{1}{2}(10)(20 + 28)$

Simplify. $= 240$

The area of the trapezoid is 240 m^2.

EXAMPLE 2

The area of a trapezoid is 36 cm^2. The bases are 4 cm and 8 cm. Find the height.

Write the formula for area of a trapezoid. $A = \frac{1}{2}h(b_1 + b_2)$

Substitute the given values. $36 = \frac{1}{2}(h)(4 + 8)$

Solve for h. $36 = \frac{1}{2}(12)h$

$36 = 6h$

$6 = h$

The height of the trapezoid is 6 cm.

Try This

Trapezoid *QRSU* is isosceles. Find the area.

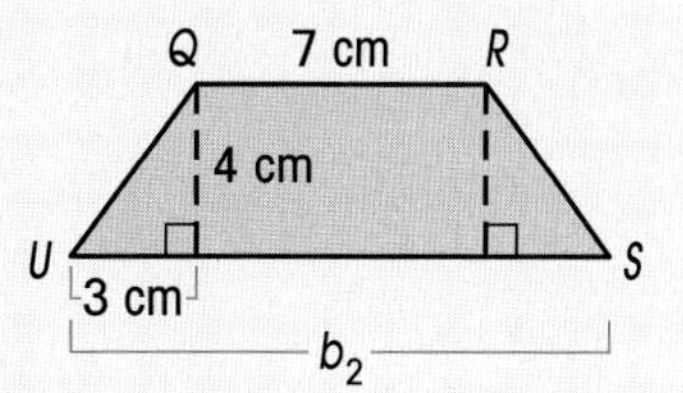

Because this is an isosceles trapezoid, you can find b_2. $b_2 = 3 + 7 + 3 = 13$

Write the formula for area of a ■. trapezoid $A = \frac{1}{2}h(b_1 + b_2)$

Substitute the values you know. $= \frac{1}{2}(■)(7 + 13)$ 4

Simplify. $= \frac{1}{2}(4)(20)$

$= ■$ 40

The area of isosceles trapezoid *QRSU* is ■. 40 cm^2

Practice

Find the area of each trapezoid.

1. 25 mm^2

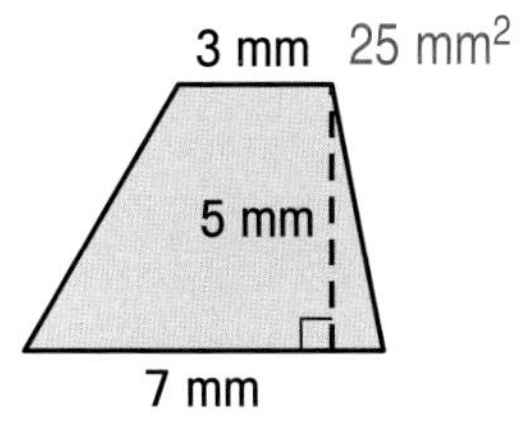

2. 25 in.^{2}

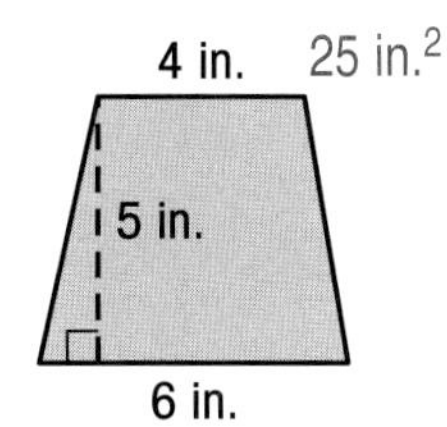

3. 120 m^2

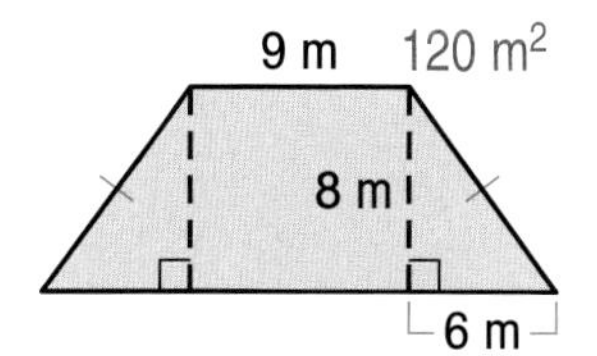

4. The area of a trapezoid is 144 ft^2. The bases are 16 ft and 32 ft. Find the height. 6 ft

5. Possible answer: Add b_1 and b_2. Then, multiply this sum and the height. Find half of this product to get the area.

Share Your Understanding

5. Explain to a partner how to find the area of a trapezoid. Use *half, height*, b_1, and b_2 in your explanation.

6. **CRITICAL THINKING** Find the perimeter of this isosceles trapezoid. (Hint: Use the Pythagorean Theorem.) 56 cm

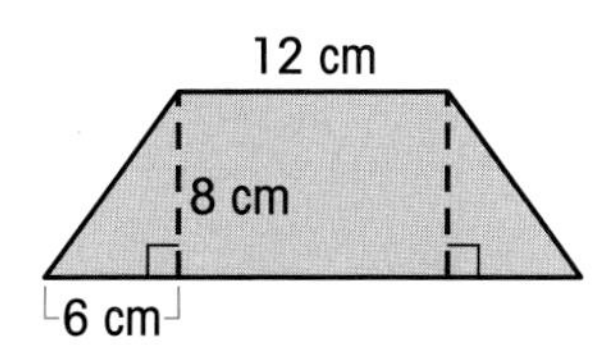

8·6 Calculator: Area of a Regular Polygon

Getting Started
Have students find the fraction key on their calculators. Have them find the product of the following problems: $\frac{2}{3} \cdot 9$ and $\frac{1}{2} \cdot 18$. [6 and 9]

A regular polygon has congruent sides and congruent angles. Look at the apothem in the regular polygon below. The **apothem** is a perpendicular line segment from the center of a regular polygon to any base.

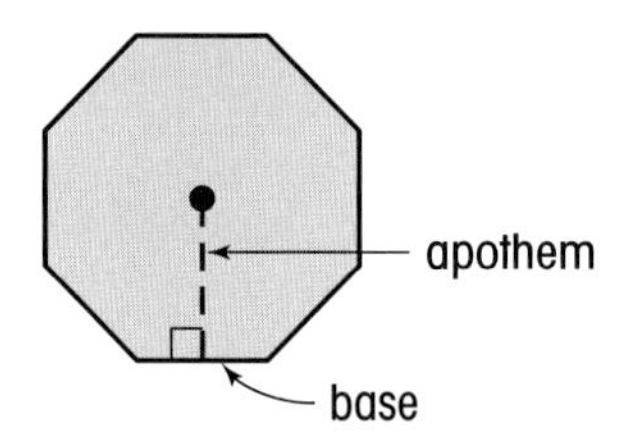

There is a formula to find the area of a regular polygon.

$$\text{Area} = \frac{1}{2}\ \text{apothem} \cdot \text{perimeter}$$

$$A = \frac{1}{2}ap$$

EXAMPLE

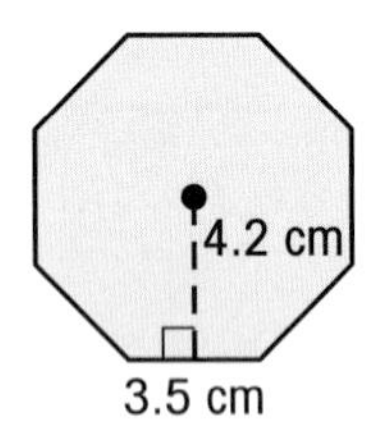

Find the area of the regular octagon on the left.

First, write the formula for perimeter. $P = ns$

Substitute the given values. $= (8)(3.5)$

Use your calculator to multiply.

PRESS: 8 × 3 . 5 =

Display: 28

The perimeter of the regular octagon is 28 cm.

Now, write the formula for area. $A = \frac{1}{2}ap$

Substitute the given values. $= \frac{1}{2}(4.2)(28)$

Use your calculator to multiply.

PRESS: 4 . 2 × 2 8 ÷ 2 =

Display: 58.8

The area of the regular octagon is $58.8\ \text{cm}^2$.

Math Fact
Multiplying by $\frac{1}{2}$ is the same as dividing by 2.

Practice

Find the area of each regular polygon.

1. 2.1 in.; 2.54 in. — 21.336 in.^{2}

2. 5.2 mm; 4.5 mm — 70.2 mm^2

3. 4.96 cm; 6 cm — 119.04 cm^2

4. 4.12 in.; 3.57 in. — 44.1252 in.^{2}

5. 1.98 m; 2.4 m — 19.008 m^2

6. 12.13 m; 10.5 m — 382.095 m

Math Connection

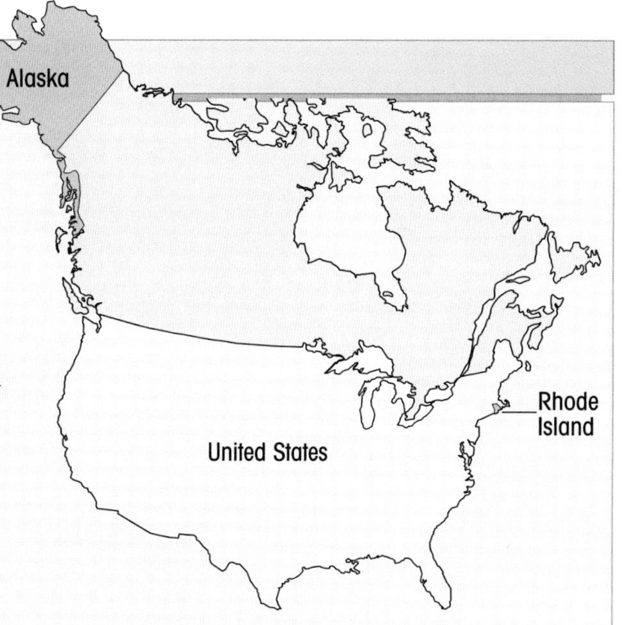

AREA OF STATES

Alaska is the largest state in the United States. It has an area of about 600,000 square miles. Rhode Island is the smallest state. It has an area of about 1,200 square miles. Rhode Island is only about 48 miles long and about 37 miles wide. If you divide the area of Alaska by the area of Rhode Island, you will find that about 500 Rhode Islands fit inside Alaska!

Alaska has one of the smallest populations in the country. There is about one person for every square mile. Rhode Island, however, is the most densely populated state in the country. It has an average of over 800 people per square mile.

Why do you think Alaska has the smallest population?

More practice is provided in the Workbook, page 46, and in the Classroom Resource Binder, page 82.

Getting Started
Have students divide a square into many polygons. Explain to them that the sum of the areas of the polygons is equal to the area of the square.

8·7 Problem-Solving Strategy: Simplify the Problem

You can estimate the area of an irregular shape. Simplify the problem. Divide the shape into polygons you know.

EXAMPLE

Estimate the area of Texas. Use two polygons.

READ **What do you need to find out?**
Find an estimate of the area of Texas.

PLAN **What do you need to do?**
Divide the shape into two polygons. Find each area. Then find the sum of the areas.

DO **Follow the plan.**

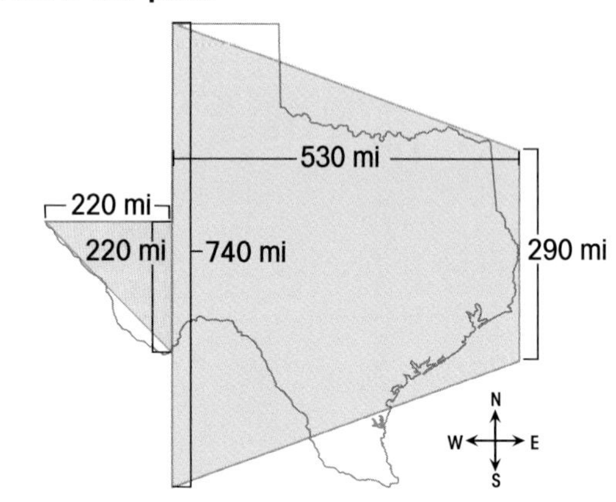

Triangle $A = \frac{1}{2}bh$

$A = \frac{1}{2}(220)(220) = 24{,}200$

Trapezoid $A = \frac{1}{2}\,h(b_1 + b_2)$

$A = \frac{1}{2}(530)(740 + 290) = 272{,}950$

Sum of the areas $272{,}950 + 24{,}200 = 297{,}150$

CHECK **Does the answer make sense?**
The greatest width on the map is 530 mi.
The greatest length is 740 mi.
$500 \cdot 700 = 350{,}000$
The estimate 297,150 mi^2 makes sense. ✓

Try This

Estimate the area of Texas. Use four polygons.

Divide Texas into four polygons.

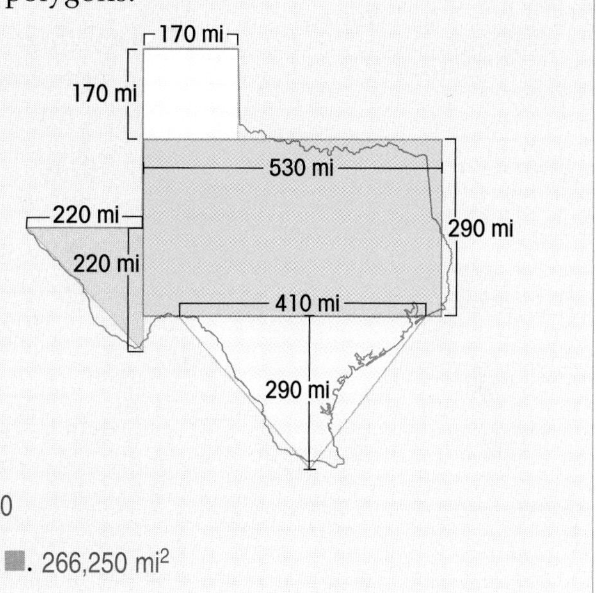

Find the area of the square.

$A = (170)(\blacksquare) = \blacksquare$ 28,900
170

Find the area of the rectangle.

$A = (530)(\blacksquare) = \blacksquare$ 153,700
290

Find the area of the right triangle.

$A = \frac{1}{2}(220)(\blacksquare) = \blacksquare$ 24,200
220

Find the area of the other triangle.

$A = \frac{1}{2}(410)(\blacksquare) = \blacksquare$ 59,450
290

Add the area of each shape.

$A = \blacksquare + \blacksquare + \blacksquare + \blacksquare = \blacksquare$ 266,250
28,900 153,700 24,200 59,450

This estimate of the area of Texas is ■. 266,250 mi^2

Practice

Solve each problem, using this map of Idaho.
Be sure to simplify each problem. Show your work.
Answers will vary. Check students' work.

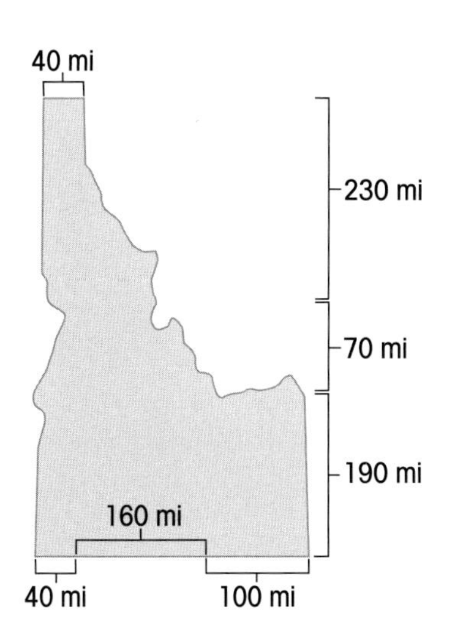

1. Estimate the area of Idaho, using two polygons.

2. Estimate the area of Idaho, using three or more polygons.

3. The actual area of Idaho is about 83,564 mi^2. Which estimate of the area is closer to the actual area of Idaho? Explain your answer.
The more polygons that are used to estimate the area, the more accurate the estimate.

More practice is provided in the Workbook, page 47, and in the Classroom Resource Binder, page 83.

Getting Started Before teaching the lesson, discuss the shape of each figure. Have students identify the length of each side.

8·8 Problem-Solving Application: Carpeting an Area

You learned how to find the area of a rectangle and a square. Sometimes you need to find the area of a shape that can be divided into rectangles and squares.

EXAMPLE 1

Carmen wants to carpet her bedroom. The shape and size of the bedroom is shown on the right. How many square feet of carpet does Carmen need?

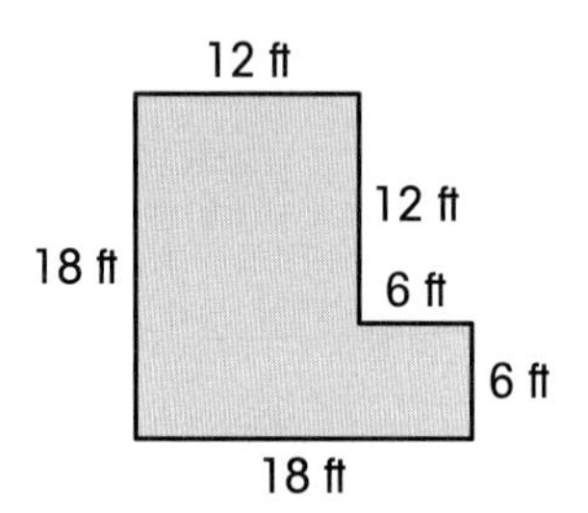

This shape can be divided into a rectangle and a square.

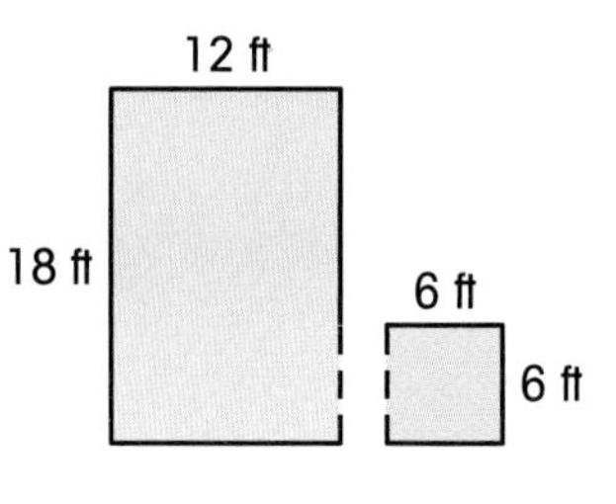

Find the area of the rectangle and the area of the square. Add the areas to find the total area of the room.

Write the formula for area of a rectangle.	$A = lw$
Substitute the given values. Multiply.	$= 18 \cdot 12 = 216$
Write the formula for area of a square.	$A = s^2$
Substitute the given values. Multiply.	$= 6 \cdot 6 = 36$
Add to find the total area of the room.	$216 + 36 = 252$

Carmen will need 252 ft^2 of carpet.

EXAMPLE 2

The carpet Carmen will use is $25 per square yard. How much will it cost to carpet her room?

First, change 252 ft^2 to yd^2.	$252 \div 9 = 28$
Then, multiply by 25.	$28 \cdot 25 = 700$

It will cost $700 to carpet the room.

Try This

Abdul wants to carpet this room. To make the carpet fit, he will cut it into two pieces. Use the diagram on the right to find out how many square feet of carpet Abdul will need for his living room.

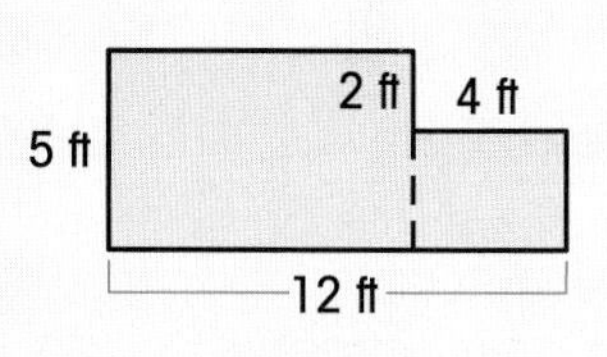

To find the area of the larger rectangle, you need to know its length and ■. width

length = 12 − ■ = ■ (4, 8)
width = ■ 5

Find the area of the larger rectangle.

$A = lw$
$= 8 \cdot 5 =$ ■ 40

To find the area of the smaller rectangle, you need to know the length and ■. width

length = ■ 4
width = 5 − ■ = ■ 3 (2)

Find the area of the smaller rectangle.

$A = lw$
$= 4 \cdot 3 =$ ■ 12

Add to find the total area of the room.

■ + ■ = ■ 52 (40, 12)

Abdul will need ■ of carpet. 52 ft^2

Practice

Use the diagram to solve each problem. Show your work.

1. How many square feet of carpet does Lee need to carpet this room? 729 ft^2

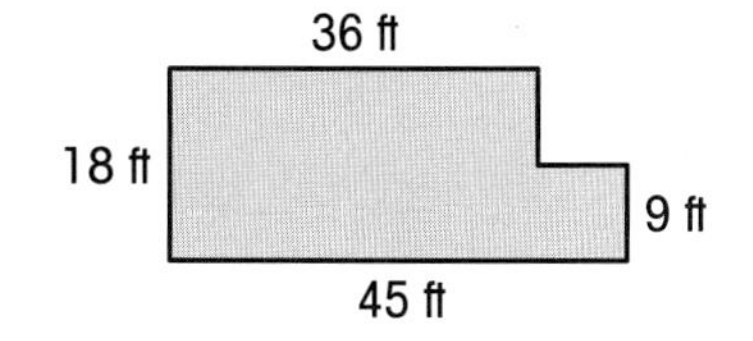

2. The carpet Lee wants to buy costs \$22 per square yard. How much will the carpet cost for this room? (Hint: 1 square yard = 9 square feet) \$1,782

3. Lee wants to put a pad under the carpet. It costs \$2 per square yard. How much will the pad cost? \$162

An alternate two-column proof lesson is provided on page 413 of the student book.

8-9 Proof: Equal Areas

Getting Started Review the triangle area formula.

THEOREM 25

Two triangles have equal areas if the heights are congruent and the bases are congruent.

You learned that triangles are congruent if they are the same size and the same shape. Triangles that are the same size but not the same shape have equal areas. If the area of $\triangle ABC$ is equal to the area of $\triangle ADC$, you can write Area $\triangle ABC$ = Area $\triangle ADC$.

You can use what you know to prove the areas of two triangles are equal.

EXAMPLE

You are given: $\overline{BD} \parallel \overline{AC}$

Prove:
Area $\triangle ABC$ = Area $\triangle ADC$

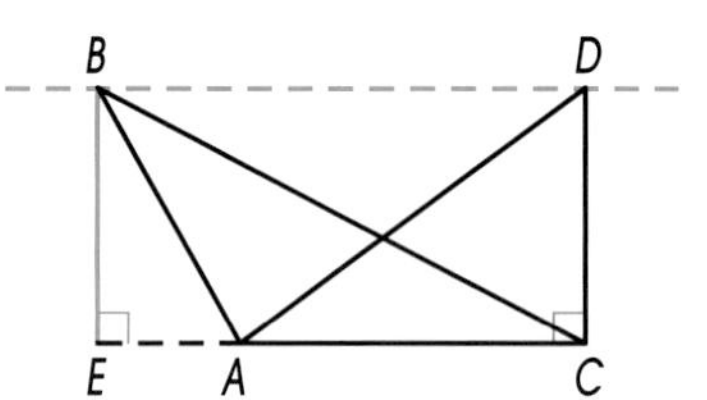

You are given $\overline{BD} \parallel \overline{AC}$. Because the distance between two parallel lines is always equal, $BE = DC$.

You know that $AC = AC$.

Then, $\frac{1}{2}(AC)(BE) = \frac{1}{2}(AC)(DC)$.

The area of $\triangle ABC$ is equal to the area of $\triangle ADC$.

So, you prove that Area $\triangle ABC$ = Area $\triangle ADC$. ✓

You proved that two triangles have equal areas if they have a common base and their vertices lie on a line parallel to the base. You proved that the area of right $\triangle ADC$ is equal to the area of obtuse $\triangle ABC$.

You learned the Addition Property of Equality. This property means that if equals are added to equals, the sums are equal. You can use this property to prove certain areas are equal.

Try This

Copy and complete the proof.

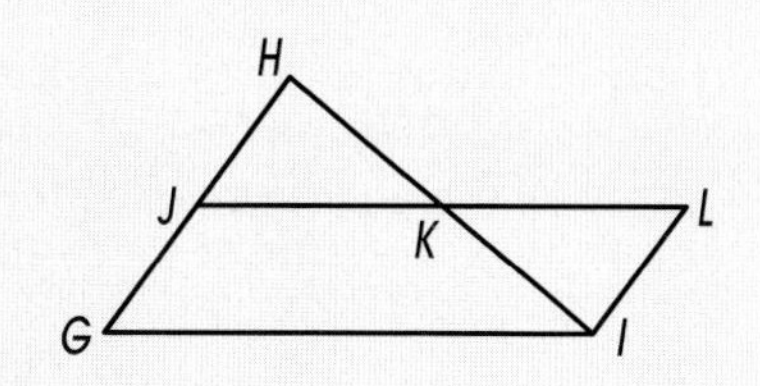

You are given:

K is the midpoint of $\overline{HI}$

$\overline{JL} \parallel \overline{GI}$

$\overline{GH} \parallel \overline{LI}$

Prove: Area $\triangle GHI$ = Area ▱$GJLI$

First, you need to prove $\triangle JHK \cong \triangle LIK$. The area of trapezoid $GJKI$ is common to both $\triangle GHI$ and ▱$GJLI$. Add the common area.

You know that K is the ■ (midpoint) of $\overline{HI}$. This is given. This means that $\overline{KH} \cong \overline{KI}$. $\angle HKJ \cong$ ■ ($\angle IKL$) because they are vertical angles. You are given $\overline{GH} \parallel$ ■ ($\overline{LI}$). This means that $\angle HJK \cong$ ■ ($\angle ILK$) because they are alternate ■ (interior) angles. Then, $\triangle JHK \cong \triangle LIK$ by ■ (AAS). Because the triangles are congruent, the areas are equal. Then, Area $\triangle JHK$ + Area trapezoid $GJKI$ = ■ (Area $\triangle LIK$) + Area trapezoid $GJKI$.

So, you prove that Area $\triangle GHI$ = Area ▱$GJLI$. ✓

Practice

Write a proof for each of the following. See Additional Answers in the back of this book.

1. You are given: trapezoid $MNOP$

Prove: Area $\triangle MNP$ = Area $\triangle POM$
(Hint: Draw a height for $\triangle MNP$. Draw a height for $\triangle POM$.)

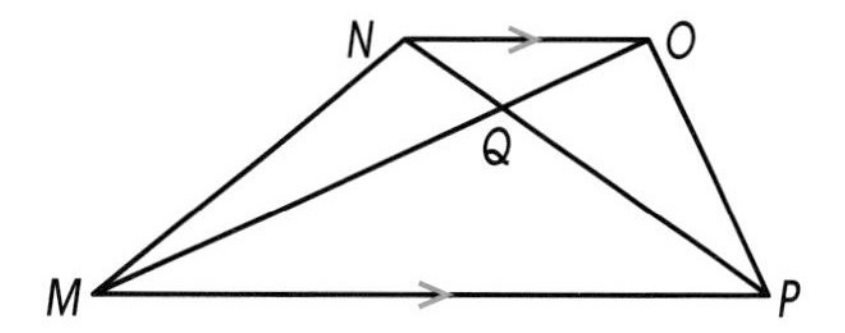

2. You are given: trapezoid $ABCD$
$\overline{BC} \cong \overline{ED}$

Prove: Area $ABCD$ = Area $\triangle ABE$

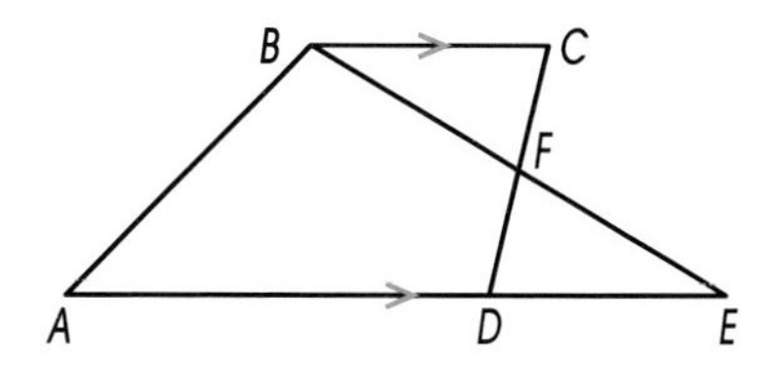

More Practice is provided in the Classroom Resource Binder.

Chapter 8 Review

Summary

To find the perimeter of a polygon, add the length of each side.
Use the formula $A = lw$ to find the area of a rectangle.
Use the formula $A = s^2$ to find the area of a square.
Use the formula $A = bh$ to find the area of a parallelogram.
Use the formula $A = \frac{1}{2}bh$ to find the area of a triangle.
Use the formula $A = \frac{1}{2}h(b_1 + b_2)$ to find the area of a trapezoid.
Use the formula $A = \frac{1}{2}ap$ to find the area of any regular polygon.
You can simplify a problem to help solve it.
You can use formulas to solve real-life problems about area.

More vocabulary review is provided in the Classroom Resource Binder, page 77.

perimeter
apothem
formula
height
area
base

Vocabulary Review

Complete the sentences with words from the box.

1. The number of square units needed to cover a surface is the ____. area

2. The ____ of a polygon is the distance around the polygon. perimeter

3. A ____ is a side of a polygon to which an altitude is drawn. base

4. The ____ is the length of the altitude of a polygon. height

5. A perpendicular line segment from the center of a regular polygon to any base is called an ____. apothem

6. A ____ is a rule that uses letters to represent measures. formula

Chapter Quiz

More assessment is provided in the Classroom Resource Binder.

Find the perimeter of each figure.

1. 12 m, 10 m, 10 m — 32 m

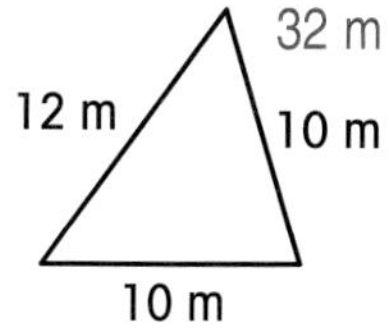

2. 53 ft, 30 ft, 3 ft, 9 ft, 56 ft, 39 ft — 190 ft

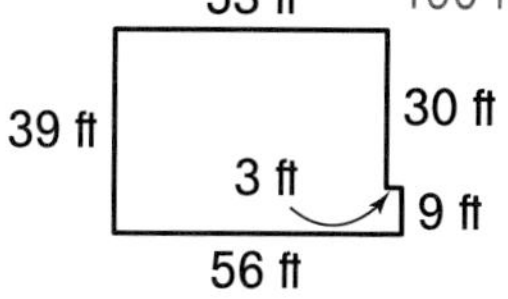

3. 15 in., 10 in., 18 in., 13 in. — 56 in.

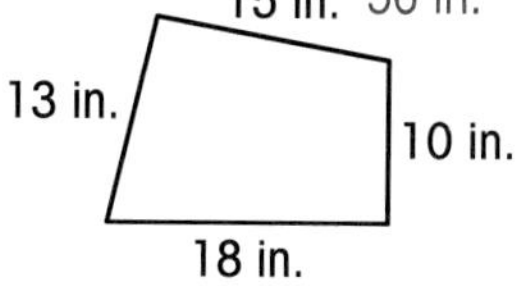

Find the area of each figure.

4. 30 mm, 8 mm — 240 mm^2

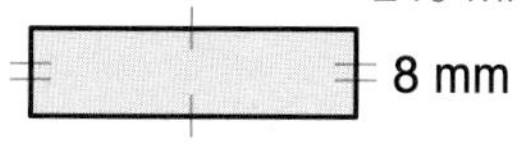

5. 56 cm — 3,136 cm^2

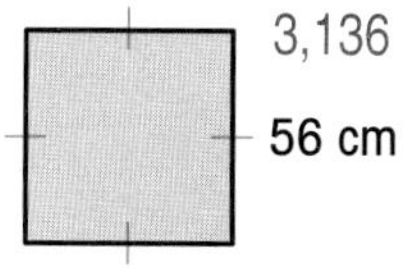

6. 6 ft, 4 ft, 19 ft — 50 ft^2

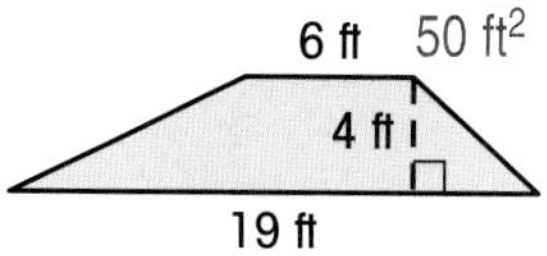

Find the value of x in each diagram.

7. 22 yd, 12 yd, 19 yd, x, $P = 73$ yd — 20 yd

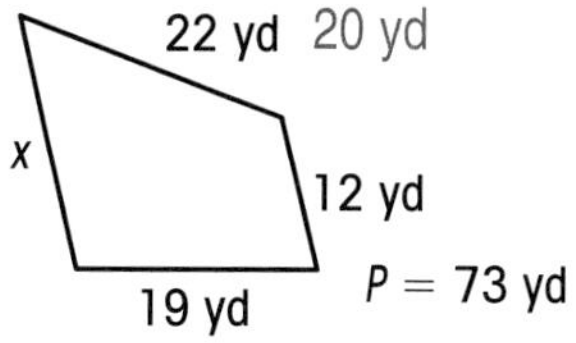

8. 21 m, x, Area $= 147$ m^2 — 7 m

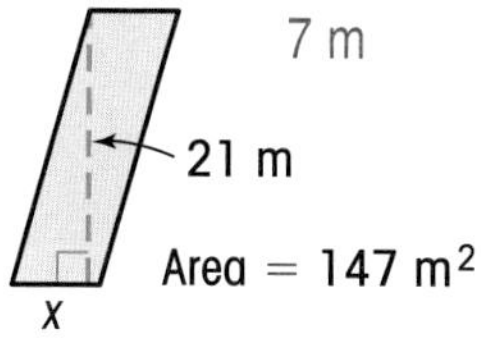

9. x, Area $= 169$ mm^2 — 13 mm

Problem Solving

Solve each problem. Show your work.

10. Angel needs a floor mat for his car. The shape of the floor is a trapezoid. The lengths of the two bases are 15 in. and 18 in. The height is 16 in. What is the area of the mat he needs? (Hint: Draw a diagram.) 264 in.^{2}

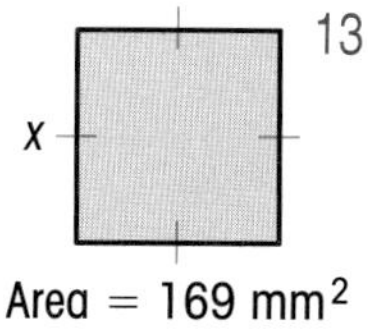

In a square, the length and width are equal. So, you are multiplying the length of one side times the length of another congruent side. $A = s^2$

Write About Math

How can you use the formula $A = lw$ to develop the formula for area of a square?

Additional Practice for this chapter is provided on page 400 of the student book.

Chapter 9 Similar Polygons

Look at the elephants in the photograph. What is similar and what is different about them?

Learning Objectives

- Write ratios and proportions.
- Identify corresponding parts in similar polygons.
- Use the Angle-Angle Similarity Postulate.
- Use the Side-Splitter Theorem.
- Find the length of an unknown side, perimeter, or area of similar polygons.
- Use a calculator to solve a proportion.
- Use proportions to solve problems.
- Use scale drawings.
- Write a proof about similar polygons.

ESL/ELL Note Explain how the terms *ratio* and *proportion* are related. Have students give examples.

Words to Know

ratio a comparison of two quantities

proportion an equation that states that two ratios are equal

cross products in a proportion, if $\frac{a}{b} = \frac{c}{d}$, then $a \bullet d = b \bullet c$

geometric mean the square root of the product of two positive numbers; in the proportion $\frac{a}{x} = \frac{x}{b}$, the geometric mean of a and b is x

similar triangles triangles that have the same shape but that may or may not be the same size

similarity ratio the ratio of the lengths of the corresponding sides of similar polygons

scale drawing a drawing showing dimensions in proportion to the object it represents; a length on the drawing is proportional to the object's actual length

Scale Drawing Project

Measure and record the length and width of a room in your home. Then, measure the length and width of the furniture in that room. Make a scale outline of the room on graph paper. Use a scale that shows 12 inches in the room as 1 unit on the graph paper. On a second sheet of graph paper, outline scale furniture (12 in. = 1 unit). Cut out the furniture and arrange it on your room outline. Try rearranging the furniture in different ways. Which arrangement do you like best?

Project See the Classroom Resource Binder for a scoring rubric to assess this project.

More practice is provided in the Classroom Resource Binder, page 88.

9-1 Algebra Review: Ratios

Getting Started
Have students count the number of girls in the class and then the number of boys. Tell them to write the numbers as a fraction and simplify it, if needed. Point out that this is a ratio.

A **ratio** is a comparison of one quantity to another. When you write a ratio, be sure to write the numbers in the order that the items are compared.

The ratio of yellow to blue circles is 4 to 3.
There are three ways to write a ratio:

4 to 3   4:3 $\frac{4}{3}$

A ratio should always be written in simplest form.

EXAMPLE 1

Write the ratio of blue triangles to yellow triangles.

Count the number of blue triangles.	3 blue
Count the number of yellow triangles.	6 yellow
Write the ratio of blue to yellow triangles.	3 to 6
Simplify. Divide each number by 3.	1 to 2

The ratio of blue to yellow triangles is 1 to 2, 1:2, or $\frac{1}{2}$.

Ratios can be used with polygons.

EXAMPLE 2

Write the ratio of the base of the triangle to its height.

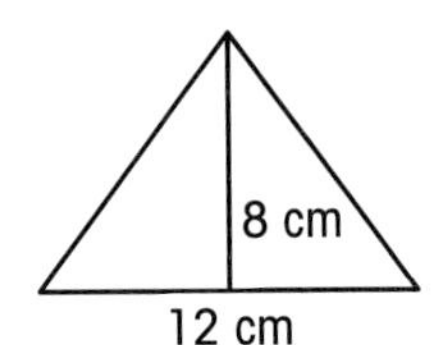

Find the length of the base.	12
Find the length of the height.	8
Write the ratio of base to height.	12 to 8
Simplify. Divide each number by 4.	3 to 2

The ratio of the base to the height is 3 to 2, 3:2, or $\frac{3}{2}$.

Try This

Write the ratio of blue parts to all parts of the pentagon.

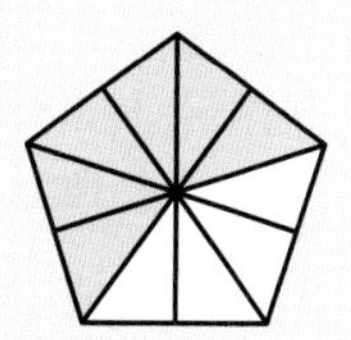

Count the number of blue parts. 6 ■ blue parts

Count all the parts in the pentagon. 10 ■ parts in all

Write the ratio. 6 to ■ 10

Simplify. 3 to 5

The ratio of blue parts to all parts of the pentagon is ■ to ■, 3:5, or $\frac{3}{5}$. (3, 5)

Practice

Write a ratio to compare the parts of this rectangle.

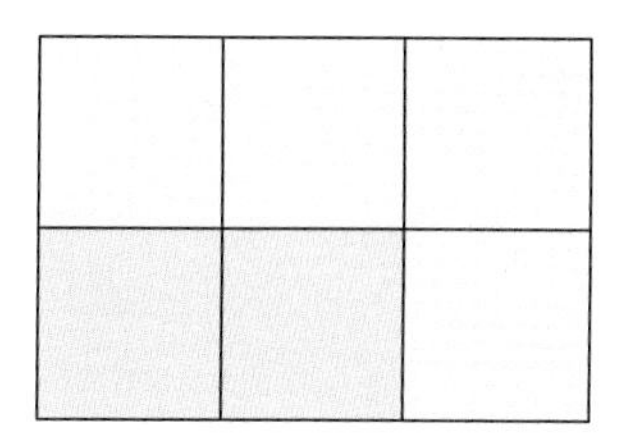

1. yellow to green
   2 to 1, 2:1, or $\frac{2}{1}$
2. green to yellow
   1 to 2, 1:2, or $\frac{1}{2}$
3. yellow to all
   2 to 3, 2:3, or $\frac{2}{3}$
4. all to green
   3 to 1, 3:1, or $\frac{3}{1}$
5. yellow to yellow
   1 to 1, 1:1, or $\frac{1}{1}$
6. blue to yellow
   0 to 4, 0:4, or $\frac{0}{4}$

Share Your Understanding

7. Explain to a partner how to find the ratio of yellow circles to blue circles. Use the words *count* and *to*. Have your partner write the ratio three different ways.

8. **CRITICAL THINKING** Identify the length (l) and the width (w) of the rectangle. Write the ratio of the width to the length. (Hint: Change feet to inches. 1 ft = 12 in.)
   1 to 3, 1:3, or $\frac{1}{3}$, for w = 8 in. and l = 2 ft.

8 in.

2 ft

7. First, count the number of yellow circles. Then, count the number of blue circles. The ratio is 4 to 3. Write 4 to 3, 4:3, or $\frac{4}{3}$.

More practice is provided in the Classroom Resource Binder, page 88.

9·2 Algebra Review: Proportions

Getting Started
Have students find the missing numbers in these equivalent fractions.

$\frac{1}{2} = \frac{5}{■}$ [10]

$\frac{3}{5} = \frac{■}{25}$ [15]

$\frac{■}{24} = \frac{5}{6}$ [20]

Math Fact
If $\frac{a}{b} = \frac{c}{d}$
and b and d are not 0,
then $ad = bc$.

A **proportion** shows that two ratios are equal. $\frac{3}{5} = \frac{9}{15}$

You can cross multiply the parts of a proportion. These products are called **cross products**. Cross products of a proportion are equal.

$\frac{3}{5} \times \frac{9}{15}$ $3 \cdot 15 = 5 \cdot 9$ cross products

$45 = 45$

Sometimes one number in a proportion is unknown. You can use cross products to find that number.

EXAMPLE 1

Solve the proportion. $\frac{8}{x} = \frac{24}{27}$

Write the cross products. $8 \cdot 27 = x \cdot 24$

Simplify. Solve for x. $216 = 24x$

$9 = x$

The unknown number in the proportion, x, is 9.

The parts of a proportion have special names.

extremes $\frac{3}{5}$ means = means $\frac{9}{15}$ extremes

When the means in a proportion are the same number, it is called the **geometric mean**.

EXAMPLE 2

Find the geometric mean of 6 and 24.

Write a proportion. $\frac{6}{x} = \frac{x}{24}$

Write the cross products. $6 \cdot 24 = x \cdot x$

Simplify. Solve for x. $144 = x^2$

$12 = x$

The geometric mean is 12.

Try This

Find the geometric mean of 4 and 9.

Write a proportion.
Let x be the geometric mean. $\frac{4}{x} = \frac{x}{9}$

Write the cross ■. products $4 \cdot ■ = x \cdot ■$ 9 x

Simplify. Solve for x.
36 ■ $= x^2$
6 ■ $= x$

The geometric mean of 4 and 9 is ■. 6

Practice

Solve each proportion.

1. $\frac{3}{x} = \frac{15}{25}$ 5
2. $\frac{5}{7} = \frac{x}{63}$ 45
3. $\frac{45}{75} = \frac{3}{x}$ 5
4. $\frac{x}{64} = \frac{15}{16}$ 60

5. $\frac{68}{x} = \frac{17}{20}$ 80
6. $\frac{2}{x} = \frac{x}{8}$ 4
7. $\frac{3}{x} = \frac{x}{27}$ 9
8. $\frac{7}{x} = \frac{x}{28}$ 14

Find the geometric mean of each pair of numbers.

9. 3 and 12 6
10. 5 and 20 10
11. 9 and 16 12
12. 5 and 45 15

Share Your Understanding

13. Explain to a partner how to solve the proportion on the right. Use the words *multiply, cross product,* and *divide*. Have your partner find the value of x. $\frac{6}{x} = \frac{24}{44}$
Write the cross products. Multiply 6 by 44 and 24 by x. Then, divide 264 by 24; $x = 11$.

14. **CRITICAL THINKING** The ratio of AB to BC is 3 to 2. If AC is 30, what is BC? 12

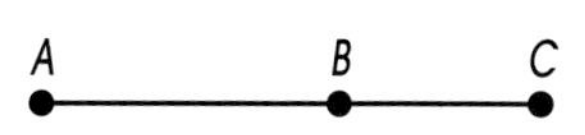

More practice is provided in the Workbook, page 48, and in the Classroom Resource Binder, page 89.

Getting Started
Draw a triangle on the board. Label two angles 35° and 75°. Ask the students to find the measure of the other angle. [70°]

9·3 Similar Triangles

You learned that congruent triangles are the same shape and the same size. **Similar triangles** are the same shape but may or may not be the same size.

Remember
Congruent angles have the same measure.

Two triangles are similar if two conditions are true. First, the corresponding angles are congruent. Second, the ratios of the lengths of the corresponding sides are equal. This ratio is called the **similarity ratio**.

EXAMPLE

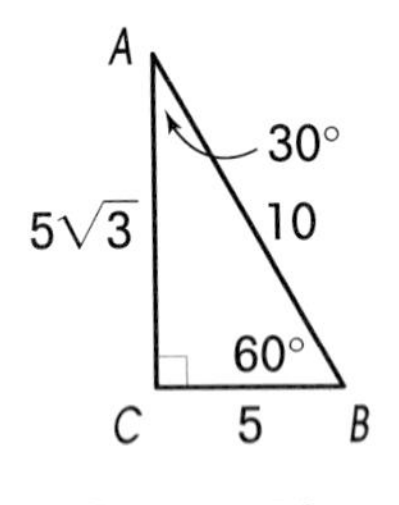

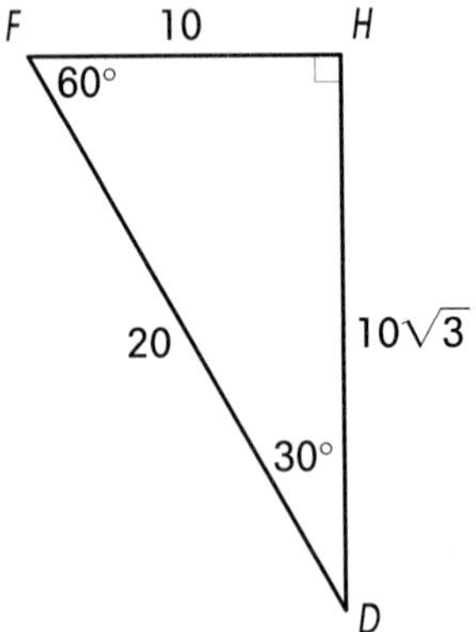

Decide if the two triangles on the left are similar. If they are, write a similarity statement. Find the similarity ratio.

First, see if all the angles are congruent.

$\angle A \cong \angle D$
$\angle B \cong \angle F$
$\angle C \cong \angle H$

Second, find the ratios of the lengths of the corresponding sides. The corresponding sides are between the corresponding angles.

$\overline{AB}$ corresponds to $\overline{DF}$. $\quad \frac{AB}{DF} = \frac{10}{20} = \frac{1}{2}$

$\overline{BC}$ corresponds to $\overline{FH}$. $\quad \frac{BC}{FH} = \frac{5}{10} = \frac{1}{2}$

$\overline{CA}$ corresponds to $\overline{HD}$. $\quad \frac{CA}{HD} = \frac{5\sqrt{3}}{10\sqrt{3}} = \frac{1}{2}$

The corresponding angles are congruent. The ratios are equal.

Write a similarity statement. Match the congruent angles.

So, $\triangle ABC$ is similar to $\triangle DFH$. The similarity ratio of $\triangle ABC$ to $\triangle DFH$ is 1 to 2.

The symbol $\sim$ means "is similar to." You can write $\triangle ABC \sim \triangle DFH$.

Try This

Decide if the two triangles are similar.
If they are, write a similarity statement.
Find the similarity ratio.

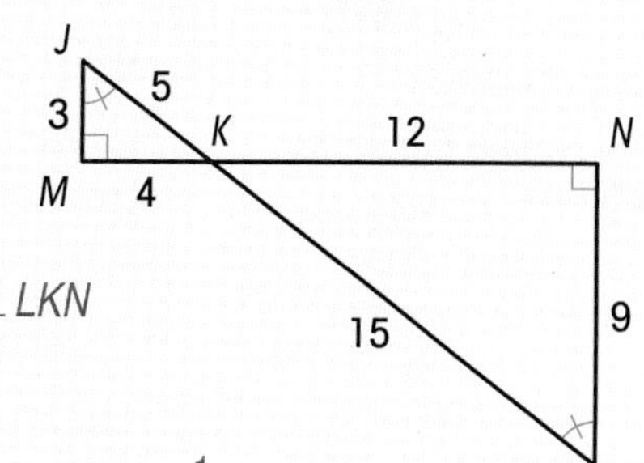

First, see if all the ■ (angles) are congruent. Remember, vertical angles are congruent.

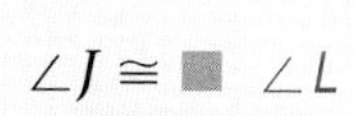

$\angle J \cong$ ■ $\angle L$

$\angle JKM \cong$ ■ $\angle LKN$

$\angle M \cong$ ■ $\angle N$

Second, write the ratios of the lengths of the ■ (corresponding) sides.

$\overline{JK}$ corresponds to ■. $\overline{LK}$ $\qquad \frac{5}{15} = \frac{■ (1)}{3}$

$\overline{KM}$ corresponds to ■. $\overline{KN}$ $\qquad \frac{4}{■ (12)} = \frac{1}{3}$

$\overline{MJ}$ corresponds to ■. $\overline{NL}$ $\qquad \frac{3}{■ (9)} = \frac{1}{3}$

The corresponding angles are ■ (congruent). The ratios are ■ (equal).

So, $\triangle JKM \sim$ ■ ($\triangle LKN$). The similarity ratio is 1 to ■. 3

Practice

Decide if the two triangles are similar. Write *yes* or *no*.
If yes, write a similarity statement. Find the similarity ratio.

1. yes; $\triangle PDK \sim \triangle LBR$; The similarity ratio is 3 to 2.

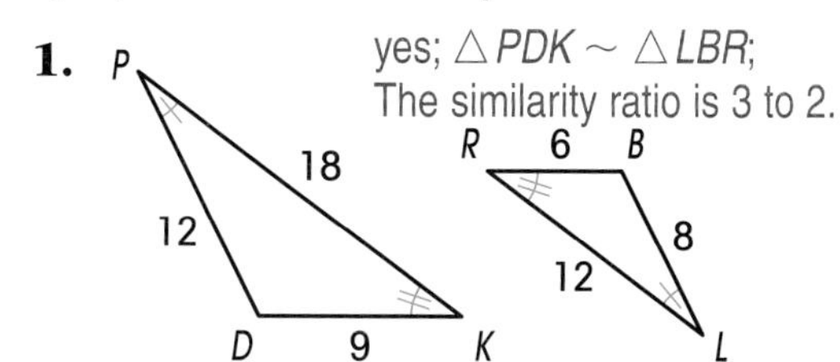

2. no

A, 10, D, 12, C, 18, B, 13, 5, E

Share Your Understanding

3. Explain to a partner how to identify each pair of corresponding angles in exercise **1**. Have your partner identify the pairs of corresponding sides. See above.

3. Look for congruent angle marks to find corresponding angles. $\overline{PK}$ corresponds to $\overline{LR}$; $\overline{DP}$ corresponds to $\overline{BL}$; and $\overline{DK}$ corresponds to $\overline{BR}$.

4. **CRITICAL THINKING** The ratio of *DE* to *EF* is 3 to 1. If *DF* is 20, what is *DE*? 15

More practice is provided in the Workbook, page 48, and in the Classroom Resource Binder, page 89.

Getting Started
Review the Reflexive Property.

9·4 Angle-Angle Similarity

> POSTULATE 12
> **Angle-Angle Similarity Postulate**
>
> If two angles of one triangle are congruent to two angles of another triangle, then the two triangles are similar.

You only need to know that two pairs of corresponding angles are congruent to know that two triangles are similar.

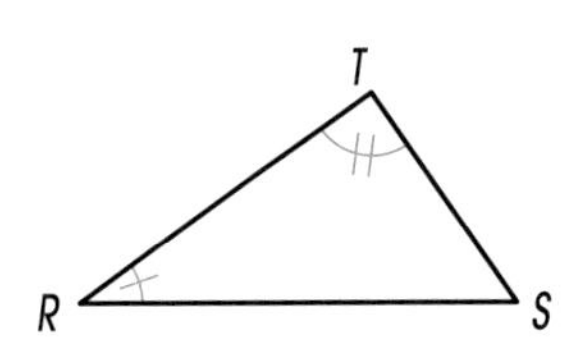

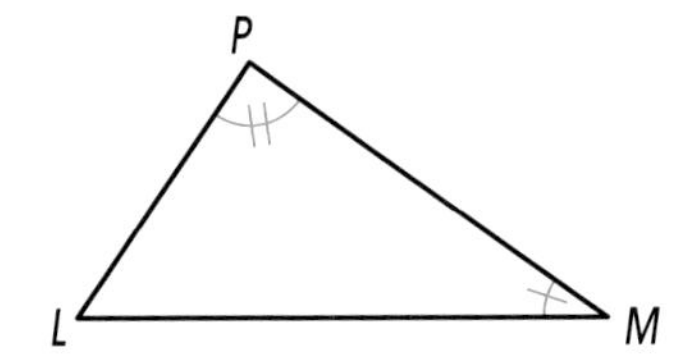

$\triangle RTS \sim \triangle MPL$ by the Angle-Angle Postulate.

If you know the triangles are similar, you can find an unknown length of a side.

EXAMPLE

Decide if $\triangle RSW$ is similar to $\triangle VSB$. If they are, find VS.

Two pairs of corresponding angles are congruent.	$\angle R \cong \angle V \quad \angle WSR \cong \angle BSV$
So, the triangles are similar.	$\triangle RSW \sim \triangle VSB$
Now, write a proportion to find VS. Look for corresponding sides.	$\frac{RW}{VB} = \frac{RS}{VS}$
Substitute the given values.	$\frac{6}{9} = \frac{8}{VS}$
Write the cross products.	$6 \bullet VS = 9 \bullet 8$
Solve for VS.	$6(VS) = 72$
	$VS = 12$

Remember
Vertical angles are congruent.

By the Angle-Angle Postulate, $\triangle RSW \sim \triangle VSB$. VS is 12.

Try This

Decide if $\triangle ARK$ is similar to $\triangle BRK$.

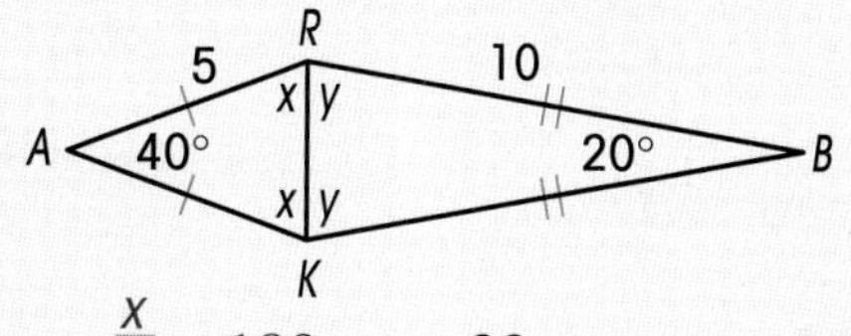

$\triangle ARK$ and $\triangle BRK$ are isosceles ■. triangles
So, the base angles are congruent.
Find the measure of each base angle. $40 + x + \blacksquare = 180$ (x) $\quad 20 + y + y = \blacksquare$ 180

The bases angles are corresponding angles.

$40 + 2x = 180 \quad 20 + 2y = 180$

$2x = \blacksquare$ 140 $\quad 2y = \blacksquare$ 160

$x = \blacksquare$ 70 $\quad y = \blacksquare$ 80

The corresponding angles are ■ (not) congruent.
So, $\triangle ARK$ is ■ (not) similar to $\triangle BRK$.

Practice

Decide if each pair of triangles is similar. Write *yes* or *no*. Explain.

1. 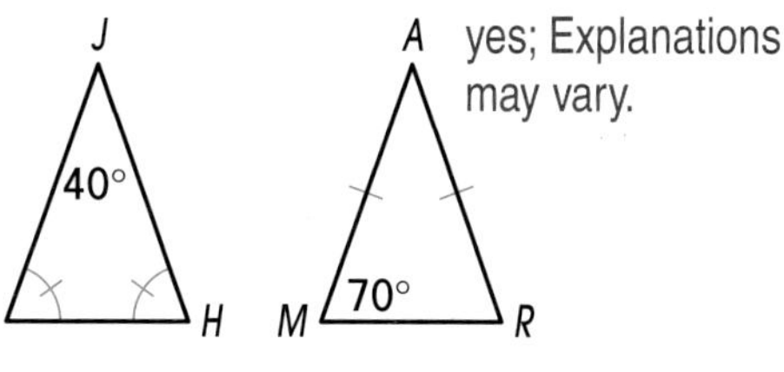

yes; Explanations may vary.

2. 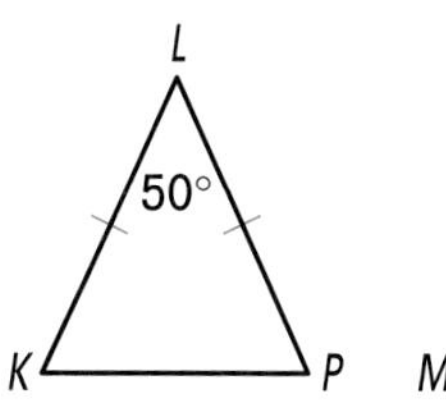

A
70°
M R

no; Explanations may vary.

Each diagram shows similar triangles. Find IN.

3. 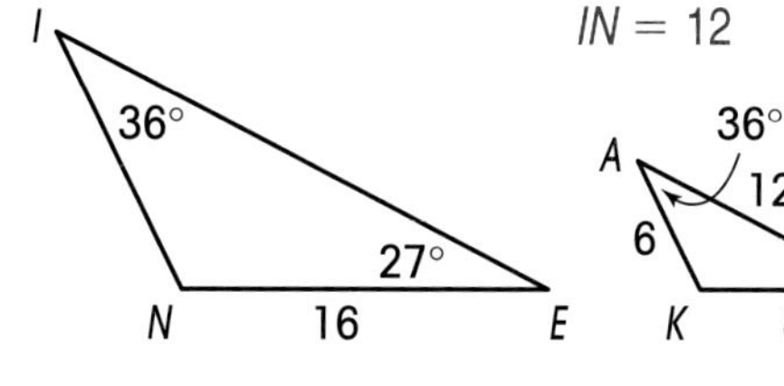

$IN = 12$

4. 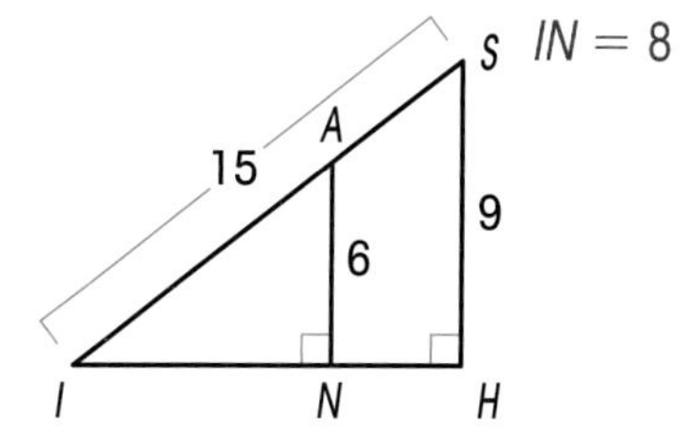

$IN = 8$

Share Your Understanding

5. Explain the Angle-Angle Similarity Postulate to a partner. Have your partner draw two triangles to illustrate your explanation. Check students' work.

6. **CRITICAL THINKING** Are congruent triangles similar? Explain why or why not? yes; Possible answer: Congruent triangles will have at least two pairs of corresponding angles that are congruent.

More practice is provided in the Workbook, page 49, and in the Classroom Resource Binder, page 90.

9-5 Altitude of a Right Triangle

Getting Started
Review simplifying radicals. Point out that $\sqrt{128} = \sqrt{64 \cdot 2} = 8\sqrt{2}$.

You can draw the altitude from the right angle of a right triangle. Then, you will form three similar triangles.

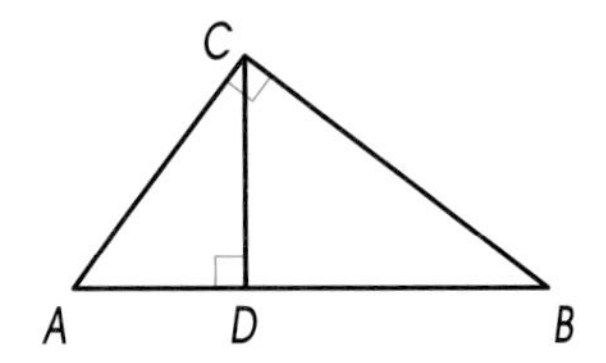

$$\triangle ABC \sim \triangle CAD \sim \triangle BCD$$

The altitude is the geometric mean between the parts of the base. $\frac{AD}{CD} = \frac{CD}{BD}$

You can use a proportion to find the measure of the altitude.

EXAMPLE

Find the length of the altitude.

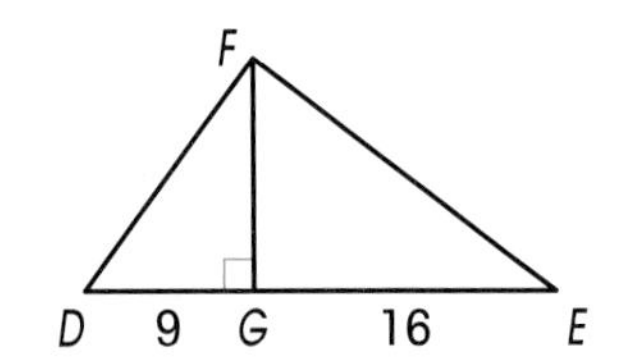

Write a proportion to find FG. FG is the geometric mean. $\frac{DG}{FG} = \frac{FG}{EG}$

Substitute the given values. $\frac{9}{FG} = \frac{FG}{16}$

Write the cross products. $9 \cdot 16 = FG \cdot FG$

Simplify. $144 = (FG)^2$

Find the square root of both sides. $\sqrt{144} = \sqrt{(FG)^2}$

$12 = FG$

The length of the altitude is 12.

Now that you know FG, you can find DF and FE. Use the Pythagorean Theorem. $DF = 15$ $FE = 20$

Try This

$\overline{JK}$ is an altitude of $\triangle HJI$. Find HK and HJ.

J 24 H K 32 I

First, find HK.

Write a ■ (proportion). JK is the ■ (geometric) mean. $\qquad \dfrac{HK}{■ \; JK} = \dfrac{■ \; JK}{IK}$

Substitute the given values. $\qquad \dfrac{HK}{24} = \dfrac{24}{■ \; 32}$

Write the cross products. $\qquad HK \cdot 32 = 24 \cdot ■ \; 24$

Solve for HK. $\qquad 32(HK) = ■ \; 576$

$HK = ■ \; 18$

Now, find HJ. Use the Pythagorean Theorem. $\qquad a^2 + b^2 = ■ \; c^2$

Substitute the given values. $\qquad 18^2 + ■^2 \; (24) = (HJ)^2$

Solve for HJ. $\qquad 900 = (HJ)^2$

$30 \; ■ = HJ$

So, HK is ■ (18) and HJ is ■ (30).

4. To find KL, write a proportion. Then, write the cross products. Find the square roots of the cross products, 5, 184 and $(KL)^2$, to get $KL = 72$. To find KI, use the Pythagorean Theorem: $54^2 + 72^2 = (KI)^2$, $(KI)^2 = 8{,}100$, $KI = 90$. To find KJ, also use the Pythagorean Theorem: $72^2 + 96^2 = (KJ)^2$, $(KJ)^2 = 14{,}400$, $KJ = 120$.

Practice

Find all three unknown lengths in each triangle.

1.

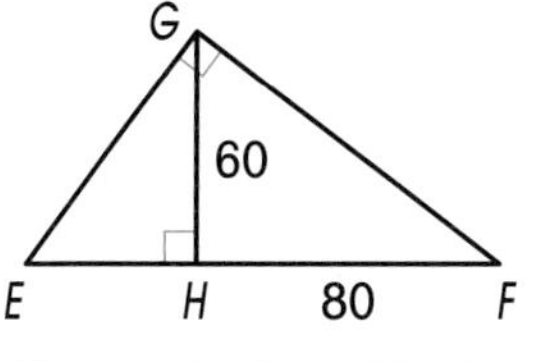

$EH = 45$; $GE = 75$; $GF = 100$

2.

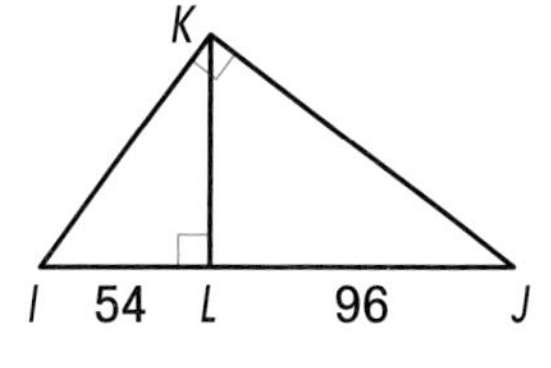

$KL = 72$; $KI = 90$; $KJ = 120$

3.

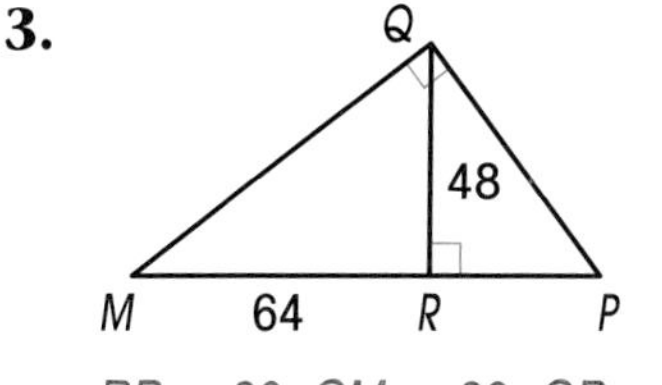

$RP = 36$; $QM = 80$; $QP = 60$

Share Your Understanding

4. Explain to a partner how to find KL in exercise **2**. Have your partner explain how to find KI and KJ. See above.

5. **CRITICAL THINKING** Find DB. $DB = 48$.

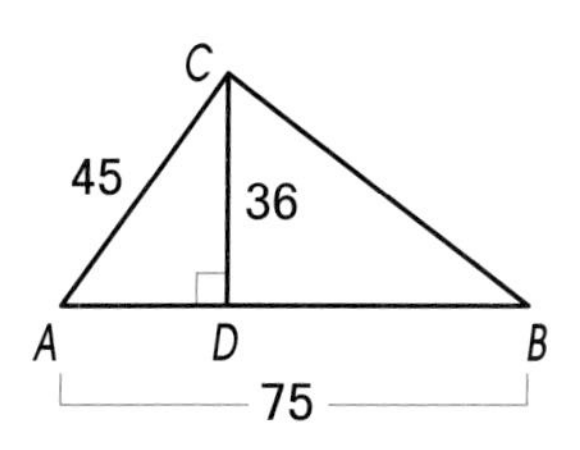

More practice is provided in the Workbook, page 50, and in the Classroom Resource Binder, page 90.

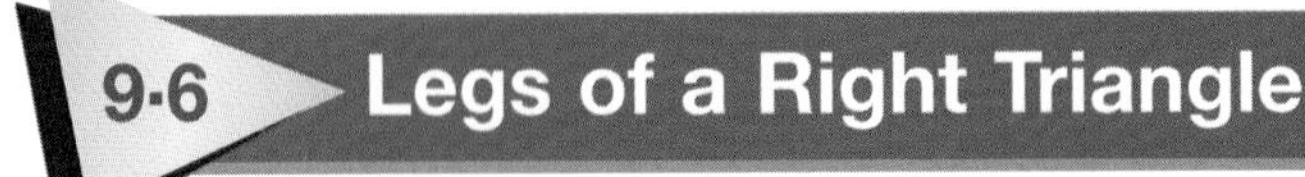

9·6 Legs of a Right Triangle

Getting Started
Review the parts of this special right triangle.

You learned that in a right triangle, the altitude from the right angle is a geometric mean between the parts of the base.

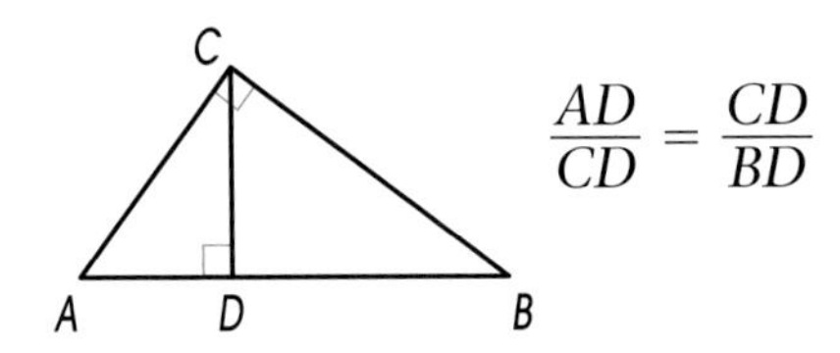

$$\frac{AD}{CD} = \frac{CD}{BD}$$

Each leg of a right triangle is also a geometric mean between part of the base and the whole base.

$$\frac{AD}{AC} = \frac{AC}{AB} \qquad \frac{BD}{BC} = \frac{BC}{BA}$$

You can use these facts to find the measure of a leg.

EXAMPLE 1

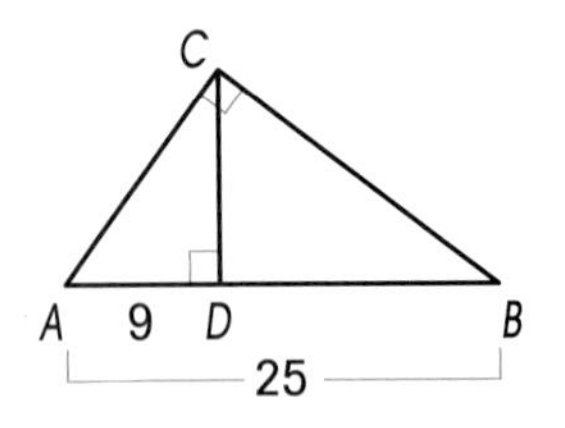

Find AC.

Write a proportion. $\frac{AD}{AC} = \frac{AC}{AB}$

Substitute the given values. $\frac{9}{AC} = \frac{AC}{25}$

Write the cross products. $9 \bullet 25 = AC \bullet AC$
Solve for AC. $225 = (AC)^2$
$15 = AC$

So, AC is 15.

EXAMPLE 2

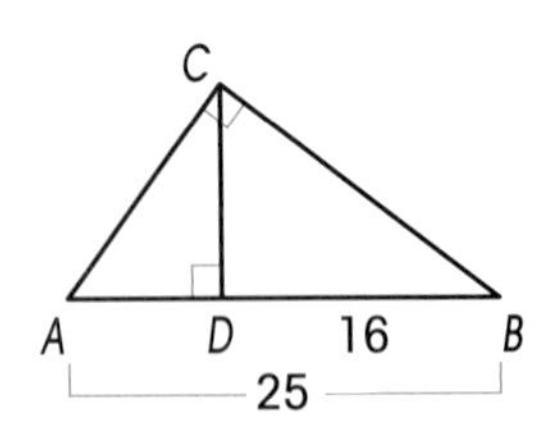

Find BC.

Write a proportion. $\frac{BD}{BC} = \frac{BC}{BA}$

Substitute the given values. $\frac{16}{BC} = \frac{BC}{25}$

Write the cross products. $16 \bullet 25 = BC \bullet BC$
Solve for BC. $400 = (BC)^2$
$20 = BC$

So, BC is 20.

Try These

1. Find *DF*.

Write a ■ to find *DF*. proportion $\frac{DG}{■} = \frac{■}{DE}$ (DF, DF)

Substitute the given values. $\frac{18}{DF} = \frac{DF}{■}$ (50)

Then, solve for *DF*. $■ = (DF)^2$ (900)

$■ = DF$ (30)

So, *DF* is ■. 30

2. Find *EF*.

Write a ■ to find *EF*. proportion $\frac{EG}{■} = \frac{■}{ED}$ (EF, EF)

Subtract to find *EG*. $ED - GD = EG$

$50 - 18 = 32$

Substitute. $\frac{■}{EF} = \frac{EF}{50}$ (32)

Then, solve for *EF*. $■ = (EF)^2$ (1,600)

$■ = EF$ (40)

So, *EF* is ■. 40

4. To find *HJ*, write and solve a proportion with *HJ*. $\frac{HK}{HJ} = \frac{HJ}{HI}$; Substitute 27 for *HK* and 75 for *HI*. *HJ* is 45. To find *IJ*, write and solve a proportion with *IJ*. $\frac{IK}{IJ} = \frac{IJ}{IH}$; Substitute 75 for *IH*. Since $75 - 27 = 48$, substitute 48 for *IK*. *IJ* is 60.

Practice

Find the length of each leg of each large triangle.

1.

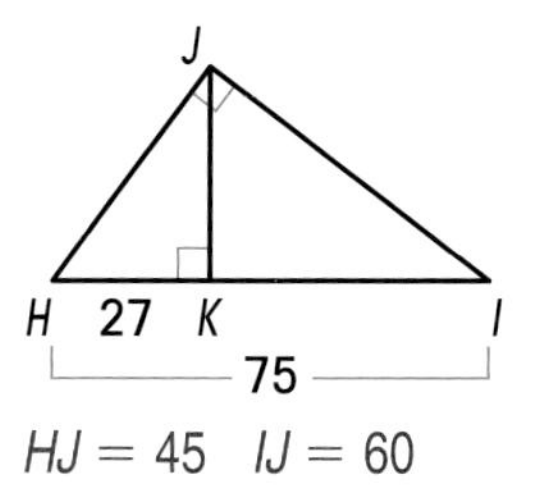

$HJ = 45$ $IJ = 60$

2.

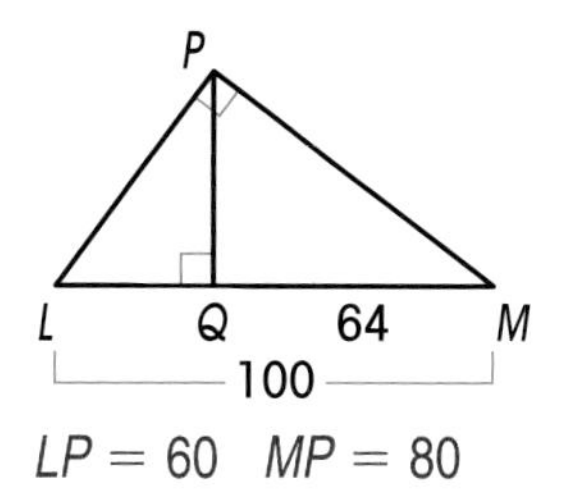

$LP = 60$ $MP = 80$

3.

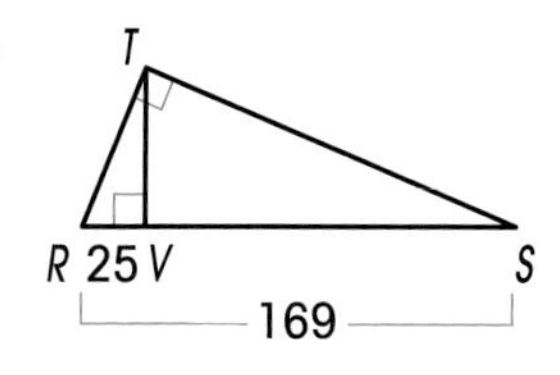

$RT = 65$ $ST = 156$

Share Your Understanding

4. Explain to a partner how to find *HJ* in exercise **1**. Have your partner explain how to find *IJ*.
See above.

5. **CRITICAL THINKING** Find *AB*. 25

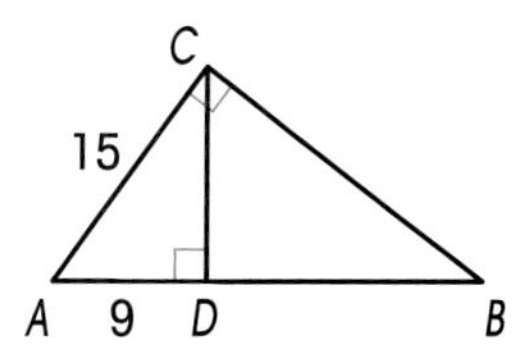

More practice is provided in the Workbook, page 51, and in the Classroom Resource Binder, page 91.

9·7 Side-Splitter Theorem

Getting Started Review the definition of a midsegment.

THEOREM 26
Side-Splitter Theorem

If a line is parallel to one side of a triangle and intersects the other two sides, then it divides the two sides proportionally.

You learned that the midsegment is parallel to a base. It connects the midpoints of two sides.

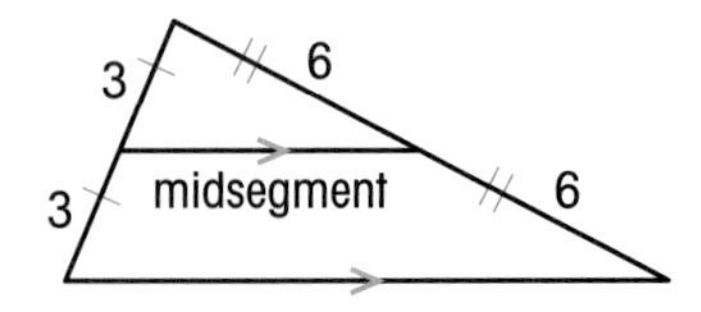

Any line that connects two sides of a triangle and is parallel to the third side divides the two sides proportionally. This is the Side-Splitter Theorem.

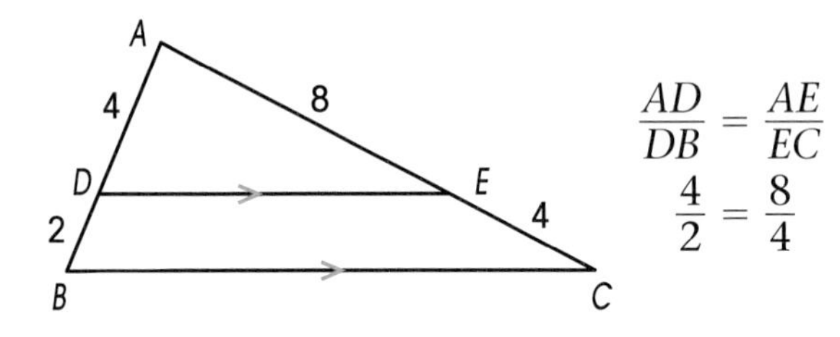

$$\frac{AD}{DB} = \frac{AE}{EC}$$
$$\frac{4}{2} = \frac{8}{4}$$

Remember

$\|$ means "is parallel to."

You can use the Side-Splitter Theorem to find the unknown measure of a side.

EXAMPLE

In the diagram, $\overline{TS} \parallel \overline{QP}$. Find RS.

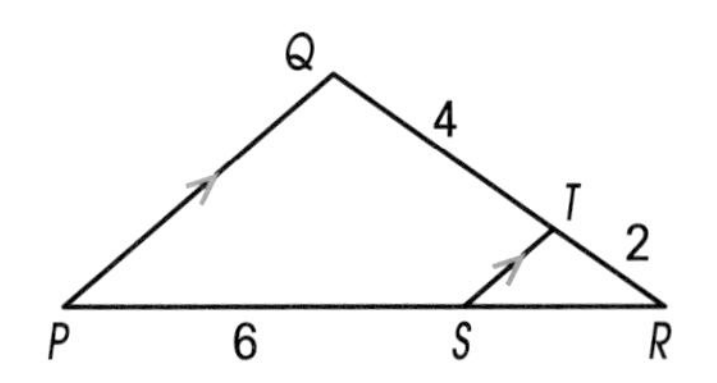

Write a proportion. Use the Side-Splitter Theorem. $\frac{RT}{TQ} = \frac{RS}{SP}$

Substitute the given values. $\frac{2}{4} = \frac{RS}{6}$

Write the cross products. $2 \cdot 6 = 4 \cdot RS$

Solve for RS.

$12 = 4(RS)$

$3 = RS$

So, RS is 3.

Try This

In the diagram, $\overline{KM} \parallel \overline{HJ}$. Find IJ.

First, you need to find MJ. Write a proportion. Use the ■ Theorem. Side-Splitter

$$\frac{IK}{■} = \frac{IM}{MJ} \quad KH$$

Substitute the given values.

$$\frac{5}{■} = \frac{15}{MJ} \quad 3$$

Write the cross products.

$$5 \cdot MJ = 3 \cdot ■ \quad 15$$

$$5(MJ) = ■ \quad 45$$

$$MJ = ■ \quad 9$$

Now, use the Segment ■ Postulate. Addition

$$IJ = IM + MJ$$

Substitute the values you know.

$$= 15 + ■ = ■ \quad 9 \quad 24$$

So, IJ is ■. 24

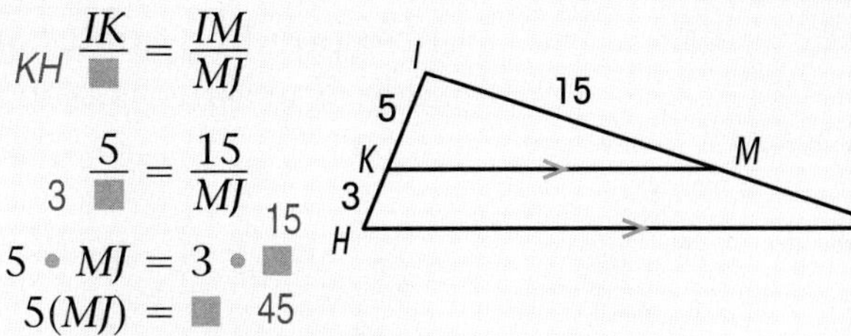

Practice

Find the value of x in each triangle.

1. 30

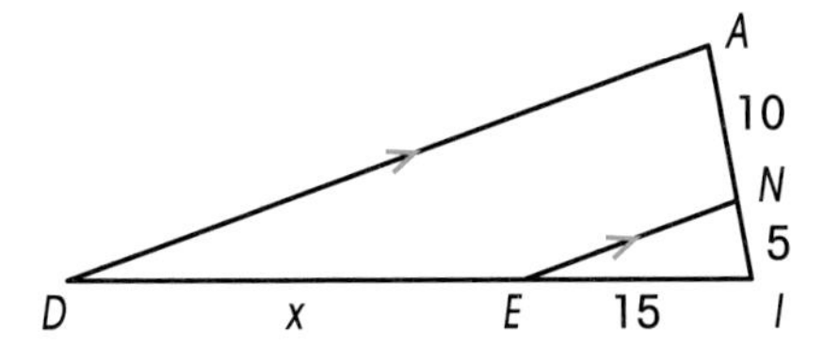

2. 4

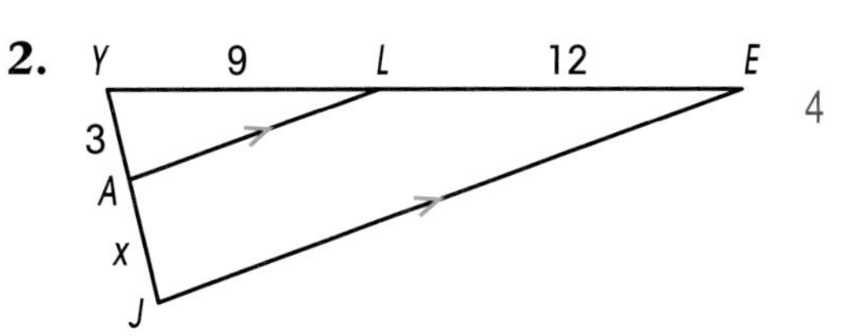

3. 20

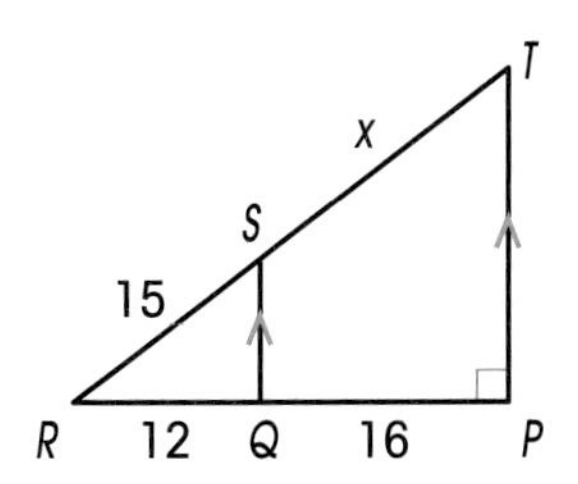

4. 25

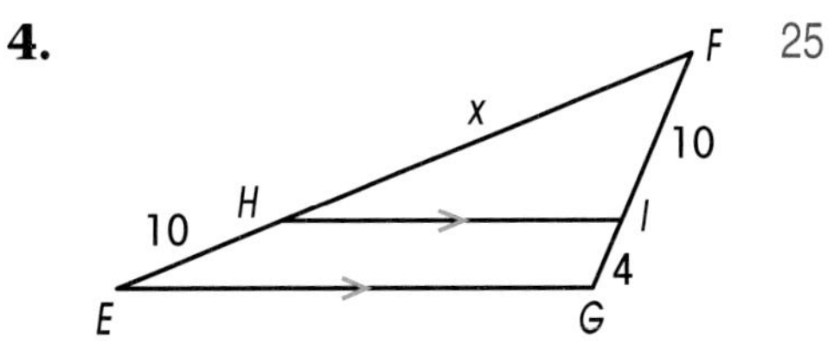

5. Write a proportion. Use the Side-Splitter Theorem. Write the cross products. Solve for x.

Share Your Understanding

5. Explain to a partner how to find the value of x in one of the exercises above. Use the words *proportion, Side-Splitter,* and *cross products* in your explanation.
See above.

6. **CRITICAL THINKING** In the diagram, $\overline{EM} \parallel \overline{JY}$. Find MY.(Hint: First, find RM. Use the Pythagorean Theorem.)
5

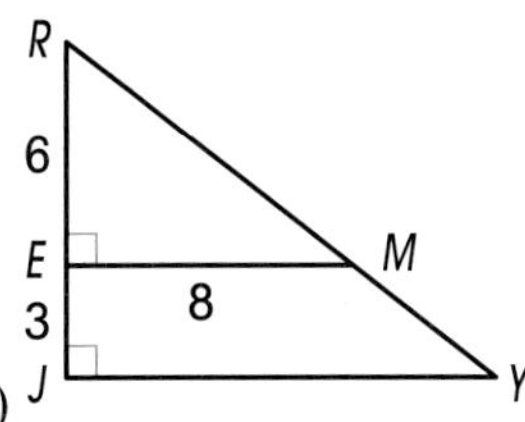

More practice is provided in the Workbook, page 52, and in the Classroom Resource Binder, page 92.

Getting Started
Ask students to list objects around the house or classroom that are similar. Call on volunteers for answers.
[Possible answers: coins, measuring cups, mixing bowls, pencils, beekers]

9·8 Similar Polygons

You learned that similar triangles have congruent corresponding angles and proportional corresponding sides. The same is true for similar polygons.

Remember
To write a similarity statement, match the congruent angles.

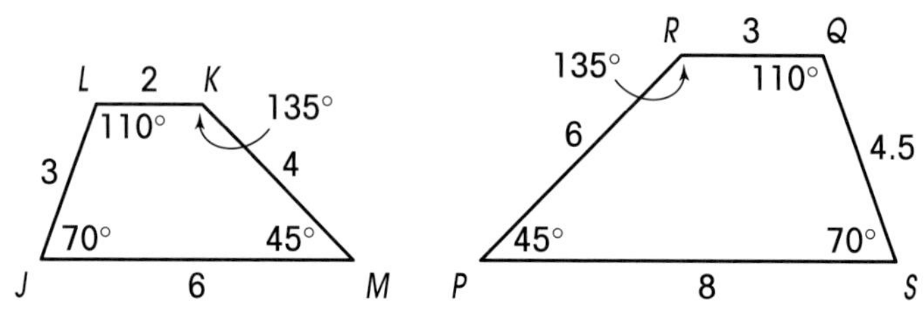

trapezoid $JLKM \sim$ trapezoid $SQRP$

The ratio of the lengths of the corresponding sides is the similarity ratio.

$\overline{LK}$ corresponds to $\overline{QR}$. $\quad \frac{LK}{QR} = \frac{2}{3}$

The similarity ratio is $\frac{2}{3}$, or 2 to 3.

If you know two polygons are similar, you can find the length of a side or the measure of an angle.

EXAMPLE

Trapezoid $AJEM \sim$ trapezoid $QBRK$. Find QK. Write the similarity ratio.

J E 4 A M 10

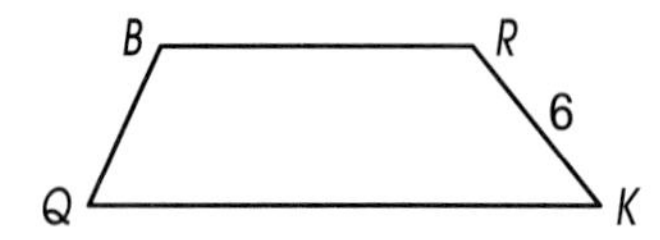

Look for corresponding sides.	$\overline{EM}$ corresponds to $\overline{RK}$. $\overline{AM}$ corresponds to $\overline{QK}$.
Write a proportion.	$\frac{EM}{RK} = \frac{AM}{QK}$
Substitute the values you know.	$\frac{4}{6} = \frac{10}{QK}$
Write the cross products. Solve for QK.	$4 \cdot QK = 6 \cdot 10$ $4(QK) = 60$ $QK = 15$

So, QK is 15.

The similarity ratio is 4 to 6, or 2 to 3.

Try This

Parallelogram $PHCT \sim$ parallelogram $DGNS$.
Find DS. Write the similarity ratio.

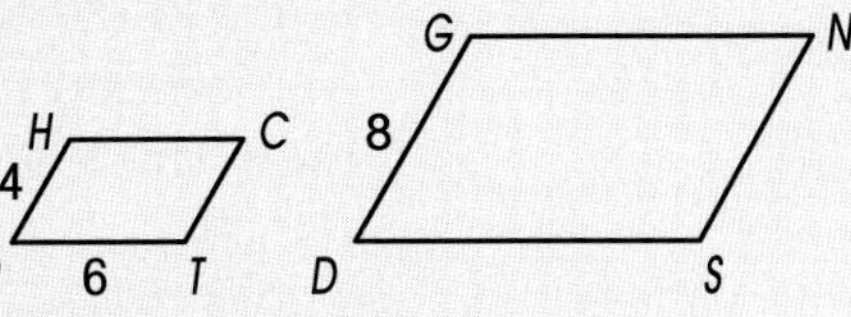

Look for corresponding sides. $\overline{PH}$ corresponds to ■. *DG*
$\overline{PT}$ corresponds to ■. *DS*

Write a proportion. $\frac{PH}{DG} = \frac{■}{DS}$ *PT*

Substitute the values you know. $\frac{4}{8} = \frac{■}{DS}$ 6

Write the cross products. $4 \bullet DS = 8 \bullet 6$

Solve for DS. $4(DS) = ■$ 48

$DS = ■$ 12

So, DS is ■. 12

The similarity ratio is 4 to 8, or 1 to ■. 2

Practice

Each pair of polygons is similar. Find the value of x. Write the similarity ratio.

1. 27

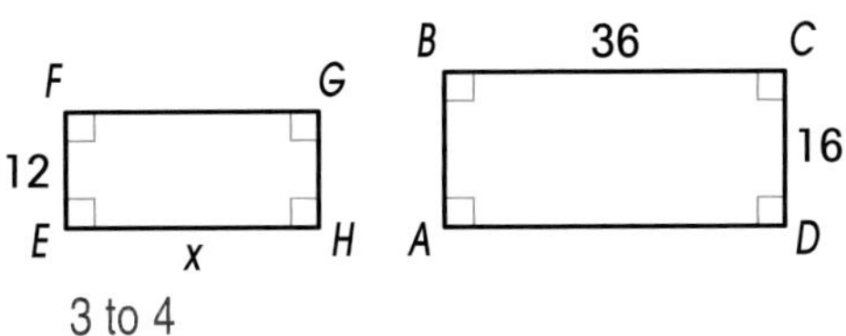

3 to 4

2. 2

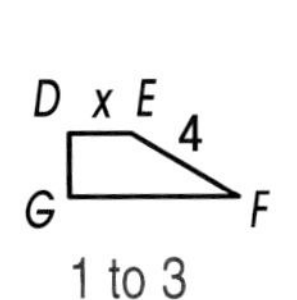

T 6 U 12 S V

1 to 3

Share Your Understanding

3. Explain to a partner how to identify similar polygons. Have your partner draw and label two similar rectangles. Check students' work. Similar polygons have congruent corresponding angles and proportional corresponding sides.

4. **CRITICAL THINKING** In the diagram, $\triangle DOV \sim \triangle DVA$. Find DO and OV. (Hint: First, find AV.)
$DO = 9 \quad OV = 12$

A V 25 15 O D

More practice is provided in the Workbook, page 52, and in the Classroom Resource Binder, page 92.

9·9 Perimeter of Similar Polygons

Getting Started Review perimeter of a polygon.

THEOREM 27

If two polygons are similar, then the ratio of their perimeters is equal to the ratio of any pair of corresponding sides.

If you know the perimeter of a polygon, you can find the perimeter of a similar polygon. You only need to know the lengths of a pair of corresponding sides.

$$\frac{\text{Perimeter}_1}{\text{Perimeter}_2} = \frac{\text{side}_1}{\text{side}_2}$$

EXAMPLE

In the diagram, $\triangle ABC \sim \triangle DEF$. Find the perimeter of $\triangle DEF$.

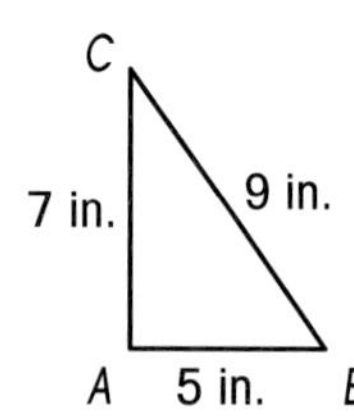

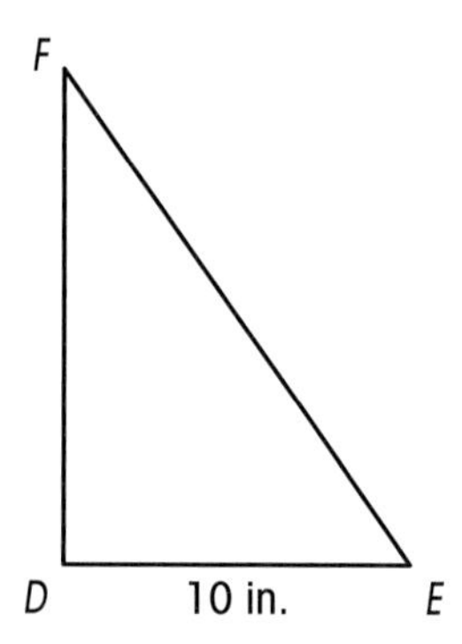

First, find the perimeter of $\triangle ABC$. Add the lengths of all the sides. $5 + 9 + 7 = 21$

$\overline{AB}$ corresponds to $\overline{DE}$. Write a proportion. $\frac{\text{Perimeter } ABC}{\text{Perimeter } DEF} = \frac{AB}{\text{DE}}$

Substitute the values you know. Let x = Perimeter DEF. $\frac{21}{x} = \frac{5}{10}$

Write the cross products. Solve for x.

$$21 \cdot 10 = x \cdot 5$$
$$210 = 5x$$
$$42 = x$$

The perimeter of $\triangle DEF$ is 42 in.

In the example above, the ratio of AB to DE is 5 to 10, or 1 to 2. The similarity ratio is 1 to 2. The ratio of the perimeters is 21 to 42, or 1 to 2. So, the ratio of the perimeters is the same as the similarity ratio. You can use the similarity ratio to write a proportion.

Try This

In the diagram, $\square GHIJ \sim \square KLMN$.
The perimeter of $\square GHIJ$ is 20 cm.
Find the perimeter of $\square KLMN$.

H I G J 4 cm L M K N 6 cm

$\overline{HG}$ corresponds to ■ $\overline{LK}$. So, the similarity ratio is 4 to 6, or 2 to ■ 3. Use it to write a proportion.

$$\frac{\text{Perimeter } GHIJ}{\text{Perimeter } KLMN} = \frac{HG}{■ \; LK}$$

Substitute the perimeter of *GHIJ*.
Let x = perimeter ■ *KLMN*.

$$\frac{20}{x} = \frac{2}{3}$$

Write the ■ cross products.
Solve for *x*.

$$20 \cdot 3 = ■ \quad x \cdot 2$$
$$60 \; ■ = 2x$$
$$30 \; ■ = x$$

The perimeter of $\square KLMN$ is ■. 30 cm

Practice

Each pair of polygons is similar. The measures of corresponding sides are given. Find the unknown perimeter.

1. 36 cm

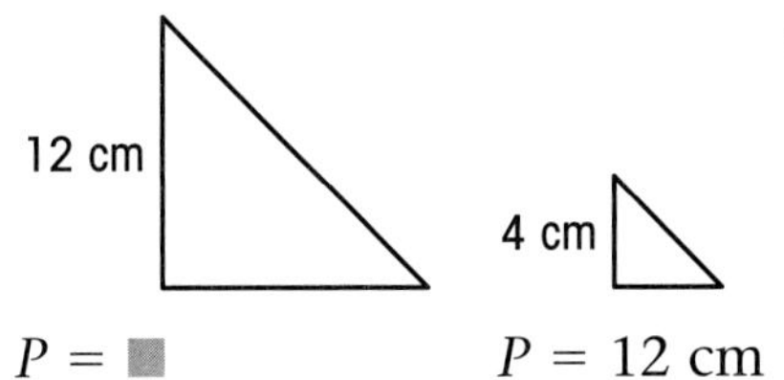

$P =$ ■ $\quad P = 12$ cm

2. 42 m

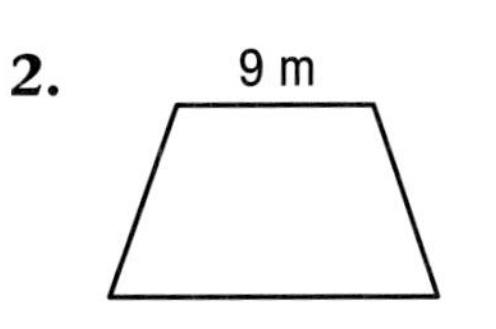

6 m

$P =$ ■ $\quad P = 28$ m

3. 12 ft

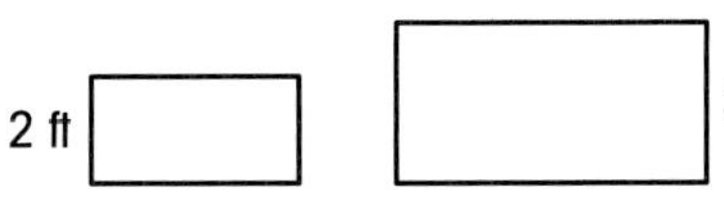

$P =$ ■ $\quad P = 18$ ft

4. 24 in.

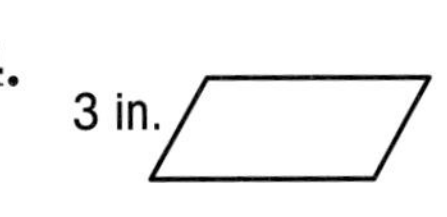

4 in.

$P = 18$ in. $\quad P =$ ■

Share Your Understanding

5. Explain to a partner how to use the similarity ratio to find the perimeter of similar polygons. Use the word *proportion* in your explanation. Write a proportion. Substitute the given values. Let x = the perimeter you want to find. Find the cross products. Solve for x.

6. **CRITICAL THINKING** $\triangle PQR \sim \triangle JKL$ The perimeter of $\triangle PQR$ is 12 in. The perimeter of $\triangle JKL$ is 18 in. What is the similarity ratio? 2 to 3

More practice is provided in the Workbook, page 53 and in the Classroom Resource Binder, page 93.

9·10 Area of Similar Polygons

Getting Started Review area of a polygon

THEOREM 28

If two polygons are similar, then the ratio of their areas is equal to the ratio of the squares of the lengths of any pair of corresponding sides.

If you know the area of one polygon, you can find the area of a similar polygon. You only need to know the lengths of a pair of corresponding sides.

$$\frac{\text{Area}_1}{\text{Area}_2} = \frac{(\text{side}_1)^2}{(\text{side}_2)^2}$$

EXAMPLE

In the diagram, rectangle *ABCD* ~ rectangle *EFGH*. The area of rectangle *ABCD* is 20 in.^{2} Find the area of rectangle *EFGH*.

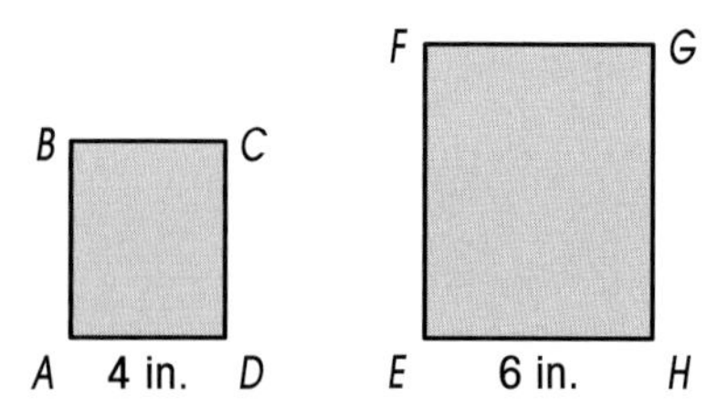

$\overline{AD}$ corresponds to $\overline{EH}$. Write a proportion.

$$\frac{\text{Area } ABCD}{\text{Area } EFGH} = \frac{(AD)^2}{(EH)^2}$$

Substitute the given values. Let x = Area *EFGH*.

$$\frac{20}{x} = \frac{4^2}{6^2}$$

Simplify.

$$\frac{20}{x} = \frac{16}{36}$$

Write the cross products. Solve for x.

$$20 \bullet 36 = x \bullet 16$$
$$720 = 16x$$
$$45 = x$$

The area of rectangle *EFGH* is 45 in.^{2}

In the example above, the ratio of *AD* to *EH* is 4 to 6, or 2 to 3. The similarity ratio is 2 to 3. The ratio of the areas is 16 to 36, or 4 to 9. So, the ratio of the area is the square of the similarity ratio.

Try This

▱$GHIJ \sim$ ▱$KLMN$. The area of ▱$GHIJ$ is 40 cm^2. Find the area of ▱$KLMN$.

$\overline{GH}$ corresponds to ■ $\overline{KL}$. The similarity ratio is 5 to 10, or 1 to 2. Write a proportion.

$$\frac{\text{Area } GHIJ}{\text{Area } KLMN} = \frac{(GH)^2}{■ \ (KL)^2}$$

Substitute the given values.
Let x = Area ■ $KLMN$.

$$\frac{40}{x} = \frac{1^2}{2^2}$$

Write the ■ cross products.

$$40 \cdot 4 = x \cdot ■ \ 1$$

Solve for x.

$$160 ■ = x$$

The area of ▱$KLMN$ is ■. 160 cm^2

Practice

Each pair of polygons is similar. The measures of corresponding sides are given. Find the unknown area.

1.

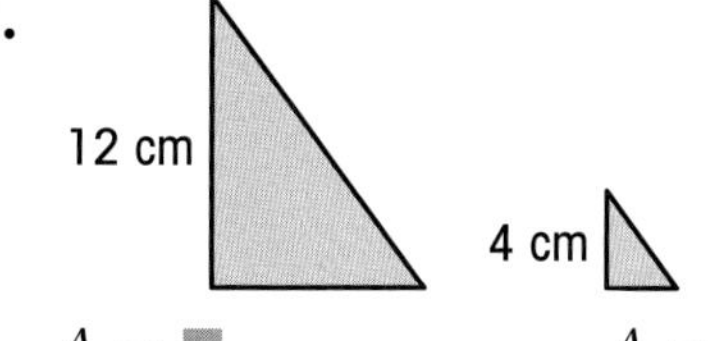

A = ■ $\quad$ $A = 6\ cm^2$ $\quad$ 54 cm^2

2.

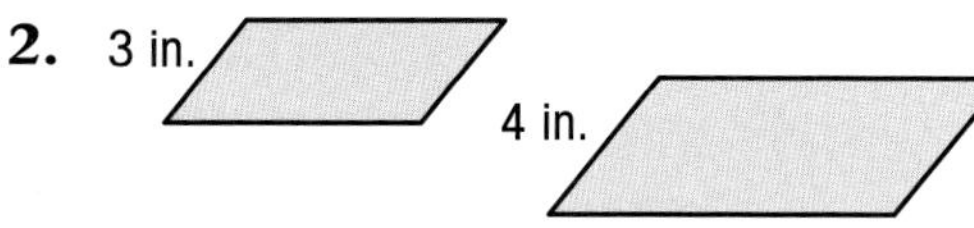

$A = 18\ in.^2$ $\quad$ A = ■ $\quad$ 32 $in.^2$

3.

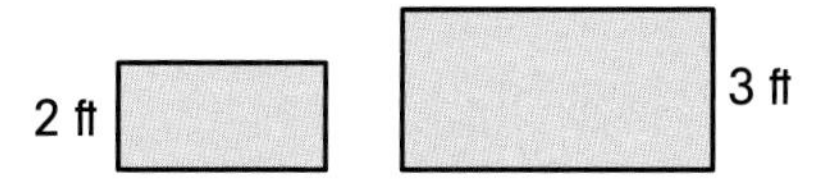

A = ■ $\quad$ $A = 18\ ft^2$ $\quad$ 8 ft^2

4.

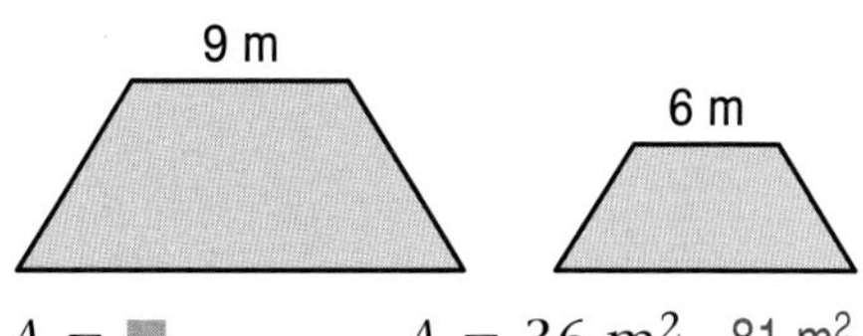

A = ■ $\quad$ $A = 36\ m^2$ $\quad$ 81 m^2

5. Write a proportion. Let x = the unknown area. Set the ratio of the areas of the similar polygons equal to the square of the similarity ratio. Solve the proportion.

Share Your Understanding

5. Explain to a partner how to use the similarity ratio to find the area of similar polygons. Use the word *proportion* in your explanation. See above.

6. **CRITICAL THINKING** $\triangle PQR \sim \triangle JKL$. The area of $\triangle PQR$ is 9 $in.^2$ The area of $\triangle JKL$ is 16 $in.^2$ What is the similarity ratio? 3 to 4

Getting Started
Tell students to use their calculators to solve this problem: A car travels 50 miles in 41 minutes. What is the car's speed? [73 mph]

9-11 Calculator: Solving Proportions

You can use a proportion to solve real-life problems. You can use a calculator to help you.

EXAMPLE 1

A plane flies 3,635 miles from New York to Paris in about 7.5 hours. About how long would it take the same plane to fly 4,933 miles from New York to Athens?

Let x = hours from New York to Athens. Write a proportion. $\frac{3,635}{7.5} = \frac{4,933}{x}$ ← miles / ← hours

Write the cross products. $3,635 \cdot x = 7.5 \cdot 4,933$

Simplify. $3,635x = 7.5 \cdot 4,933$

Use your calculator to multiply. **Display**

PRESS: 7 . 5 × 4 9 3 3 = → 36997.5

Do not clear the calculator. Solve for x. $\frac{3,635x}{3,635} = \frac{36,997.5}{3,635}$

Use your calculator to divide. **Display**

PRESS: ÷ 3 6 3 5 = → 10.178129

Round 10.178129 to the nearest hour.

It would take the plane about 10 hours.

EXAMPLE 2

Decide if the ratios 6:8 and 55:75 form a proportion.

Write the ratios in fractional form. Use ? for =. $\frac{6}{8} \ ? \ \frac{55}{75}$

Write the cross products. $6 \cdot 75 \ ? \ 8 \cdot 55$

Use your calculator to multiply the left side. **Display**

PRESS: 6 × 7 5 = → 450

Use your calculator to multiply the right side. **Display**

PRESS: 8 × 5 5 = → 440

The cross products are not equal. So, the ratios do not form a proportion.

Practice

Solve each problem, using a proportion. Show your work.

1. Daniel drives 125 miles to visit his grandmother. The trip takes 2.5 hours. Daniel's uncle lives 150 miles away from Daniel. Assuming Daniel always travels at the same speed, how long will it take Daniel to get to his uncle's house?
   3 hours

2. It takes Daniel 4 hours to travel 200 miles. How long would it take Daniel to travel 360 miles? 7.2 hours

Decide if each pair of ratios forms a proportion. Write *yes* or *no*.

3. 3 to 2 and 4 to 5 no
4. 16 to 8 and 4:2 yes
5. 3:1 and 9:3 yes
6. 5:4 and 12:15 no
7. $\frac{11}{1}$ and $\frac{22}{2}$ yes
8. 25 to 5 and 36:6 no

Math Connection

THE GOLDEN RECTANGLE

The Golden Rectangle is thought to be the most beautiful rectangle. It is often used in the construction of buildings. The ratio of the length to the width is about 1.6:1. This is called the Golden Ratio.

The Parthenon, built by ancient Greeks, is an example of the Golden Rectangle.

A Golden Rectangle can be divided into a square and a rectangle that is similar to the original rectangle. Draw a Golden Rectangle.

More practice is provided in the Workbook, page 54, and in the Classroom Resource Binder, page 94.

Getting Started
Tell students that a proportion can be written in many ways to solve a word problem. Point out that once the first ratio is written, the second ratio must be written in the same order.

9·12 Problem-Solving Strategy: Write an Equation

You can use similar triangles and proportions to measure distances.

EXAMPLE

A 5-ft-tall woman standing near a flagpole casts a shadow 12 ft long. The flagpole casts a shadow 24 ft long. What is the height of the flagpole?

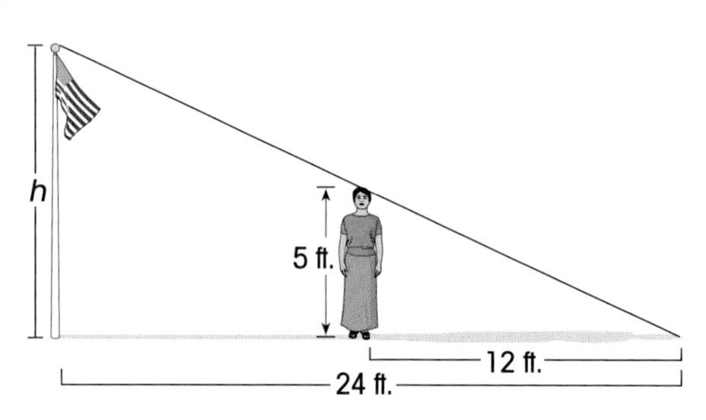

READ **What do you need to find out?**
Find the height of the flagpole.

PLAN **What do you need to do?**
Set up a proportion and solve it.

DO **Follow the plan.**
Let h = height of the flagpole.

$$\frac{\text{height of flagpole}}{\text{height of woman}} = \frac{\text{length of flagpole's shadow}}{\text{length of woman's shadow}}$$

$$\frac{h}{5} = \frac{24}{12}$$

$$h \cdot 12 = 5 \cdot 24$$

$$h = 120 \div 12 = 10$$

The height of the flagpole is 10 ft.

CHECK **Does your answer make sense?**
Yes, the flagpole's shadow is longer than the woman's shadow. ✓

Try This

Look at the drawing on the right.
What is the height of the cactus?

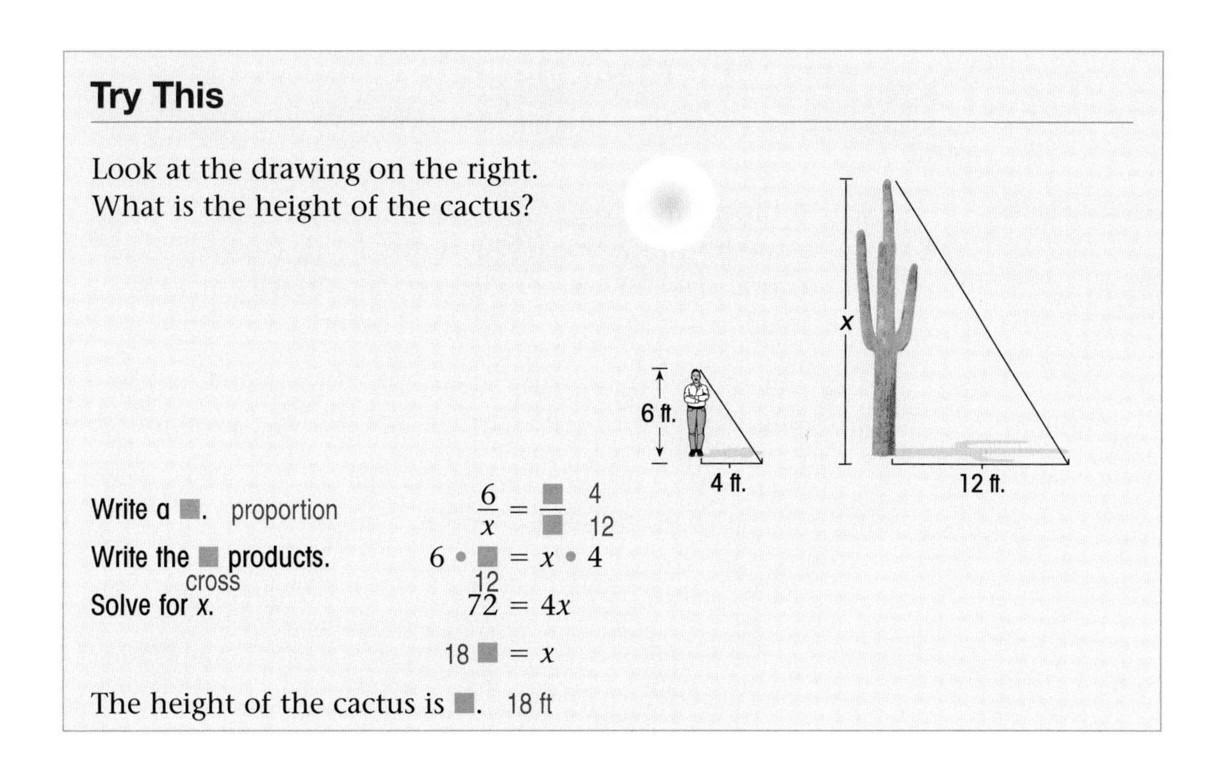

Write a ■. proportion $\quad \frac{6}{x} = \frac{■}{■}$ 4 12

Write the ■ products. cross $\quad 6 \bullet ■ = x \bullet 4$ 12

Solve for x. $\quad 72 = 4x$

18 ■ $= x$

The height of the cactus is ■. 18 ft

Practice

Solve each problem. Show your work.

1. A radio tower casts a shadow that is 14 m long. At the same time, a 2-m post casts a shadow 0.4 m long. What is the height of the radio tower? 70 m

2. A 6-ft-tall woman casts a shadow 5 ft long. At the same time, a tree casts a shadow 30 ft long. How tall is the tree? 36 ft

3. A 4-ft-tall girl casts a shadow 3 ft long. At the same time, a telephone pole casts a shadow 18 ft long. How tall is the telephone pole? 24 ft

4. A 5-ft-tall boy standing near a tree casts a shadow 12 ft long. At the same time, the tree casts a shadow 48 ft long. What is the height of the tree? 20 ft

More practice is provided in the Workbook, page 55, and in the Classroom Resource Binder, page 95.

Problem-Solving Application: Scale Drawings

Getting Started
Review with students how to change a mixed number to a fraction.

A blueprint is an example of a **scale drawing**. A scale drawing shows a real object smaller or larger than its actual size. The dimensions on the drawing are in proportion to those of the real object.

This is a scale drawing of an apartment. The scale is $\frac{1}{8}$ inch to 1 foot. This means that $\frac{1}{8}$ inch in the drawing represents 1 foot in the apartment.

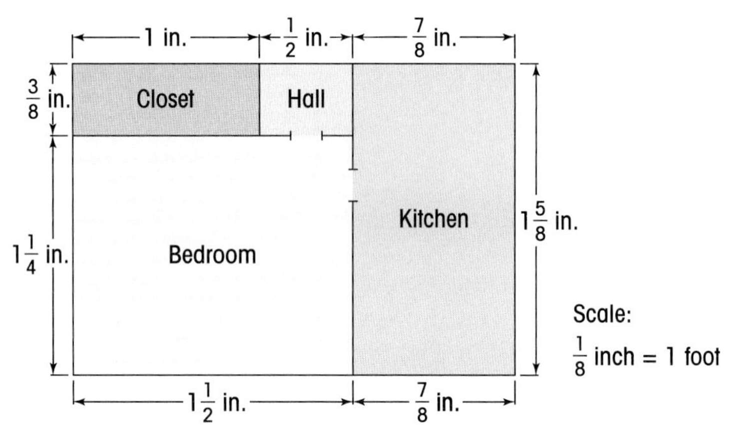

You can find the actual measures of each room in the apartment.

EXAMPLE

The length of the bedroom is $1\frac{1}{2}$ in. on the drawing. What is the actual measure?

Write a proportion. $\quad \dfrac{\frac{1}{8}}{1} = \dfrac{1\frac{1}{2}}{x}$

Write the cross products. $\quad \frac{1}{8} \bullet x = 1 \bullet 1\frac{1}{2}$

$$\frac{1}{8}x = 1\frac{1}{2}$$

Solve for x. $\quad \frac{1}{8}x = \frac{3}{2}$

$$x = \frac{3}{2} \bullet \frac{8}{1}$$

$$x = 12$$

Math Fact
$1\frac{1}{2} = \frac{3}{2}$

The bedroom is 12 feet long.

Try This

What is the width of the bedroom in the scale drawing on page 258?

Write a ■. proportion $\quad \dfrac{\frac{1}{8}}{1} = \dfrac{■}{x}$ $1\frac{1}{4}$

Write the ■ products. cross $\quad ■ \bullet x = 1 \bullet 1\frac{1}{4}$ ($■ = \frac{1}{8}$)

Solve for x. $\quad \frac{1}{8}x = ■$ $1\frac{1}{4}$

Change $1\frac{1}{4}$ to $\frac{5}{4}$. $\quad \frac{1}{8}x = \frac{5}{4}$

Multiply both sides by $\frac{8}{1}$. $\quad x = ■ \bullet \frac{8}{1}$ $\frac{5}{4}$

$x = ■$ $\frac{40}{4}$ or 10

The bedroom is ■ wide. 10 ft

Practice

Use the scale drawing on page 258 to solve each problem. Show your work.

1. What is the length of the kitchen? 13 ft

2. What is the width of the kitchen? 7 ft

3. What is the length of the closet? 8 ft

4. What is the width of the closet? 3 ft

5. What is the total area of the apartment? (Hint: Find the length and width.) 247 ft^2

Use a proportion to solve the problem. Show your work.

6. On a map, the distance from Portland, Oregon, to San Francisco, California, is about 4.6 cm. The scale of the map is 1 cm = 230 km. Estimate the actual distance between the two cities. about 1,058 km

An alternate two-column proof lesson is provided on page 414 of the student book.

Getting Started
Have students list postulates and theorems about triangles that they already know.

9-14 Proof: Angle-Angle Similarity Postulate

> POSTULATE 12
> **Angle-Angle Similarity Postulate**
>
> If two angles of a triangle are congruent to two angles of another triangle, then the two triangles are similar.

You can use the Angle-Angle Similarity Postulate to prove two triangles are similar. You only need to show that two pairs of corresponding angles are congruent. Then, the triangles are similar. You can write a paragraph proof.

EXAMPLE 1

You are given: $\overline{AD} \parallel \overline{CB}$
$\overline{AB}$ intersects $\overline{CD}$

Prove: $\triangle ADE \sim \triangle BCE$

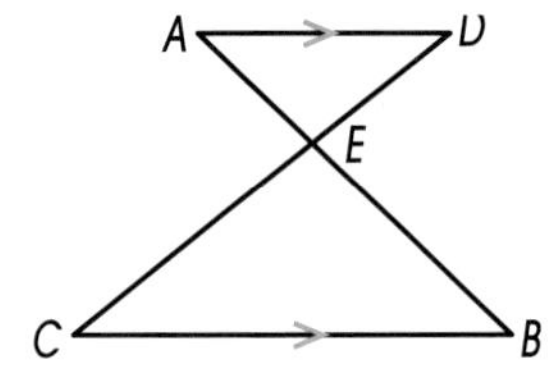

You are given that $\overline{AD} \parallel \overline{CB}$. This means that the alternate interior angles are congruent. Then, $\angle A \cong \angle B$ and $\angle D \cong \angle C$. By the Angle-Angle Similarity Postulate, the two triangles are similar.

So, you prove that $\triangle ADE \sim \triangle BCE$. ✓

You can use this theorem to prove triangles similar.

EXAMPLE 2

You are given: $\overline{JK} \parallel \overline{GI}$
Prove: $\triangle GHI \sim \triangle JHK$

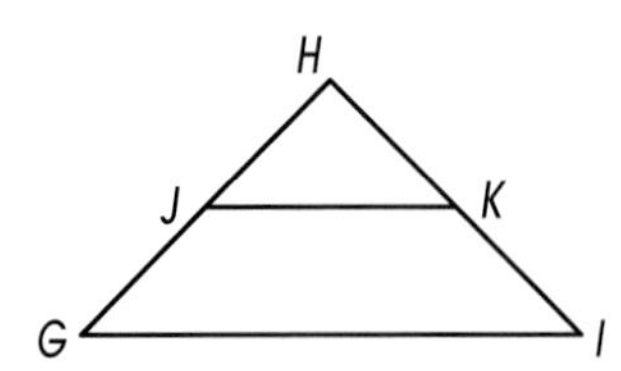

You are given that $\overline{JK} \parallel \overline{GI}$. This means that the corresponding angles are congruent. Then, $\angle HJK \cong \angle HGI$. You also know that $\angle H \cong \angle H$. Because of the Angle-Angle Similarity Postulate, the two triangles are similar. ✓

So, you prove that $\triangle GHI \sim \triangle JHK$. ✓

Try This

Copy and complete the proof.

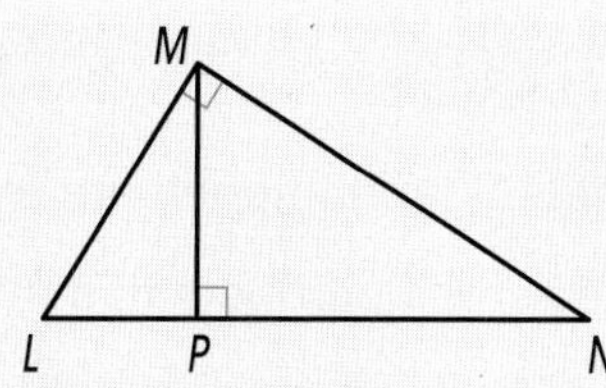

You are given: $\triangle LMN$ is a right triangle.
$\overline{MP} \perp \overline{LN}$
Prove: $\triangle LMN \sim \triangle LPM$

Look at the diagram and at what you need to prove. To prove $\triangle LMN \sim \triangle LPM$, you need to show that two pairs of corresponding angles are congruent.

You know that $\overline{MP} \perp \overline{LN}$. This is ■ (given). So, $\angle LPM$ is a right angle by the definition of ■ (perpendicular) lines. You know that $\angle LMN$ is a right angle. Right angles are congruent, so $\angle LPM \cong \angle LMN$. $\angle L \cong \angle L$ because an angle is ■ (congruent) to itself. $\triangle LMN \sim \triangle LPM$ by the Angle-Angle ■ (similarity) Postulate.

So, you prove that $\triangle LMN \sim$ ■. ✓ $\triangle LPM$

Practice

Write a proof for each of the following. See Additional Answers in the back of this book.

1. You are given: $\overline{AD}$ intersects $\overline{BE}$.
$\angle B$ and $\angle E$ are right angles.
Prove: $\triangle ABC \sim \triangle DEC$

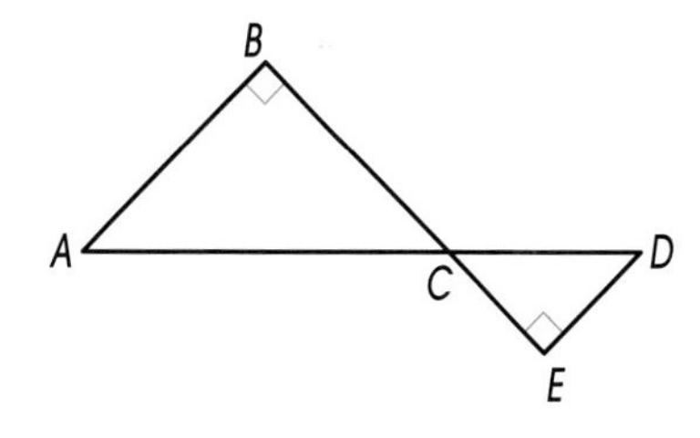

2. You are given: $\triangle FGH$ is a right triangle.
$\overline{GI}$ is an altitude.
Prove: $\triangle FGH \sim \triangle GIH$

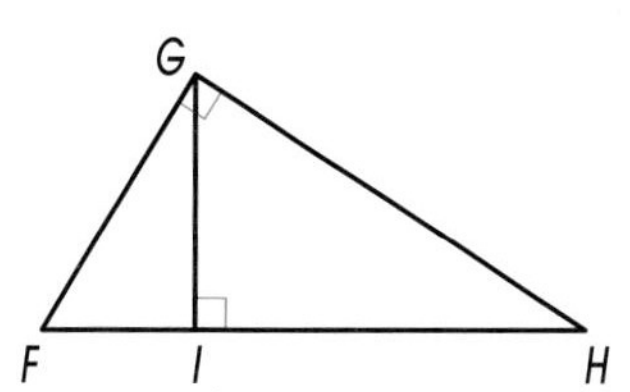

3. You are given: $\overline{MN} \parallel \overline{JL}$.
Prove: $\triangle MKN \sim \triangle JKL$

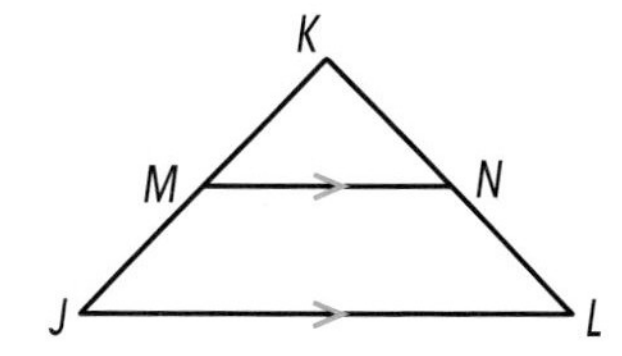

Chapter 9 Review

Summary

You can write a ratio to compare two quantities.
You can use cross products to solve a proportion.
When two triangles are similar, you can use proportions to find the unknown length of a side of a triangle.
If a line cuts two sides of a triangle and is parallel to the third side, the two sides are cut proportionally.
The altitude from the right angle of a right triangle makes similar triangles.
The perimeter of similar triangles has the same ratio as the corresponding sides.
The area of similar triangles has the same ratio as the squares of the corresponding sides.
You can use similar triangles to find distances you cannot measure.
You can use a proportion to find the actual size from a scale drawing.
You can write a proof to show that two triangles are similar.

More vocabulary review is provided in the Classroom Resource Binder, page 87.

ratio
proportion
geometric mean
similar triangles
similarity ratio

Vocabulary Review

Complete the sentences with the words from the box.

1. A ____ is a comparison of two quantities. ratio
2. The ratio of the lengths of the corresponding sides of similar polygons is the ____. similarity ratio
3. An equation that states that two ratios are equal is a ____. proportion
4. The ____ of two positive numbers, a and b, is the positive number x such that $\frac{a}{x} = \frac{x}{b}$. geometric mean
5. Triangles that have the same shape but that may not be the same size are ____. similar triangles

Chapter Quiz **More assessment** is provided in the Classroom Resource Binder.

1. Write the ratio of blue circles to white circles.
   3 to 2, 3:2, or $\frac{3}{2}$

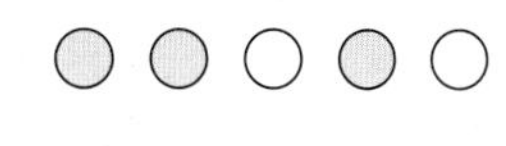

Solve each proportion.

2. $\frac{1}{3} = \frac{15}{x}$ 45
3. $\frac{x}{24} = \frac{5}{8}$ 15
4. $\frac{9}{x} = \frac{x}{25}$ 15

Each pair of triangles is similar. Find the value of x.

5. 3

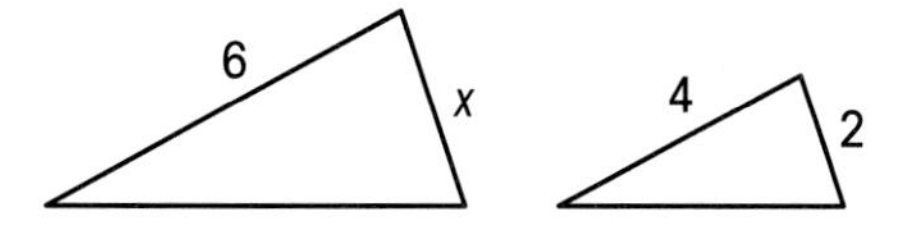

6. 6

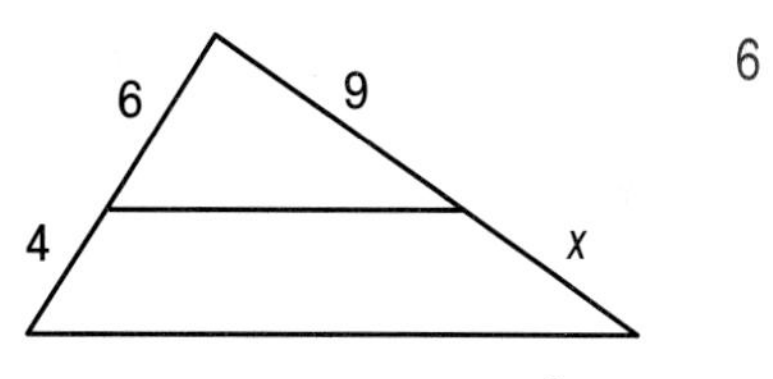

7.

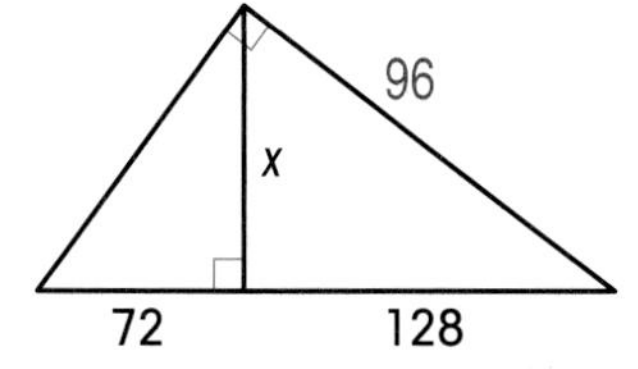

8.

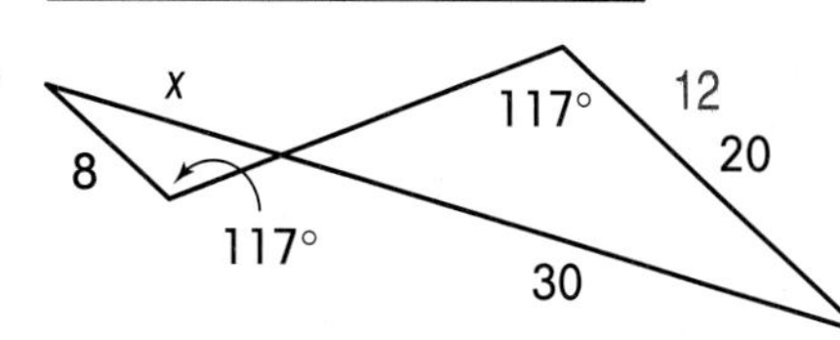

Find the perimeter or the area.

9. ▱$ABCD \sim$ ▱$EFGH$. The ratio of a pair of corresponding sides is 2:3. The perimeter of ▱$ABCD$ is 24 cm. Find the perimeter of ▱$EFGH$. 36 cm

10. $\triangle JKL \sim \triangle MNP$. The ratio of a pair of corresponding sides is 1 to 2. The area of $\triangle JKL$ is 10 cm^2. Find the area of $\triangle MNP$. 40 cm^2

Write About Math Yes. The triangles will have corresponding angles congruent, and because the sides are the same, the ratios of the lengths of the corresponding sides will also be congruent.

Problem Solving

Solve each problem. Show your work.

11. A 6-ft-tall man casts a 2-ft shadow. At the same time, a tree casts a 3-ft shadow. How tall is the tree? 9 ft

12. The scale on a scale drawing is $\frac{1}{4}$ in. = 1 ft. A room on the drawing is $1\frac{1}{2}$ in. long. What is the actual length of the room? 6 ft

Write About Math
Are two congruent triangles also similar? Why or why not?
See above.

Additional Practice for this chapter is provided on page 401 of the student book.

Chapter 10 Circles

These circles show star tracks. Why do you think some circles are smaller and some circles are larger?

Learning Objectives

- Find the circumference and area of a circle.
- Find the measure of a central angle, an inscribed angle, and the intercepted arc.
- Find the length of an arc and the area of a sector.
- Find the measure of angles and line segments formed by tangents, secants, and chords.
- Use a calculator to find area and circumference.
- Solve problems about circles.
- Write proofs about chords and circles.

ESL/ELL Note On index cards, have students write a math vocabulary word in both English and a second language on one side, and an illustrative diagram on the other side. Students can then use the index cards as flashcards for review, or as a reference dictionary.

Words to Know

circle	a plane figure made up of points all the same distance from a point called the center
center	the point inside a circle that is the same distance from each point on the circle
radius	a line segment from the center to any point on the circle
diameter	a line segment that passes through the center of a circle with endpoints on the circle
circumference	the distance around a circle
central angle	an angle with its vertex at the center of the circle and sides that are radii
arc	a curve between any two points on a circle
sector	a region formed by two radii and the intercepted arc
inscribed angle	an angle with its vertex on the circle and sides that are chords
tangent	a line that touches a circle at only one point
secant	a line that intersects a circle at two points
chord	a line segment with endpoints on a circle

Mobile Project

Circles with the same center in the same plane but with different radii are concentric. You can create a mobile of concentric circles. Place a compass in the center of a sheet of paper. Draw a small circle. Now, without moving the compass from the center, draw nine different-size circles. Cut around the edge of each circle to make nine rings. Place five of them inside one another. Connect them with a 12-inch piece of string. Your mobile is now ready to display in school or at home!

Project See the Classroom Resource Binder for a scoring rubric to assess this project.

More practice is provided in the Workbook, page 56, and in the Classroom Resource Binder, page 100.

10-1 Circumference of a Circle

Getting Started
Have students use string to measure the circumference of different-sized cans. Then, have them measure the diameter of each can and find the ratio of the circumference to the diameter. [about 3]

Math Fact
In all circles, $\pi = \frac{\text{circumference}}{\text{diameter}}$.

Every point on a **circle** is the same distance from the center. A circle is named by its **center**. The symbol ⊙ means "circle." You can write ⊙P for the circle on the right.

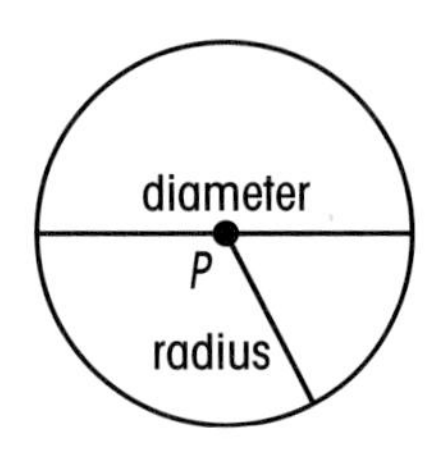

A **radius** is a line segment from the center of a circle to any point on the circle. A **diameter** is a line segment that connects two points on the circle and passes through the center. The length of the diameter is two times the length of the radius. So, $d = 2r$.

The **circumference** of a circle is the distance around the circle. You can use a formula to find the circumference.

$$\text{Circumference} = \pi \cdot \text{diameter, or } C = \pi d$$
$$\text{Circumference} = \pi \cdot 2 \cdot \text{radius, or } C = 2\pi r$$

The symbol π is read as *pi*. π is about 3.14 or about $\frac{22}{7}$.

EXAMPLE 1

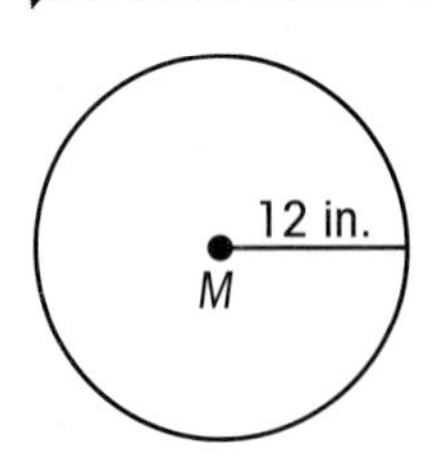

Find the circumference of ⊙M. Use 3.14 for π.

Write the formula that uses the radius. $C = 2\pi r$

Substitute 3.14 for π and 12 for r. $C \approx 2(3.14)(12)$

Simplify. ≈ 75.36

The circumference of ⊙M is about 75.36 inches.

EXAMPLE 2

The circumference of a circle is 12π inches. Find the length of the radius.

Write the formula that uses the radius. $C = 2\pi r$

Substitute 12π for C. $12\pi = 2\pi r$

Solve for r. Divide both sides of the equation by 2π. $\frac{12\pi}{2\pi} = \frac{2\pi r}{2\pi}$

$6 = r$

The length of the radius is 6 inches.

Try This

The diameter of ⊙Q is 14 yards. Find the circumference. Use $\frac{22}{7}$ for π.

Write the formula that uses the ■. diameter — $C = \pi d$

Substitute ■ for π and 14 for d. $\frac{22}{7}$ — $C \approx \left(\frac{22}{7}\right)(■)$ 14

Simplify. — ≈ 44

The circumference of the circle is about ■. 44 yards

Practice

Find the circumference of each circle. Use 3.14 for π in exercises **1–4.** Use $\frac{22}{7}$ for π in exercises **5–6.**

1.

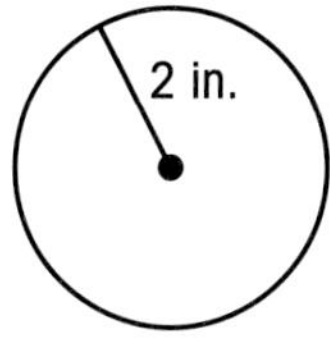

about 12.56 in.

2.

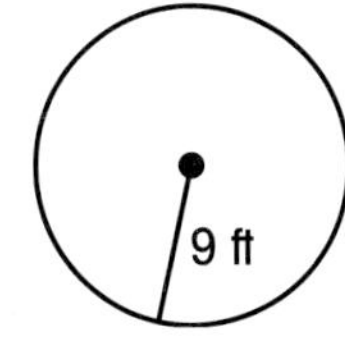

about 56.52 ft

3.

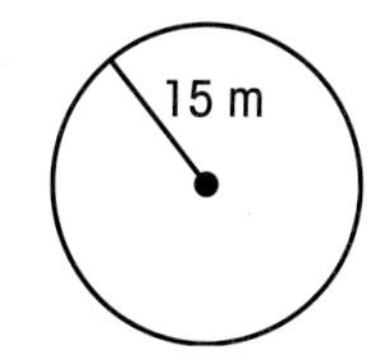

about 94.2 m

4.

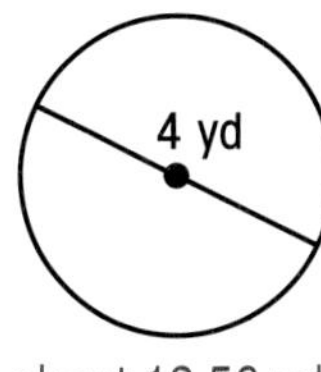

about 12.56 yd

5.

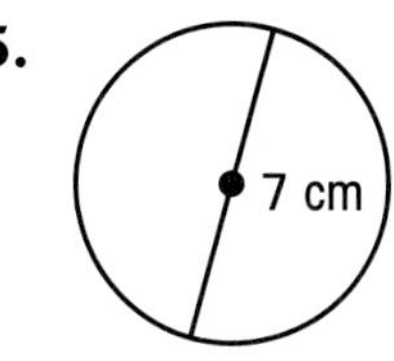

about 22 cm

6.

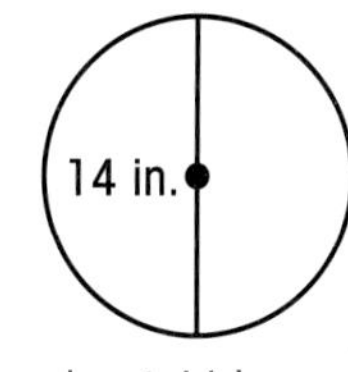

about 44 in.

The circumference of each circle is given. Find the length of the radius.

7. $C = 16\pi$ ft
8 ft

8. $C = 32\pi$ cm
16 cm

9. $C = 30\pi$ in.
15 in.

10. $C = 24\pi$ m
12 m

Share Your Understanding

11. Explain to a partner how to find the circumference of one of the circles above. Answers will vary. Students should demonstrate an understanding of the formulas.

12. **CRITICAL THINKING** Find the circumference of the circle inside the square. about 18.84 in. for $\pi = 3.14$; about 18.86 in. for $\pi = \frac{22}{7}$

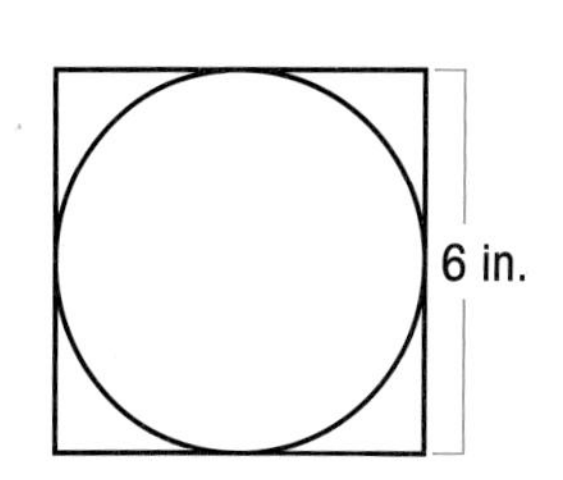

More practice is provided in the Workbook, page 56, and in the Classroom Resource Binder, page 101.

Getting Started Tell students that polygons with the same area are equal polygons, but are not always congruent. Discuss why circles with the same area are always congruent. [All circles are similar.]

Remember
π is about 3.14.

10-2 Area of a Circle

You used π and the length of the radius in a formula to find the circumference of a circle. You can also use these values in a formula to find the area of a circle.

$$\text{Area} = \pi \cdot (\text{radius})^2$$
$$A = \pi r^2$$

You can use the area formula if you know the radius.

EXAMPLE 1

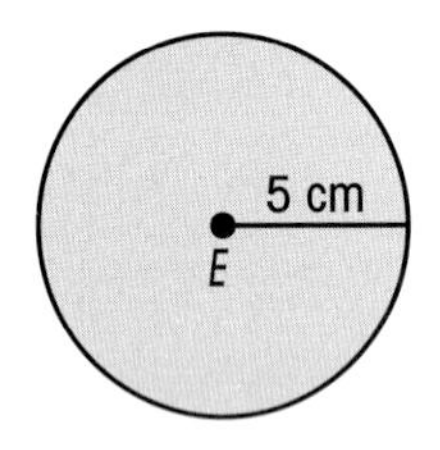

Find the area of $\odot E$. Use 3.14 for π.

Write the formula for area of a circle.	$A = \pi r^2$
Substitute 3.14 for π and 5 for r.	$A \approx 3.14(5^2)$
Simplify.	$A \approx 3.14 \cdot 25$
	≈ 78.5

The area of $\odot E$ is about 78.5 cm^2.

If you know the diameter, you can find the radius. Just divide the diameter by 2.

EXAMPLE 2

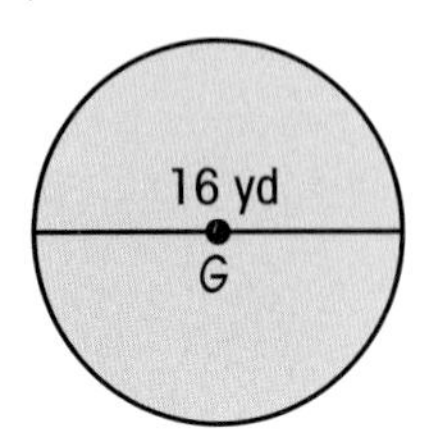

Find the area of $\odot G$. Use 3.14 for π.

Write the formula for area of a circle.	$A = \pi r^2$
Because the diameter is 16, the radius is 8. Substitute 3.14 for π and 8 for r.	$A \approx 3.14(8^2)$
Simplify.	$A \approx 3.14 \cdot 64$
	≈ 200.96

The area of $\odot G$ is 64π yd^2, or about 200.96 yd^2.

You can also use the area formula to find the radius. You need to know the area of the circle in terms of π.

Try This

The area of a circle is 36π in.^{2} Find the length of the radius.

Write the formula for ■ of a circle. (area)	$A = \pi r^2$
Substitute ■ for A. Solve for r. (36π)	$36\pi = \pi r^2$
Divide both sides of the equation by π.	$\frac{36\pi}{\pi \text{ ■}} = \frac{\pi r^2}{\pi}$
Simplify.	36 ■ $= r^2$
Find the square root of both sides of the equation.	6 ■ $= r$

The length of the radius is ■. 6 in.

Practice

Find the area of each circle. Use 3.14 for π.

1.

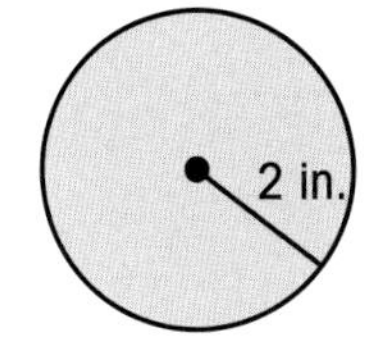

about 12.56 in.^{2}

2.

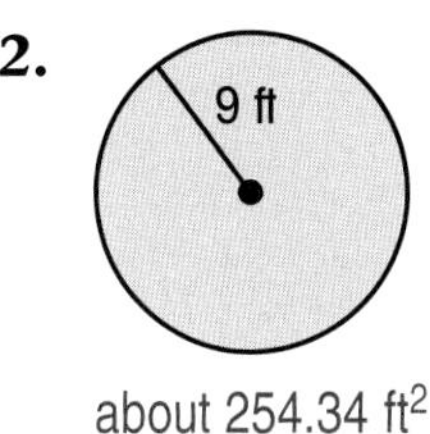

about 254.34 ft^2

3.

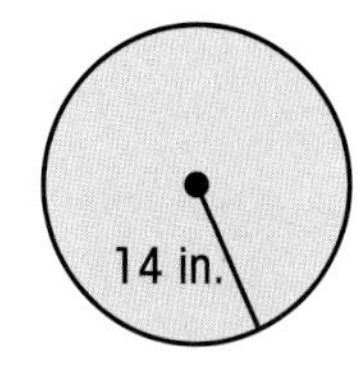

about 615.44 in.^{2}

4.

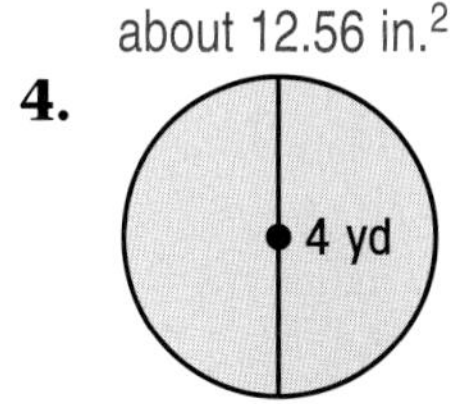

about 12.56 yd^2

5.

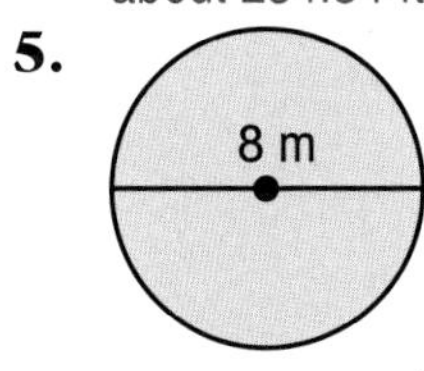

about 50.24 m^2

6.

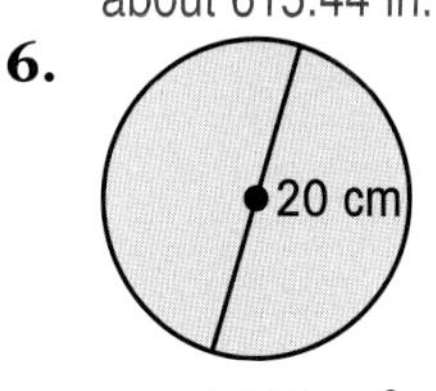

about 314 cm^2

The area of each circle is given. Find the length of the radius.

7. $A = 9\pi$ cm^2 3 cm

8. $A = 25\pi$ ft^2 5 ft

9. $A = 81\pi$ m^2 9 m

10. $A = 100\pi$ in.^{2} 10 in.

Share Your Understanding

11. Explain to a partner how to find the area of any circle in exercises **1–6**. Answers will vary. Students should demonstrate understanding of the formula.

12. **CRITICAL THINKING** Look at the diagram on the right. Find the area of the shaded part. about 84.78 m^2

3 m
3 m

More practice is provided in the Workbook, page 57, and in the Classroom Resource Binder, page 102.

10·3 Arcs and Central Angles

Getting Started
Explain that every circle has a measure of 360° around the center.

An angle with a vertex at the center of a circle is called a **central angle**. $\angle ABC$ is a central angle.

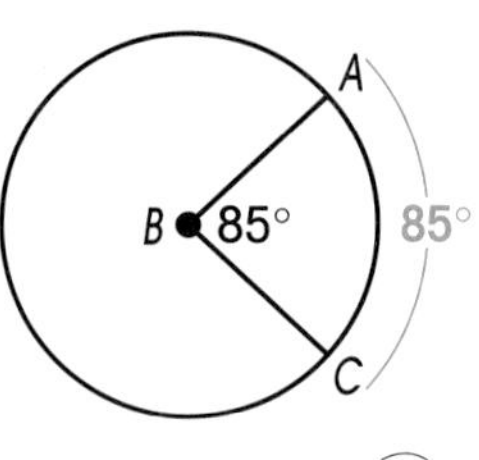

$m\angle ABC = m\widehat{AC}$

The curve of the circle cut by the sides of a central angle is an **arc**. The symbol ⌒ means "arc." Use the two points on the circle to name the arc. $\widehat{AC}$ means arc AC.

The measure of a central angle is equal to the degree measure of its arc. $\angle ABC$ is 85°. $\widehat{AC}$ is 85°.

You can find the measure of a central angle if you know the degree measure of its arc.

EXAMPLE 1

Find the measure of $\angle LMN$ in $\odot M$.

In $\odot M$, $\angle LMN$ is a central angle. $\widehat{LN}$ is its arc. $\widehat{LN}$ is 35°.

$$m\angle LMN = m\widehat{LN}$$
$$= 35$$

So, $\angle LMN$ is 35°.

$\widehat{LN}$ is a minor arc on $\odot M$. A minor arc is less than 180°. A major arc of $\odot M$ is $\widehat{LPN}$. You need three letters to name a major arc. A major arc is greater than 180° and less than 360°.

You can find the measure of a central angle if you know the degree measure of the major arc. Use the fact that there are 360 degrees in a circle.

EXAMPLE 2

Find the measure of $\angle KIJ$ in $\odot I$.

$\widehat{KHJ}$ is a major arc. $m\widehat{KHJ}$ is 250. Subtract 250 from 360 to find $m\widehat{KJ}$.

$$m\widehat{KJ} = 360 - 250$$
$$= 110$$

In $\odot I$, $\angle KIJ$ is a central angle. $\widehat{KJ}$ is its arc. $\widehat{KJ}$ is 110°.

$$m\angle KIJ = m\widehat{KJ}$$
$$= 110$$

So, $\angle KIJ$ is 110°.

Try This

Find the degree measure of $\overset{\frown}{AC}$.

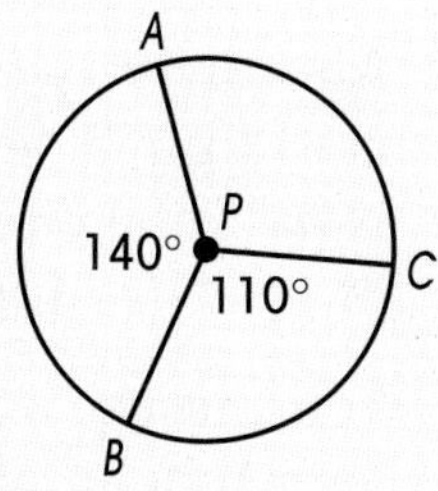

Since $m\overset{\frown}{AC} = m\angle APC$, first find $m\angle APC$.

There are 360 degrees in a ■. circle

Use this fact to write an equation. $m\angle APB + m\angle BPC + m\angle APC =$ ■ 360

Substitute the given values. $140 +$ ■ $+ m\angle APC = 360$ 110

Solve for $m\angle APC$. $250 + m\angle APC =$ ■ 360

$m\angle APC =$ ■ 110

Substitute $m\overset{\frown}{AC}$ for $m\angle APC$. $m\overset{\frown}{AC} =$ ■ 110

So, $\overset{\frown}{AC}$ is ■. 110°

Practice

Find the value of x in each circle.

1. 90°

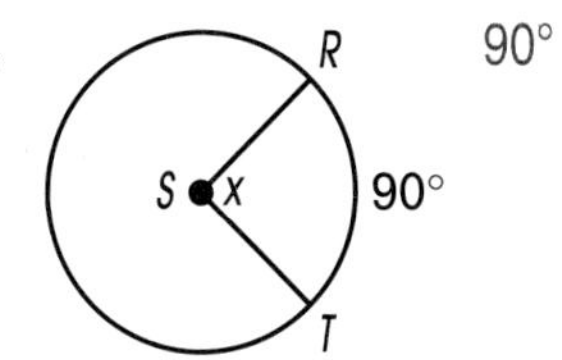

2. 130°

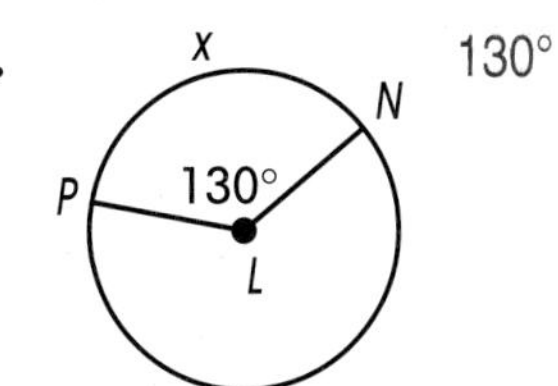

3. 140°

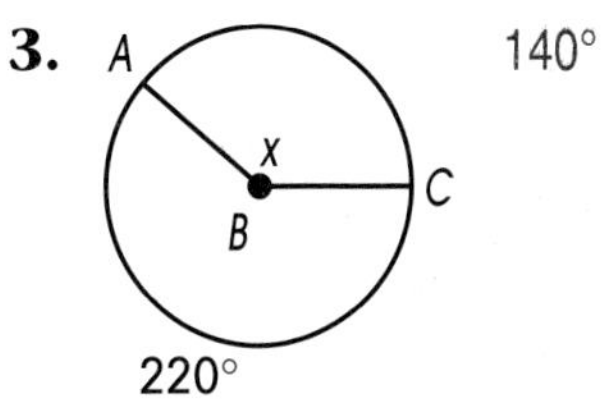

4. 180°

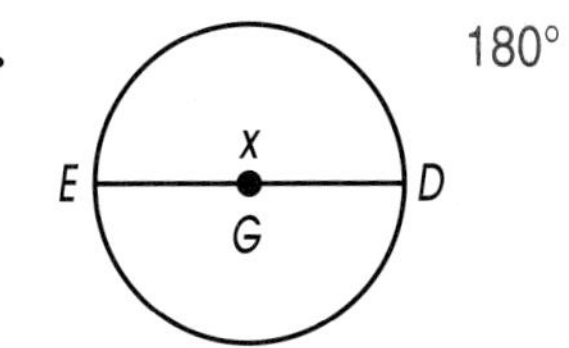

5. 130°

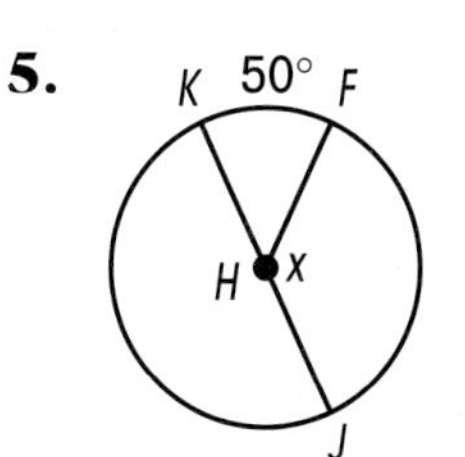

6. 150°

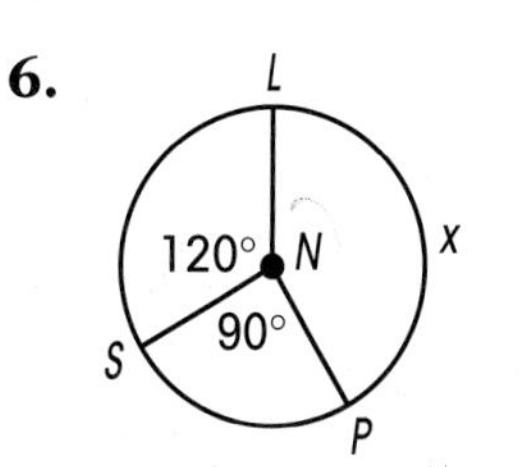

7. A minor arc is less than 180° and is labeled by two points on the arc. A major arc is greater than 180° and less than 360°. You need three letters to name a major arc.

Share Your Understanding

7. Explain to a partner the difference between a minor arc and a major arc. See above.

8. **CRITICAL THINKING** $\overset{\frown}{AC}$ in $\odot B$ is 90°. $\overline{BD}$ bisects $\angle ABC$. What is $m\angle ABD$? (Hint: Draw a diagram.) 45°

More practice is provided in the Workbook, page 57, and in the Classroom Resource Binder, page 102.

10·4 Arc Length and Sectors

Getting Started
Draw a pie chart on the board. Call on volunteers to estimate and assign a degree measure to each section. Remind them that the sum of the angles must be 360°.

Another way to measure an arc is to find its length. Because an arc is part of a circle, its length is part of the circumference.

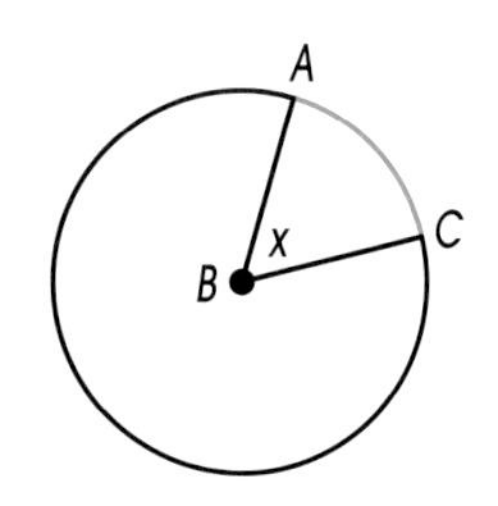

You can use a formula to find the length of an arc.

$$\text{length of } \widehat{AC} = \frac{m\angle x}{360} \cdot 2\pi r$$

EXAMPLE 1

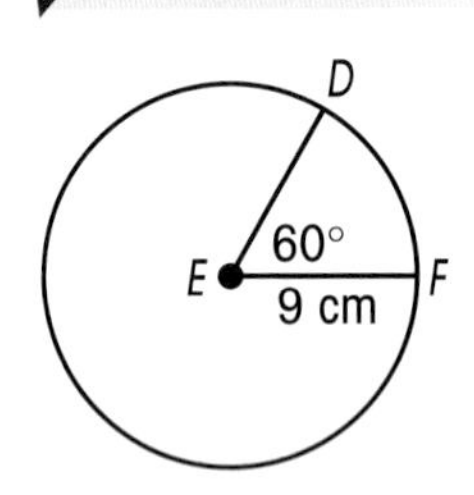

Find the length of $\widehat{DF}$ in $\odot E$. Use 3.14 for π.

Write the formula for length of $\widehat{DF}$. $\quad \text{length of } \widehat{DF} = \frac{m\angle E}{360} \cdot 2\pi r$

Substitute the values you know. $\quad \text{length of } \widehat{DF} \approx \frac{60}{360} \cdot 2(3.14)(9)$

Simplify. $\quad \approx \frac{1}{6} \cdot 18(3.14)$

≈ 9.42

The length of $\widehat{DF}$ is about 9.42 centimeters.

Math Fact
The plural of *radius* is *radii*.

A **sector** is a region formed by two radii and their arc. It is part of the area of a circle. You can use a formula to find the area of a sector.

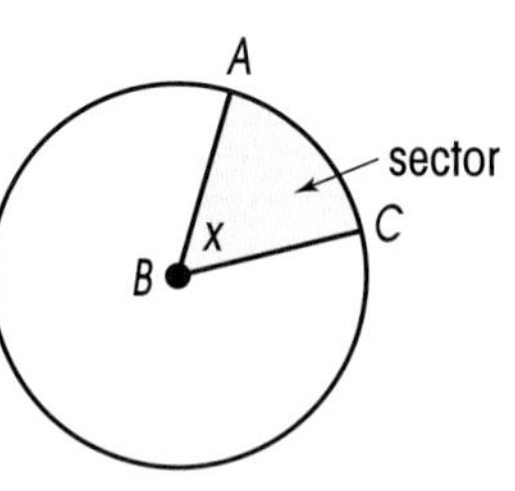

$$\text{Area of sector } ABC = \frac{m\angle x}{360} \cdot \pi r^2$$

EXAMPLE 2

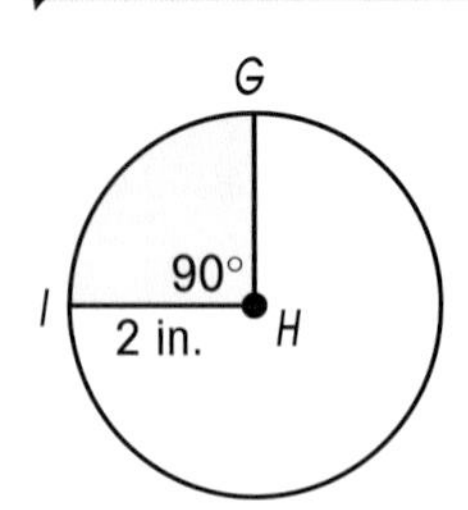

Find the area of sector GHI in $\odot H$. Use 3.14 for π.

Write the area formula of sector *GHI*. $\quad A = \frac{m\angle H}{360} \cdot \pi r^2$

Substitute the values you know. $\quad A \approx \frac{90}{360} \cdot (3.14)(2^2)$

Simplify. $\quad \approx \frac{1}{4} \cdot (3.14)(4)$

≈ 3.14

The area of the sector is about 3.14 in.^{2}

Try This

Find the length of $\widehat{LM}$. Use 3.14 for π.

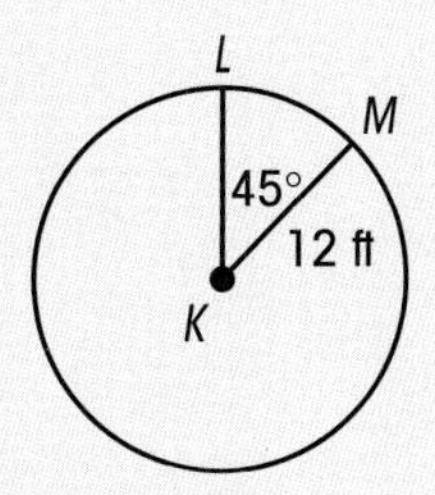

Write the formula for length of $\widehat{LM}$. length of $\widehat{LM} = \frac{m\angle K}{360} \bullet 2\pi r$

Substitute the values you know. $\approx \frac{45}{360} \bullet 2(3.14) \bullet 12$

Simplify. ≈ 9.42

The length of $\widehat{LM}$ is about 9.42 ft.

Practice

Find the length of each minor arc. Find the area of each sector. Use 3.14 for π.

1. 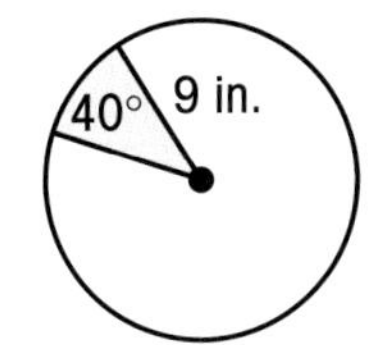

Length of minor arc ≈ 6.28 in.; Area of sector ≈ 28.26 in.^{2}

2. 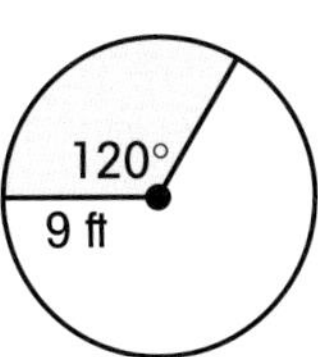

Length of minor arc ≈ 18.84 ft; Area of sector ≈ 84.78 ft^2

3. 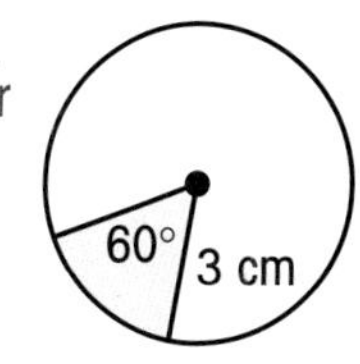

Length of minor arc ≈ 3.14 cm; Area of sector ≈ 4.71 cm^2

4. 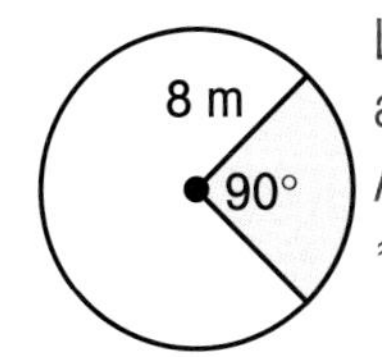

Length of minor arc ≈ 12.56 m; Area of sector ≈ 50.24 m^2

5. 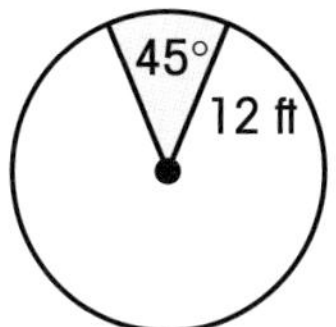

Length of minor arc ≈ 9.42 ft; Area of sector ≈ 56.52 ft^2

6. 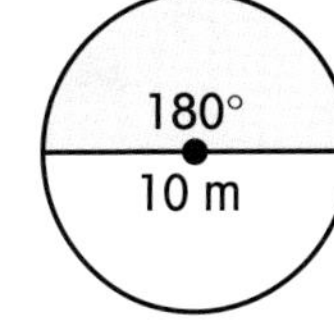

Length of minor arc ≈ 15.7 m; Area of sector ≈ 39.25 m^2

7. Possible answer: To find the length of an arc, first find the circumference using the formula. Next, divide the measure of the central angle by 360. Then, multiply the circumference by the quotient.

Share Your Understanding

7. Explain to a partner how to find the length of an arc.

8. **CRITICAL THINKING** A semicircle is half of a circle. Explain how to find the area of a semicircle. Give an example. Find the area of a circle and divide it by 2.

More practice is provided in the Workbook, page 58, and in the Classroom Resource Binder, page 103.

Inscribed Angles

THEOREM 29

The measure of an angle inscribed in a circle is equal to one-half the measure of its intercepted arc.

Getting Started

Tell students to draw an acute angle with a protractor. Then, have them draw a circle around the angle, making sure the vertex is on the circle.

You learned that an angle whose vertex is the center of a circle is a central angle.

An angle whose vertex is on the circle is an **inscribed angle**.

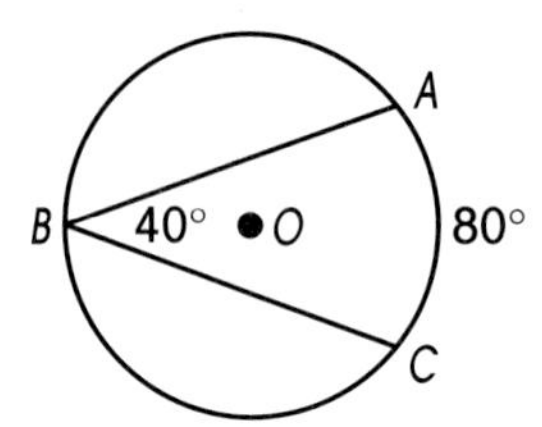

$\angle ABC$ is an inscribed angle of $\odot O$. The arc cut by this inscribed angle is $\widehat{AC}$. The measure of an inscribed angle is equal to one-half the degree measure of its arc.

$m\angle ABC = \frac{1}{2}m\widehat{AC}$

$$m\angle ABC = \frac{1}{2}(80) = 40$$

You can find the measure of an inscribed angle if you know the degree measure of its arc.

EXAMPLE 1

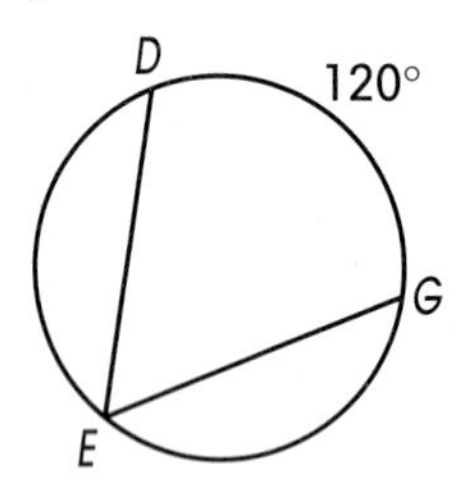

Find the measure of $\angle DEG$.

Write an equation.	$m\angle DEG = \frac{1}{2}m\widehat{DG}$
Substitute the given value.	$m\angle DEG = \frac{1}{2}(120)$
Simplify.	$m\angle DEG = 60$

So, $\angle DEG$ is 60°.

You can find the degree measure of an arc if you know the measure of its inscribed angle.

EXAMPLE 2

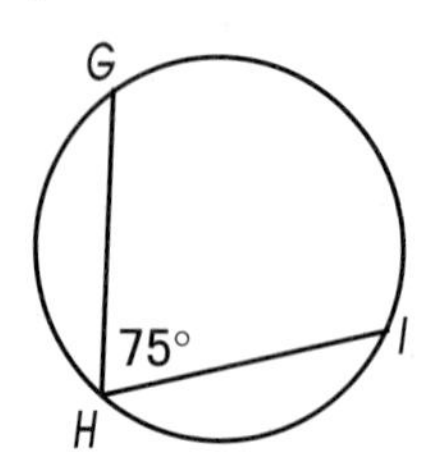

Find the degree measure of $\widehat{GI}$.

Write an equation.	$m\angle GHI = \frac{1}{2}m\widehat{GI}$
Substitute the given value.	$75 = \frac{1}{2}m\widehat{GI}$
Solve for $m\widehat{GI}$.	$150 = m\widehat{GI}$

So, $\widehat{GI}$ is 150°.

Try This

Find the measure of $\angle LMP$.

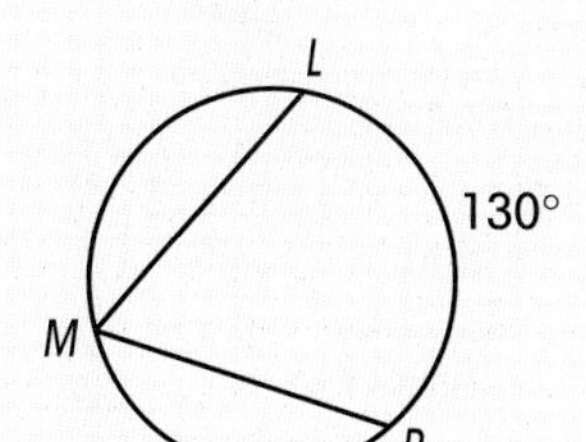

The measure of $\angle LMP$ is equal to ■ the measure of $\widehat{LP}$. Write an equation.	$m\angle LMP = \frac{1}{2}m■$	one-half; $\widehat{LP}$
Substitute the given value.	$m\angle LMP = \frac{1}{2}(■)$	130
Simplify.	$m\angle LMP = ■$	65

So, $\angle LMP$ is ■. 65°

Practice

Find the value of x in each circle.

1.

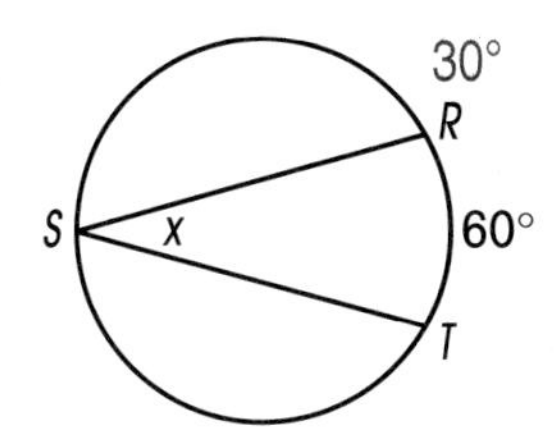

2.

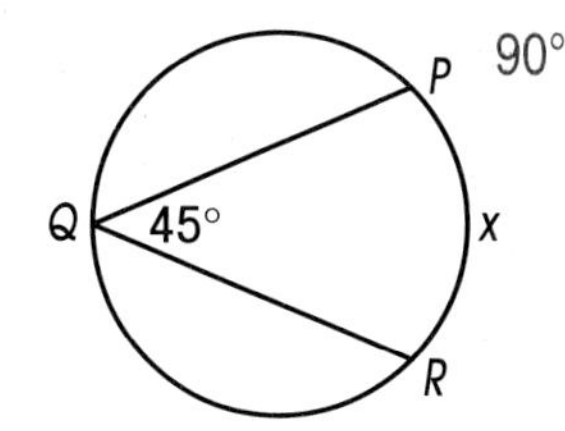

3.

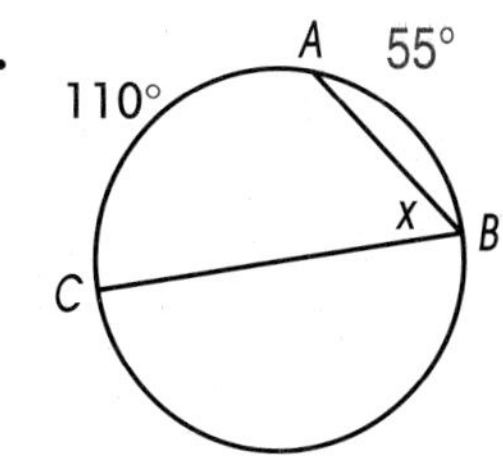

4.

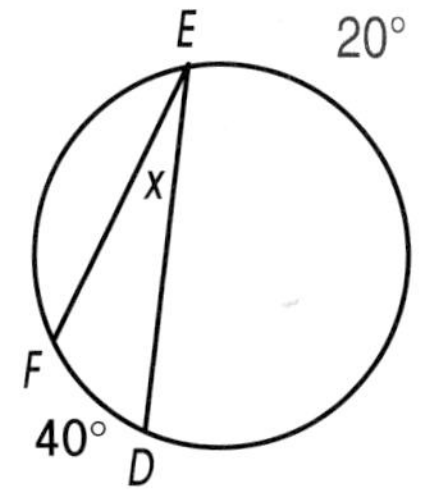

5.

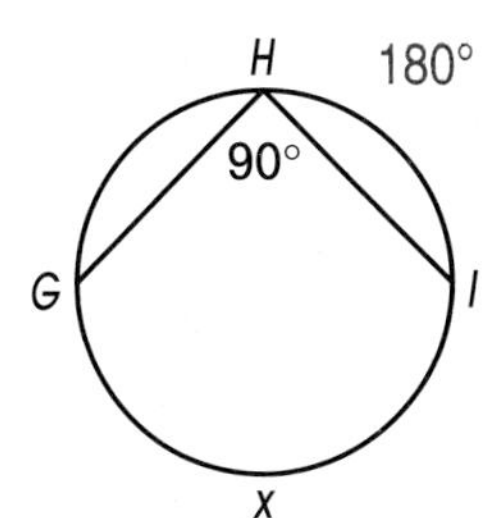

6.

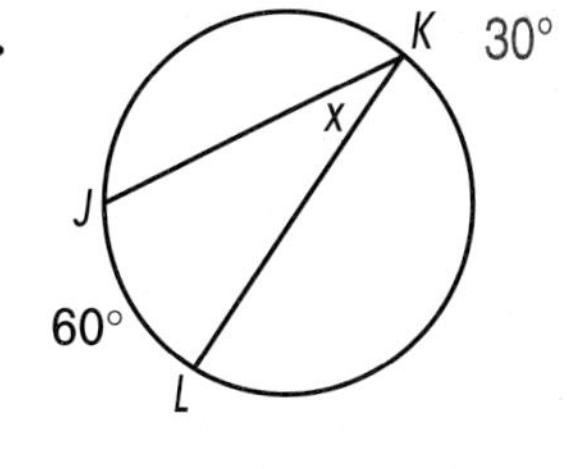

7. To find the measure of an inscribed angle, divide the degree measure of the arc in half. To find the degree measure of its arc, double the measure of its corresponding inscribed angle.

Share Your Understanding

7. Explain to a partner how to find the measure of an inscribed angle or the degree measure of its arc. Use the words *inscribed angle* and *arc* in your explanation. See above.

central angle

8. **CRITICAL THINKING** Explain the difference between an inscribed angle and a central angle. Draw a diagram of each.

   The vertex of the inscribed angle is on the circle, and the vertex of the central angle is the center of the circle.

inscribed angle

More practice is provided in the Workbook, page 59, and in the Classroom Resource Binder, page 104.

10·6 Tangents

THEOREM 30

If a line is tangent to a circle, then it is perpendicular to the radius drawn to the point of tangency.

Getting Started

Tell students that *tangent* comes from the Latin word *tangere*, which means "to touch." Draw a tangent on the board. Discuss how many tangents a circle can have. [an infinite number]

A **tangent** is a line that intersects a circle at only one point. $\overleftrightarrow{PN}$ is tangent to $\odot O$ at point T. The point where the tangent meets the circle is called the point of tangency. All tangents are perpendicular to the radius at the point of tangency.

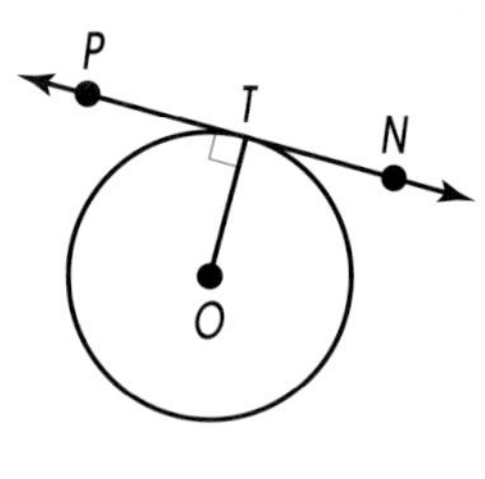

A tangent can be shown as a line, a line segment, or a ray.

Because the tangent and radius form a right angle, they can be the legs of a right triangle. Given the lengths of two sides, you can find the length of the third side.

EXAMPLE 1

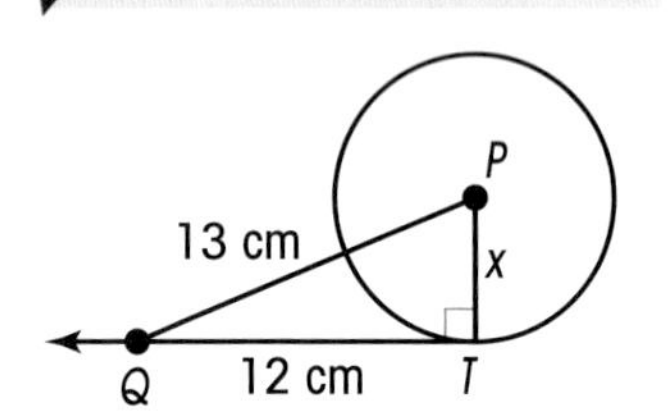

$\overrightarrow{TQ}$ is tangent to $\odot P$ at point T. Find the length of the radius.

The radius is leg x of the right triangle QPT.
Use the Pythagorean Theorem to find x. $\quad a^2 + b^2 = c^2$

Substitute x for a, 12 for b, and 13 for c. $\quad x^2 + 12^2 = 13^2$

Solve for x. $\quad x^2 + 144 = 169$
$x^2 = 25$
$x = 5$

The length of the radius is 5 centimeters.

The right triangle may be a special right triangle.

EXAMPLE 2

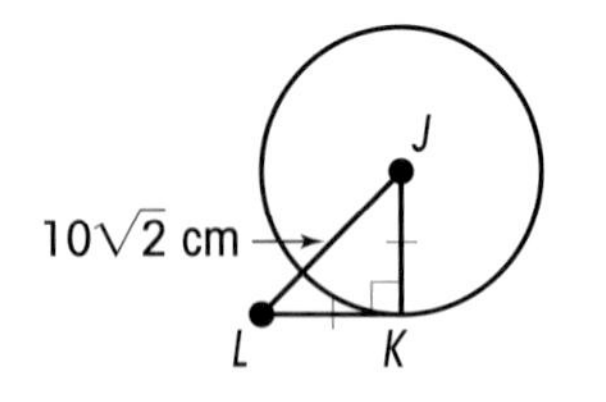

$\overline{KL}$ is tangent to $\odot J$ at point K. Find the length of $\overline{KL}$.

$\triangle LJK$ is a 45°-45°-90° right triangle.
Write the formula for the hypotenuse. $\quad \text{hypotenuse} = \text{leg}\sqrt{2}$
Substitute the given value. $\quad 10\sqrt{2} = \text{leg}\sqrt{2}$
Find the length of the leg. $\quad 10 = \text{leg}$

The length of $\overline{KL}$ is 10 centimeters.

Try This

$\overleftrightarrow{LN}$ is tangent to $\odot D$ at point M. Find the length of the radius.

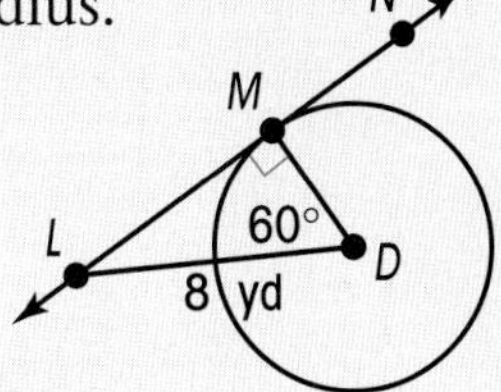

$\angle LMD$ is a 90° angle. $\angle D$ is a ■ angle. 60° Then, $\triangle LMD$ is a 30°-60°-90° ■ right triangle. The radius of $\odot D$ is the short leg of this triangle. You are given the hypotenuse.

Write the formula for the hypotenuse of this triangle. hypotenuse = 2 • short leg

Substitute the given value. 8 ■ = 2 • short leg

Solve for the short leg. Divide both sides of the equation by 2. 4 ■ = short leg

The length of the radius is ■. 4 yd

Practice

The given line is tangent to each circle. Find the value of x for each circle.

1. 6 cm

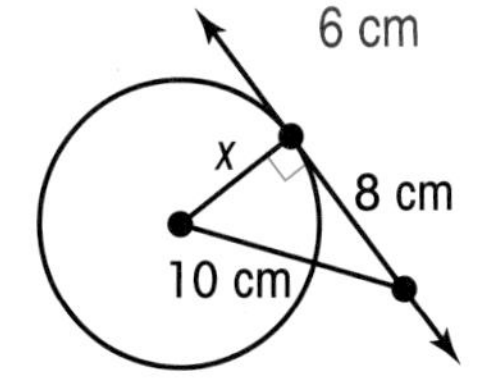

2. 9 ft

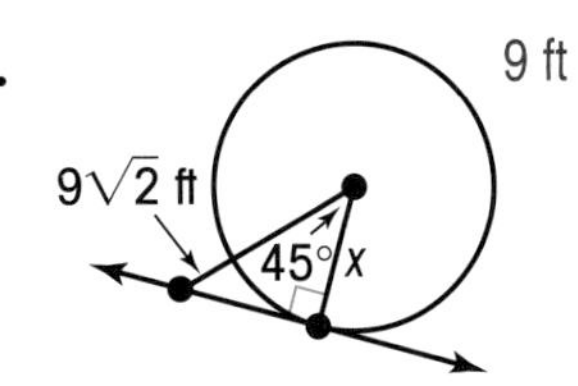

3. 12 in. 6 in.

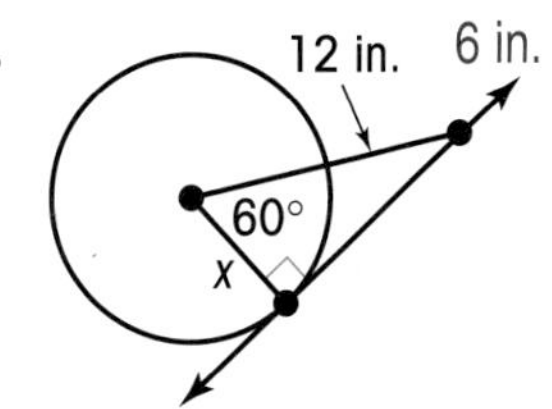

4. 14 m

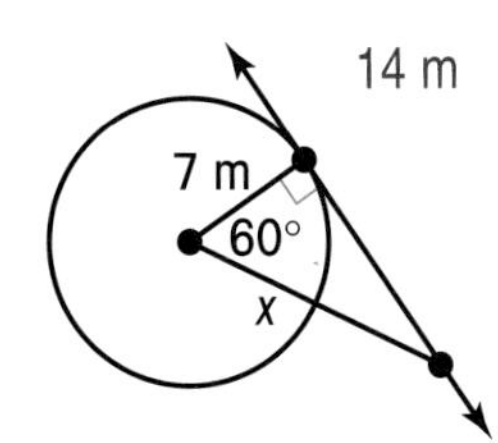

5. $5\sqrt{2}$ cm

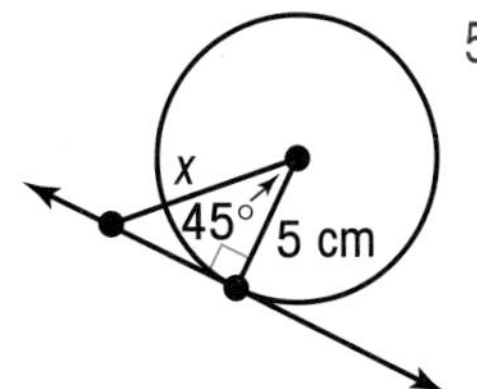

6. 12 ft

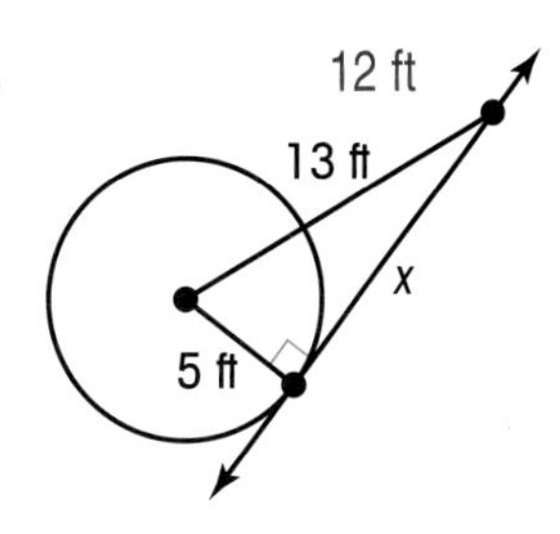

Share Your Understanding

7. Explain to a partner how to find the value of x in any exercise above. Answers will vary.

8. **CRITICAL THINKING** $\overline{RP}$ is tangent to $\odot T$ at point P. $\overline{RT}$ is 13 m and $\overline{RP}$ is 12 m. Find the length of the diameter of $\odot T$. (Hint: Draw a diagram.) 10 m

More practice is provided in the Workbook, page 60, and in the Classroom Resource Binder, page 105.

10·7 Tangents, Secants, and Angles

Getting Started Review intercepted arcs.

> THEOREM 31
> **Tangent-Tangent Angle Theorem**
>
> The measure of the angle formed by two tangents is one-half the difference of the degree measures of the intercepted arcs.

When two tangents to the same circle intersect outside the circle, they form a tangent angle. In $\odot S$, $\angle PFQ$ is a tangent angle.

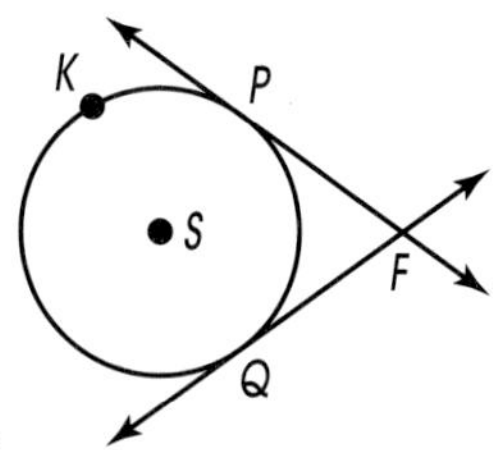

The measure of a tangent angle is one-half the difference of the degree measures of the intercepted arcs.

$$m\angle PFQ = \frac{1}{2}(m\overset{\frown}{PKQ} - m\overset{\frown}{PQ})$$

EXAMPLE 1

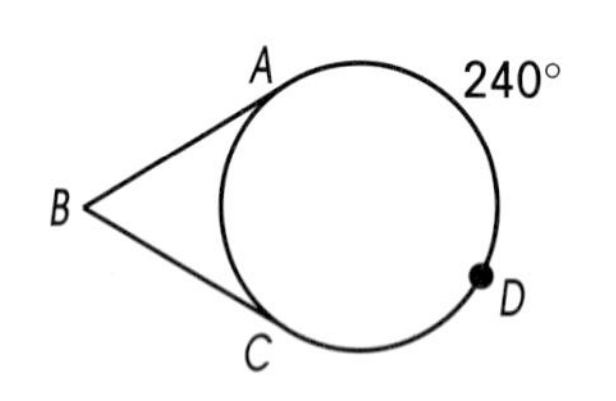

Find the measure of $\angle ABC$.

Write an equation. Use the Tangent-Tangent Angle Theorem.	$m\angle ABC = \frac{1}{2}(m\overset{\frown}{ADC} - m\overset{\frown}{AC})$
Find $m\overset{\frown}{AC}$. You know a circle has 360°. Subtract 240 from 360.	$m\overset{\frown}{AC} = 360 - 240 = 120$
Substitute the values you know in the first equation. Simplify.	$m\angle ABC = \frac{1}{2}(240 - 120)$ $= \frac{1}{2}(120) = 60$

So, $\angle ABC$ is 60°.

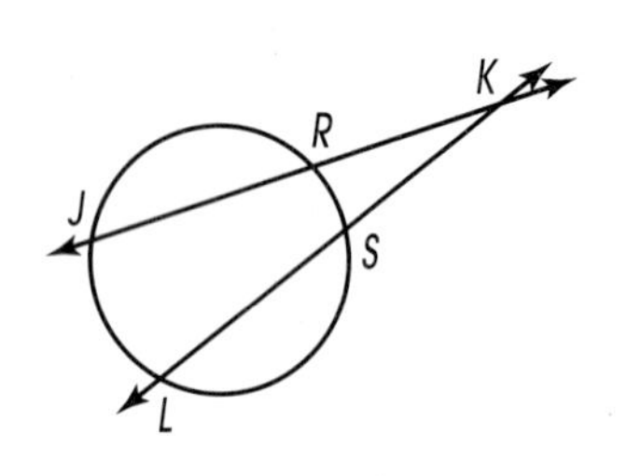

A **secant** is a line that intersects a circle at two points. Two secants that intersect outside a circle form a secant angle. The measure of a secant angle is one-half the difference of the measures of the intercepted arcs.

$$m\angle JKL = \frac{1}{2}(m\overset{\frown}{JL} - m\overset{\frown}{RS})$$

EXAMPLE 2

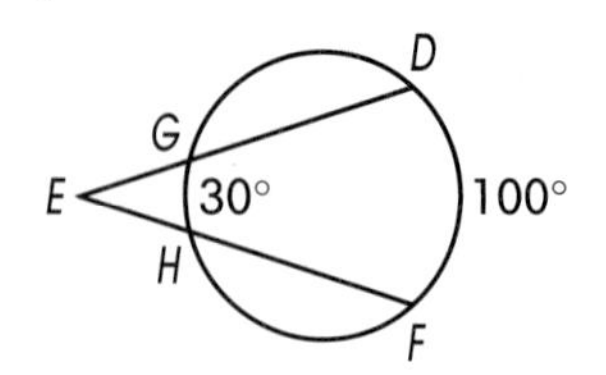

Find the measure of $\angle DEF$.

Write an equation.	$m\angle DEF = \frac{1}{2}(m\overset{\frown}{DF} - m\overset{\frown}{GH})$
Substitute the given values.	$m\angle DEF = \frac{1}{2}(100 - 30)$
Simplify.	$= \frac{1}{2}(70) = 35$

So, $\angle DEF$ is 35°.

Try This

Find the measure of $\angle IJK$.

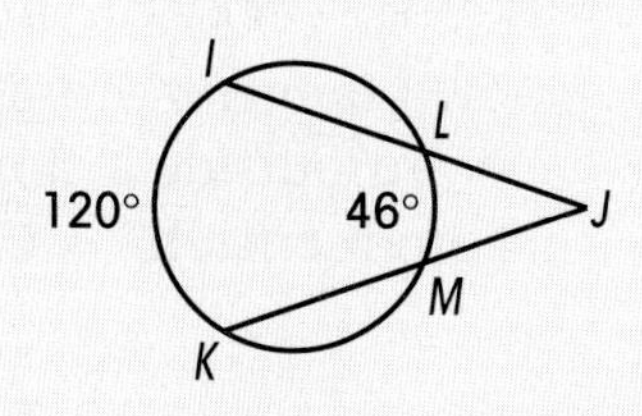

Write an equation. $m\angle IJK = \frac{1}{2}(m\widehat{IK} - m\widehat{LM})$

Substitute 120 for $\widehat{IK}$ and ■ for $\widehat{LM}$. 46 $m\angle IJK = \frac{1}{2}(120 - ■)$ 46

Simplify. $m\angle IJK = \frac{1}{2}■$ 74

$m\angle IJK = ■$ 37

So, $\angle IJK$ is ■. 37°

Practice

Find the measure of the tangent angle or secant angle for each circle.

1. 90°

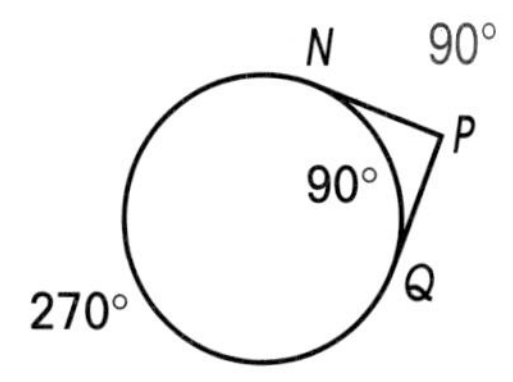

2. 45°

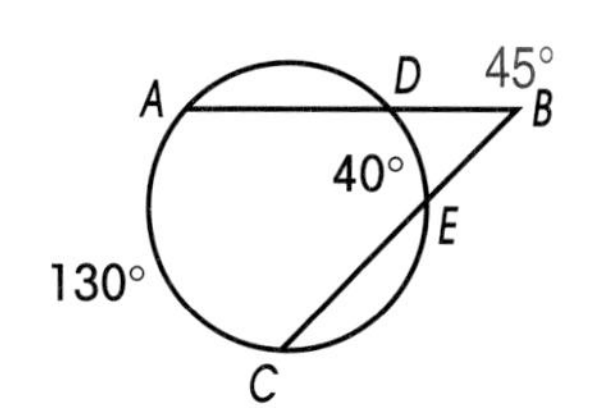

3. 20°

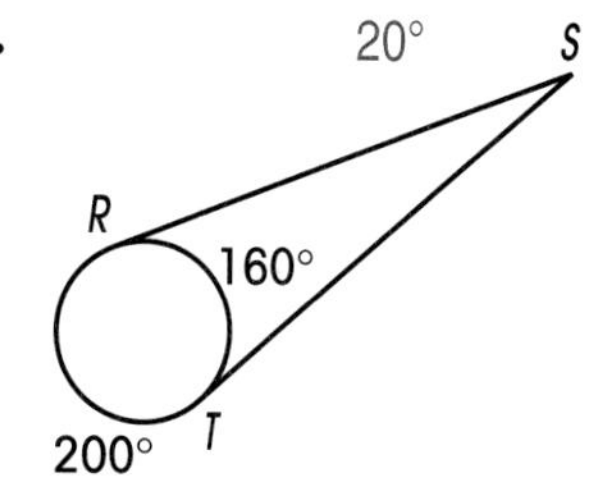

4. 30°

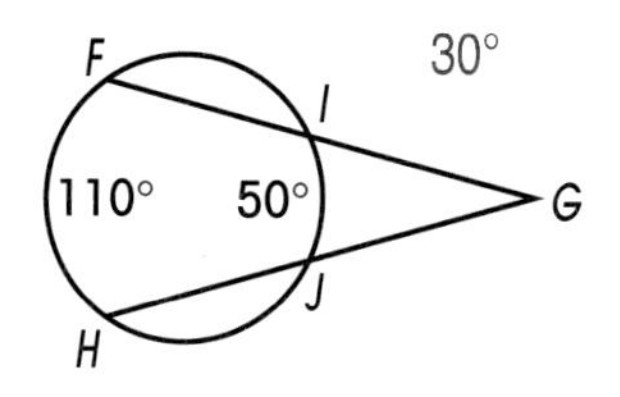

5. 40°

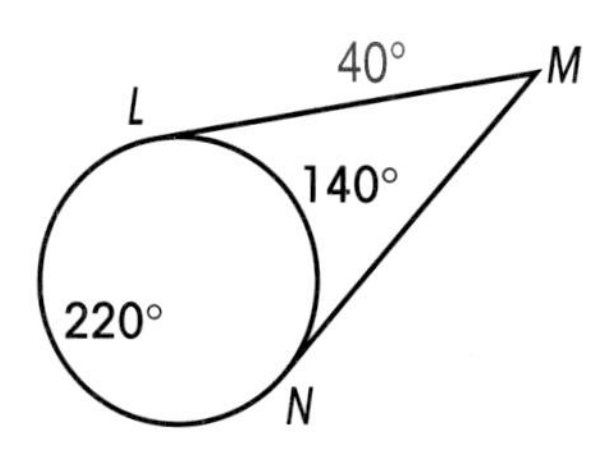

6. 35°

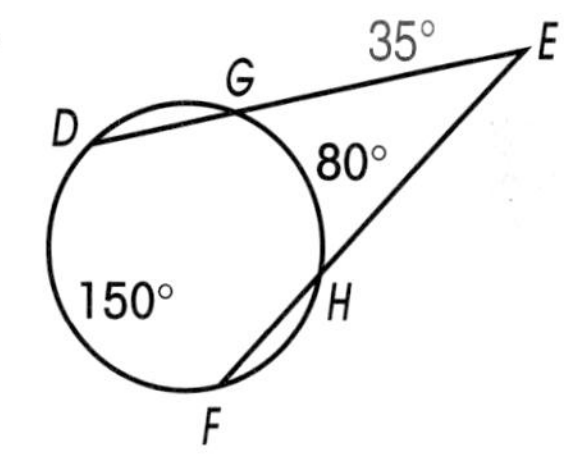

Share Your Understanding

7. Explain to a partner how to find the tangent angle or the secant angle in any exercise above. Answers will vary.

8. **CRITICAL THINKING** Copy the diagram in exercise 1. Draw the center C. Then draw $\overline{CN}$, $\overline{CP}$, and $\overline{CQ}$. Explain why $\overline{NP} \cong \overline{QP}$.

$\overline{NC} \cong \overline{QC}$ because both are radii. $\angle PNC$ and $\angle PQC$ are right triangles because a radius is perpendicular to a tangent at the point of tangency. $\overline{CP} \cong \overline{CP}$. So, by the Hypotenuse-Leg Theorem, $\triangle NPC \cong \triangle QPC$, and $\overline{NP} \cong \overline{QP}$.

More practice is provided in the Workbook, page 61, and in the Classroom Resource Binder, page 106.

10·8 Tangents and Segments

> THEOREM 32
> **Tangent-Segment Theorem**
>
> If two segments from the same exterior point are tangent to a circle, then they are congruent.

Two tangents from the same point outside a circle are congruent. $\overline{AB} \cong \overline{CB}$ in $\odot O$.

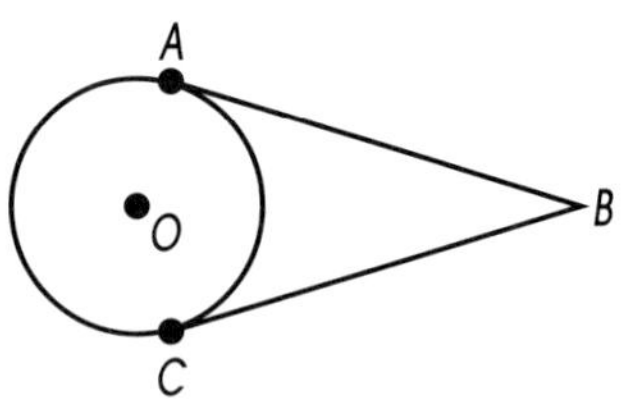

You can use what you know about congruent line segments to solve problems about tangents.

EXAMPLE 1

$\overline{DE}$ and $\overline{FE}$ are tangent to $\odot P$. Find the value of x.

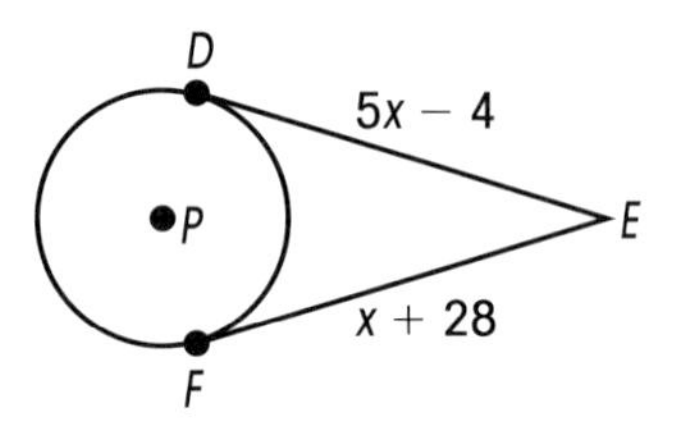

Getting Started
Have students trace a quarter and then draw two tangents from a single point outside the circle. Then, have them measure each tangent from the point where they touch the circle to the point outside the circle.

Write an equation. Use the Tangent-Segment Theorem. $5x - 4 = x + 28$

Solve for x.

$$4x - 4 = 28$$
$$4x = 32$$
$$x = 8$$

The value of x is 8.

EXAMPLE 2

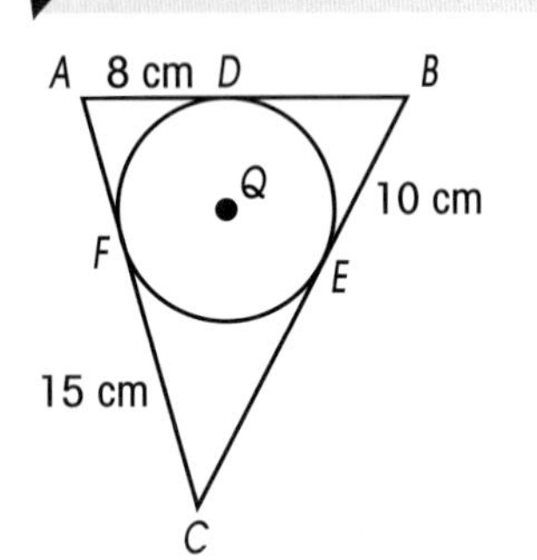

In the diagram on the left, each line segment is tangent to $\odot Q$. Find the perimeter of $\triangle ABC$.

First, identify the congruent line segments. Find the measures.

$$AD = AF = 8$$
$$BD = BE = 10$$
$$CF = CE = 15$$

Write an equation for the perimeter. $P = AD + AF + BD + BE + CF + CE$

Substitute the values you know. Simplify.

$$P = 8 + 8 + 10 + 10 + 15 + 15$$
$$= 66$$

The perimeter of $\triangle ABC$ is 66 centimeters.

Try This

In the diagram, each line segment is tangent to $\odot P$. Find the perimeter of $\triangle NTR$.

T, 3 in., L, K, P, N, 4 in., M, 9 in., R

First, identify the ■ line segments. (congruent) $TK = TL =$ ■ 3

Find the measures. $RL = RM =$ ■ 9

$NK =$ ■ $=$ ■ 4 (NM)

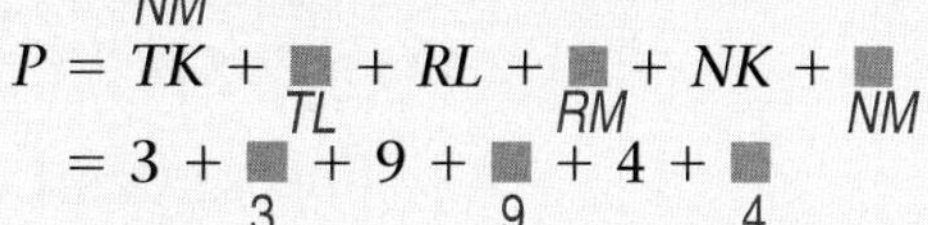

Write an equation for the perimeter. $P = TK +$ ■ $+ RL +$ ■ $+ NK +$ ■ (TL, RM, NM)

Substitute the values you know. $= 3 +$ ■ $+ 9 +$ ■ $+ 4 +$ ■ (3, 9, 4)

Simplify. $=$ ■ (32)

The perimeter of $\triangle NTR$ is ■. 32 in.

Practice

In each diagram, the line segments are tangent to the circle. Find the value of x.

1.

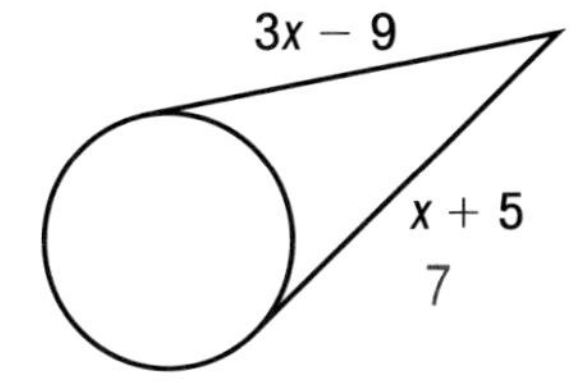

2.

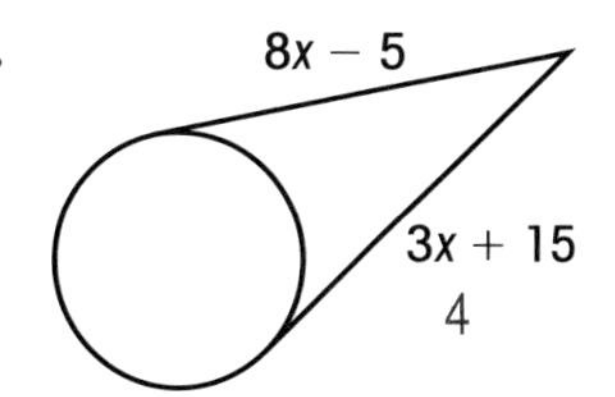

3.

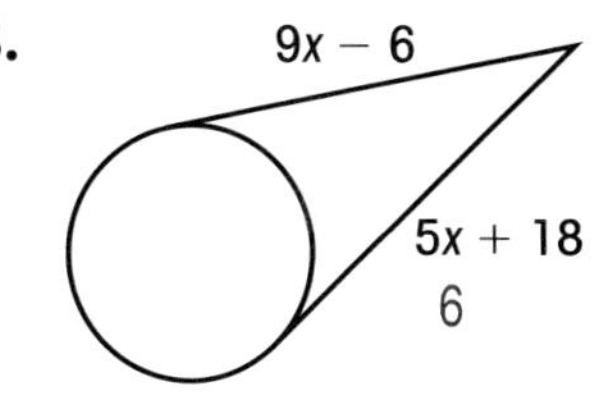

4. In the diagram, each line segment is tangent to $\odot T$. Find the perimeter of $\triangle PQR$. 88 cm

Q, 12 cm, X, Y, T, 15 cm, P, Z, 17 cm, R

5. Set both expressions equal to each other and solve for x. Then, substitute the value of x into one expression and simplify. This will give you the length of both tangents.

Share Your Understanding

5. Explain to a partner how to find the length of any tangent in exercises **1–3**. See above.

6. **CRITICAL THINKING** Each side of an equilateral triangle is tangent to a circle. The perimeter of the triangle is 24 cm. What is the length of one tangent segment from a vertex to the point of tangency? (Hint: Draw a diagram.) 4 cm

More practice is provided in the Workbook, page 62, and in the Classroom Resource Binder, page 107.

10-9 Chords

Getting Started
Review the definition of diameter. Draw a circle on the board with a diameter and another chord that is not a diameter. Ask students what the difference is between the two segments.

A **chord** is a line segment with endpoints on a circle. $\overline{AB}$ is a chord of $\odot M$. ME is the distance that chord AB is from the center of $\odot M$.

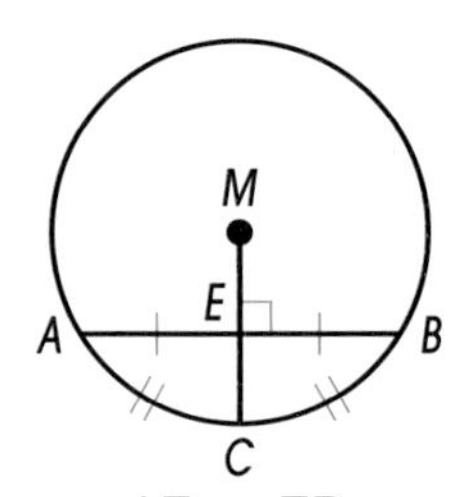

If a radius of a circle is perpendicular to a chord, the radius bisects the chord and its arc.

$AE = EB$
$m\overset{\frown}{AC} = m\overset{\frown}{CB}$

Any line segment from the center of a circle that is perpendicular to a chord bisects the chord.

If you know the length of a chord and its distance from the center, you can find the length of the radius.

EXAMPLE

Find the length of radius ER of $\odot R$.

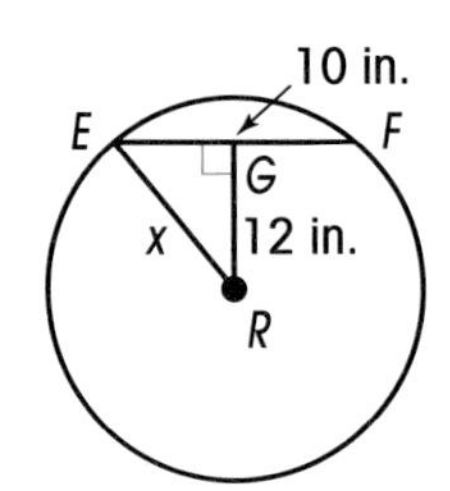

$\triangle EGR$ is a right triangle. $\overline{ER}$ is the hypotenuse and a radius of $\odot R$. Use the Pythagorean Theorem.

First, find EG. Because $\overline{RG}$ is perpendicular to chord EF, $\overline{RG}$ bisects $\overline{EF}$. Divide EF by 2.

$EG = GF$
$EG = EF \div 2$

Substitute 10 for EF.

$EG = 10 \div 2$
$= 5$

Now, use the Pythagorean Theorem.

$a^2 + b^2 = c^2$
$(EG)^2 + (GR)^2 = (ER)^2$

Substitute the values you know.

$5^2 + 12^2 = x^2$

Solve for x.

$25 + 144 = x^2$
$169 = x^2$
$13 = x$

The length of radius ER is 13 inches.

Try This

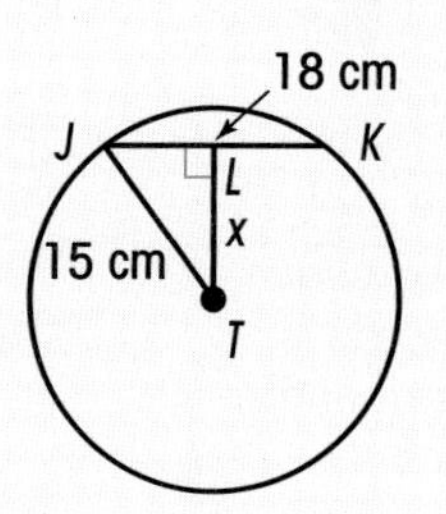

Radius JT is 15 centimeters.
Chord JK is 18 centimeters. Find the value of x.

Use the ■ Theorem. Pythagorean $a^2 + b^2 = c^2$

Because $JL = LK$, $JL =$ ■. 9 $(JL)^2 + (LT)^2 = (JT)^2$

Substitute the values you know. 9^2 ■ $+ x^2 = 15^2$

81 ■ $+ x^2 = 225$

Solve for x. $x^2 =$ ■ 144

$x =$ ■ 12

The value of x is ■. 12 cm

Practice

Find the value of x in each circle.

1. 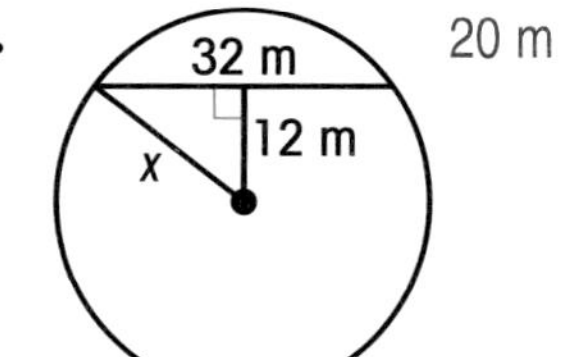

20 m

2. 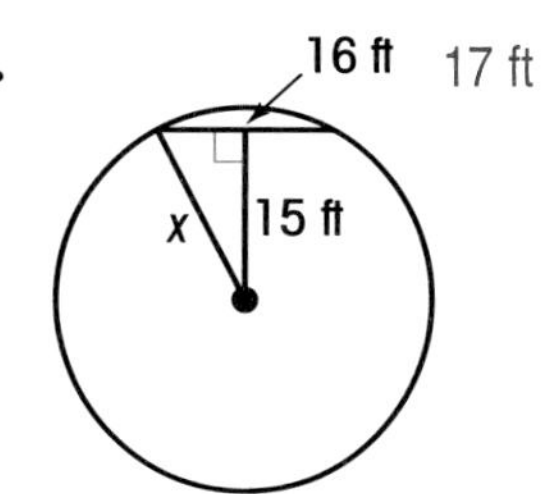

17 ft

3. 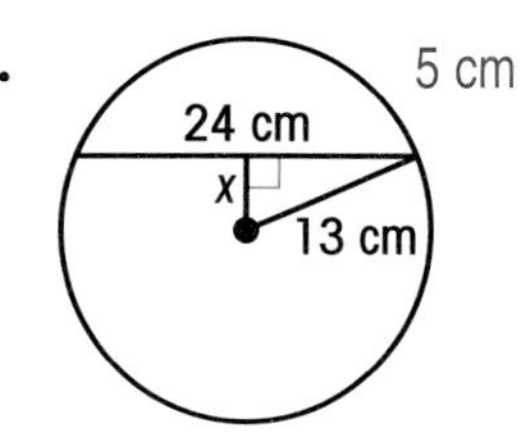

5 cm

4. 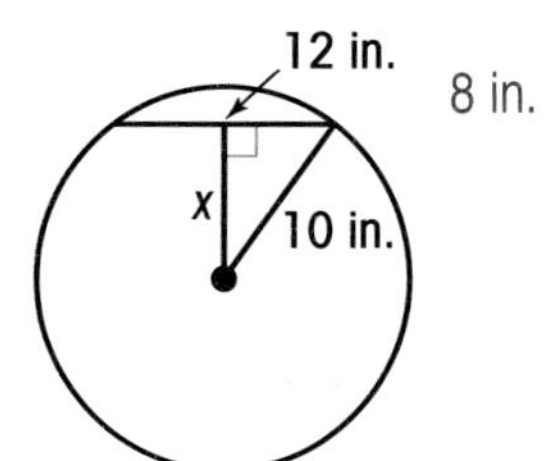

8 in.

5. 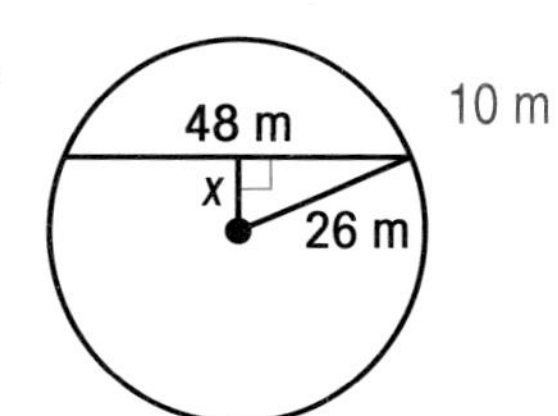

10 m

6. 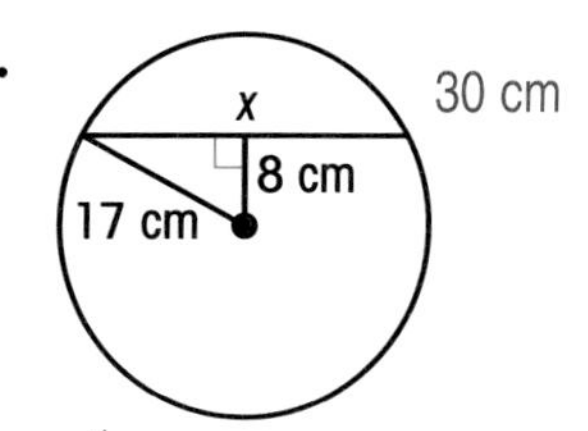

30 cm

7. Write the Pythagorean Theorem. Find $\frac{1}{2}$ of $32 = 16$. So, $(16)^2 + (12)^2 = x^2$. Square 16 and 12; then add to get $400 = x^2$. Find the square root of both sides to get $x = 20$.

Share Your Understanding

7. Explain to a partner how to find the length of the radius in exercise **1**. See above.

8. **CRITICAL THINKING** The diameter of $\odot P$ is 20 in. Chord AB is 12 in. How far is $\overline{AB}$ from the center of $\odot P$? (Hint: Draw a circle with chord AB perpendicular to the diameter.) 8 in.

More practice is provided in the Workbook, page 63, and in the Classroom Resource Binder, page 108.

10·10 Chords and Angles

Getting Started Review the method for finding an average.

THEOREM 33

The measure of an angle formed by two intersecting chords is one-half the sum of the intercepted arcs.

Two chords that intersect form vertical angles. You can find the measure of these angles if you know the measures of the arcs.

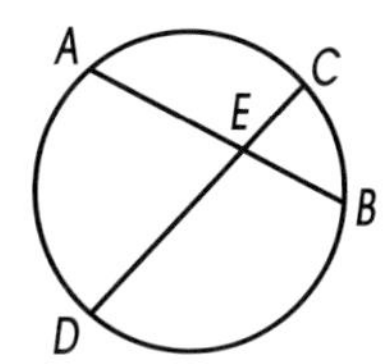

$$m\angle AED = \frac{1}{2}(m\overset{\frown}{AD} + m\overset{\frown}{CB})$$

EXAMPLE 1

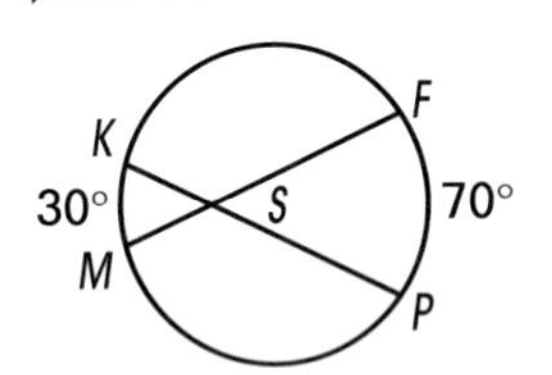

Find the measure of $\angle KSM$.

Write an equation. $m\angle KSM = \frac{1}{2}(m\overset{\frown}{KM} + m\overset{\frown}{FP})$

Substitute the given values. $m\angle KSM = \frac{1}{2}(30 + 70)$

Simplify. $= \frac{1}{2}(100)$

$= 50$

So, $\angle KSM$ is 50°.

You can also find the measure of an intercepted arc.

EXAMPLE 2

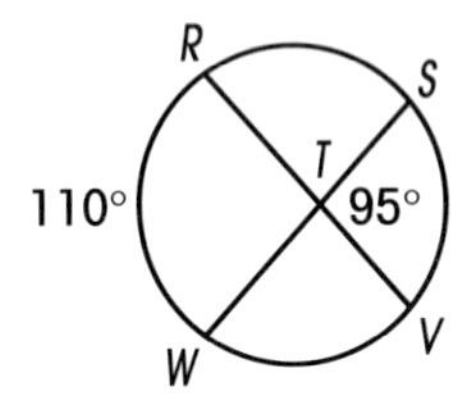

Find the measure of $\overset{\frown}{SV}$.

Write an equation. $m\angle STV = \frac{1}{2}(m\overset{\frown}{SV} + m\overset{\frown}{RW})$

Substitute the given values. $95 = \frac{1}{2}(m\overset{\frown}{SV} + 110)$

Solve for $m\overset{\frown}{SV}$. $190 = m\overset{\frown}{SV} + 110$

$80 = m\overset{\frown}{SV}$

So, $\overset{\frown}{SV}$ is 80°.

Try This

Find the measure of $\angle JGM$.

$\angle JGM$ and $\angle JGE$ are supplementary.
First, find the measure of $\angle JGE$.

Write an equation. $\quad m\angle JGE = \frac{1}{2}(m\widehat{JE} + \blacksquare)$ $\quad m\widehat{MK}$

Substitute the given values. $\quad m\angle JGE = \frac{1}{2}(\blacksquare + 72)$ $\quad 80$

Simplify. $\quad = \frac{1}{2}(\blacksquare)$ $\quad 152$

$= \blacksquare$ $\quad 76$

Now, find the measure of $\angle JGM$. $\quad m\angle JGM = 180 - \blacksquare$ $\quad 76$

So, $\angle JGM$ is $\blacksquare$. 104° $\quad = \blacksquare$ $\quad 104$

Practice

Find the value of x in each circle.

1.

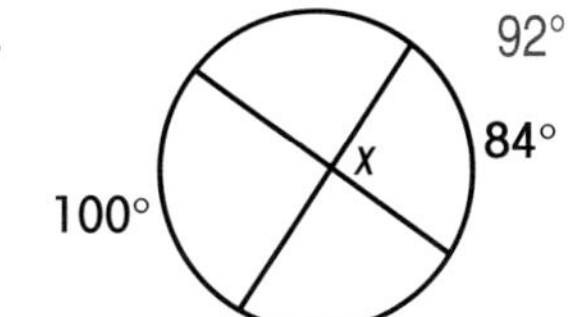

2.

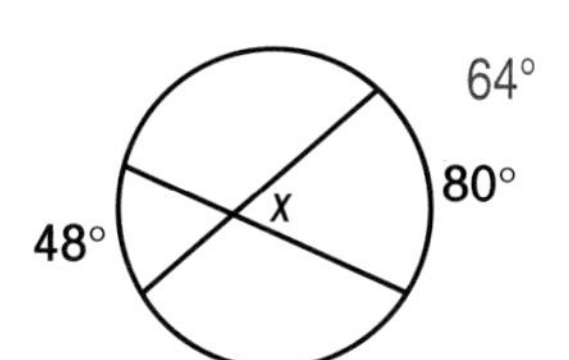

3.

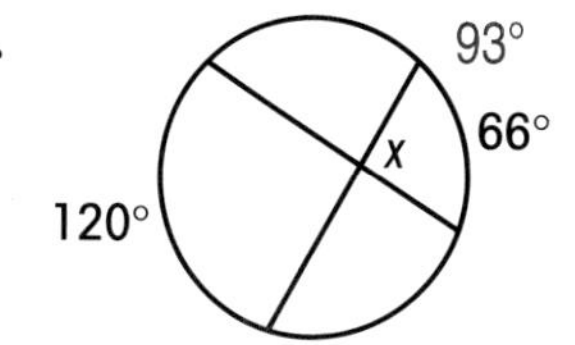

4.

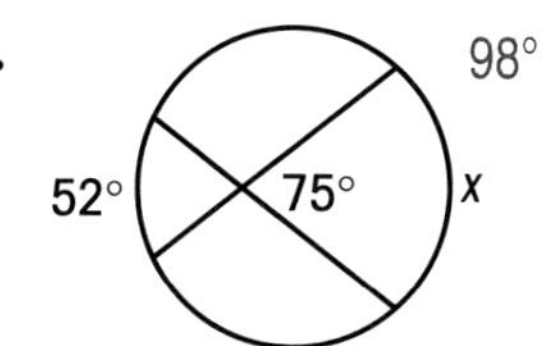

5.

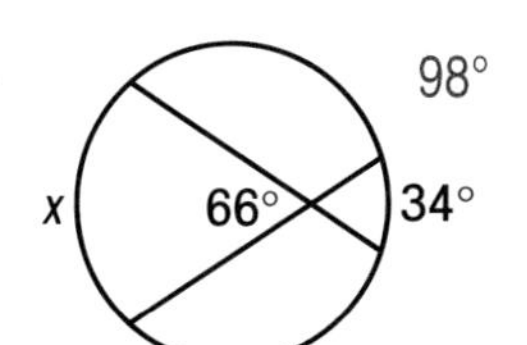

6.

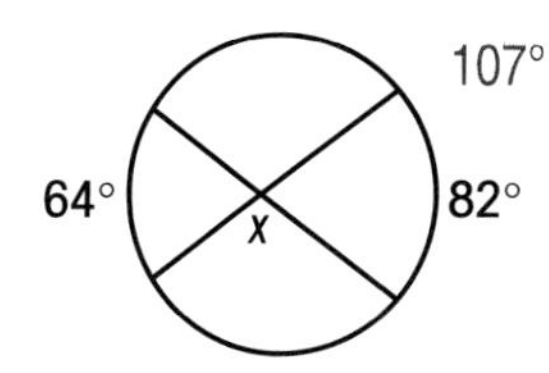

8. Yes, the intercepted arcs will be congruent because the diameters form vertical angles and vertical angles are congruent. Congruent angles form congruent arcs when they are central angles.

Share Your Understanding

7. Explain to a partner how to find the value of x in any of the exercises above. Answers will vary.

8. **CRITICAL THINKING** Two diameters intersect in the center of a circle. Will the intercepted arcs be congruent? Tell why or why not. See above.

More practice is provided in the Workbook, page 64, and in the Classroom Resource Binder, page 109.

10-11 Chords and Segments

Getting Started
Review the definition of a chord.

When two chords intersect, the segments have a special relationship.

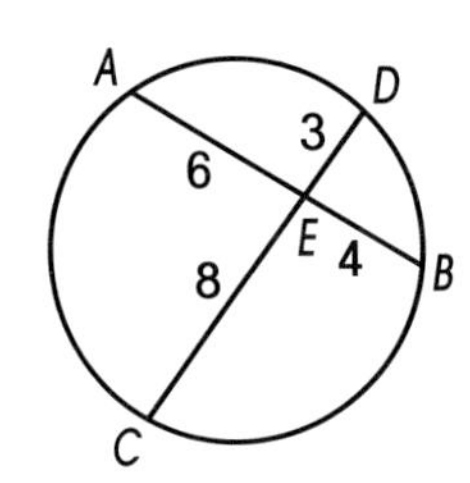

$$AE \cdot EB = CE \cdot ED$$
$$3 \cdot 8 = 6 \cdot 4$$
$$24 = 24$$

You can find the length of segments formed by intersecting chords.

EXAMPLE 1

Find the value of x.

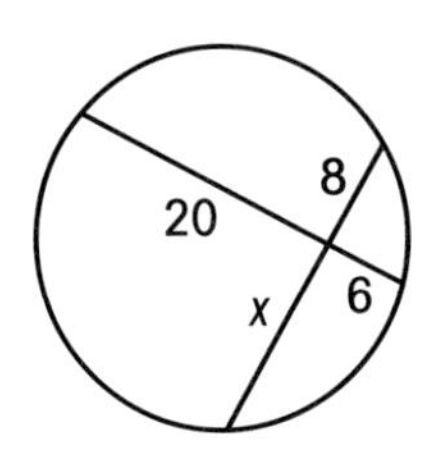

Write an equation. $20 \cdot 6 = x \cdot 8$

Solve for x. $120 = 8x$

$15 = x$

The value of x is 15.

EXAMPLE 2

Find the value of x.

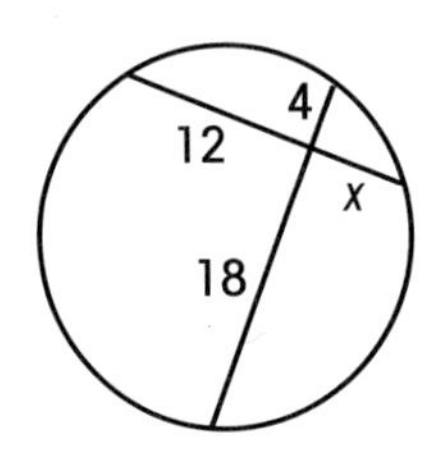

Write an equation. $12 \cdot x = 18 \cdot 4$

Solve for x. $12x = 72$

$x = 6$

The value of x is 6.

Try This

Find the value of x.

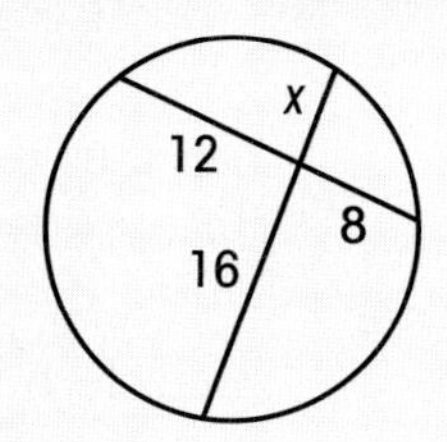

Write an equation. $\quad 12 \bullet \blacksquare = \blacksquare \bullet x$ (8, 16)

Solve for x. $\quad 96 \blacksquare = 16x$

$6 \blacksquare = x$

The value of x is ■. 6

Practice

Find the value of x in each circle.

1. 24

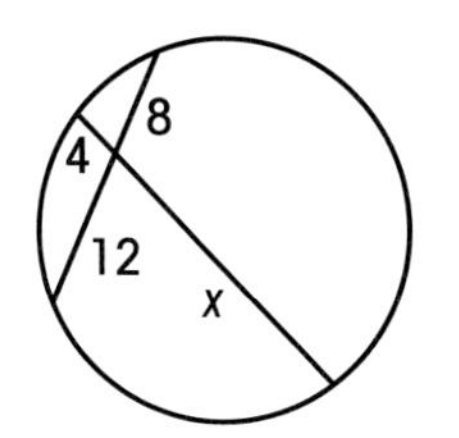

2. 8

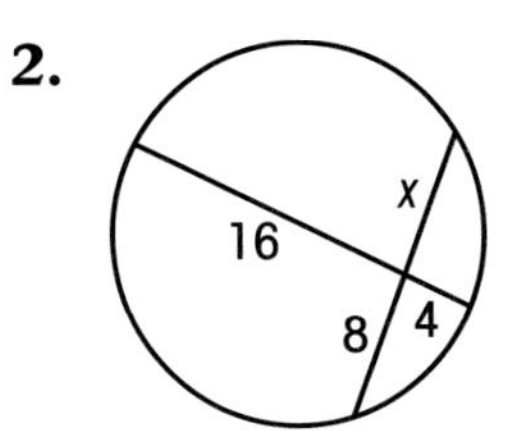

3. 9

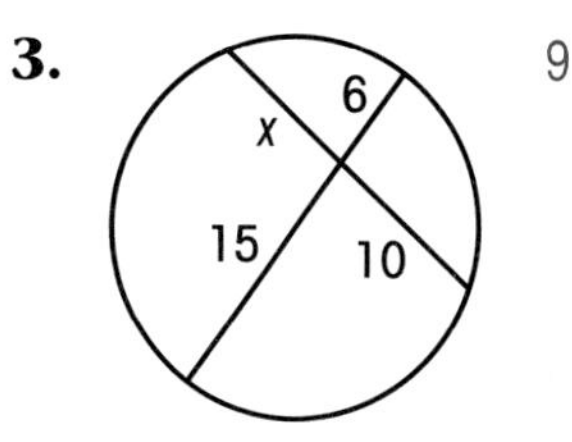

4. 14

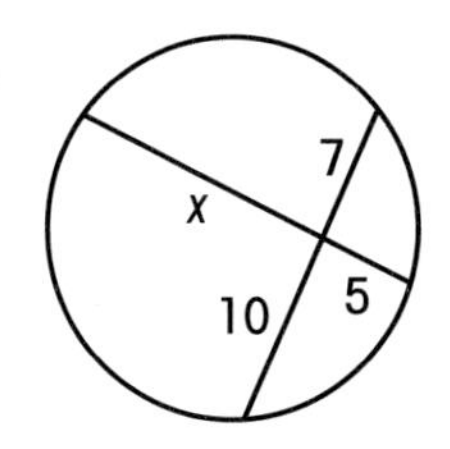

5. 22

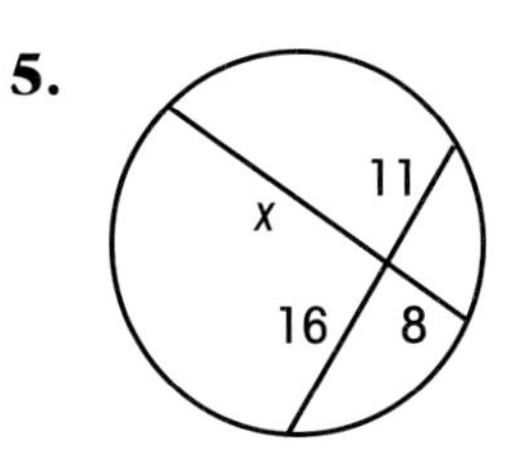

6. 6

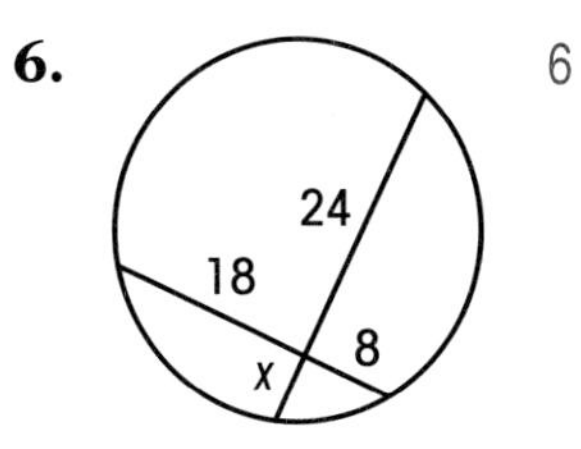

Share Your Understanding

7. Explain to a partner how to find the value of x in any of the exercises above. Use the words *equation* and *multiply* in your explanation. Answers will vary.

8. **CRITICAL THINKING** Find the value of x in $\odot P$ in the diagram on the right. 2

10-12 Calculator: Circumference and Area of a Circle

Getting Started
Remind students that 3.14 is an estimation of pi.

You can use a calculator to find the circumference and the area of a circle.

EXAMPLE 1

A circle has a diameter of 28 ft. Find the circumference of the circle. Round your answer to the nearest whole number.

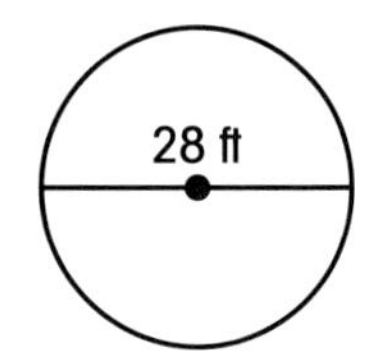

Math Fact
Some calculators have a special π key. If you multiply by the π key, you will need to round your answer.

Write the formula for the circumference of a circle that uses diameter. $C = \pi d$

Substitute 3.14 for π and 28 for d. $C \approx (3.14)(28)$

Use your calculator to multiply. **Display**

PRESS: [3] [.] [1] [4] [×] [2] [8] [=] 87.92

The circumference of the circle is about 88 ft.

EXAMPLE 2

A compact disc has a diameter of 12 cm. Find the area of the compact disc. Round your answer to the nearest whole number.

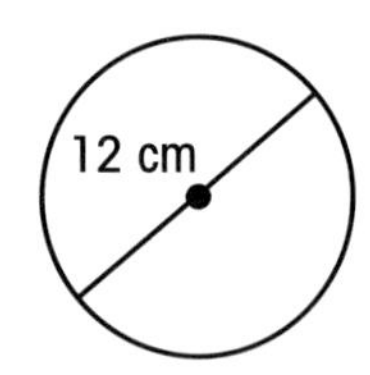

Write the formula for the area of a circle. $A = \pi r^2$

The radius of the disc is half the diameter. Half of 12 is 6. Substitute 3.14 for π and 6 for r. $A \approx (3.14)(6^2)$

Use your calculator to multiply. **Display**

PRESS: [3] [.]  [4]  [6] [×]  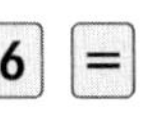 113.04

The area of the disc is about 113 cm^2.

Practice

Find the area of each circle. Use 3.14 for π. Round your answer to the nearest hundredth.

1. diameter = 41.5 cm
about 1,351.97 cm^2

2. diameter = 62 in.
about 3,017.54 $in.^2$

3. radius = 10 mm
about 314.00 mm^2

Find the circumference of each circle. Use 3.14 for π. Round your answer to the nearest whole number.

4.

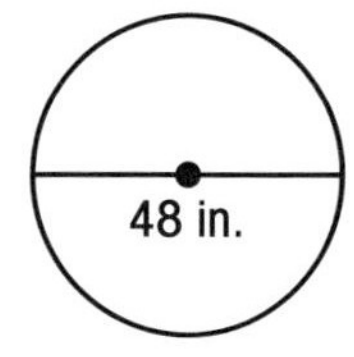

about 151 in.

5.

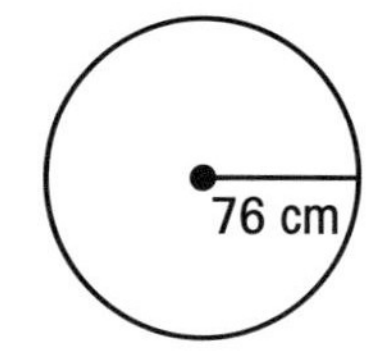

about 477 cm

6.

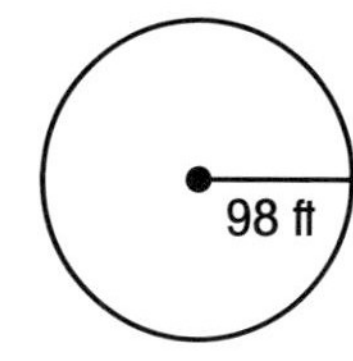

about 615 ft

Find the area and circumference of each circle. Use 3.14 for π. Round your answer to the nearest tenth.

7. radius = 24 ft
$A \approx 1{,}808.6\ ft^2$; $C \approx 150.7$ ft

8. diameter = 51 cm
$A \approx 2{,}041.8\ cm^2$; $C \approx 160.1$ cm

9. diameter = 38 in.
$A \approx 1{,}133.5\ in.^2$; $C \approx 119.3$ in.

Math in Your Life

CONCENTRIC CIRCLES IN NATURE

Circles that have the same center and lie in the same plane are called concentric circles. Concentric circles are found in flowers, dart boards, and gardens.

Sometimes when it rains, you can see concentric circles. When a raindrop hits a pool of water, it becomes the center of concentric circles. Many circles with radii of different lengths form inside each other. These circles lie in the same plane and have the same center.

You can make your own concentric circles. Dip your finger in a cup of water. Lift it out and let one drop of water fall from your fingertip into the water. How many concentric circles do you see?

More practice is provided in the Workbook, page 65, and in the Classroom Resource Binder, page 110.

Problem-Solving Skill: Inscribed and Circumscribed Circles

Getting Started
Review with students the meaning of the prefixes *in-* and *circum-*. (*in-* means "inside" and *circum-* means "around.")

When a circle is inscribed in a square, the sides of the square are tangent to the circle. The diameter of the inscribed circle is equal to the length of a side of the square.

You can find the perimeter or the area of a square if you know the length of the diameter or the radius of the inscribed circle.

EXAMPLE 1

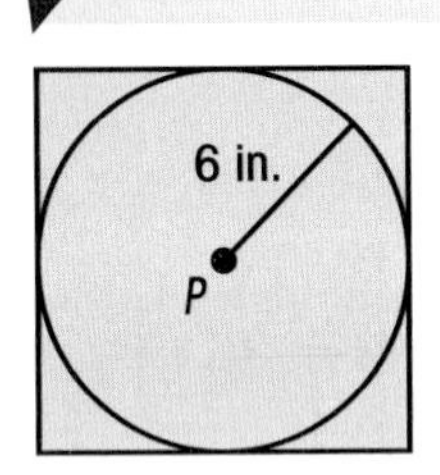

$\odot P$ is inscribed in a square. Find the area of the square.

Find the diameter.

$$d = 2 \bullet r$$
$$= 2 \bullet 6 = 12$$

Find the area. The side of the square is equal to the diameter.

$$A = s^2$$
$$= 12 \bullet 12 = 144$$

The area of the square is 144 in.^{2}

When a circle is circumscribed about a square, it touches the four vertices of the square. The diameter of the circle is a diagonal of the square.

You can find the circumference or the area of the circumscribed circle if you know the length of a side of the square.

EXAMPLE 2

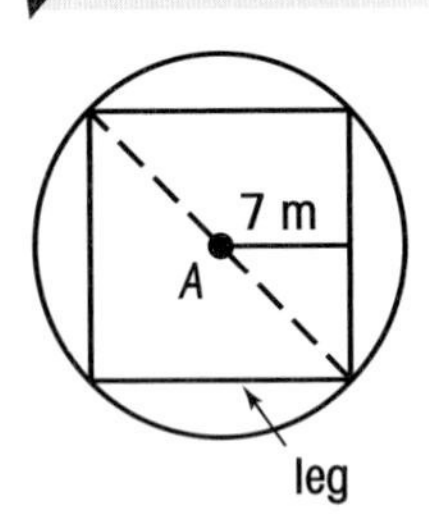

$\odot A$ is circumscribed about a square. Find the circumference of $\odot A$. Leave π in the answer.

Find the length of the side of the square. $s = 2 \bullet 7 = 14$

The diagonal is the hypotenuse of a 45°-45°-90° right triangle. Find the length of the diagonal.

$$\text{hypotenuse} = \text{leg} \bullet \sqrt{2}$$
$$\text{diagonal} = \text{leg} \bullet \sqrt{2}$$
$$= 14\sqrt{2}$$

The diagonal of the square is equal to the diameter. Find the circumference.

$$C = \pi d$$
$$= 14\sqrt{2}\,\pi$$

The circumference of the circle is $14\sqrt{2}\pi$ inches.

Try This

$\odot N$ is circumscribed about a square. Find the area of $\odot N$. Leave π in your answer.

Find the ■ length of the diagonal of the square.
$$\text{diagonal} = \text{leg} \cdot \sqrt{2}$$
$$= ■ \sqrt{2} \quad 6$$

The diagonal of the square is equal to the ■ diameter; of the circle. Divide the diameter by ■ 2 to find the radius.
$$\text{radius} = \text{diameter} \div ■ \quad 2$$
$$= 6\sqrt{2} \div ■ \quad 2$$
$$= 3\sqrt{2}$$

Find the area.
$$A = \pi r^2$$
$$= ■ (3\sqrt{2})^2 \quad \pi$$
$$= \pi \cdot 3 \cdot 3 \cdot 2$$
$$= ■\pi \quad 18$$

The area of $\odot N$ is ■. 18π in.^{2}

Practice

Solve each problem. Show your work.

1. $\odot K$ is inscribed in a square. Find the area of the square.
100 cm^2

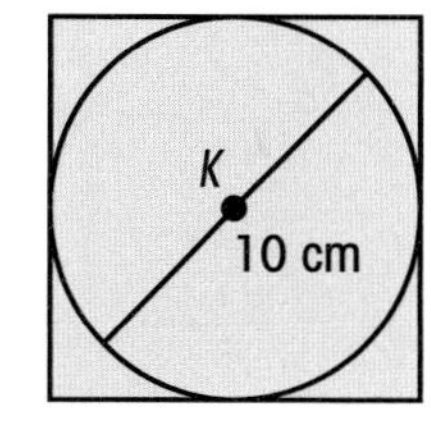

2. $\odot R$ is inscribed in a square. Find the perimeter of the square.
64 ft

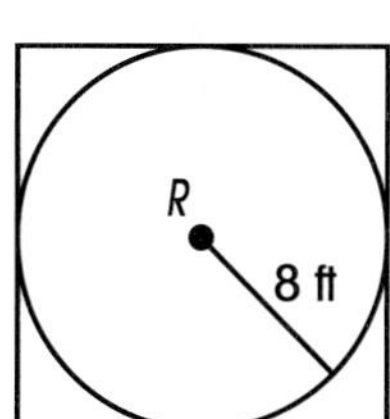

3. $\odot T$ is circumscribed about a square. Find the circumference of $\odot T$. Leave π in your answer.
$10\sqrt{2}\,\pi$ in.

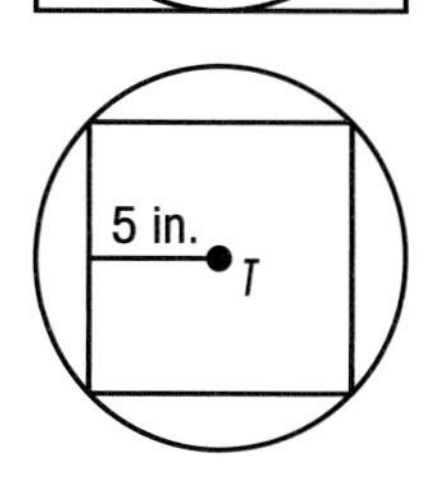

More practice is provided in the Workbook, page 66, and in the Classroom Resource Binder, page 111.

10-14 Problem-Solving Application: Revolutions of a Circle

Getting Started
Have students cut a length of string to fit around a tire. They should also measure the diameter of the wheel. Have students use the diameter in a circumference. The length of the string should be the same as the circumference found with the formula.

The distance a wheel travels in one complete turn is equal to the circumference of the wheel. One complete turn of a wheel is called a revolution.

You can use circumference to solve distance problems.

EXAMPLE 1

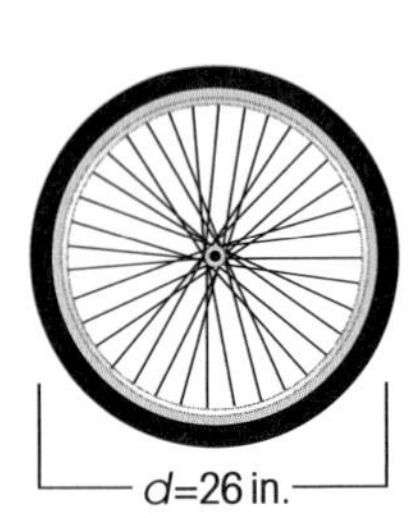

Aaron rides a bicycle with 26-inch wheels. This means that the diameter of a wheel is 26 inches. What is the distance in feet he will travel in 1,000 revolutions?

First, find the distance traveled in one revolution. Use the formula for circumference.

$$C = \pi d \approx (3.14)(26) \approx 81.64$$

Then, multiply the circumference by 1,000 revolutions to find the distance in inches.

$$\text{distance} \approx 81.64 \cdot 1{,}000 \approx 81{,}640$$

Divide the distance by 12 to find the number of feet. 12 in. = 1 ft

$$\approx 81{,}640 \div 12 \approx 6{,}803.3$$

Aaron can travel about 6,803 feet.

EXAMPLE 2

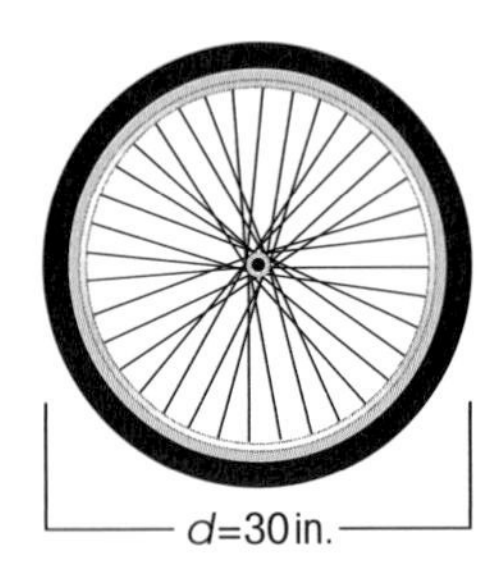

Mari rides a bicycle with 30-inch wheels. She travels 7,850 feet. How many revolutions of the wheel is that?

First, multiply by 12 to find the distance traveled in inches.

$$\text{distance} = 7{,}850 \cdot 12 = 94{,}200$$

Then, find the circumference.

$$C = \pi d \approx (3.14)(30) \approx 94.2$$

Divide the distance by the circumference to find the number of revolutions.

$$\text{revolutions} \approx 94{,}200 \div 94.2 \approx 1{,}000$$

There were about 1,000 revolutions.

Try This

April rides a bicycle with 22-inch wheels. She can do 250 revolutions in one minute. At this rate, what is the distance in feet April can travel in 5 minutes?

First, find the distance traveled in one ■. revolution
Use the formula for ■. circumference

$C = \pi d$
$\approx (3.14)(■)$ 22
$\approx ■$ 69.08

Then, find the distance traveled in 250 ■. revolutions

distance $\approx (69.08)\ (■)$ 250
$\approx ■$ 17,270

Then, find the distance traveled in 5 minutes.

distance $\approx (17{,}270)\ (■)$ 5
$\approx ■$ 86,350

Now, divide by 12 to find the number of feet.

distance $\approx ■ \div 12$ 86,350
$\approx ■$ 7,195.8

April can travel about ■ in 5 minutes. 7,196 feet

Practice

Solve each problem. Show your work.

1. Carmen rides a bicycle with 24-inch wheels. What is the distance in feet she can travel in 2,000 revolutions?
   about 12,560 ft

2. Pedro rides a bicycle with 26-inch wheels. He travels 4,900 feet. About how many revolutions of the wheel is that? about 720 revolutions

3. Dennis rides a bicycle with 30-inch wheels. He can do 300 revolutions in one minute. At this rate, what is the distance in feet he can travel in 4 minutes?
   about 9,420 ft

4. Kim rides a bicycle with 24-inch wheels. She can do 300 revolutions in one minute. At this rate, what is the distance in feet she can travel in 4 minutes?
   about 7,536 ft

5. Heather rides a bicycle with 22-inch wheels. Sarah rides a bicycle with 30-inch wheels. What is the distance in feet each girl can travel in 1,000 revolutions? How much farther can Sarah travel?
   Heather can ride about 5,757 feet. Sarah can ride about 7,850 feet. Sarah can travel about 2,093 feet farther.

An alternate two-column proof lesson is provided on page 415 of the student book.

Getting Started
Remind students that an angle inscribed in a circle is formed by 2 chords. Ask if an inscribed angle can be formed with one radius and one chord. Encourage students to explain their reasoning.

THEOREM 34

If two inscribed angles of a circle intercept the same arc, then the angles are congruent.

10-15 Proof: Circles

You learned that the measure of an inscribed angle is equal to one-half the measure of its arc.

$$m\angle ABC = \frac{1}{2}m\widehat{AC}$$

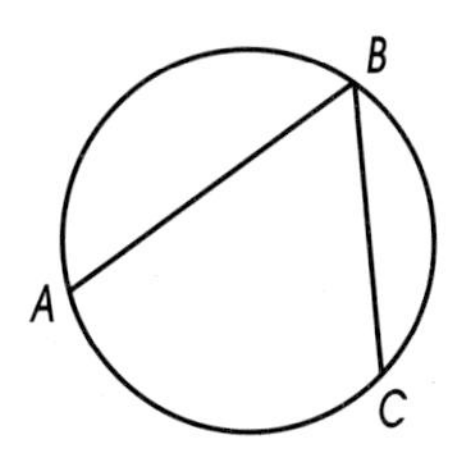

You can use this fact to prove that if two inscribed angles intercept the same arc, they are congruent.

EXAMPLE

You are given:
$\angle DEF$ intercepts $\widehat{DF}$.
$\angle DGF$ intercepts $\widehat{DF}$.

Prove: $\angle DEF \cong \angle DGF$

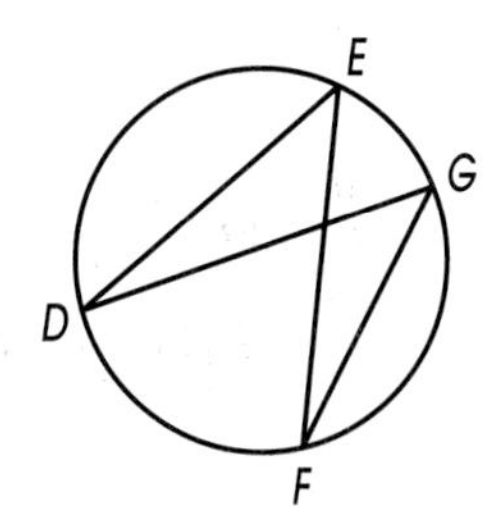

You are given that $\angle DEF$ intercepts $\widehat{DF}$. This means that $m\angle DEF = \frac{1}{2}m\widehat{DF}$. You are also given that $\angle DGF$ intercepts $\widehat{DF}$. This means that $m\angle DGF = \frac{1}{2}m\widehat{DF}$.

Because the measures of both angles are equal to $\frac{1}{2}m\widehat{DF}$, the measures are equal to each other. Since $m\angle DEF = m\angle DGF$, then $\angle DEF \cong \angle DGF$.

So, you prove that $\angle DEF \cong \angle DGF$. ✓

Try This

Copy and complete the paragraph proof.

You are given: $\overline{HI} \cong \overline{JK}$

Prove: $\overline{LP} \cong \overline{MP}$

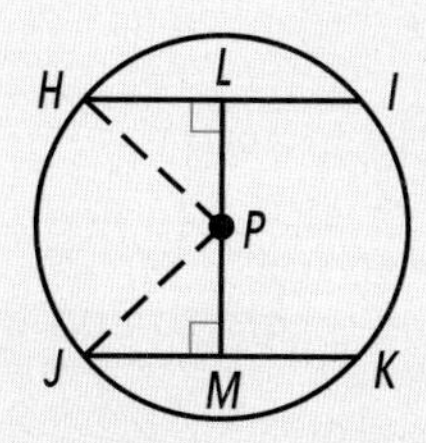

First, you can prove two triangles congruent. You can draw dashed lines to make right triangles. These lines are radii of the circle.

You are given that $\overline{HI} \cong$ ■ ($\overline{JK}$). *You know that* $\overline{LM}$ *bisects* $\overline{HI}$ *and* $\overline{JK}$. *Then,* $\overline{HL} \cong$ ■ ($\overline{JM}$). *You know that* $\overline{HP} \cong$ ■ ($\overline{JP}$) *because radii of the same circle are congruent. Then, by the Hypotenuse -*■ (Leg) *Theorem* $\triangle HLP \cong$ ■ ($\triangle JMP$). *So,* $\overline{LP} \cong$ ■ ($\overline{MP}$) *by CPCTC.*

So, you prove that ■. Congruent chords are the same distance from the center of a circle. ✓ $\overline{LP} \cong \overline{MP}$

Practice

Write a proof for each of the following.

1. You are given: $\angle RQS \cong \angle TQU$
   Prove: $\overset{\frown}{RS} \cong \overset{\frown}{TU}$
   See Additional Answers for exercises 1–3 in the back of this book.

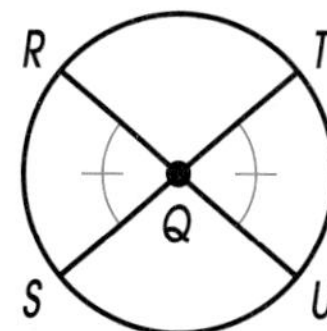

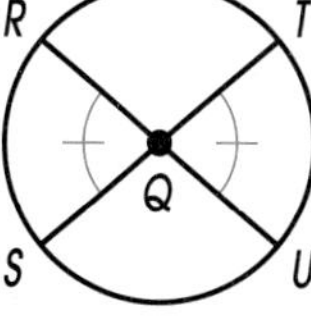

2. You are given: $\overline{AB} \cong \overline{CD}$
   Prove: $\overset{\frown}{AB} \cong \overset{\frown}{CD}$

3. You are given: $\overline{EF} \parallel \overline{GH}$
   Prove: $\overset{\frown}{EG} \cong \overset{\frown}{FH}$

More Practice is provided in the Classroom Resource Binder.

Chapter 10 Review

Summary

Summary
Use the formula $C = 2\pi r$ to find the circumference of a circle.
Use the formula $A = \pi r^2$ to find the area of a circle.
Use the formula length $= \frac{m\angle x}{360} \bullet 2\pi r$ to find the length of an arc.
Use the formula $A = \frac{m\angle x}{360} \bullet \pi r^2$ to find the area of a sector.
The measure of a central angle is equal to its intercepted arc.
The measure of an inscribed angle is one-half its intercepted arc.
You can use a calculator to find the area and circumference of a circle.
The revolution of a wheel is equal to its circumference.
You can use theorems and postulates you know about circles to prove a statement.

More vocabulary review is provided in the Classroom Resource Binder, page 99.

arc
circle
central angle
tangent
circumference
diameter
chord

Vocabulary Review

Complete the sentences with words from the box.

1. The ____ is the distance around a circle. circumference
2. A ____ is a plane figure made up of points all the same distance from the center. circle
3. An angle with its vertex at the center of a circle and sides that are radii is a ____. central angle
4. A curve between any two points on the circle is called an ____. arc
5. The ____ is a line segment that passes through the center of a circle. diameter
6. A line segment connecting two points on a circle is a ____. chord
7. A ____ is a line that touches a circle at only one point. tangent

Chapter Quiz

More assessment is provided in the Classroom Resource Binder.

Find the circumference and area of each circle. Use 3.14 for π. Round your answers to the nearest whole number.

1.

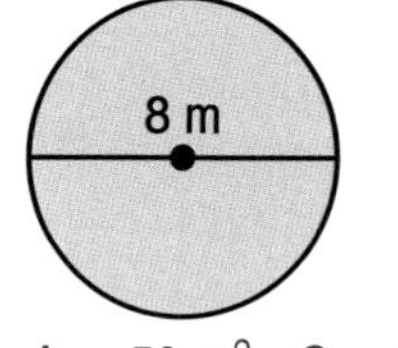

$A \approx 50$ m^2; $C \approx 25$ m

2.

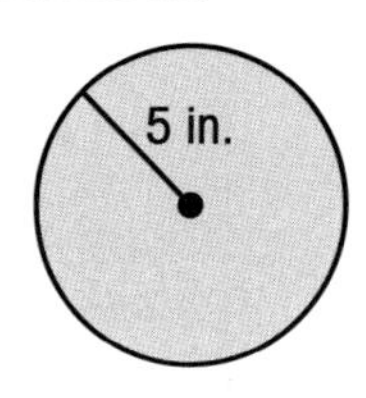

$A \approx 79$ in.^{2}; $C \approx 31$ in.

3.

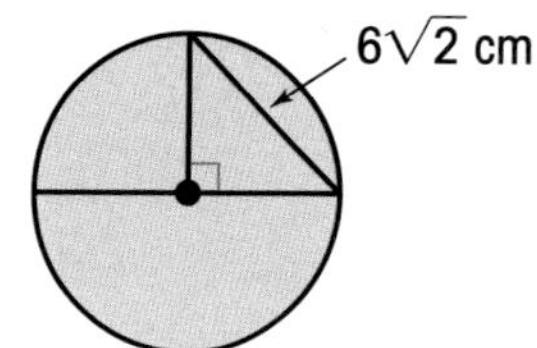

$A \approx 113$ cm^2; $C \approx 38$ cm

Find the value of x in each figure.

4.

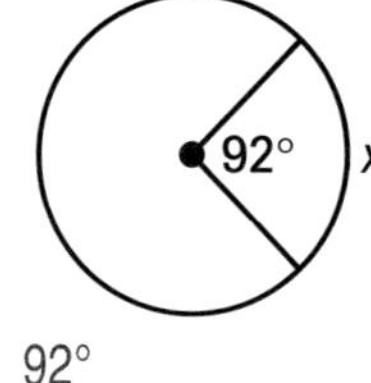

92°

5.

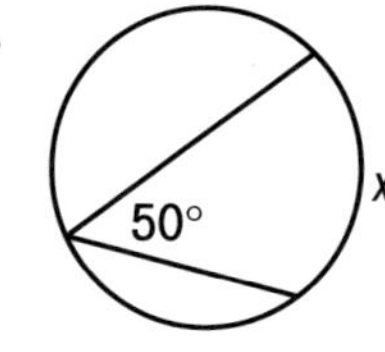

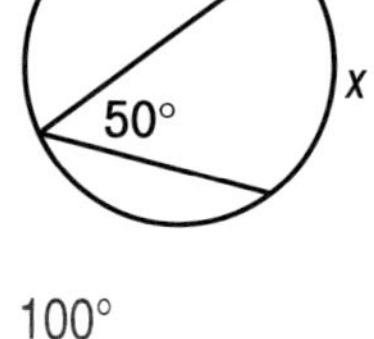

100°

6.

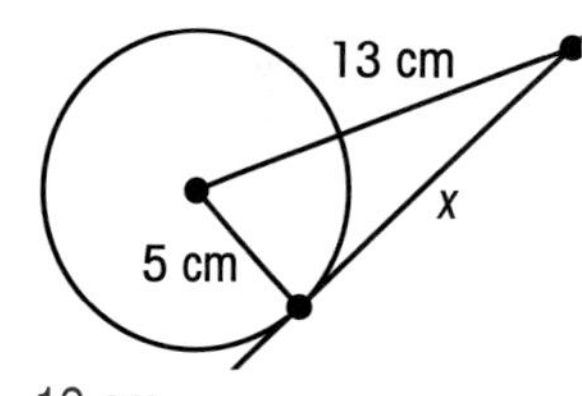

12 cm

7.

x 120° 240°

60°

8.

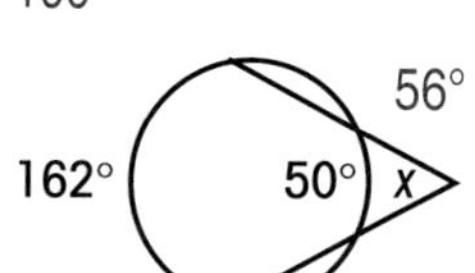

56°

9.

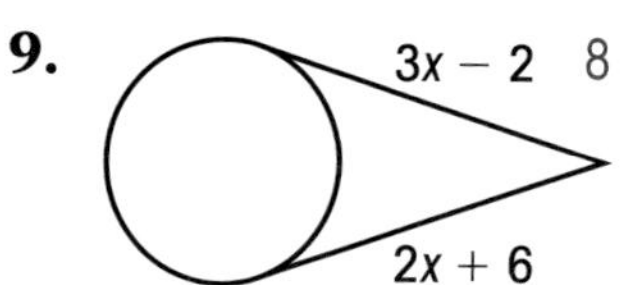

8

A chord and a diameter both touch two points on the circle. A diameter always passes through the center of the circle; not all chords do. All diameters are chords, but a chord can only be a diameter if it passes through the center of the circle.

Problem Solving

Solve each problem. Show your work. Round your answer to the nearest whole number

10. Andre rides a bicycle with 28-inch wheels. He travels 5,600 feet. About how many revolutions of the wheel is that?
about 764 revolutions

11. A circle is circumscribed about a square. The diagonal of the square is 10 feet. What is the area of the circle?
Area of the circle ≈ 79 ft^2

> **Write About Math**
> What is the difference between a chord and a diameter? Can a chord be a diameter? Can a diameter be a chord? Explain why or why not.

Additional practice for this chapter is provided on page 402 of the student book.

Unit 3 Review

Standardized Test Preparation This unit review follows the format of many standardized tests. A Scantron sheet is provided in the Classroom Resource Binder.

Write the letter of the correct answer.

1. Find the perimeter of a rectangle with a length of 4 cm and a width of 7 cm.
A. 28 cm
B. 28 cm^2
(C.) 22 cm
D. 15 cm

2. What is the area of the figure?
A. 360 ft^2
B. 76 ft
(C.) 300 ft^2
D. 360 ft

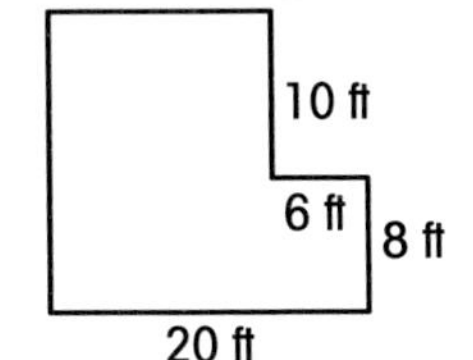

3. You have 5 black beans and 7 green beans. What is the ratio of green beans to total beans?
A. 5:7
(B.) 7:12
C. 5:12
D. 12:7

4. These polygons are similar. Find the area of the larger polygon.
(A.) 144 m^2
B. 48 m^2
C. 64 m^2
D. 192 m^2

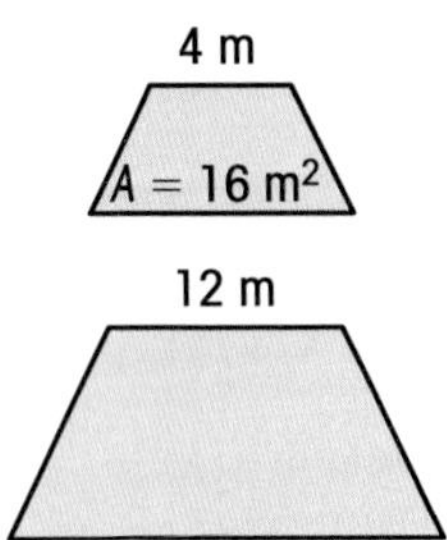

5. The area of a circle is 36π cm^2. Find the circumference in terms of π.
A. 9π cm
B. 3π cm
(C.) 12π cm
D. 6π cm

6. $\overleftrightarrow{TL}$ is a ____.
A. chord
B. diameter
C. radius
(D.) tangent

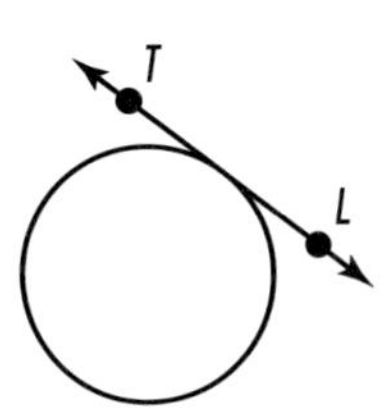

Critical Thinking

Janet has a tricycle. The front wheel has a diameter of 18 inches. The back wheels each have a diameter of 6 inches. Janet travels 1,000 feet. Will each of the wheels have the same number of revolutions? Explain why or why not.

CHALLENGE Find the actual number of revolutions for each wheel.

large wheel, about 212 rotations; small wheels, about 637 rotations

Critical Thinking No. The small wheels will make more revolutions. The small wheels have a smaller circumference. The quotient is larger when you divide the distance by the circumference of the small wheels than when you divide the distance by the circumference of the large wheel.

See the Classroom Resurce Binder for a scoring rubric to assess the Critical Thinking question.

Unit Four

Chapter 11 Surface Area and Volume

City skylines include buildings of all shapes and sizes. What geometric shapes do you see in this photograph?

Learning Objectives

- Identify space figures and their nets.
- Find the surface area of a prism, cylinder, and sphere.
- Find the volume of a prism, cylinder, cone, sphere, and pyramid.
- Compare the volumes of similar figures.
- Use a calculator to find the volume of a pyramid.
- Solve problems by writing an equation.
- Apply concepts and skills to real-life problems.
- Write proofs about volume.

ESL/ELL Note Have students add this chapter's vocabulary to their vocabulary list. Have students use models to identify each figure.

Words to Know

space figure a three-dimensional figure

prism a space figure with two congruent parallel bases and rectangular lateral faces

pyramid a space figure with one base and triangular lateral faces that meet at a vertex

cube a prism with six faces that are all congruent squares

cylinder a space figure with two congruent, parallel bases that are circles

cone a space figure with one circular base and one vertex

sphere a space figure with all points the same distance from the center

net a two-dimensional pattern that can be folded to create a space figure

surface area the sum of the areas of all the surfaces of a space figure

volume the number of cubic units that fill a space figure

cubic unit a cube with sides one unit long that is used to measure volume

Volume of a Prism Project

On cardboard, draw a net for a prism. Make all the measurements in millimeters. Cut out the net. Fold and tape the edges to create the prism.

First, calculate the surface area of your prism to find out how much contact paper you will need to cover it.

Then, calculate the volume of your prism, using a formula. Do the same for other prisms. See how they compare.

Project Students can begin work on the project after completing Lesson 11.2. See the Classroom Resource Binder for a scoring rubric to assess this project.

More practice is provided in the Workbook, page 67, and in the Classroom Resource Binder, page 116.

11·1 Space Figures

Getting Started
Bring in models of space figures. Ask a volunteer to explain the difference between a square and a cube.

Space figures are three-dimensional solid shapes. **Prisms** and **pyramids** are space figures with flat surfaces called faces.

A prism has two congruent faces, or bases, that are opposite each other. The other faces are called lateral faces. A pyramid has only one base.

Rectangular Prism	Cube	Square Pyramid
base; base; lateral face	base; lateral face; base	vertex; lateral face; base
Cylinder	**Cone**	**Sphere**
base; lateral face; base	vertex; base	

A **cube** is a special type of prism. The six faces are all congruent squares.

A **sphere** has no base or flat surface.

A prism or a pyramid is named by the shape of its base.

EXAMPLE

Name the space figure.

There are two bases.
So, the figure is a prism.
The bases are triangles.

The space figure is a triangular prism.

Try This

Name the space figure.

There is one ■. The lateral faces are ■. triangles
base

So, this is a pyramid.

The base is a ■. triangle

The space figure is a triangular ■. pyramid

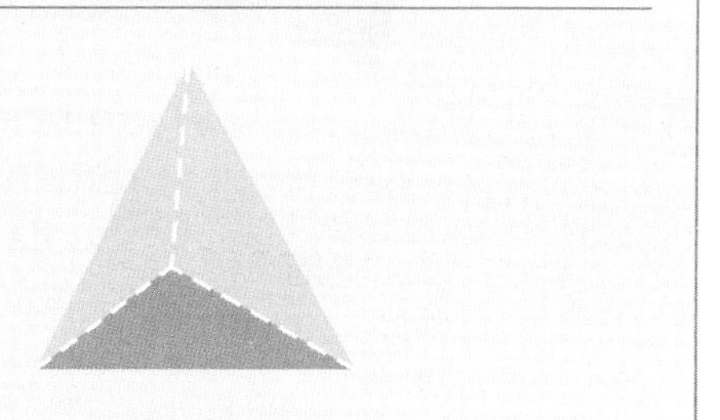

Practice

Name each space figure.

1.

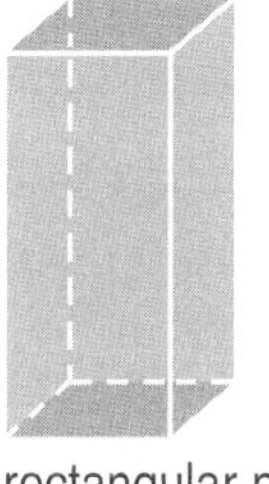

rectangular prism

2.

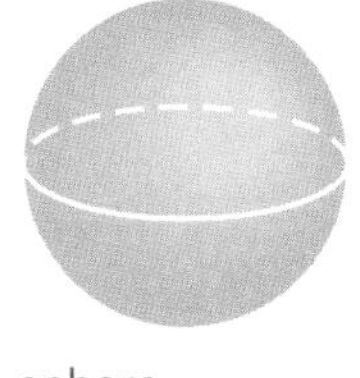

sphere

3.

cube

4.

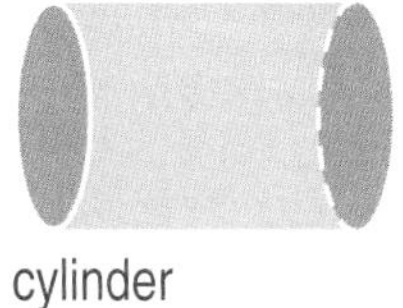

cylinder

5.

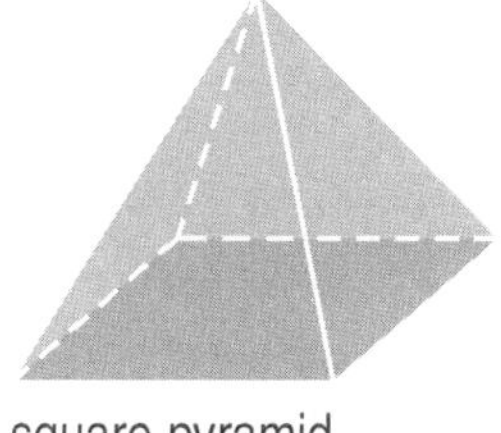

square pyramid

6.

cone

Share Your Understanding

7. Work with a partner. List examples of space figures you can find in your classroom. Explain your choices. Possible answers: A soda can is a cylinder; a basketball is a sphere; a door is a rectangular prism.

8. **CRITICAL THINKING** How is a pyramid like a cone? How is it different? Possible answer: The lateral faces of the pyramid and the curved surface of the cone meet at a single vertex. Both figures have one base. The base of a pyramid is a polygon, and the base of a cone is circular. The pyramid has triangular lateral faces, and the cone has one curved surface.

More practice is provided in the Workbook, page 67, and in the Classroom Resource Binder, page 116.

11·2 Nets of Space Figures

Getting Started
Discuss with students how a flat pattern can be used to make a three-dimensional shape. Show examples such as dress patterns and collapsible boxes.

You can use a pattern called a **net** to make a three-dimensional figure. Do this by folding the net along the dashed lines. The net below can be used to make a cube.

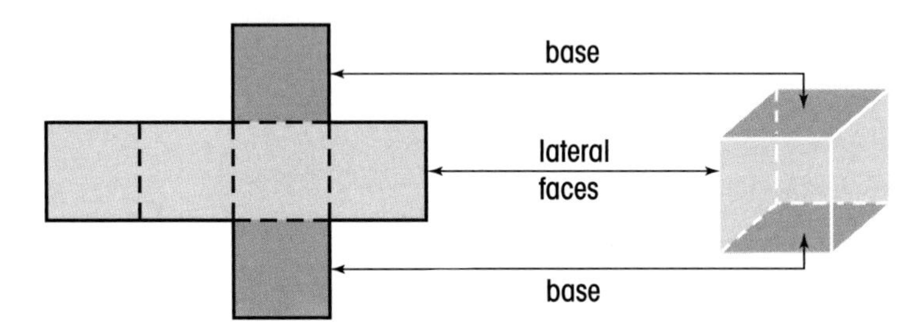

EXAMPLE 1

Name the space figure you can make from the net.

There are two congruent triangles.
These can be the bases.

There are three rectangles.
These can be the lateral faces.

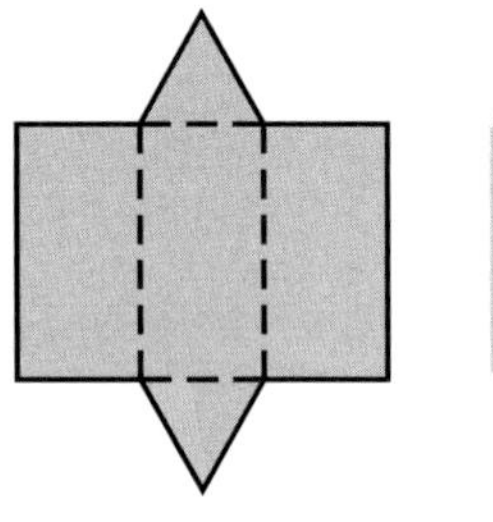

You can make a triangular prism from the net.

EXAMPLE 2

Name the space figure you can make from the net.

There are two congruent circles.
These can be the two bases.

There is one rectangle.
This can be a curved surface.

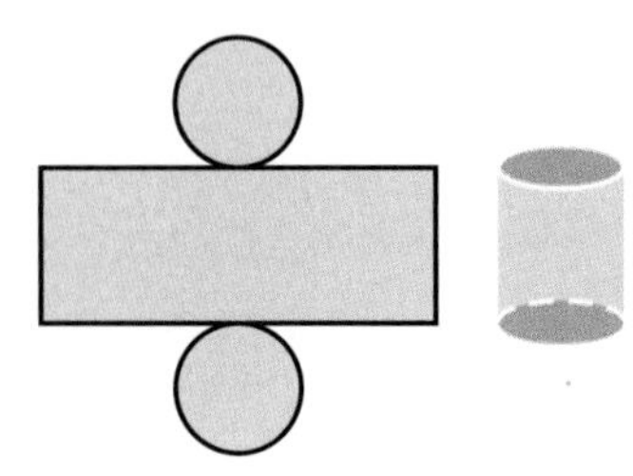

You can make a cylinder from the net.

Try This

Draw a net for the space figure.

First, draw a ■ to represent the base. square

Then, draw four ■ to represent the ■. sides triangles

The space figure has 5 faces, and the net has 5 parts.

So, you have drawn a net for the figure.

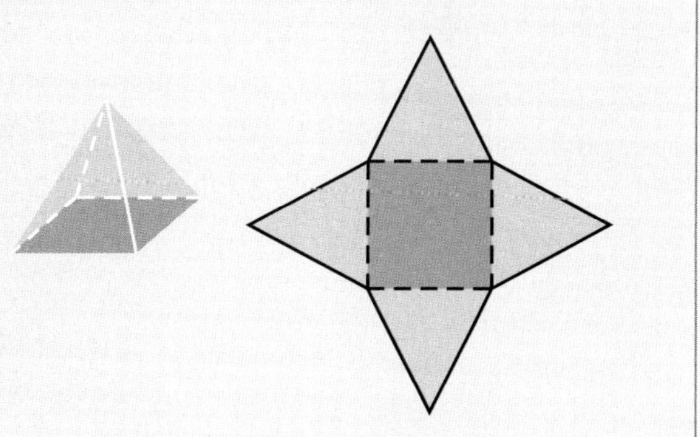

Practice

Name the space figure you can make from each net.

1.

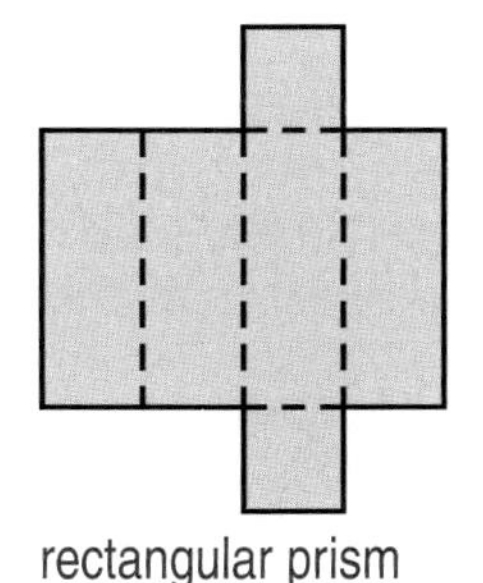

rectangular prism

2.

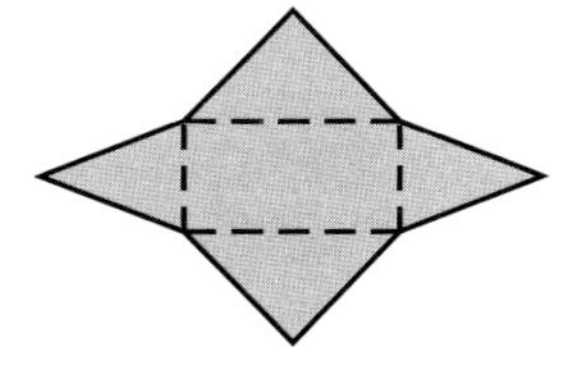

rectangular pyramid

3.

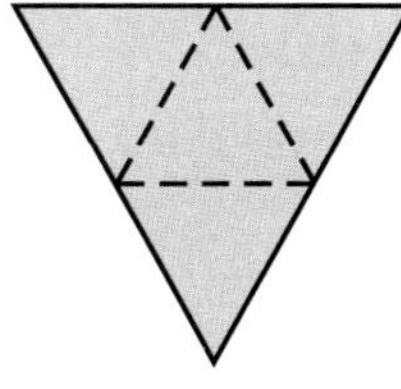

triangular pyramid

Draw a net for each space figure. Possible answers are given.

4.

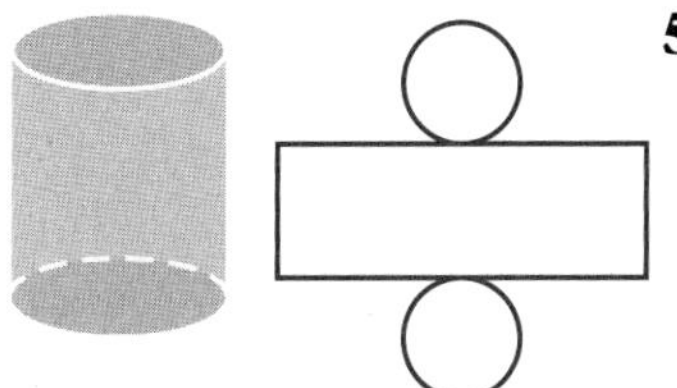

5.

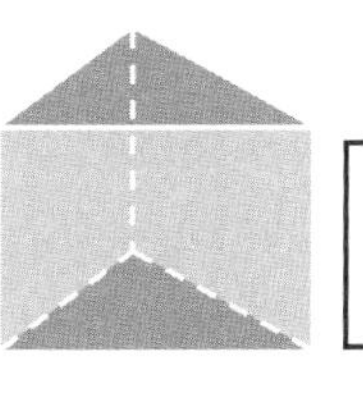

6.

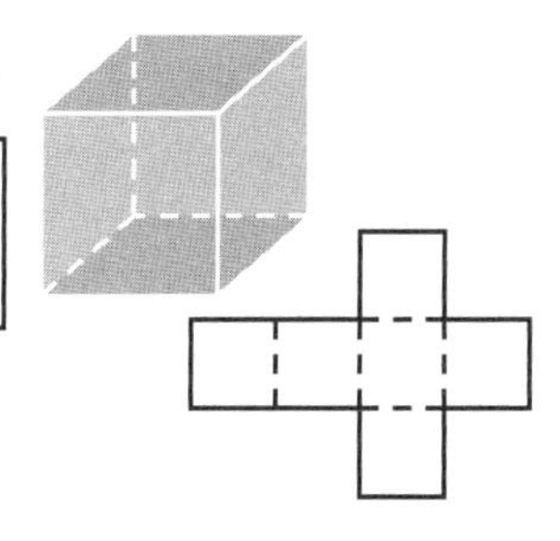

Share Your Understanding

7. Choose a net from the exercises above. Explain to a partner how you know which space figure you could make from the net. Answers will vary.

8. **CRITICAL THINKING** Can this diagram be a net for a cube? Explain your thinking. No. A cube has 6 faces and the diagram has only 5 parts.

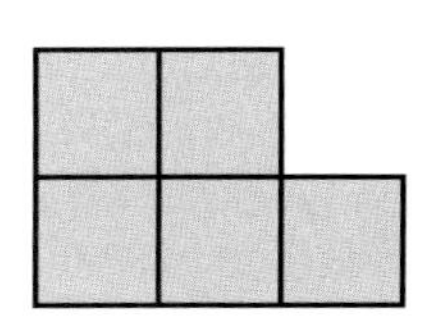

More practice is provided in the Workbook, page 68, and in the Classroom Resource Binder, page 117.

11-3 Surface Area of a Prism

Getting Started Review how to find the area of a rectangle and a triangle.

> **THEOREM 35**
>
> The surface area of a prism is the sum of the lateral area and the area of each of the two bases.

You learned that a net shows all the faces of a prism. You can use a net to help you find the **surface area** of a prism. The surface area is the sum of the areas of all the faces. You can write *SA* to mean *surface area*.

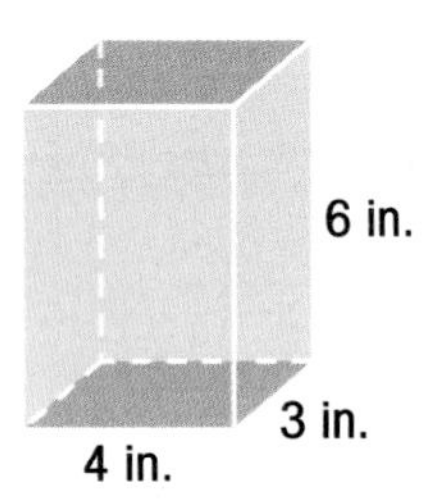

EXAMPLE

Find the surface area of the prism above.

First, make a net to show all the faces of the prism. Each face of this prism is a rectangle.

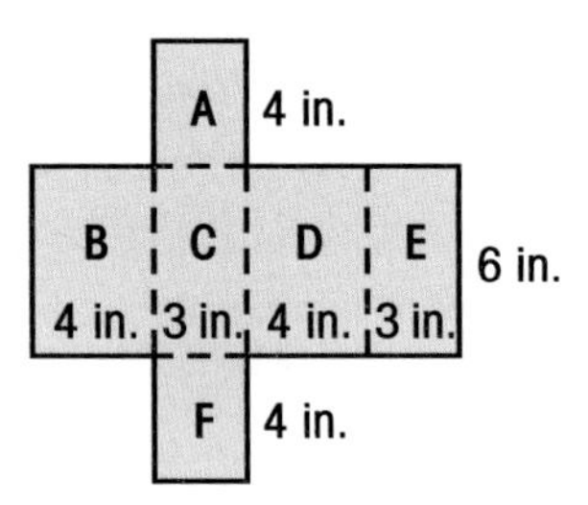

Write the formula for area of a rectangle. Then find the area of each rectangle in the net.

$$A = lw$$
$$\text{Area A} = 3 \bullet 4 = 12$$
$$\text{Area B} = 4 \bullet 6 = 24$$
$$\text{Area C} = 3 \bullet 6 = 18$$
$$\text{Area D} = 4 \bullet 6 = 24$$
$$\text{Area E} = 3 \bullet 6 = 18$$
$$\text{Area F} = 3 \bullet 4 = 12$$

Then, add to find the total surface area.

$$SA = 12 + 24 + 18 + 24 + 18 + 12 = 108$$

The surface area of the prism is 108 in.^{2}

Remember
Area is written in square units, such as square inches.

Try This

Find the surface area of the prism.

First, make a net to show all the faces. Three faces are rectangles and two faces are triangles.

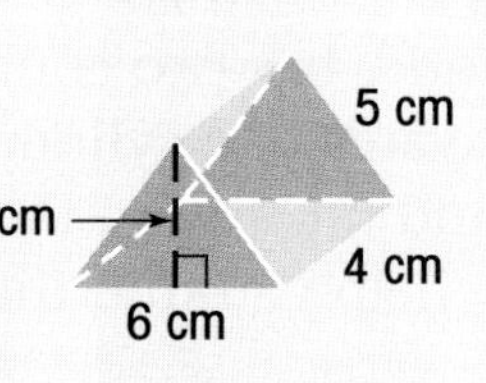

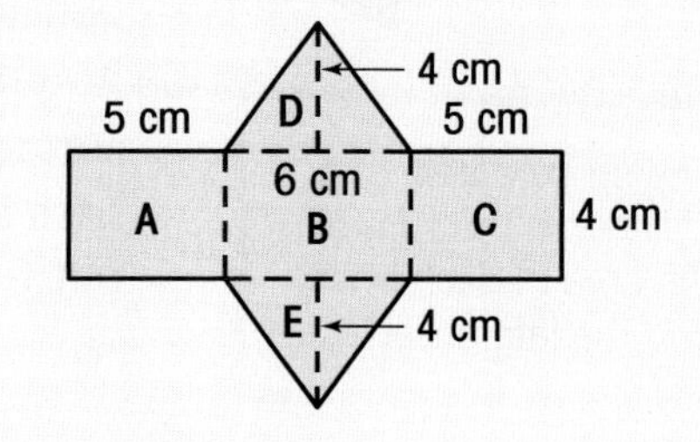

Find the area of each face.

The formula for area of a rectangle is lw.

Area A = 5 • 4 = ■ 20

Area B = 6 • ■ 4 = ■ 24

Area C = 5 • ■ 4 = ■ 20

The formula for area of a triangle is $\frac{1}{2}bh$.

Area D = $\frac{1}{2}$ (■ 6)(■ 4) = 12

Area E = ■ $\frac{1}{2}$ (■ 6)(■ 4) = ■ 12

Then, add to find the total surface area.

SA = ■ 88

The surface area of the prism is ■. 88 cm²

Practice

Find the surface area of each prism.

1.

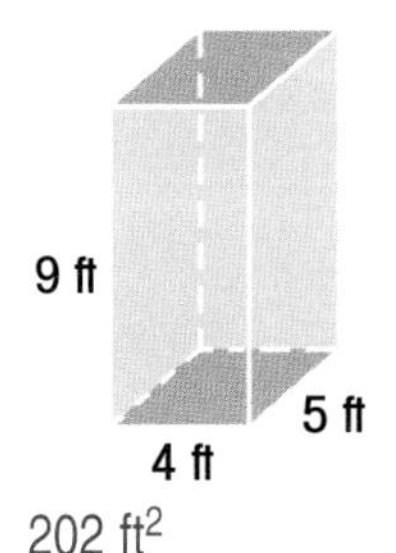

202 ft²

2.

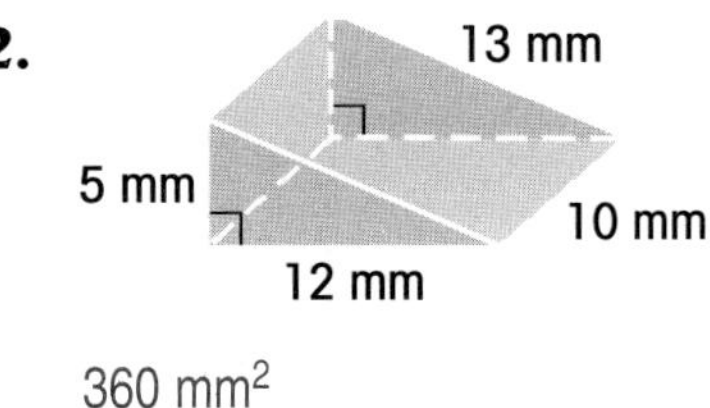

360 mm²

3.

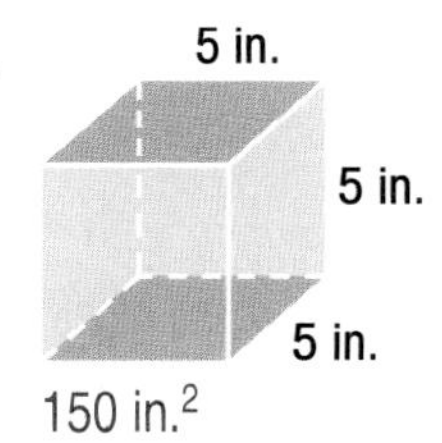

150 in.²

Share Your Understanding

4. Explain to a partner how you found the surface area of one of the prisms in the exercises above. Answers will vary.

5. **CRITICAL THINKING** Describe an easy way to find the surface area of a cube. (Hint: What do you know about all the faces?)
All faces are congruent. Find the area of one lateral face, a square.
Then, multiply this area by 6, the total number of faces.

More practice is provided in the Workbook, page 69, and in the Classroom Resource Binder, page 118.

11·4 Surface Area of a Cylinder

THEOREM 36

The surface area of a cylinder is the sum of the area of each base and the lateral surface.

Getting Started Review the circumference and area of a circle.

A cylinder has two congruent bases that are circles. The surface area of a cylinder is the sum of the area of each base and the area of the lateral surface. You can use a net to help you find the surface area.

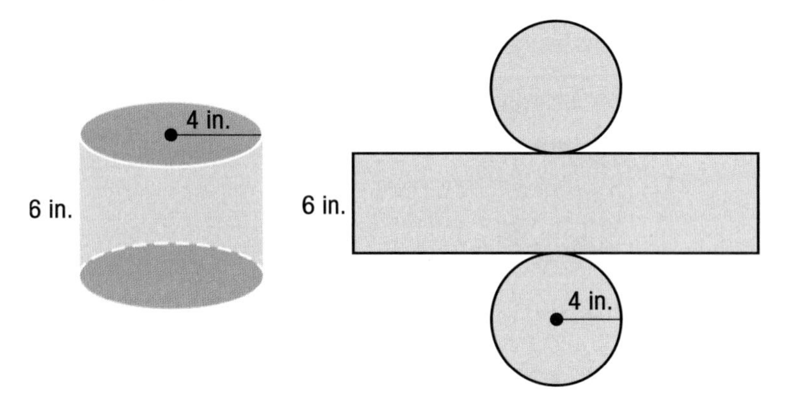

EXAMPLE

Find the surface area of the cylinder above, to the nearest square unit.

Find the area of each circular base. Write the formula for area of a circle. Use 3.14 for π. Substitute 4 for r.

$$A = \pi r^2 \approx 3.14(4^2) \approx 50.24$$

Remember
The area of a circle is πr^2.
The circumference of a circle is $2\pi r$, or πd.

Find the area of the lateral surface. The lateral face of a cylinder is a rectangle. To find the length of the lateral surface, find the circumference of the base.

$$C = 2\pi r \approx 2(3.14)(4) \approx 25.12$$

Write the formula for area of a rectangle. Substitute 25.12 for l and 6 for w.

$$A = lw = 25.12 \cdot 6 = 150.72$$

The total surface area is the sum of the areas of the two bases and the lateral surface.

$$SA \approx 2(50.24) + 150.72 \approx 251.20$$

The surface area is about 251 in.^{2}

Try This

Find the surface area of the cylinder, to the nearest square unit.

5 cm

10 cm

First, find the area of each base. Write the formula for area of a circle. Use 3.14 for π.

$A = \pi r^2$
$\approx 3.14(5^2)$
$\approx$ ■ 78.5

To find the area of the lateral face, first find the circumference of the base.

$C = 2\pi r$
$\approx 2(3.14)($■$)$ 5
$\approx$ ■ 31.4

Write the formula for area of a rectangle. The circumference is the length of the rectangle. Substitute ■ 31.4 for l and 10 for w.

$A = lw$
$\approx$ ■ 31.4 $\cdot$ 10
$\approx$ ■ 314

Find the total surface area.

$SA \approx 2 \cdot$ ■ $+$ ■ $=$ ■ 471
78.5 314

The surface area of the cylinder is about ■. 471 cm²

Practice

Find the surface area of each cylinder, to the nearest square unit. Use 3.14 for π.

1. 10 cm, 10 cm

about 1,256 cm²

2. 20 yd, 17 yd

about 4,647 yd²

3.

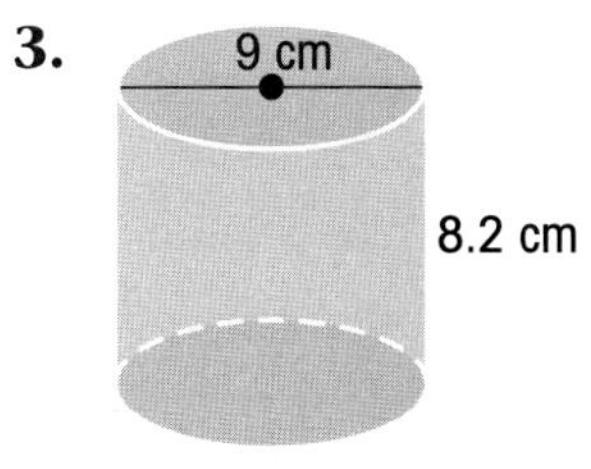

about 359 cm²

Share Your Understanding

4. Draw a cylinder and label its dimensions. Ask a partner to explain how to find the surface area. Answers will vary.

5. **CRITICAL THINKING** How many square centimeters of plastic are needed to make a drinking straw 30 cm long with a diameter of 0.5 cm? Round your answer to the nearest whole number. Use 3.14 for π. about 47 cm²

More practice is provided in the Workbook, page 69, and in the Classroom Resource Binder, page 118.

11·5 Surface Area of a Sphere

Getting Started Review the radius and the diameter of a circle.

A sphere is like a ball. It has no base and no flat surfaces. All points of a sphere are the same distance from the center. The radius of a sphere is the distance from the center of the sphere to any point on the sphere.

You can use the radius to find the surface area. The surface area is the square of the radius multiplied by 4π.

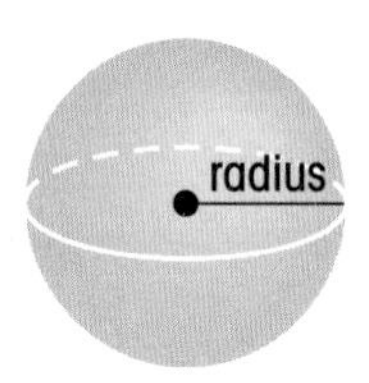

$$\text{Surface Area} = 4 \bullet \pi \bullet (\text{radius})^2$$
$$SA = 4\pi r^2$$

EXAMPLE

Find the surface area of the sphere. Write your answer in terms of π.

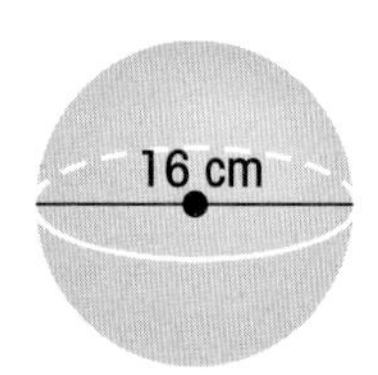

Math Fact
The radius of a sphere is half the diameter.

First, find the radius of the sphere.

$$r = d \div 2 = 16 \div 2 = 8$$

Then, write the formula for surface area. Substitute 8 for r. Simplify.

$$SA = 4\pi r^2 = 4\pi(8^2) = 256\pi$$

The surface area is $256\pi \text{ cm}^2$.

Try This

The surface area of a sphere is 36π yd^2. Find the radius.

Write the formula for surface area.	$SA = 4\pi r^2$
Substitute 36π for the surface area.	36π ■ $= 4\pi r^2$
Solve for r.	$36\pi \div 4\pi = 4\pi r^2 \div$ ■π 4
	9 ■ $= r^2$
	3 ■ $= r$

The radius of the sphere is ■. 3 yd

Practice

Find the surface area of each sphere. Write your answer in terms of π.

1. 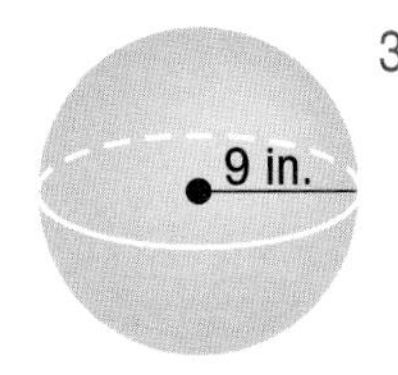

324π in.^{2}

2. 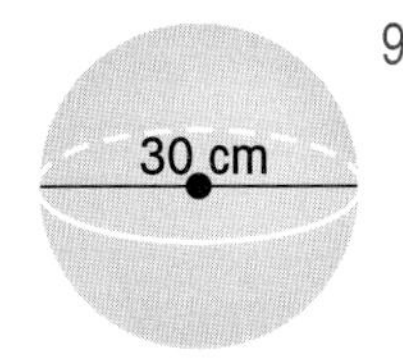

900π cm^2

3. 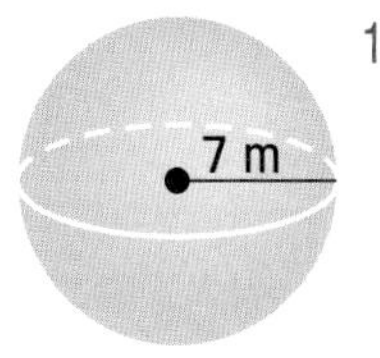

196π m^2

Find the radius of each sphere. You are given the surface area.

4. $SA = 64\pi$ m^2 4 m

5. $SA = 196\pi$ ft^2 7 ft

6. $SA = 100\pi$ cm^2 5 cm

Share Your Understanding

7. Draw a sphere and label its diameter or radius. Have a partner explain how to find the surface area of the sphere. Answers will vary.

8. **CRITICAL THINKING** Explain how to find the surface area of this grain silo. (Hint: Think about how the shape of the top is like a sphere.)
Possible answer: Find the surface area of the lateral face of the cylinder. Then find the surface area of a sphere with a radius of 10 ft. Divide the surface area of the sphere by 2 and add to the surface area of the lateral face of the cylinder.

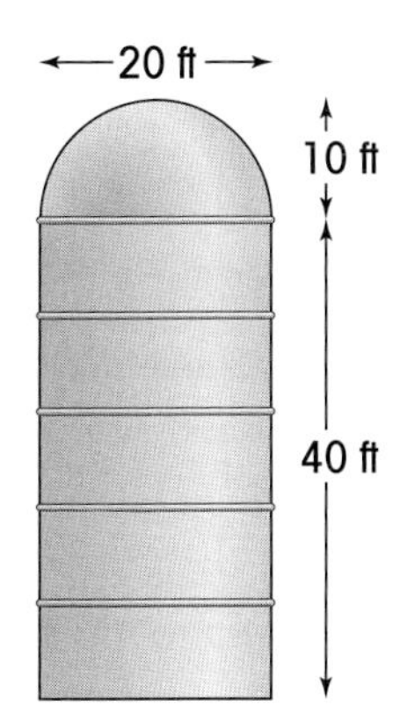

More practice is provided in the Workbook, page 70, and in the Classroom Resource Binder, page 119.

11·6 Volume of a Prism

Getting Started Use models to review square units and cubic units. Compare surface area and volume by asking volunteers to give real-life examples of when each measure is used.

The **volume** of a prism is the number of **cubic units** that fill it. You can find the volume of a prism. Multiply the area of the base (B) by the height (h) of the prism.

1 cubic unit

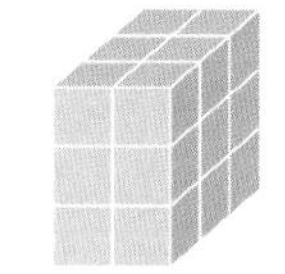

Volume = 18 cubic units

$$\text{Volume} = \text{Area of base} \cdot \text{height}$$
$$V = Bh$$

EXAMPLE 1

Find the volume of the prism.

Write the formula for volume of a prism. $V = Bh$

The area of the base of a rectangular prism is lw. $= (lw)h$

Substitute 4 for l, 3 for w, and 6 for h. Simplify. $= (4)(3)(6)$ $= 72$

6 m
4 m
3 m

The volume of the prism is 72 cubic meters, or 72 m^3.

There is a special formula for finding the volume of a cube. Remember that all the sides are congruent. So the length, width, and height are all equal.

$V = l \cdot w \cdot h$

Volume of a cube equals the length of side cubed.

$V = s \cdot s \cdot s = s^3$

EXAMPLE 2

Find the volume of the cube.

Write the formula for volume of a cube. $V = s^3$

Substitute 4 for s. $= 4^3$

Simplify. $= 64$

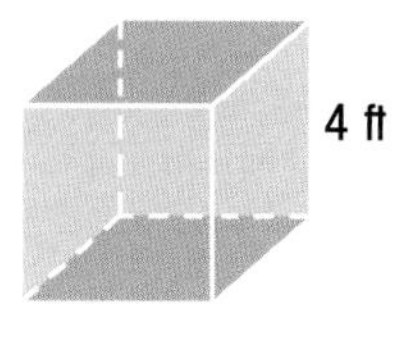

The volume of the cube is 64 ft^3.

Try This

Find the volume of the triangular prism.

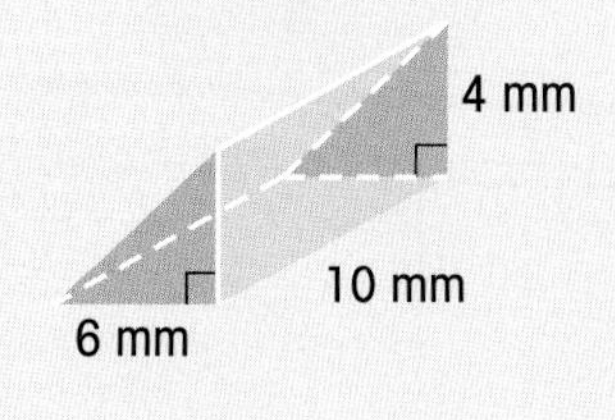

Write the formula for volume of a prism. $V = Bh$

Find the area of the triangular base (B). $B = \frac{1}{2}bh$

Substitute the given values. $= \frac{1}{2}(\blacksquare)(4)$ 6

Simplify. $= \blacksquare$ 12

Write the formula for volume of a prism. $V = Bh$

Substitute the given values. $= \blacksquare(10)$ 12

Simplify. $= \blacksquare$ 120

The volume of the triangular prism is ■. 120 mm^3

Practice

Find the volume of each prism.

1. 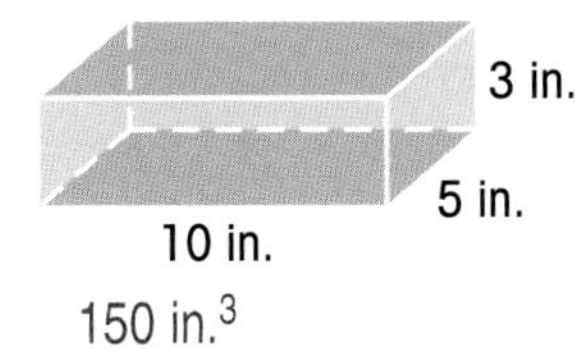

150 in.^{3}

2. 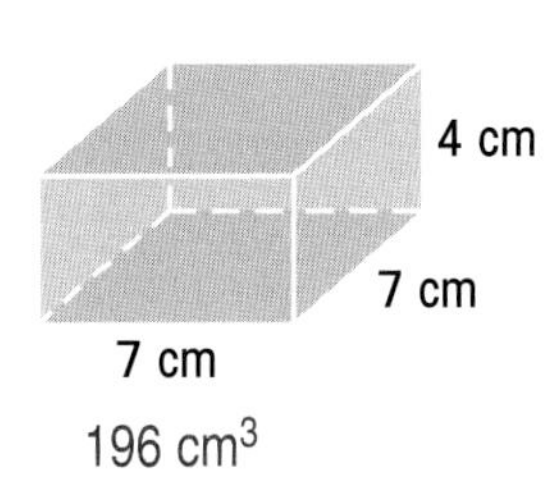

196 cm^3

3. 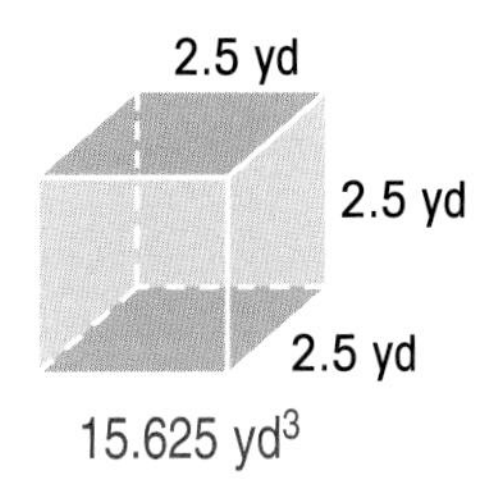

15.625 yd^3

4. 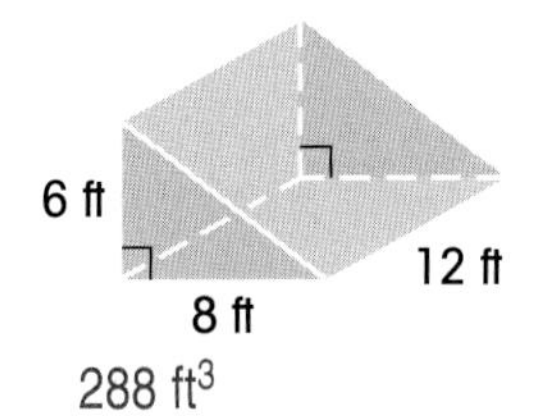

288 ft^3

5. 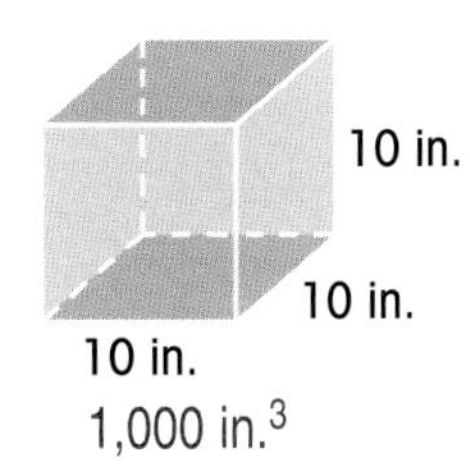

1,000 in.^{3}

6. 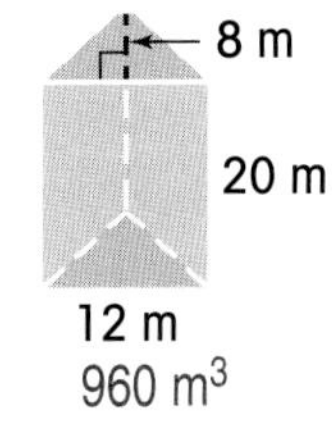

960 m^3

Share Your Understanding

7. Find an object in the classroom that is shaped like a rectangular prism. Explain to a partner how to find the volume. Have your partner find the volume.
The volume is the product of length times width times height.

8. **CRITICAL THINKING** The volume of a cube is 2,197 cm^3. Is the length of one side more or less than 10 cm? Explain your thinking.
More than 10 cm; a cube with 10 cm sides will have a volume of 1,000 cm^3.

More practice is provided in the Workbook, page 71, and in the Classroom Resource Binder, page 120.

11·7 Volume of a Cylinder

Getting Started Review how to find the area of a circle.

Finding the volume of a cylinder is like finding the volume of a prism. You multiply the area of the base by the height.

A cylinder has two congruent circular bases. So, the area of the base is πr^2.

$$\text{Volume} = \text{Area of base} \bullet \text{height}$$
$$V = Bh$$
$$= \pi r^2 h$$

Find the volume of the cylinder, to the nearest cubic unit.

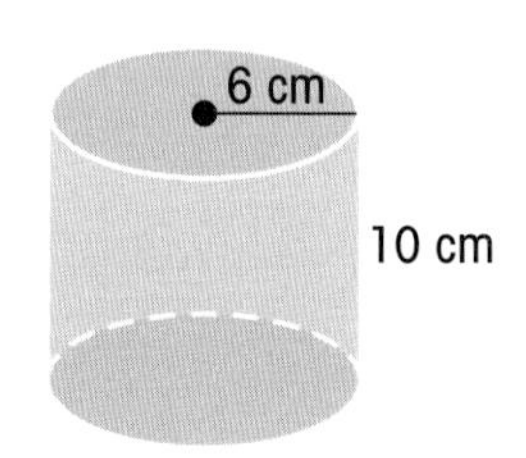

Write the formula for volume of a cylinder. $V = \pi r^2 h$

Use 3.14 for π. Substitute 6 for r and 10 for h. Simplify. $\approx (3.14)(6^2)(10)$
$\approx 1{,}130.4$

The volume of the cylinder is about 1,130 cm^3.

You can also find the volume if you know the diameter.

EXAMPLE 2

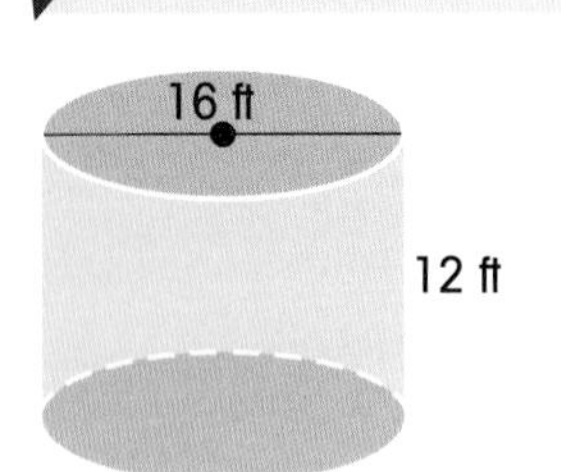

Find the volume of the cylinder. Write your answer in terms of π.

Find the radius of the cylinder. Divide the diameter by 2. $r = d \div 2$
$= 16 \div 2 = 8$

Write the formula for volume of a cylinder. Substitute 8 for r and 12 for h. $V = \pi r^2 h$
$= \pi(8^2)(12)$
$= 768\pi$

The volume of the cylinder is 768π ft^3.

Try This

The volume of the cylinder is 100π m^3.
Find the height of the cylinder.

5 m

Write the formula for volume of a cylinder.	$V = (\blacksquare)h$ πr^2
Substitute the given values.	$100\pi = \pi(\blacksquare)h$ 5^2
Solve for h.	$4\ \blacksquare = h$

The height is ■. 4 m

Practice

Find the volume of each cylinder. Write your answer in terms of π.

1. 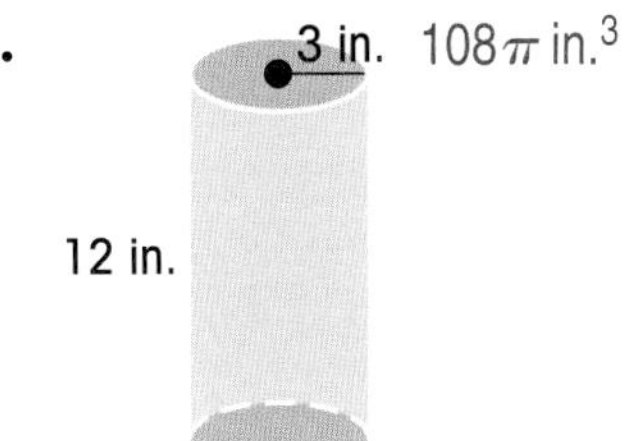

108π in.^{3}

2. 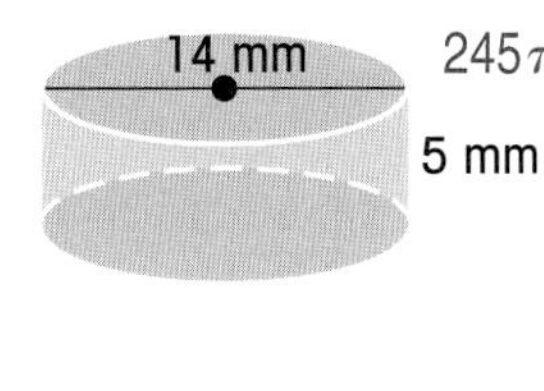

245π mm^3

3. 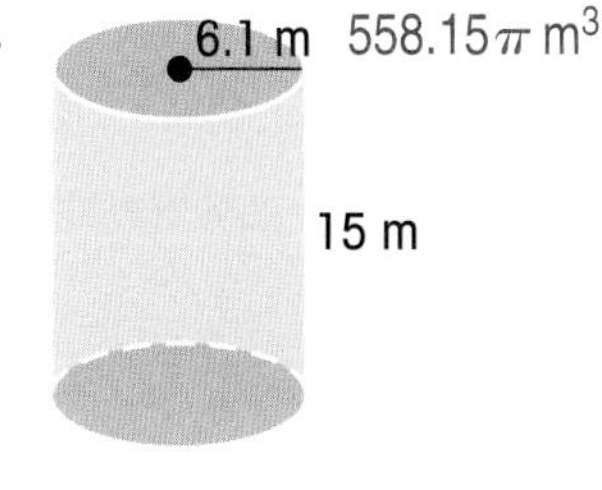

558.15π m^3

Find the height of each cylinder. You are given the volume and the radius of the base.

4. $V = 2{,}916\pi$ ft^3
$r = 18$ ft 9 ft

5. $V = 2{,}057\pi$ cm^3
$r = 11$ cm 17 cm

6. $V = 810\pi$ yd^3
$r = 9$ yd 10 yd

Share Your Understanding

7. Explain to a partner how to find the volume of a cylinder. Use the words *radius*, *base*, and *height* in your explanation.
Multiply the radius times itself. Multiply this product times pi to find the area of the base. Then, multiply the area times the height.

8. **CRITICAL THINKING** Does the volume of a cylinder double when you double the radius? Does it double when you double the height? No; yes

More practice is provided in the Workbook, page 72, and in the Classroom Resource Binder, page 121.

11·8 Volume of a Cone

Getting Started Review how to find the volume of a cylinder.

A cone has a circular base and one vertex. The volume of a cone is one-third the volume of a cylinder with the same base and the same height.

Cylinder

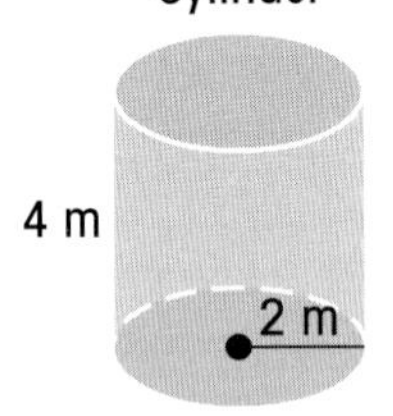

Cone

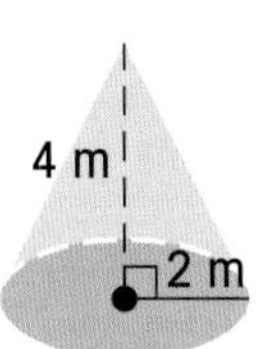

$V = \pi \bullet (\text{radius})^2 \bullet \text{height}$

$= \pi r^2 h$

$V = \frac{1}{3} \bullet \pi \bullet (\text{radius})^2 \bullet \text{height}$

$= \frac{1}{3}\pi r^2 h$

EXAMPLE 1

Find the volume of the cone, to the nearest cubic unit.

Remember
The base of a cone is a circle.

Write the formula for volume of a cone. Substitute 3 for *r* and 8 for *h*. Use 3.14 for π	$V = \frac{1}{3}\pi r^2 h$
	$\approx \frac{1}{3}(3.14)(3^2)(8)$
Simplify.	≈ 75.36

The volume of the cone is about 75 m^3.

You can also find the volume if you know the diameter.

EXAMPLE 2

Find the volume of the cone.
Write your answer in terms of π.

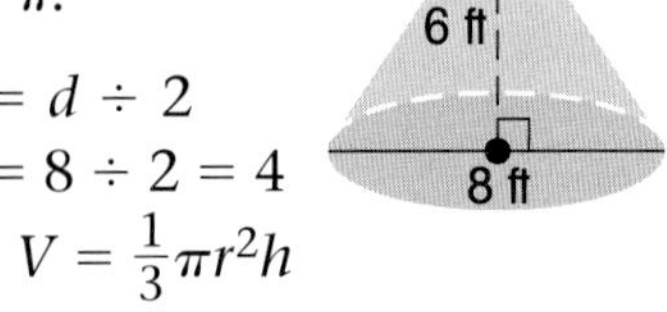

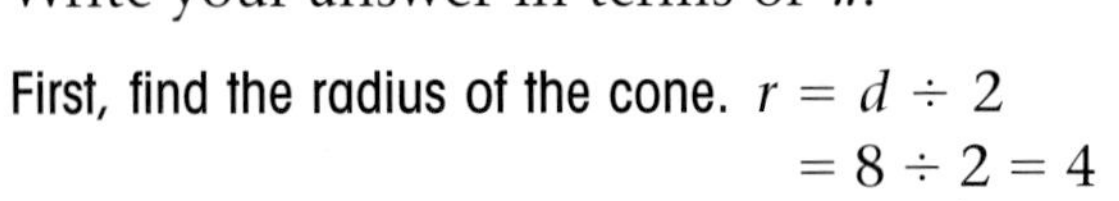

First, find the radius of the cone. $r = d \div 2$
$= 8 \div 2 = 4$

Write the formula for volume of a cone. Substitute 4 for *r* and 6 for *h*. Simplify.	$V = \frac{1}{3}\pi r^2 h$
	$= \frac{1}{3}(\pi)(4^2)(6)$
	$= 32\pi$

The volume of the cone is 32π ft^3.

Try This

The volume of the cone is 540π cm^3. Find the height.

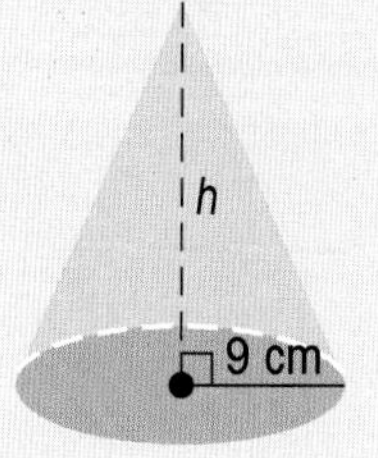

Write the formula for volume of a cone. $V = \frac{1}{3}\pi(\blacksquare)h$ r^2

Substitute the given values. $540\pi = \frac{1}{3}\pi(\blacksquare^2)h$ 9

Solve for *h*. $20\ \blacksquare = h$

The height of the cone is ■. 20 cm

Practice

Find the volume of each cone, to the nearest cubic unit. Use 3.14 for π.

1. 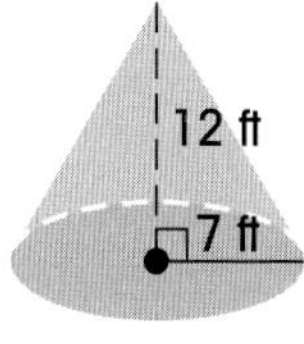

about 615 ft^3

2. 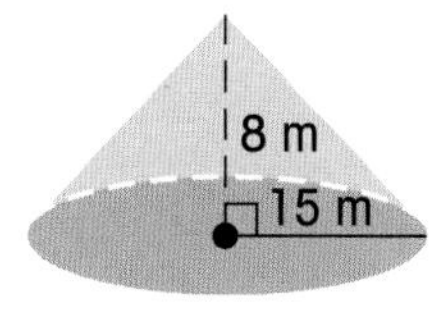

about 1,884 m^3

3. 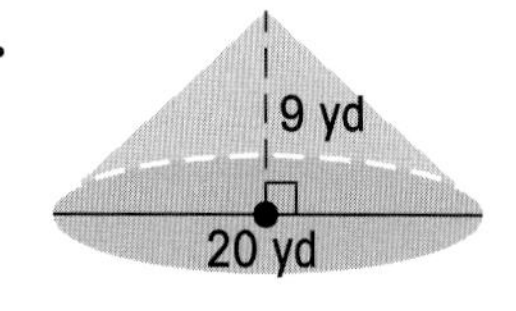

about 942 yd^3

Find the height of each cone.

4. $V = 540\pi$ cm^3
$r = 3$ cm
180 cm

5. $V = 240\pi$ m^3
$r = 12$ m
5 m

6. $V = 48\pi$ ft^3
$r = 6$ ft
4 ft

Share Your Understanding

7. Work with a partner. Draw a cone and label its height and radius. Have your partner explain how to find the volume of the cone. The volume of the cone is $\frac{1}{3}\pi$ times the radius squared times the height.

8. **CRITICAL THINKING** A cone has a volume of 188.4π in.^{3} The area of the base is 28.26π in.^{2} What is the height?
20 in.

More practice is provided in the Workbook, page 72, and in the Classroom Resource Binder, page 121.

11•9 Volume of a Sphere

Getting Started Ask volunteers to find the cubes of 2, 5, and 10.

You learned how to find the surface area of a sphere. You can also find the volume of a sphere. The volume is $\frac{4}{3}\pi$ times the cube of the radius.

$$\text{Volume} = \frac{4}{3} \cdot \pi \cdot (\text{radius})^3$$

$$V = \frac{4}{3}\pi r^3$$

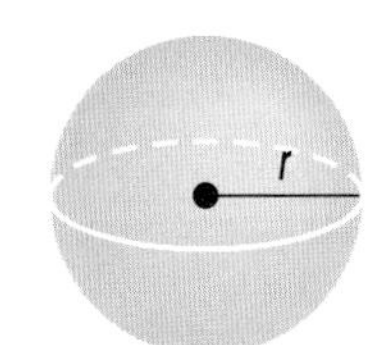

EXAMPLE 1

Find the volume of the sphere, to the nearest cubic unit.

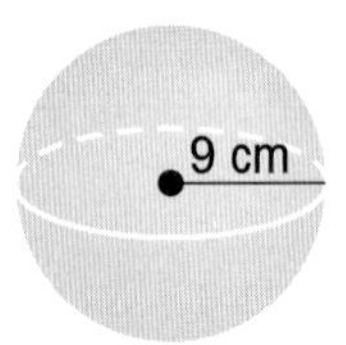

Write the formula for volume of a sphere. $V = \frac{4}{3}\pi r^3$

Use 3.14 for π. Substitute 9 for r. $\approx \frac{4}{3}(3.14)(9^3)$

Simplify. $\approx \frac{4}{3}(3.14)(729)$

$\approx 3{,}052.08$

Remember
$9^3 = 9 \cdot 9 \cdot 9 = 729$

The volume of the sphere is about 3,052 cm^3.

If you know the diameter, you can still find the volume.

EXAMPLE 2

Find the volume of the sphere, to the nearest cubic unit.

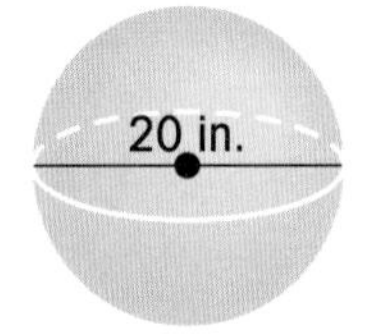

Use the diameter to find the radius. $r = d \div 2$

$= 20 \div 2 = 10$

Write the formula for volume. $V = \frac{4}{3}\pi r^3$

Use 3.14 for π. Substitute 10 for r. $\approx \frac{4}{3}(3.14)(10^3)$

$\approx \frac{4}{3}(3.14)(1{,}000)$

Simplify. $\approx 4{,}186.66$

The volume of the sphere is about 4,187 $in.^3$

Try This

Find the volume of the sphere, to the nearest cubic unit.

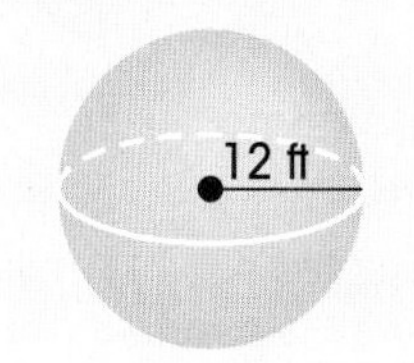

Write the formula for volume of a sphere. $V = \frac{4}{3}\pi r^3$

Use 3.14 for π. Substitute 12 for r. $\approx \frac{4}{3}(■)(12^3)$ 3.14

$\approx$ ■ 7,234.56

The volume is about ■. 7,235 ft^3

Practice

Find the volume of each sphere, to the nearest cubic unit.
Use 3.14 for π.

1.

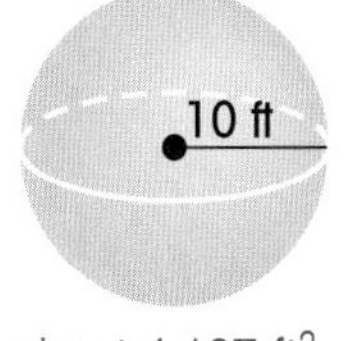

about 4,187 ft^3

2.

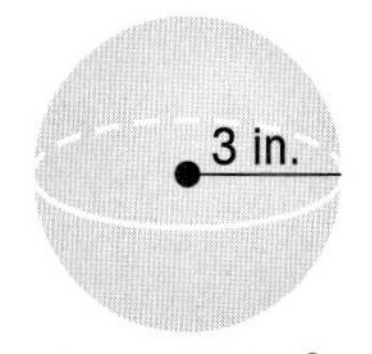

about 113 in.^{3}

3.

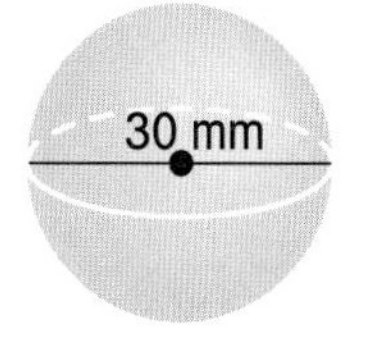

about 14,130 mm^3

4.

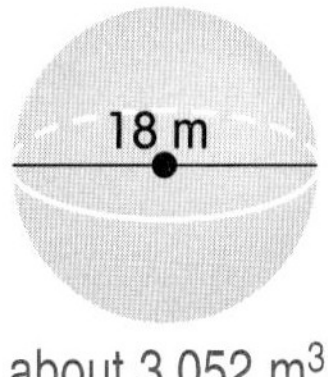

about 3,052 m^3

5.

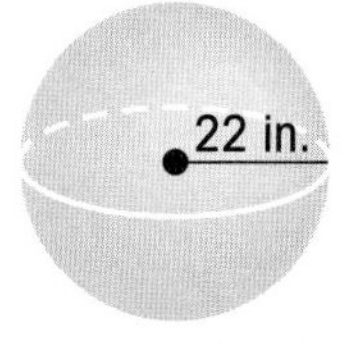

about 44,580 in.^{3}

6.

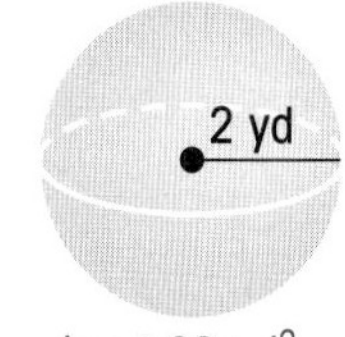

about 33 yd^3

Share Your Understanding

7. Choose a sphere from the exercises above. Explain to a partner how to find the volume.
Volume is the product of $\frac{4}{3}\pi$ times the radius cubed.

8. **CRITICAL THINKING** How does the volume of a sphere change when the radius is doubled? It becomes 8 times as large.

More practice is provided in the Workbook, page 73, and in the Classroom Resource Binder, page 122.

11·10 Volume of Similar Figures

Getting Started Review the definition of similar shapes.

> **THEOREM 37**
>
> If the similarity ratio of the sides of two similar figures is a:b, then the ratio of the volumes of the figures is a^3:b^3.

You learned that the areas of similar figures are related. The volumes of similar figures are also related.

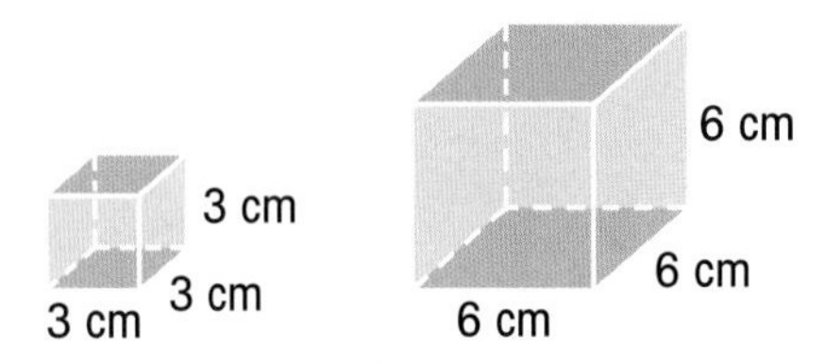

These cubes are similar. You can write a ratio to compare the sides. The similarity ratio of the sides is 1:2. So, you know that the ratio of the surface areas is 1^2:2^2, or 1:4.

The volume of the smaller cube is 3^3, or 27 cm^3. The volume of the larger cube is 6^3, or 216 cm^3.

So, the ratio of the volumes is 1^3:2^3, or 1:8.

Suppose two figures are similar. If you know the ratio of the sides and the volume of one figure, you can find the volume of the other. Use a proportion.

EXAMPLE

Prism A and prism B are similar. Their similarity ratio is 1:3. The volume of prism A is 30 m^3. Find the volume of prism B.

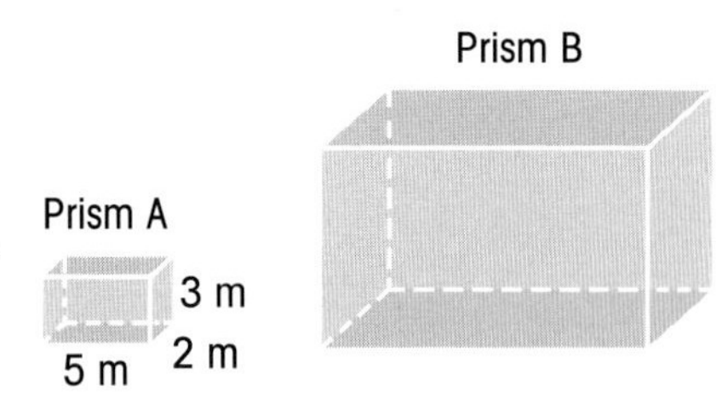

Use the similarity ratio to write a volume ratio. $1^3:3^3 = 1:27$

Then, write a proportion. $\frac{\text{Volume A}}{\text{Volume B}} = \frac{1}{27}$

Substitute 30 for Volume A. $\frac{30}{\text{Volume B}} = \frac{1}{27}$

Write the cross products. $\text{Volume B} = 30 \bullet 27$

Simplify. $= 810$

Volume B is 810 m^3.

Try This

These two spheres are similar. Their similarity ratio is 1:4. Find the volume of sphere A. Express your answer in terms of π.

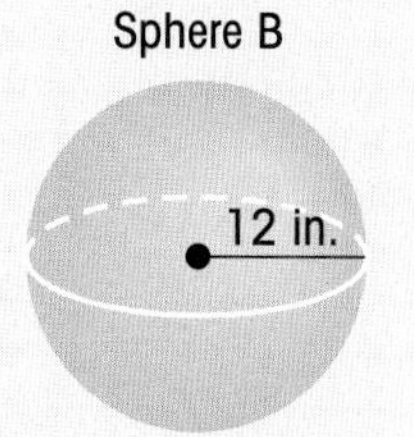

Write a similarity ratio for the volumes. $1^3:4^3 = 1:64$

Write a proportion. $\frac{\text{Volume A}}{\text{Volume B}} = \frac{1}{■}$ 64

Substitute the given values. Solve. $\frac{\text{Volume A}}{2{,}304\pi} = \frac{1}{■}$ 64

64 Volume A (■) = ■ $2{,}304\pi$

Volume A = ■ 36π

The volume of sphere A is 36π in.^{3}

Practice

Each pair of figures is similar. Find the volume of the larger figure. Use 3.14 for π as needed.

1.

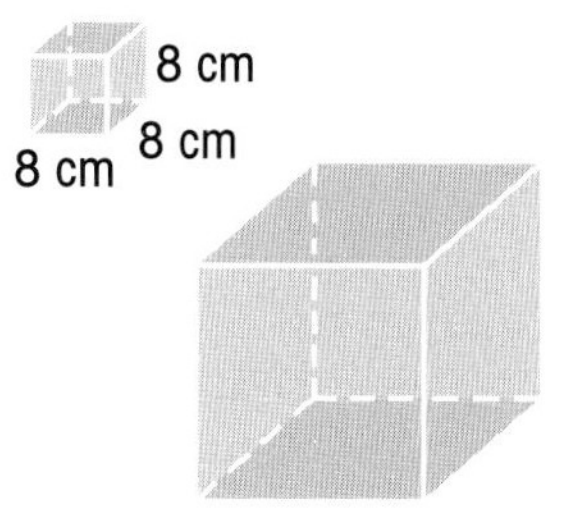

Similarity ratio 1:3
13,824 cm^3

2.

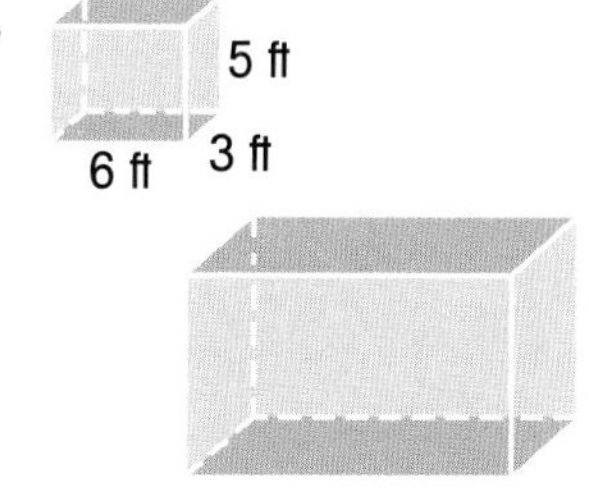

Similarity ratio 1:2
720 ft^3

3.

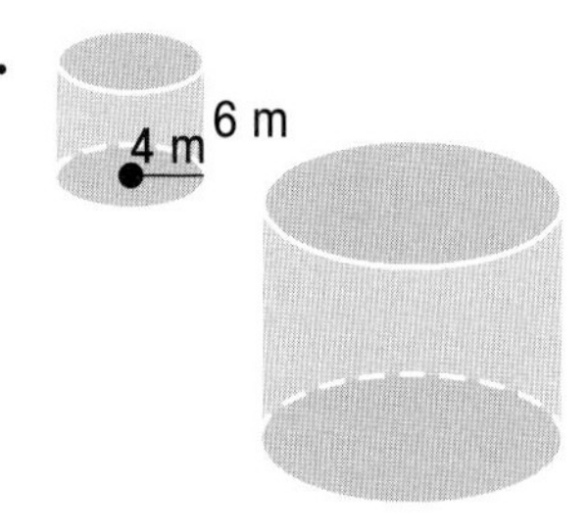

Similarity ratio 1:4
about 19,292 m^3

Share Your Understanding

4. Choose a pair of figures from the exercises above. Explain to a partner how you found the volume of the larger figure. Answers will vary.

5. **CRITICAL THINKING** A cube measures 5 cm on a side. The volume of a larger, similar cube is 1,000 cm^3. What is the similarity ratio of the sides? 1:2

More practice is provided in the Workbook, page 74, and in the Classroom Resource Binder, page 123.

11-11 Calculator: Volume of a Pyramid

Getting Started Use models to compare and contrast a pyramid and a prism.

Remember that a pyramid is a space figure. The volume of a pyramid is one-third the volume of a prism with the same base and height.

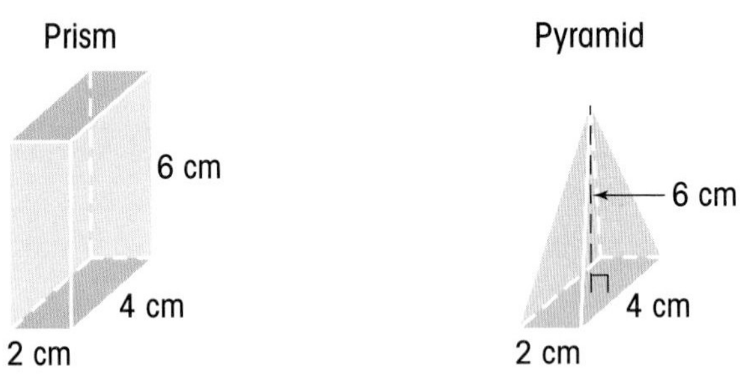

$V = \text{Area of base} \cdot \text{height}$ $\quad$ $V = \frac{1}{3} \cdot \text{Area of base} \cdot \text{height}$

$= Bh$ $\quad$ $= \frac{1}{3}Bh$

You can use a calculator to find the volume of a pyramid.

EXAMPLE

Find the volume of this pyramid, to the nearest cubic meter.

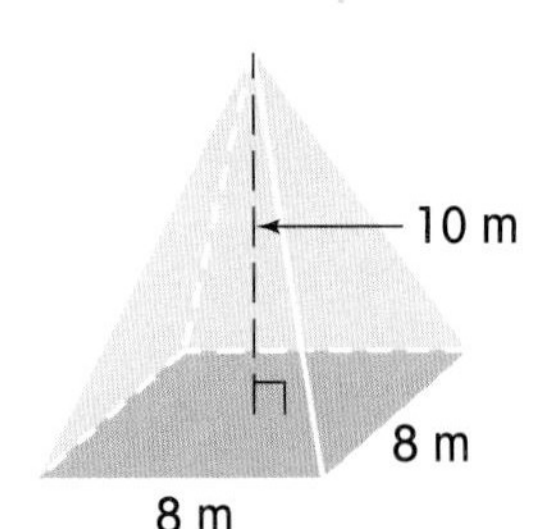

Write the formula for volume. $\quad V = \frac{1}{3}Bh$

The area of the square base is s^2. $\quad = \frac{1}{3}s^2h$

Substitute 8 for s and 10 for h. $\quad = \left(\frac{1}{3}\right)(8^2)(10)$

Now, use your calculator to find the volume. **Display**

PRESS: 8 × 8 × 1 0 ÷ 3 = $\quad$ 213.3333333

Round to the nearest whole number. $\quad 213.3333333 \approx 213$

The volume of the pyramid is 213 m^3, to the nearest cubic meter.

Practice

Find the volume of each pyramid, to the nearest cubic unit.

1.

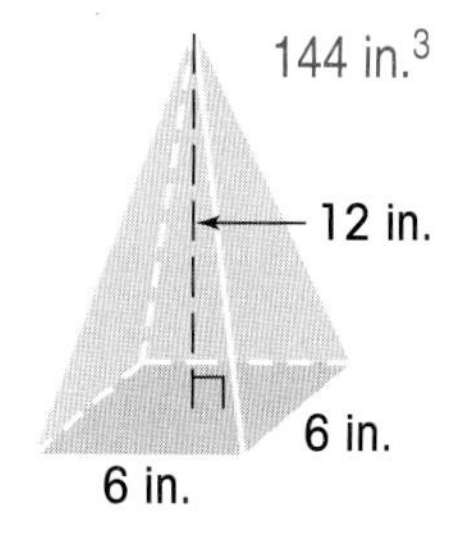

2.

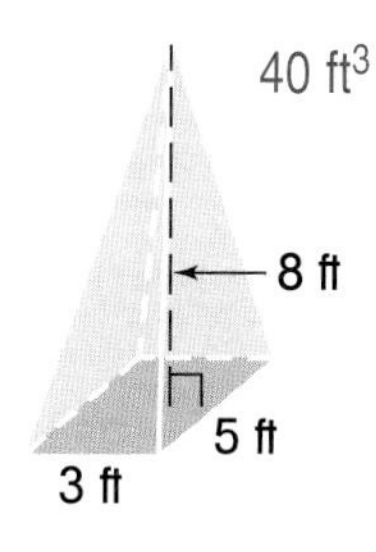

3.

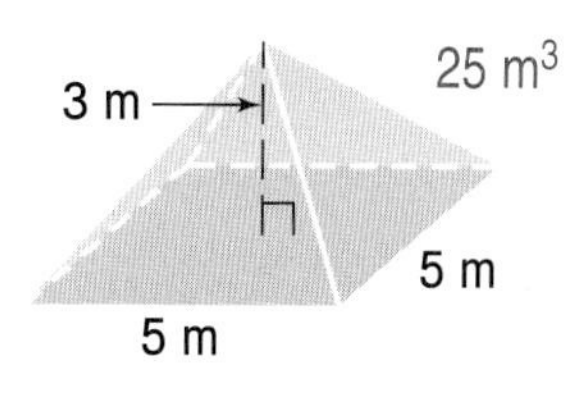

4.

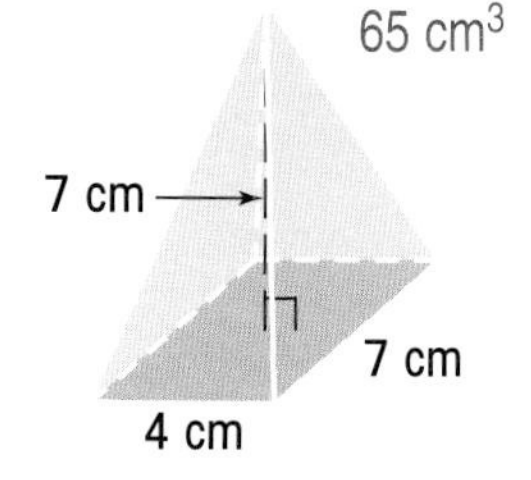

5.

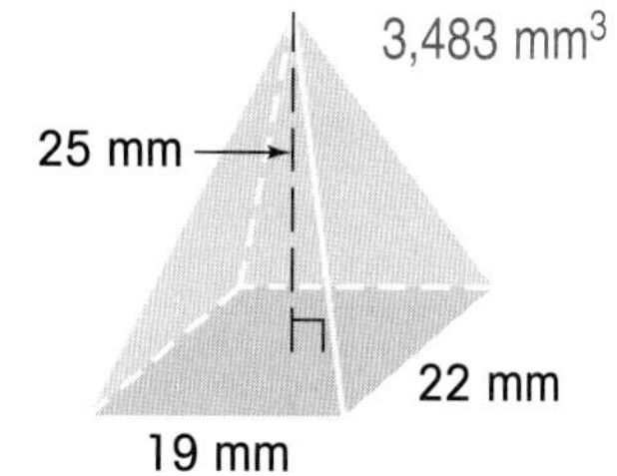

6.

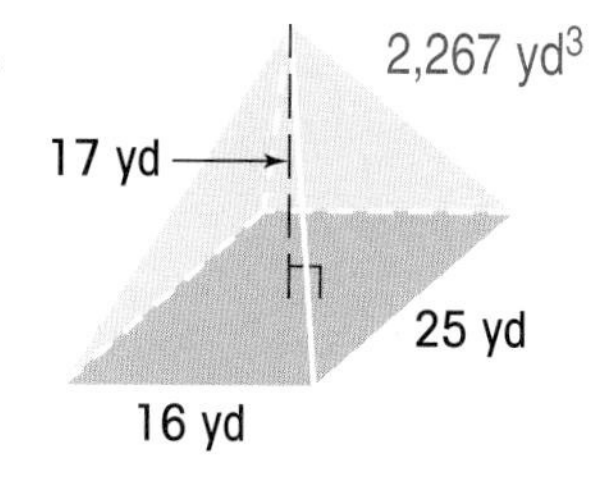

Math Connection

ARCHIMEDES AND VOLUME

Archimedes (287–212 B.C.) was one of the greatest mathematicians of all time. He discovered and proved formulas for the surface area and the volume of a sphere.

Legend says that Archimedes made a great discovery when he got into his bath and saw the water overflow. He realized that if he put a solid object into a full container of water, the volume of water that overflowed would equal the volume of the object. He also knew that if two solid objects were made of the same material and had the same weight, they should have the same volume. Archimedes used what he knew to prove that a king's crown was not pure gold.

You can use Archimedes' method to measure volume. Sink an unopened can of soup into a full container of water. The volume of the water that flows out is equal to the volume of the can.

More practice is provided in the Workbook, page 74, and in the Classroom Resource Binder, page 124.

Problem-Solving Strategy: Write an Equation

Getting Started Review how to find the surface area of a rectangular prism.

Writing an equation can help you solve a problem.

EXAMPLE

Dena wants to paint the walls of a room 10 ft wide and 7 ft long. The walls are 8 ft high. A glass patio door is 6 ft wide by 7 ft high. Another door is 3 ft wide by 7 ft high. How many square feet of walls will Dena paint?

You can use a net to show the dimensions of the room. Shade the areas Dena will paint.

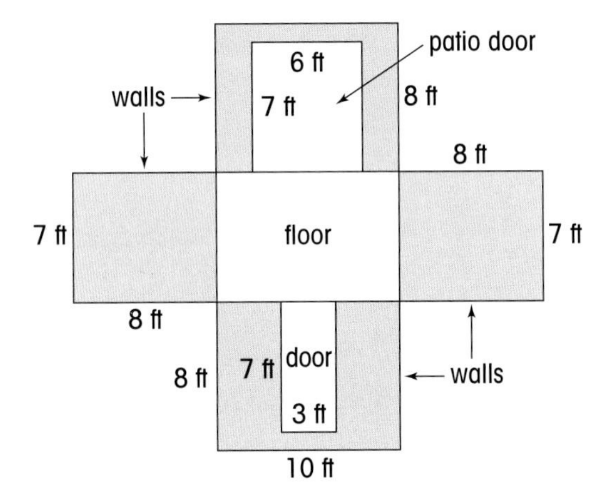

Write an equation to find the area of the 2 pairs of opposite walls.
Substitute the given values.

$$\text{Area} = 2lh + 2wh$$
$$= 2 \cdot 7 \cdot 8 + 2 \cdot 10 \cdot 8$$
$$= 272$$

Find the area of each door.
Substitute the given values.

$$\text{Area of patio door} = wh = 6 \cdot 7 = 42$$

$$\text{Area of other door} = wh = 3 \cdot 7 = 21$$

Subtract the area of the doors from the area of the walls.

$$\text{Area to paint} = 272 - 63 = 209$$

Dena will paint 209 ft^2.

Try This

Joe uses three coats of paint on the walls of a shed. The total surface area of the walls is 300 ft^2. One gallon of paint covers 350 ft^2. The paint is only sold in gallon cans. How much paint must Joe buy?

Write an equation to find the number of gallons.	Number of gallons $= 3x \div 350$
Let x equal the surface area of the walls of the shed.	300 $= 3(\blacksquare) \div 350$
Multiply to find the amount needed for 3 coats.	$= \blacksquare$ 2.6
Divide by the area that one gallon covers.	

Joe cannot buy part of a gallon, so he must buy ■ gallons. 3

Practice

Write an equation to solve each problem.

1. Caleb wants to paint a room that measures 18 ft by 22 ft. The walls are 12 ft high. The doors and windows have a total area of 155 ft^2. How many square feet of walls will Caleb paint? 805 ft^2

2. Sheanna is painting the outside walls of a dollhouse. The dollhouse is 48 in. long, 36 in. wide, and 24 in. high. It has two doors that are 3 in. by 6 in. It has 16 square windows that are each 3 in. on a side. How many square inches of walls will Sheanna paint? (Hint: Make a net.) 3,852 in.^{2}

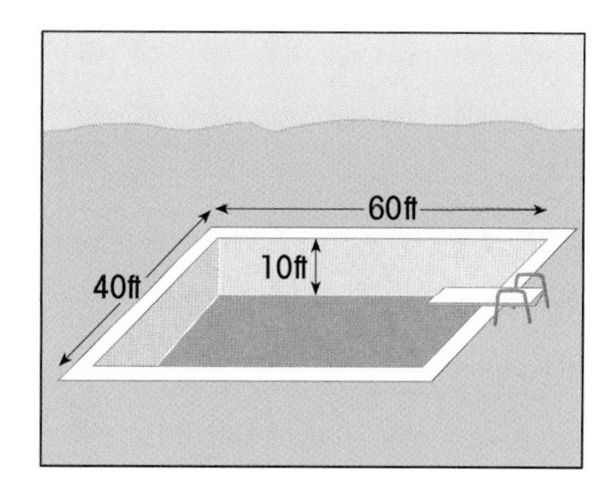

3. Kim's Construction is refinishing the inside walls and bottom of a swimming pool. Use the diagram at the right. How many square feet will be refinished? 4,400 ft^2

4. A gallon of coating material covers 125 ft^2 when it is used to refinish a pool. It costs \$25 per gallon. How much will it cost to buy enough to refinish a pool 10 ft deep, 25 ft long, and 12 ft wide? \$225.00

More practice is provided in the Workbook, page 75, and in the Classroom Resource Binder, page 125.

11·13 Problem-Solving Application: Air Conditioning

Getting Started Review how to find the volume of a rectangular prism.

Remember
The volume of a prism is $V = Bh$.

The cooling capacity of an air conditioner is measured in British thermal units (BTUs). To find the cooling capacity you need for a room with an 8-ft-high ceiling, use this formula. V is the volume of the room.

$$\text{BTU} = 3V$$

EXAMPLE 1

Ken wants to cool a room that is 20 ft by 17 ft with an 8-ft-high ceiling. What cooling capacity should the air conditioner have?

Find the volume of the room.
$$\begin{aligned} V &= Bh \\ &= (lw)h \\ &= 20 \cdot 17 \cdot 8 \\ &= 2{,}720 \end{aligned}$$

Use the formula for finding cooling capacity. $\text{BTU} = 3V$
Substitute the values you know. $= 3 \cdot 2{,}720$
Simplify. $= 8{,}160$

The air conditioner needs a cooling capacity of 8,160 BTUs.

EXAMPLE 2

The price of an air conditioner depends on its cooling capacity. The table shows the prices for air conditioners with different cooling capacities. Use the cooling capacity from Example 1. Which model should Ken buy?

Model	BTU	Price
A	5,000	$199.99
B	6,000	$229.99
C	8,000	$299.99
D	10,000	$349.99
E	12,000	$449.99

Look for the model with the smallest cooling capacity that is still large enough to meet Ken's needs. Ken needs an air conditioner with a capacity of 8,160 BTUs. Model C does not have enough cooling capacity.

Ken will need to buy Model D.

Try These

A family room is 15 ft by 15 ft with an 8-ft-high ceiling.

1. How many BTUs are needed to cool the room?

Find the volume of the room.

$V = Bh$
$= (lw)h$
15 15 $= (■)(■)(8)$
$= ■$ 1,800

Use the formula for finding cooling capacity

$BTU = 3V$
$= 3(■)$ 1,800
$= ■$ 5,400

To cool the room, ■ BTUs are needed. 5,400

2. Use the table on page 326. How much would an air conditioner with the correct cooling capacity cost?

Model ■ B has a cooling capacity of ■ 6,000 and costs ■. $229.99

Model ■ B is the correct size and costs ■. $229.99

Practice

Solve each problem. Use the table on page 326 if necessary.

1. Tyrone wants to buy an air conditioner for his living room. The room measures 15 ft by 18 ft and has an 8-ft-high ceiling. How many BTUs are needed to cool this room? 6,480 BTUs

2. Jamil does woodworking in a shop that is 20 ft by 16 ft with an 8-ft-high ceiling. Which model air conditioner should he purchase for his shop? Model C

3. Celine is buying an air conditioner for two connecting rooms. One room measures 12 ft by 14 ft, and the other measures 12 ft by 15 ft. Both have 8-ft-high ceilings. How much will an air conditioner cost that cools both rooms? $349.99

An alternate two-column proof lesson is provided on page 416 of the student book.

11-14 Proof: Volume of Figures

Getting Started
Have students compare a rectangular prism with a cube.

You can use what you know about the volume of a prism to prove that the volume of a cube is s^3.

EXAMPLE 1

Remember
For a prism, $V = Bh$.

You are given:
a cube with side s

Prove: Volume $= s^3$

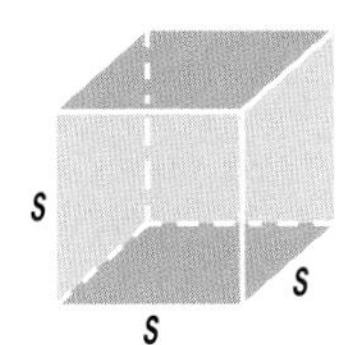

You know that the length of each side of the cube is s. The base is a square with side s. So, the area of the base is s^2. The height of the cube is also s. You know that the volume of a prism is equal to the area of the base times the height, $V = Bh$. Using substitution, the volume of a cube is $s^2 \cdot s$ or s^3.

So, you prove that for a cube with side s, Volume $= s^3$. ✓

EXAMPLE 2

Math Fact
For a cone, $V = \frac{1}{3}Bh$.

You are given:
a cone with radius r
and height h

Prove: Volume $= \frac{1}{3}\pi r^2 h$

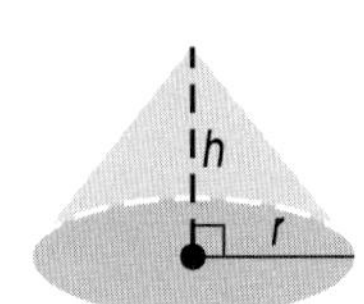

You know that the base of the cone is a circle with radius r. So, the area of the base is πr^2. The height of the cone is h. The volume is $\frac{1}{3}$ the area of the base times the height, $V = \frac{1}{3}Bh$. Using substitution, $V = \frac{1}{3}\pi r^2 h$.

So, you prove that Volume $= \frac{1}{3}\pi r^2 h$. ✓

Try This

Copy and complete the proof.

You are given:

a hemi-sphere with radius r

Prove: Volume $= \frac{2}{3}\pi r^3$

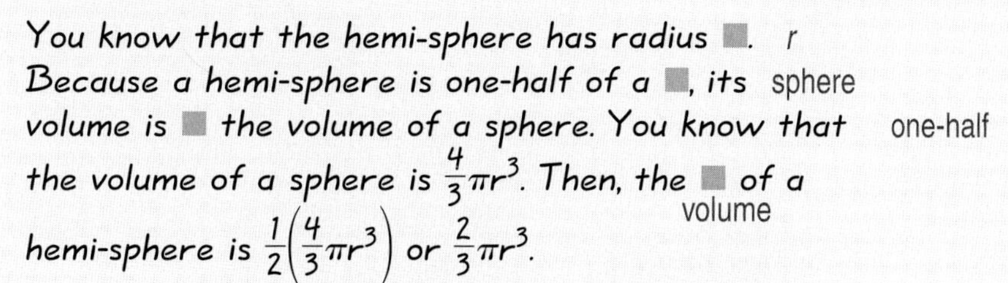
You know that the hemi-sphere has radius ■. r

Because a hemi-sphere is one-half of a ■, its sphere

volume is ■ the volume of a sphere. You know that one-half

the volume of a sphere is $\frac{4}{3}\pi r^3$. Then, the ■ of a volume

hemi-sphere is $\frac{1}{2}\left(\frac{4}{3}\pi r^3\right)$ or $\frac{2}{3}\pi r^3$.

So, you prove that for a hemi-sphere, Volume $= \frac{2}{3}\pi r^3$.

Practice

Write a proof for each of the following. See Additional Answers in the back of this book.

1. You are given:

a cylinder with radius r

and height h

Prove: Volume $= \pi r^2 h$

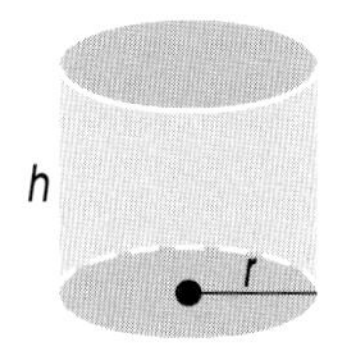

2. You are given:

a square pyramid

with side s for the

base and height h

Prove: Volume $= \frac{1}{3}s^2 h$

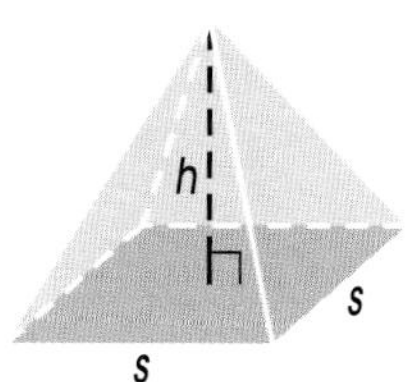

More practice is provided in the Classroom Resource Binder.

Chapter 11 Review

Summary

Prisms, pyramids, cylinders, cones, and spheres are space figures.
A net is a pattern that can be folded to make a space figure.
The surface area of a prism or a cylinder is the sum of the areas of all the faces.
Use the formula $SA = 4\pi r^2$ to find the surface area of a sphere.
The volume of a space figure is the number of cubic units needed to fill it.
Use the formula $V = Bh$ to find the volume of a prism or cylinder.
Use the formula $V = \frac{4}{3}\pi r^3$ to find the volume of a sphere.
Use the formula $V = \frac{1}{3}Bh$ to find the volume of a cone or a pyramid.
You can use a similarity ratio to compare the volumes of similar figures.
You can write an equation to solve a problem about surface area.
You can use volume formulas to solve measurement problems.
You can use facts you know about volume to prove a statement.

More vocabulary review is provided in the Classroom Resource Binder, page 115.

volume
surface area
cone
prism
pyramid
cylinder

Vocabulary Review

Complete the sentences with words from the box.

1. ____ is the number of cubic units that fill a space figure. Volume
2. A ____ is a space figure with two congruent parallel bases that are circles. cylinder
3. The sum of the areas of all the surfaces of a space figure is called the ____. surface area
4. A ____ has a circular base and one vertex. cone
5. A ____ is a space figure with faces that are polygons and two bases that are parallel and congruent. prism
6. A space figure with one base and triangular lateral faces that meet in a vertex is called a ____. pyramid

Chapter Quiz

More assessment is provided in the Classroom Resource Binder.

Name the space figure you can make from each net.

1. 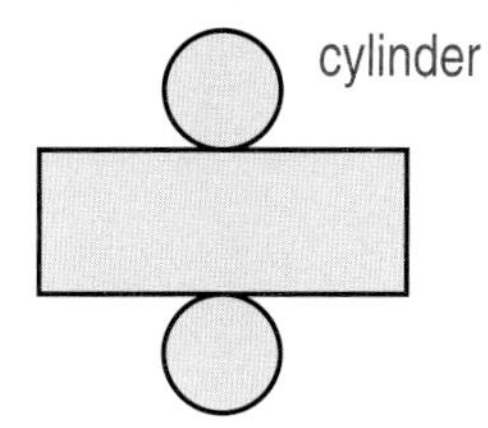
cylinder

2. 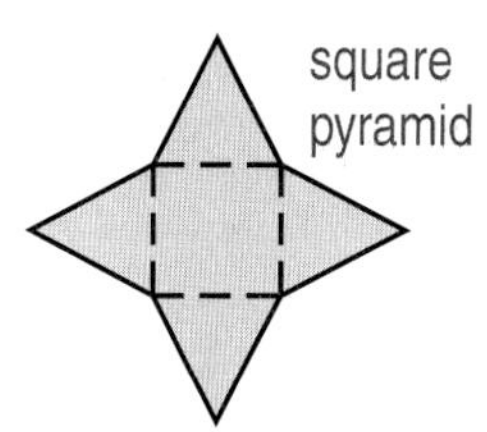
square pyramid

3. 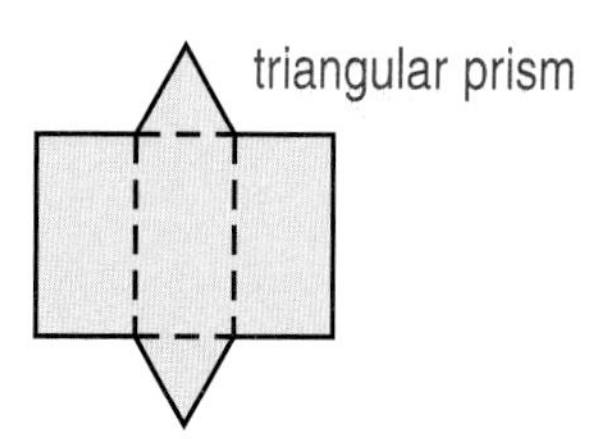
triangular prism

Find the surface area of each space figure. Use 3.14 for π as needed.

4. 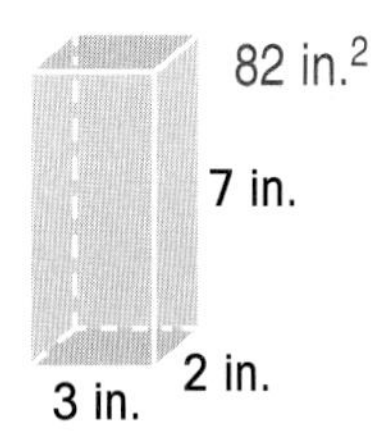

82 in.^{2}

5. 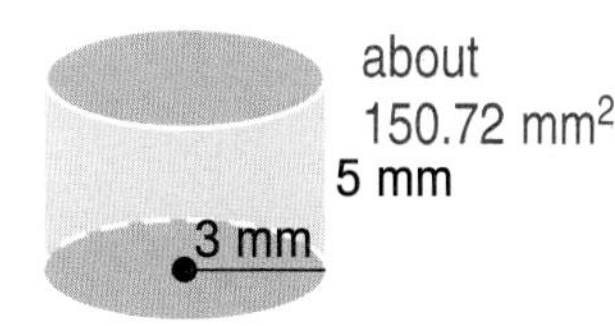

about 150.72 mm^2

6. 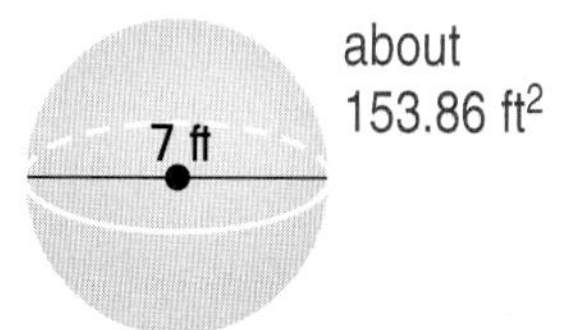

about 153.86 ft^2

Find the volume of each space figure. Use 3.14 for π as needed.

7. 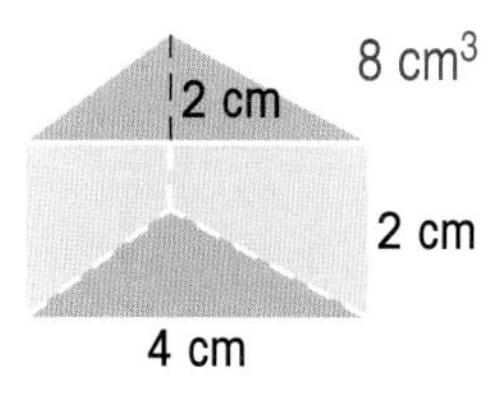

8 cm^3

8. 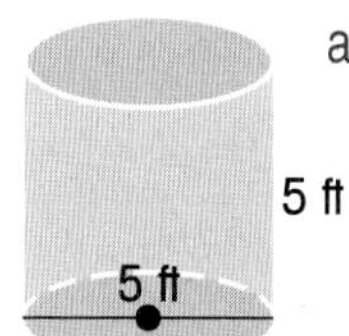

about 98.13 ft^3

9. 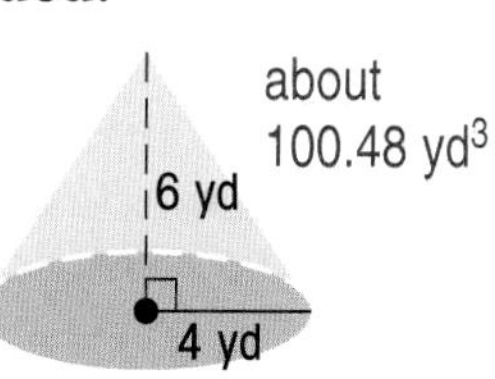

about 100.48 yd^3

Problem Solving

Solve each problem. Show your work.

10. Two prisms have a similarity ratio of 1 to 3. The volume of the larger one is 135 ft^3. What is the volume of the smaller one? 5 ft^3

11. How much paper is needed to cover all sides of a cube that measures 3 ft on a side?
54 ft^2

Write About Math

Can the same unit of measure be used to describe the surface area of a prism and its volume? Explain why or why not.

No. For volume, you multiply three dimensions, so the result is given in cubic units. For surface area, you multiply two dimensions, so the result is given in square units.

Additional Practice for this chapter is provided on page 403 of the student book.

Chapter 12 Coordinate Geometry and Transformations

Have you ever made a mirror drawing? It is similar to the reflection of the mountain in the lake. In geometry you can also create a mirror image of a shape. How do you think you can do that?

Learning Objectives

- Graph points on a coordinate plane and find the distance between two points.
- Determine the midpoint and the slope of a line.
- Map a transformation.
- Graph points in three-dimensional space and find the distance between two points.
- Use a calculator to find the magnitude of a vector.
- Solve problems about resultant vectors by drawing a diagram.
- Apply concepts and skills to vectors in the coordinate plane.
- Write a coordinate proof.

ESL/ELL Note To help students understand reflections, use a mirror to show a line of reflection. Introduce and discuss the terms *preimage* and *image*.

Words to Know

ordered pair	two numbers that locate a point on a coordinate plane
coordinate plane	a grid with two axes where points are graphed
coordinates	the ordered numbers that locate a point on a plane or in space
origin	the point where the x- and y-axes meet
slope	the steepness of a line; the ratio of vertical change to horizontal change
transformation	a change in the position or size of a figure
preimage	the original figure before a transformation
image	the figure that results from a transformation
translation	a transformation in which a figure slides to a new position
reflection	a transformation in which a figure is flipped over a line
line of reflection	the line over which a figure is flipped in a reflection
rotation	a transformation in which a figure is turned about a point
center of rotation	the point about which a figure is turned
dilation	a transformation in which the image is similar to the original figure
vector	any quantity that has size and direction
magnitude	the length of a vector
resultant vector	the sum of two vectors

Transformation Project

Draw a large coordinate plane. Draw a design in the first quadrant by graphing three or more points and connecting them. Then use rotations, translations, and/or reflections to create a design in all four quadrants.

Project
See the Classroom Resource Binder for a scoring rubric to assess this project.

More practice is provided in the Workbook, page 76, and in the Classroom Resource Binder, page 130.

Algebra Review: Points on the Coordinate Plane

Getting Started
Review locating points on a number line.

Math Fact
The regions of the coordinate plane are called quadrants. They are labeled as shown.

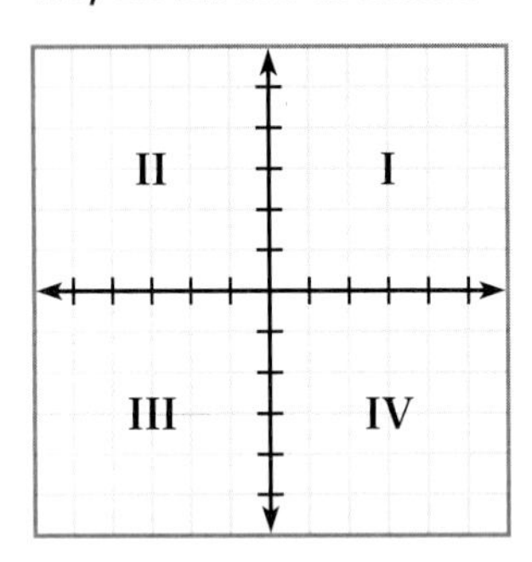

You can locate a point on the **coordinate plane** with an **ordered pair** of numbers. The ordered pair (2, 3) shows the location of point P. The numbers 2 and 3 are the **coordinates** of point P.

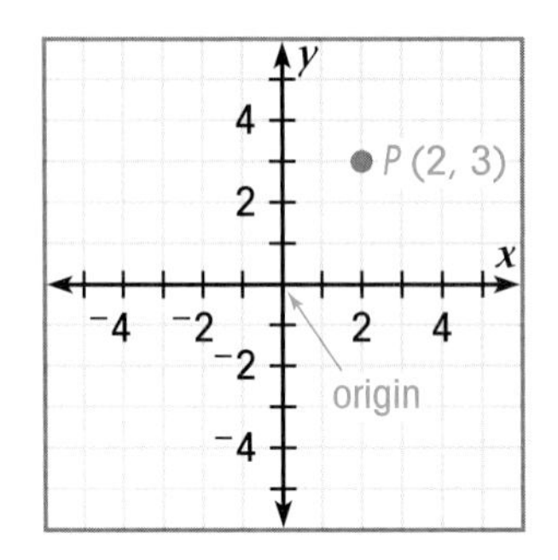

To locate or graph a point (x, y), begin at the **origin**. The origin is where the x- and y- axes meet. If the x-coordinate is positive, move to the right. If it is negative, move to the left. Then, if the y-coordinate is positive, move up. If it is negative, move down.

EXAMPLE 1

Graph point F at $(^-3, 2)$.

Begin at the origin.

Move 3 units to the left.

Then, move 2 units up.

Label the point F $(^-3, 2)$.

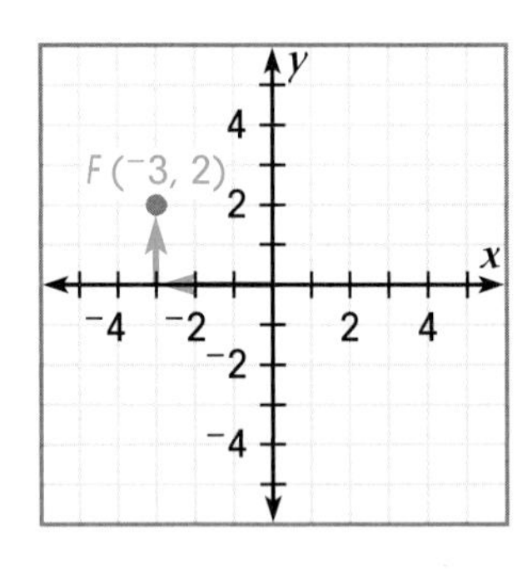

EXAMPLE 2

Write the coordinates of point A.

Begin at the origin.

Move 3 units to the right.

Then, move 4 units down.

The coordinates of point A are $(3, ^-4)$.

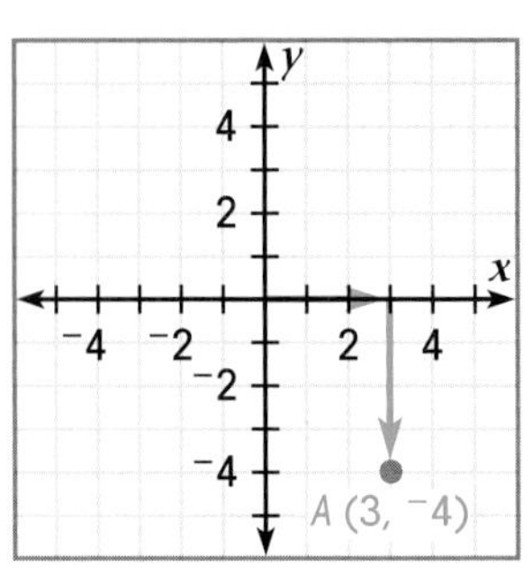

Try This

Write the coordinates of point R.

Begin at the origin.
Move ■ units to the left. 3
Then, move ■ units down. 0
The coordinates of point R are $(^-3, ■)$. 0

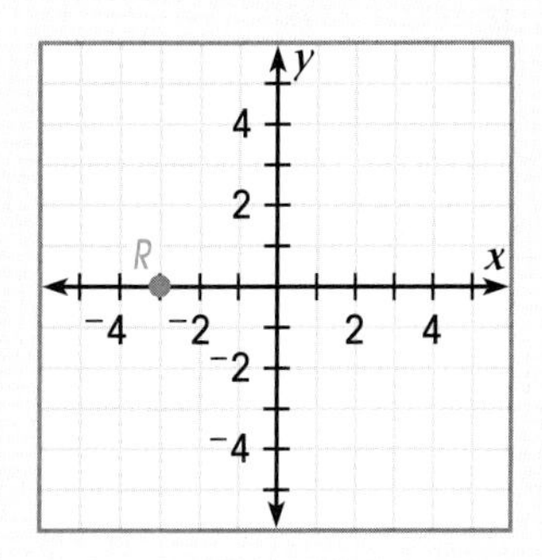

Practice

Draw a coordinate plane on grid paper. Graph each point. See below.

1. A (3, 2)
2. B (2, ⁻5)
3. C (⁻5, 2)
4. D (0, 4)
5. E (5, ⁻2)
6. F (⁻2, ⁻4)
7. G (2, 0)
8. H (⁻4, 2)

Write the coordinates of each point.

9. R (5, 5)
10. S (0, ⁻5)
11. T (⁻5, ⁻1)
12. U (2, 0)
13. V (⁻3, 3)
14. W (4, ⁻3)

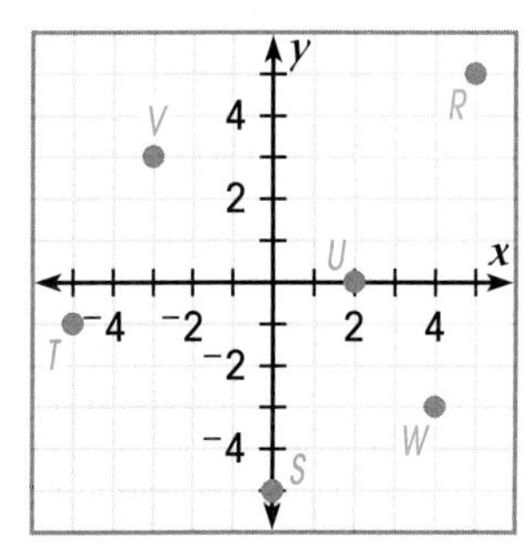

Share Your Understanding

15. Draw a coordinate plane on grid paper. Graph four points. Ask a partner to name the coordinates of each point. Answers will vary.

16. **CRITICAL THINKING** Does (8, 4) represent the same point as (⁻8, 4) on a coordinate plane? Explain.
No; (8, 4) is to the right of the y-axis and above the x-axis. (⁻8, 4) is above the x-axis, but it is to the left of the y-axis.

1–8

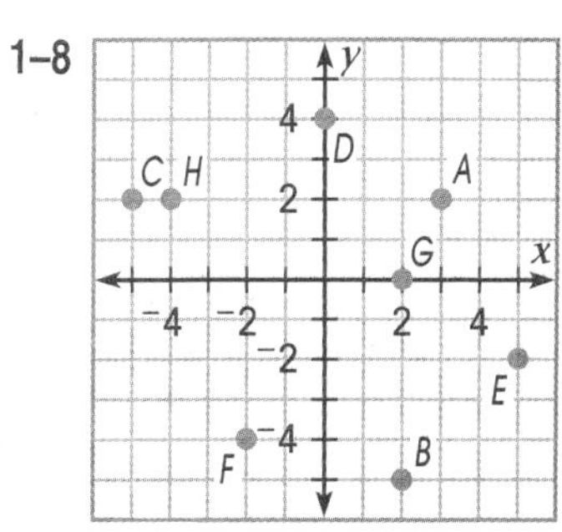

More practice is provided in the Workbook, page 76, and in the Classroom Resource Binder, page 130.

12·2 Algebra Review: Finding Distance

Getting Started
Review order of operations.

You learned to graph two points on the coordinate plane. You can find the distance between any two points, (x_1, y_1) and (x_2, y_2). Use the distance formula.

$$d = \sqrt{(x_2 - x_1)^2 + (y_2 - y_1)^2}$$

EXAMPLE 1

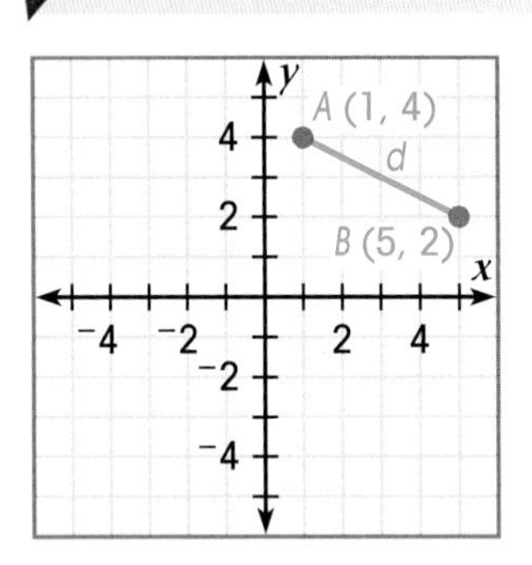

Find the distance between points A (1, 4) and B (5, 2).

Let (x_1, y_1) represent one point and (x_2, y_2) the other.
$A\ (1, 4) = (x_1, y_1)$
$B\ (5, 2) = (x_2, y_2)$

Write the formula for distance. $d = \sqrt{(x_2 - x_1)^2 + (y_2 - y_1)^2}$

Substitute the given values. $= \sqrt{(5 - 1)^2 + (2 - 4)^2}$

Simplify. $= \sqrt{(4)^2 + (^-2)^2}$

$= \sqrt{16 + 4}$

$= \sqrt{20}$

$= \sqrt{4 \cdot 5}$

$= 2\sqrt{5}$

The distance between points A (1, 4) and B (5, 2) is $2\sqrt{5}$.

EXAMPLE 2

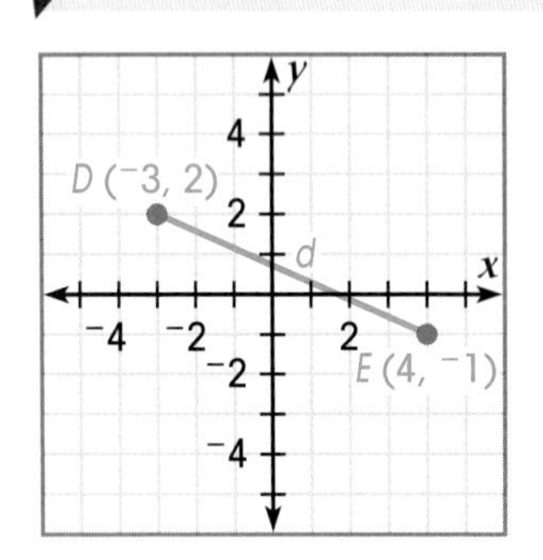

Find the distance between points D (⁻3, 2) and E (4, ⁻1).

Let (x_1, y_1) represent one point and (x_2, y_2) the other.
$D\ (^-3, 2) = (x_1, y_1)$
$E\ (4, ^-1) = (x_2, y_2)$

Write the formula for distance. $d = \sqrt{(x_2 - x_1)^2 + (y_2 - y_1)^2}$

Substitute the given values. $= \sqrt{[4 - (^-3)]^2 + (^-1 - 2)^2}$

Simplify. $= \sqrt{(7)^2 + (^-3)^2}$

$= \sqrt{49 + 9}$

$= \sqrt{58}$

The distance between points D (⁻3, 2) and E (4, ⁻1) is $\sqrt{58}$.

Try This

Find the distance between points F (0, 5) and G ($^-3$, 1).

Let (x_1, y_1) represent one point	$F\ (0, 5) = (x_1, y_1)$
and (x_2, y_2) the other.	$G\ (^-3, 1) = ■$ (x_2, y_2)
Write the formula for ■. distance	$d = \sqrt{(x_2 - x_1)^2 + (y_2 - y_1)^2}$
Substitute the given values.	$= \sqrt{(^-3 - ■)^2 + (■ - 5)^2}$ 0, 1
Solve for d.	$= \sqrt{(^-3)^2 + (^-4)^2}$
	$= \sqrt{■ + ■}$ 9, 16
	$= \sqrt{25}$
	$= ■$ 5

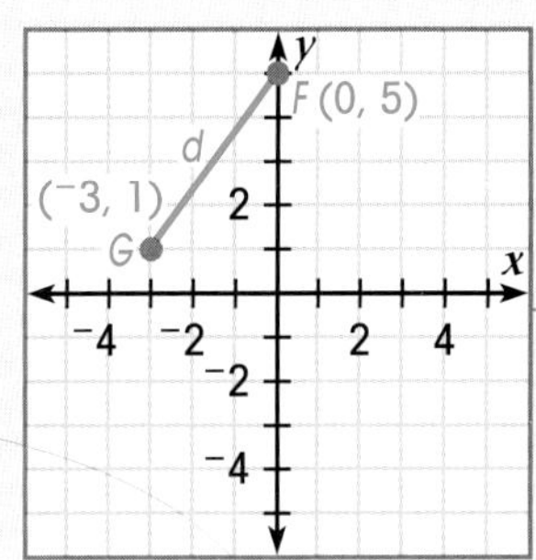

The distance between F (0, 5) and G ($^-3$, 1) is ■. 5

Practice

Find the distance between each pair of points.

1. H (1, 6) and K (3, 9) $\sqrt{13}$

2. L (4, 3) and M (8, 6) 5

3. N (4, 6) and P ($^-2$, $^-2$) 10

4. R ($^-3$, 1) and T (1, $^-3$) $4\sqrt{2}$

5. D (0, $^-2$) and E ($^-2$, 0) $2\sqrt{2}$

6. G ($^-4$, $^-6$) and H (2, 3) $3\sqrt{13}$

Share Your Understanding

7. Draw a coordinate plane on grid paper. Choose two points. Explain to a partner how to find the distance between them. Have your partner find this distance. Answers will vary. Check students' work. Students should understand how to use the distance formula.

8. **CRITICAL THINKING** Find the length of each side of $\triangle ABC$. Describe two ways to find the lengths. $AC = 3$, $CB = 5$, $AB = \sqrt{34}$; students may use the Pythagorean Theorem or the distance formula.

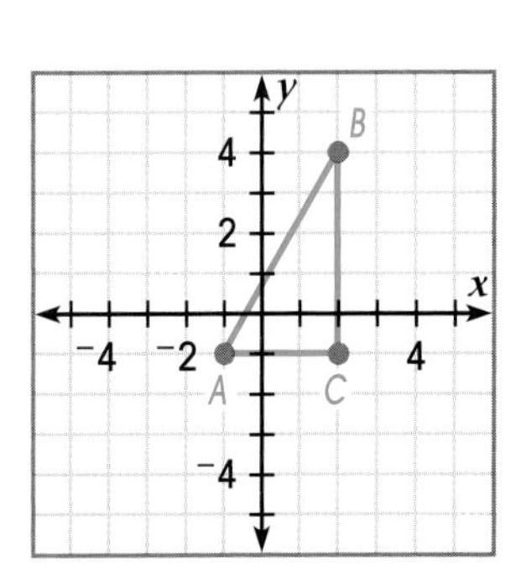

More practice is provided in the Workbook, page 77, and in the Classroom Resource Binder, page 131.

12-3 Midpoint of a Line Segment

Getting Started
Review finding the average of two numbers, for example, 2 and 6.

$$\text{Average} = \frac{2+6}{2} = \frac{8}{2} = 4$$

Remember
The average is the number that is halfway between two numbers on the number line.

You learned that the midpoint of a line segment is halfway between the endpoints. You can find the coordinates of the midpoint if you know the coordinates of the endpoints. You need to find the average of the x-coordinates and of the y-coordinates. Use the midpoint formula.

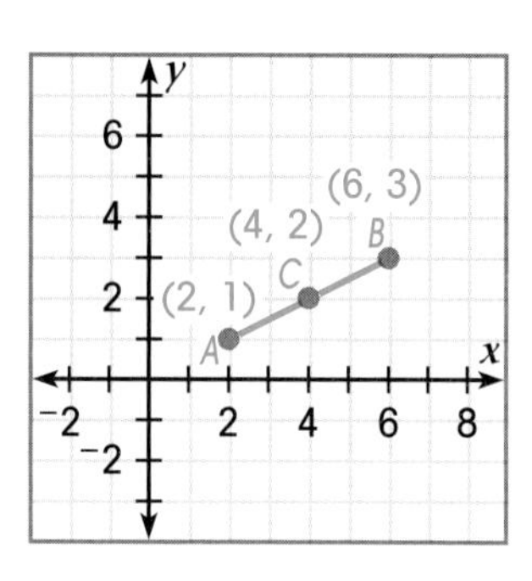

$$\text{midpoint} = \left(\frac{x_1 + x_2}{2}, \frac{y_1 + y_2}{2}\right)$$

$$\text{midpoint of } \overline{AB} = \left(\frac{2+6}{2}, \frac{1+3}{2}\right) = (4, 2)$$

To use the midpoint formula, substitute the coordinates of the endpoints of the line segment.

EXAMPLE

The endpoints of $\overline{MQ}$ are $M\ (^-3, 1)$ and $Q\ (5, ^-3)$. Find the midpoint.

Let (x_1, y_1) represent one point and (x_2, y_2) the other.
$M\ (^-3, 1) = (x_1, y_1)$
$Q\ (5, ^-3) = (x_2, y_2)$

Write the midpoint formula. $\text{midpoint} = \left(\frac{x_1 + x_2}{2}, \frac{y_1 + y_2}{2}\right)$

Substitute the given values. $= \left(\frac{^-3 + 5}{2}, \frac{1 + (^-3)}{2}\right)$

Simplify. $= \left(\frac{2}{2}, \frac{^-2}{2}\right)$

$= (1, ^-1)$

The midpoint of $\overline{MQ}$ is $(1, ^-1)$.

Try This

The endpoints of $\overline{DE}$ are D ($^-10$, 6) and E (4, 2). Find the midpoint.

Let (x_1, y_1) represent one point and (x_2, y_2) the other.	$D\,(^-10, 6) = (x_1, y_1)$ $E\,(4, 2) = (x_2, y_2)$
Write the midpoint formula.	$\text{midpoint} = \left(\frac{x_1 + x_2}{2}, \frac{y_1 + y_2}{2}\right)$
Substitute the given values.	$\text{midpoint} = \left(\frac{^-10 + 4}{2}, \frac{\blacksquare + 2}{2}\right)$ 6
Simplify.	$= \left(\frac{^-6}{2}, \frac{\blacksquare}{2}\right)$ 8 $= (^-3, \blacksquare)$ 4

The midpoint of $\overline{DE}$ is ■. ($^-3$, 4)

Practice

You are given the endpoints of a line segment. Find each midpoint.

1. A (1, 1) and B (5, 7) (3, 4)

2. Q (0, 4) and T (6, 6) (3, 5)

3. Y (5, 2) and J (1, 6) (3, 4)

4. W ($^-2$, 1) and B (6, 3) (2, 2)

5. V (4, $^-2$) and M (0, 8) (2, 3)

6. N ($^-3$, 0) and P (3, 0) (0, 0)

7. R ($^-8$, $^-2$) and A ($^-2$, $^-2$) ($^-5$, $^-2$)

8. H ($^-3$, $^-4$) and K (1, 2) ($^-1$, $^-1$)

9. K (2, 0) and L ($^-4$, $^-8$) ($^-1$, $^-4$)

10. R (0, 0) and P (4, 4) (2, 2)

Share Your Understanding

Answers will vary. Students should understand how to find the average of the x-coordinates and the y-coordinates.

11. Choose a line segment from the exercises above. Explain to a partner how to find the midpoint.

12. **CRITICAL THINKING** You are given the endpoints of a line segment. The x-coordinate of each point is 0. The y-coordinates are opposite integers. What is the midpoint of the line segment? (Hint: Draw a coordinate plane.)
the origin at (0,0)

More practice is provided in the Workbook, page 77, and in the Classroom Resource Binder, page 131.

Slope of a Line

Getting Started

Draw a line on the coordinate plane. Discuss how x and y change as you move from left to right along the line.

Slope describes the steepness of a line. It compares the change in y to the change in x. The change in y is called rise. The change in x is called run. Use this formula to find slope.

$$\text{slope} = \frac{y_2 - y_1}{x_2 - x_1} = \frac{\text{rise}}{\text{run}}$$

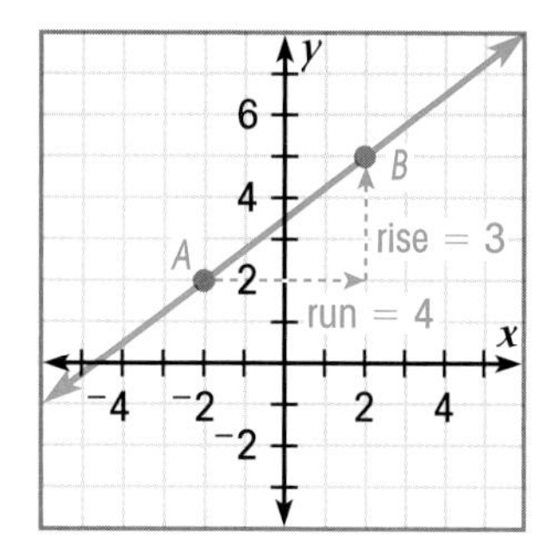

The slope can be positive or negative.

EXAMPLE 1

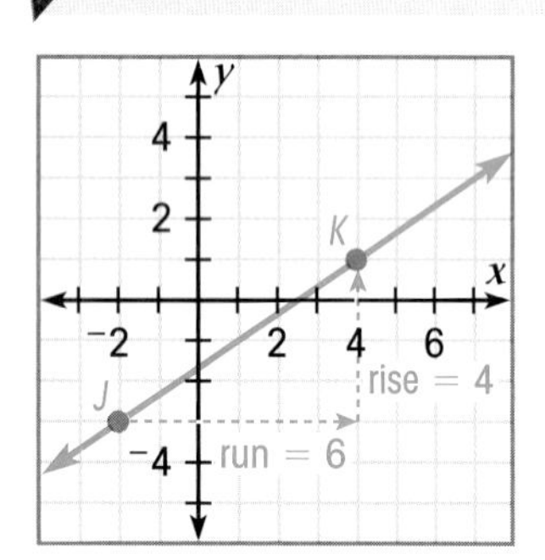

Find the slope of a line that contains points J and K.

Let (x_1, y_1) represent one point and (x_2, y_2) the other.

$J(^-2, ^-3) = (x_1, y_1)$
$K(4, 1) = (x_2, y_2)$

Write the formula for slope. $\text{slope} = \frac{y_2 - y_1}{x_2 - x_1}$

Substitute the given values. $\text{slope} = \frac{1 - (^-3)}{4 - (^-2)}$

Simplify. $= \frac{4}{6} = \frac{2}{3}$

The slope of this line is $\frac{2}{3}$.

EXAMPLE 2

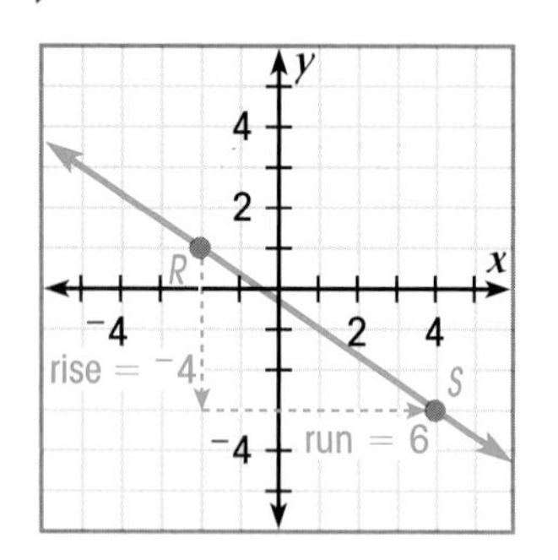

Find the slope of a line that contains points R and S.

Let (x_1, y_1) represent one point and (x_2, y_2) the other.

$R(^-2, 1) = (x_1, y_1)$
$S(4, ^-3) = (x_2, y_2)$

Write the formula for slope. $\text{slope} = \frac{y_2 - y_1}{x_2 - x_1}$

Substitute the given values. $\text{slope} = \frac{^-3 - 1}{4 - (^-2)}$

Simplify. $= \frac{^-4}{6} = \frac{^-2}{3}$

The slope of this line is $\frac{^-2}{3}$.

Try This

Find the slope of a line that contains points Q ($^-5$, 1) and R (1, $^-7$).

Let (x_1, y_1) represent one point and (x_2, y_2) the other.	$Q(^-5, 1) = (x_1, y_1)$ $R(1, ^-7) = (x_2, y_2)$
Write the formula for slope.	$\text{slope} = \frac{y_2 - y_1}{x_2 - x_1}$
Substitute the given values.	$\text{slope} = \frac{■ - 1}{■ - (^-5)}$ $^-7$; 1
Simplify.	$= \frac{^-8}{■} = ■$ 6; $\frac{^-4}{3}$

The slope of this line is ■. $\frac{^-4}{3}$

Practice

Find the slope of each line that contains the given points.

1. T (2, 4) and R (5, 1) $^-1$

2. N (1, $^-2$) and M (4, $^-3$) $\frac{^-1}{3}$

3. B (8, $^-9$) and Z (2, $^-4$) $\frac{^-5}{6}$

4. E (0, 6) and C (3, 1) $\frac{^-5}{3}$

5. W ($^-7$, $^-8$) and A ($^-3$, $^-3$) $\frac{5}{4}$

6. P (0, 0) and L ($^-5$, $^-5$) 1

7. M (2, 0) and F (1, 6) $^-6$

8. S (6, 4) and V (2, 8) $^-1$

9. D ($^-2$, 2) and C (2, 9) $\frac{7}{4}$

10. J (3, $^-3$) and K (1, $^-1$) $^-1$

Share Your Understanding

11. Work with a partner. Draw two lines on a coordinate plane. Find each slope. Answers will vary. Check students' work.

12. **CRITICAL THINKING** (3, 1) and ($^-2$, 1) are two points on a line. What is the slope of the line? 0

More practice is provided in the Workbook, page 78, and in the Classroom Resource Binder, page 132.

12·5 Parallel and Perpendicular Lines

Getting Started
Ask volunteers to identify parallel and perpendicular lines in the classroom.

You can use slope to identify parallel lines and perpendicular lines.

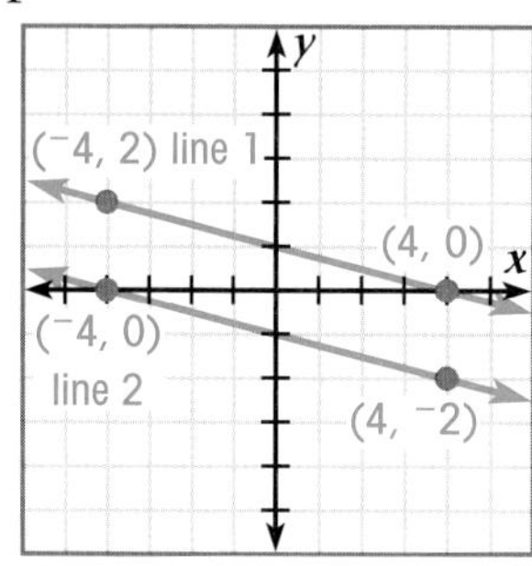

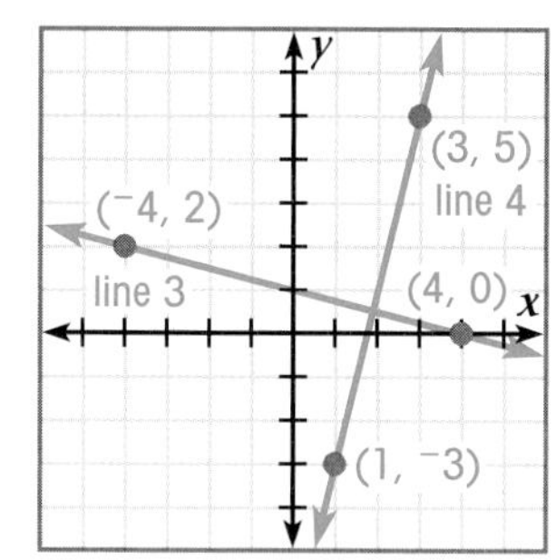

slope of line 1: $\frac{-1}{4}$

slope of line 2: $\frac{-1}{4}$

slope of line 3: $\frac{-1}{4}$

slope of line 4: $\frac{4}{1}$

Math Fact
$\frac{1}{4}$ and $\frac{4}{1}$ are reciprocals.
$-\frac{1}{4}$ and $\frac{4}{1}$ are negative reciprocals.

Lines 1 and 2 are parallel. They have the same slope. Lines 3 and 4 are perpendicular. Their slopes are negative reciprocals of each other.

EXAMPLE 1

Line 1 contains points (⁻4, 1) and (2, 5). Line 2 contains points (0, 4) and (3, 6). Decide if the lines are parallel.

Find the slope of line 1. $\text{slope} = \frac{y_2 - y_1}{x_2 - x_1}$

$= \frac{5 - 1}{2 - (^-4)} = \frac{4}{6} = \frac{2}{3}$

Find the slope of line 2. $\text{slope} = \frac{y_2 - y_1}{x_2 - x_1} = \frac{6 - 4}{3 - 0} = \frac{2}{3}$

The slopes are equal. So, the lines are parallel.

EXAMPLE 2

Line 1 contains points (5, ⁻3) and (⁻2, 1). Line 2 contains points (⁻5, ⁻3) and (⁻1, 4). Decide if the lines are perpendicular.

Find the slope of line 1. $\text{slope} = \frac{y_2 - y_1}{x_2 - x_1} = \frac{1 - (^-3)}{^-2 - 5} = \frac{4}{^-7} = \frac{^-4}{7}$

Find the slope of line 2. $\text{slope} = \frac{y_2 - y_1}{x_2 - x_1} = \frac{4 - (^-3)}{^-1 - (^-5)} = \frac{7}{4}$

$\frac{^-4}{7}$ and $\frac{7}{4}$ are negative reciprocals. So, the lines are perpendicular.

Try This

Line 1 contains points ($^-1$, 3) and (4, 6). Line 2 contains points ($^-1$, 3) and ($^-2$, 4). Decide if the lines are perpendicular.

Find the slope of line 1. slope $= \frac{y_2 - y_1}{x_2 - x_1} = \frac{6 - 3}{4 - (^-1)} = \frac{3}{■}$ 5

Find the slope of line ■. 2 slope $= \frac{y_2 - y_1}{x_2 - x_1} = \frac{4 - 3}{^-2 - ■} = \frac{1}{^-1} = {^-1}$ −1

$\frac{3}{5}$ and ■ −1 are not negative reciprocals. So, the lines are not ■. perpendicular

Practice

Decide if the lines are parallel. Write *parallel* or *not parallel.*

1. Line 1 points: (5, 3) and (4, 1)
Line 2 points: (3, 2) and (5, 6)
parallel

2. Line 1 points: (4, $^-1$) and ($^-2$, 3)
Line 2 points: (5, 2) and ($^-1$, 4)
not parallel

3. Line 1 points: (3, 6) and (2, $^-2$)
Line 2 points: (1, 4) and (3, $^-1$)
not parallel

4. Line 1 points: (4, 2) and ($^-1$, 2)
Line 2 points: (4, −2) and (6, $^-2$)
parallel

Decide if the lines are perpendicular. Write *perpendicular* or *not perpendicular.*

5. Line 1 points: (1, $^-1$) and (3, 2)
Line 2 points: (4, 1) and (5, $^-2$)
not perpendicular

6. Line 1 points: (3, 6) and ($^-4$, 2)
Line 2 points: (1, 3) and (5, $^-4$)
perpendicular

7. Line 1 points: (6, $^-3$) and (1, 2)
Line 2 points: (4, 1) and (3, $^-2$)
not perpendicular

8. Line 1 points: (5, 0) and (3, 1)
Line 2 points: (2, 4) and (0, 1)
not perpendicular

9. Find the slope of the two lines. If the slopes are equal, then the lines are parallel. If the slopes are negative reciprocals, then the lines are perpendicular. If the slopes are not equal and are not negative reciprocals, then the lines are neither.

Share Your Understanding

9. Explain to a partner how to decide if two lines are perpendicular, parallel, or neither. See above.

10. **CRITICAL THINKING** You are given the coordinates of the vertices of a parallelogram. How can you decide if the parallelogram is a rectangle? For the figure to be a rectangle, the slopes of the consecutive sides must be negative reciprocals. This means the consecutive sides are perpendicular, creating a right angle in each corner.

More practice is provided in the Workbook, page 79, and in the Classroom Resource Binder, page 133.

12-6 Translations in the Coordinate Plane

Getting Started
Draw two points on a coordinate plane. Label one point *A* and the other point *A′*. Ask a volunteer to describe how you can move from point *A* to point *A′*.

Math Fact
$ABCD \longrightarrow A'B'C'D'$ can be read "*ABCD* becomes *A* prime *B* prime *C* prime *D* prime."

A **transformation** is a change in the position or size of a figure on the coordinate plane. The original figure is called the **preimage**. The figure that results from the transformation is called the **image**.

One kind of transformation is a **translation**, or slide. The preimage and the transformation image are congruent.

Rectangle *ABCD* slides 5 units to the right. The translation image is rectangle $A'B'C'D'$. Write $ABCD \longrightarrow A'B'C'D'$. The symbol $\longrightarrow$ points from the preimage to the image.

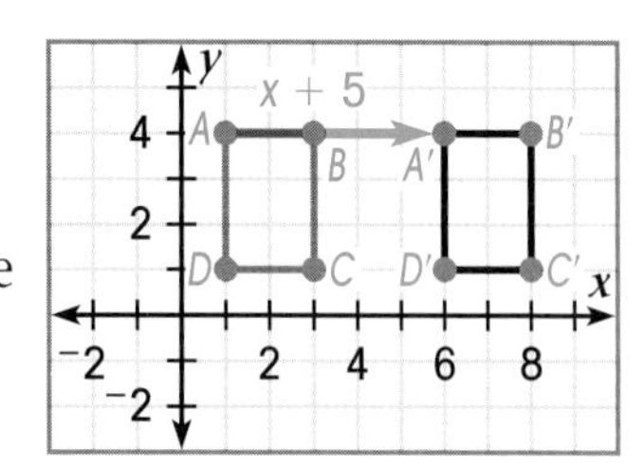

Each point in rectangle *ABCD* moved 5 units to the right. So, each x-coordinate increased by 5. Each y-coordinate stayed the same. So, the rule for this translation is $T(x, y) \longrightarrow (x + 5, y)$.

EXAMPLE

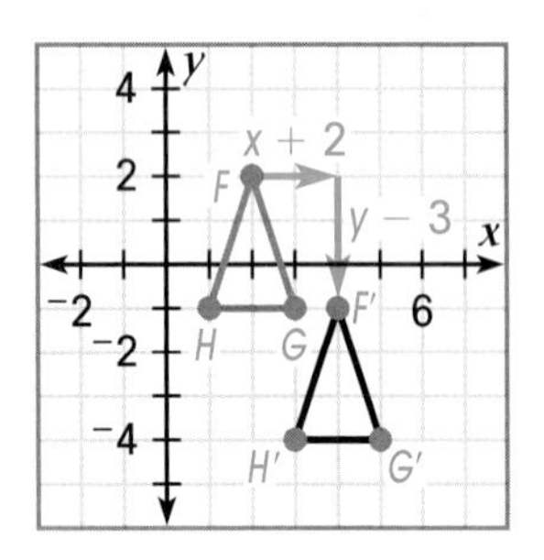

Translate $\triangle FGH$. Use the rule $T(x, y) \longrightarrow (x + 2, y - 3)$. Draw the translation image $\triangle F'G'H'$.

Copy $\triangle FGH$ onto grid paper.

From point F (2, 2), move 2 units to the right and 3 units down. Graph point F' here. $F = (2, 2)$, $F' = (4, ^{-}1)$

From point G (3, $^{-}1$), move 2 units to the right and 3 units down. Graph point G' here. $G = (3, ^{-}1)$, $G' = (5, ^{-}4)$

From point H (1, $^{-}1$), move 2 units to the right and 3 units down. Graph point H' here. $H = (1, ^{-}1)$, $H' = (3, ^{-}4)$

Connect points F', G', and H' to create the translation image $\triangle F'G'H'$.

So, $\triangle FGH \longrightarrow \triangle F'G'H'$ using the rule $T(x, y) \longrightarrow (x + 2, y - 3)$.

Try This

Write the rule to describe the translation $\triangle DEG \longrightarrow \triangle D'E'G'$.

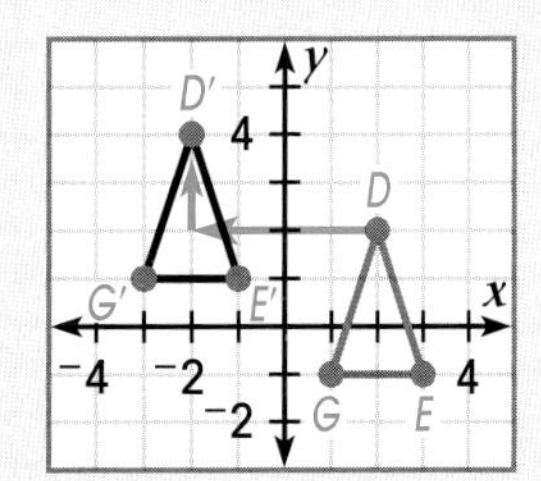

Find the change in the coordinates. Start with points D and D'.

Find the change in the x-coordinate. Count the units across from point D to the vertical line passing through point D'. — Move 4 units left. The x-coordinate decreases by ■. 4

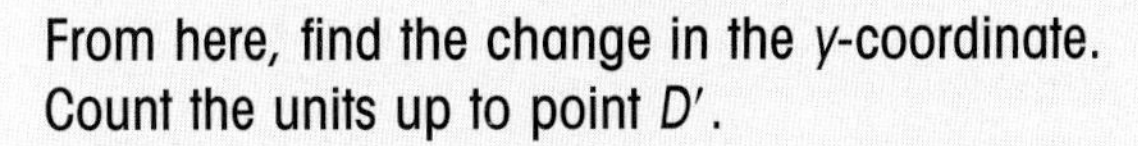

From here, find the change in the y-coordinate. Count the units up to point D'. — Move 2 units up. The y-coordinate ■ by 2. increases

Check to see if points E and G translate in the same way as point D. — yes

So, $\triangle DEG \longrightarrow \triangle D'E'G'$ using the rule $T(x, y) \longrightarrow (x - ■, y + ■)$. 4 2

Practice

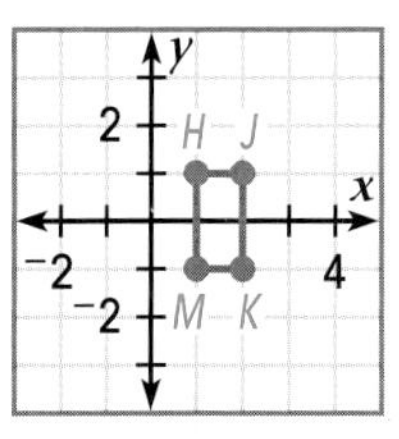

Use grid paper. Use each rule to translate rectangle $HJKM$. Draw each translation image.
See Additional Answers in the back of this book.

1. $T(x, y) \longrightarrow (x - 3, y)$
2. $T(x, y) \longrightarrow (x - 3, y + 2)$
3. $T(x, y) \longrightarrow (x, y - 3)$
4. $T(x, y) \longrightarrow (x + 1, y - 3)$

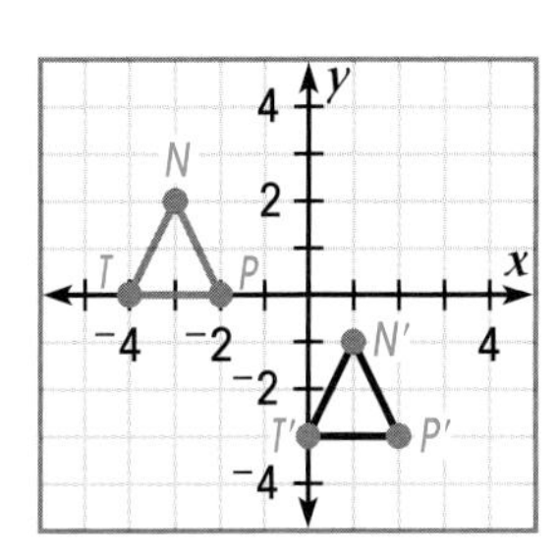

Use the diagram on the right.

5. Write the rule to describe the translation $\triangle NPT \longrightarrow \triangle N'P'T'$.
   $T(x, y) \longrightarrow (x + 4, y - 3)$

Share Your Understanding

6. Draw a square on a coordinate plane and write a translation rule. Ask a partner to use your rule to draw the image. Check students' work.

7. **CRITICAL THINKING** Draw a coordinate plane. Draw a square in the first quadrant. Have a partner draw a congruent square in the third quadrant. Write the translation rule. Check students' work.

More practice is provided in the Workbook, page 80, and in the Classroom Resource Binder, page 133.

Getting Started
After students have completed the example, have them fold their papers on the line of reflection. Point out that the preimage is congruent to the image and that the line of reflection is also a line of symmetry.

Math Fact
The line of reflection can be any line on the coordinate plane. $\triangle ABC$ is reflected over the y-axis. The equation for the y-axis is $x = 0$.

12-7 Reflections in the Coordinate Plane

Another type of transformation is called a **reflection**, or flip. The preimage is reflected over the **line of reflection**, or the line of symmetry. The preimage and the reflection image are congruent.

In the diagram, $\triangle ABC$ is reflected over the y-axis. The y-axis is the line of reflection. The reflection image is $\triangle A'B'C'$. You can write $\triangle ABC \longrightarrow \triangle A'B'C'$.

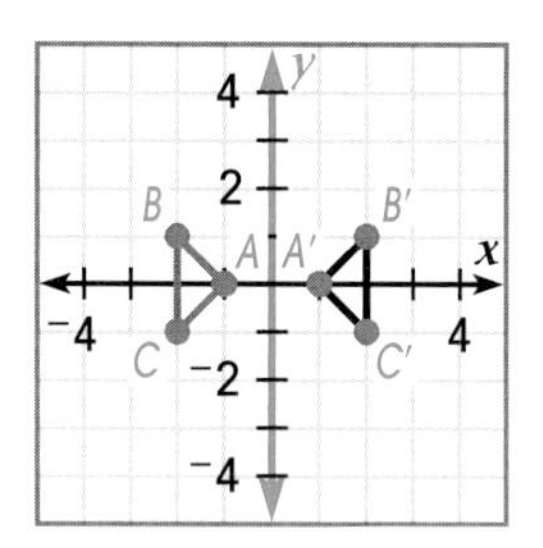

EXAMPLE

Reflect rectangle $DEGH$ over the line $x = 1$. Draw the reflection image rectangle $D'E'G'H'$.

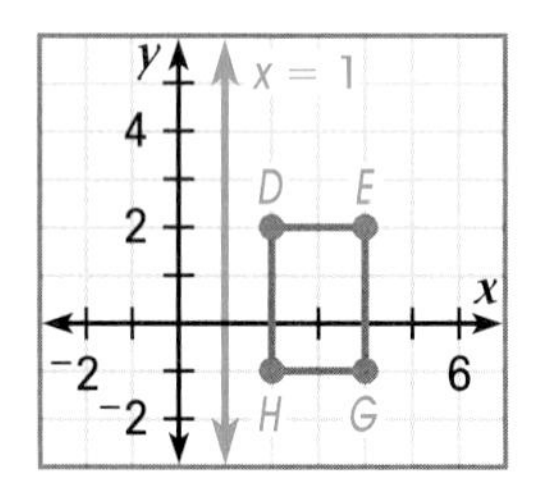

Copy rectangle $DEGH$ onto grid paper.
Graph the line of reflection $x = 1$.

Look at point D (2, 2). Point D is 1 unit to the right of $x = 1$. Stay on the same horizontal line. Graph point D' 1 unit to the left of $x = 1$.

$D = (2, 2)$
$D' = (0, 2)$

Look at point E (4, 2). Point E is 3 units to the right of $x = 1$. Stay on the same horizontal line. Graph point E' 3 units to the left of $x = 1$.

$E = (4, 2)$
$E' = (^-2, 2)$

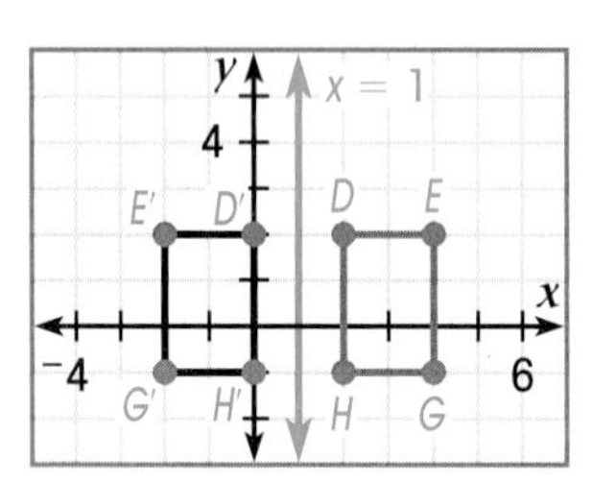

Look at point G (4, $^-1$). Point G is 3 units to the right of $x = 1$. Stay on the same horizontal line. Graph point G' 3 units to the left of $x = 1$.

$G = (4, ^-1)$
$G' = (^-2, ^-1)$

Look at point H (2, $^-1$). Point H is 1 unit to the right of $x = 1$. Stay on the same horizontal line. Graph point H' 1 unit to the left of $x = 1$.

$H = (2, ^-1)$
$H' = (0, ^-1)$

Connect points D', E', G', and H' to create the reflection image rectangle $D'E'G'H'$.

So, $DEGH \longrightarrow D'E'G'H'$ after reflecting rectangle $DEGH$ over the line $x = 1$.

Try This

Reflect rectangle $JKMP$ over the line $y = {}^{-}2$. Draw the reflection image rectangle $J'K'M'P'$.

Copy rectangle *JKMP* onto grid paper. Draw the line of reflection.

Point J is 3 units above $y = {}^{-}2$. Stay on the same vertical line. Graph point J' 3 units below $y = {}^{-}2$. $J = ({}^{-}4, 1)$ $J' = (■, {}^{-}5)$ ${}^{-}4$

Point K is 3 units above $y = {}^{-}2$. Stay on the same vertical line. Graph point K' ■ units below $y = {}^{-}2$. 3 $K = ({}^{-}1, 1)$ $K' = ({}^{-}1, ■)$ ${}^{-}5$

Point M is 1 unit above $y = {}^{-}2$. Stay on the same vertical line. Graph point M' ■ unit below $y = {}^{-}2$. 1 $M = ({}^{-}1, {}^{-}1)$ $M' = ({}^{-}1, ■)$ ${}^{-}3$

Point P is 1 unit above $y = {}^{-}2$. Stay on the same vertical line. Graph point P' ■ unit below $y = {}^{-}2$. 1 $P = ({}^{-}4, {}^{-}1)$ $P' = ({}^{-}4, ■)$ ${}^{-}3$

Connect points J', K', M', and P' to create the reflection image ■. rectangle $J'K'M'P'$

So, $JKMP \longrightarrow$ ■ after reflecting rectangle $JKMP$ over $y =$ ■. ${}^{-}2$ $J'K'M'P'$

Practice

Use grid paper. Reflect $\triangle DNS$ over each given line of reflection. Draw each reflection image.
See Additional Answers in the back of this book.

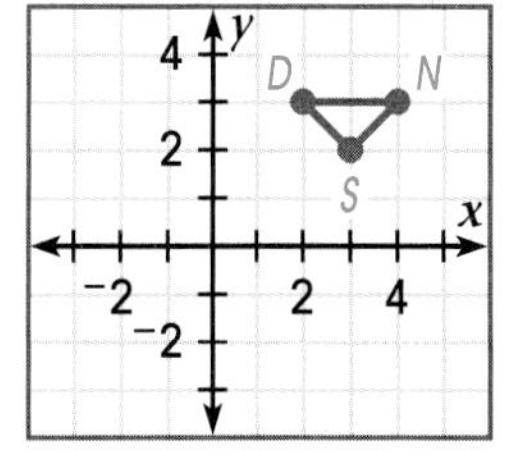

1. x-axis **2.** y-axis **3.** $x = {}^{-}1$

4. $x = 1$ **5.** $y = 1$ **6.** $y = 2$

Share Your Understanding

7. Explain to a partner how to draw one of the reflections in the exercises above. Use the words *units* and *line of reflection* in your explanation. See below.

8. **CRITICAL THINKING** Draw a coordinate plane. Draw a triangle in the first quadrant. Have a partner reflect this triangle over the y-axis. Then, you reflect this image over the x-axis. In which quadrant is the final reflection image? Quadrant III

7. Copy $\triangle DNS$ onto grid paper. Draw the line of reflection. Then count the number of units from D to the line of reflection and place D' the same number of units away from the line on the opposite side. Do the same for points N and S. Then connect the points of the image to draw $\triangle D'N'S'$.

More practice is provided in the Workbook, page 81, and in the Classroom Resource Binder, page 134.

12-8 Rotations in the Coordinate Plane

Getting Started Have students draw a square on a coordinate plane. Have them put a finger in the center of the origin and turn the paper to the right until the right-hand edge faces downward. Explain that they just rotated the square 90° clockwise about the origin. Show them a 180° rotation as two 90° turns and then a 270° rotation as three 90° turns. Explain that a 360° rotation made with four 90° turns puts the figure back in the original position.

Another type of transformation is called a **rotation**, or turn. The preimage is rotated around a **center of rotation**. The preimage and the rotation image are congruent.

Rectangle *ABCD* is rotated 90° counterclockwise about the origin. The rotation image is rectangle *A′B′C′D′*. You can write *ABCD* ⟶ *A′B′C′D′*.

90° counterclockwise

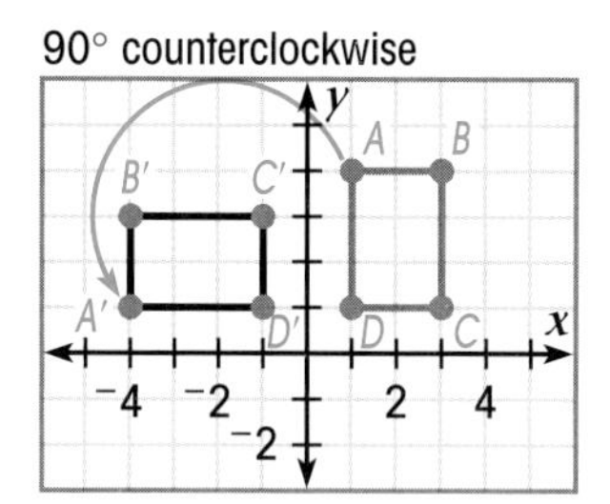

Any figure can be rotated any number of degrees clockwise or counterclockwise about any point.

EXAMPLE

Rotate △*GHJ* 270° clockwise about the origin. Draw the rotation image △*G′H′J′*.

Copy △*GHJ* onto grid paper.

Turn the grid paper to the right three 90° turns to make a 270° turn. Think of the new horizontal line as the *x*-axis. Think of the new vertical line as the *y*-axis.

Now, find point *G* after the rotation. Record the coordinates. These coordinates become point *G′*. $G' = (^-3, 2)$

Find point *H* after the rotation. Record the coordinates. These coordinates become point *H′*. $H' = (^-1, 3)$

Find point *J* after the rotation. Record the coordinates. These coordinates become point *J′*. $J' = (^-1, 1)$

Turn the paper back to its original position.

Graph points *G′*, *H′*, and *J′*. Connect these points to create the rotation image △*G′H′J′*.

So, △*GHJ* ⟶ △*G′H′J′* after rotating △*GHJ* 270° clockwise about the origin.

Try This

Dilate $\triangle JKL$ with a similarity ratio of $\frac{1}{2}$.
Draw the dilation image $\triangle J'K'L'$.

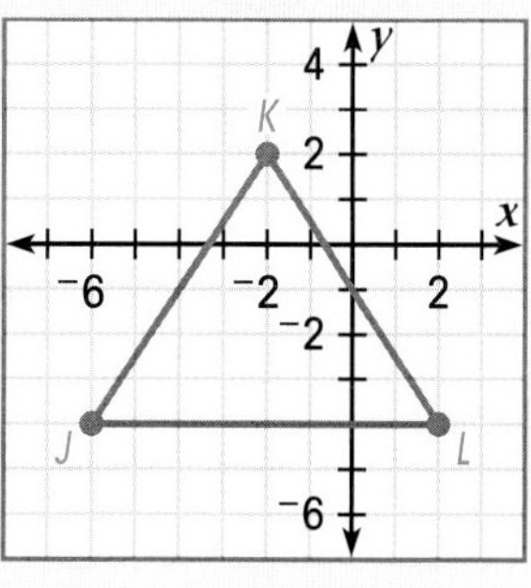

Copy $\triangle JKL$ onto grid paper. Record the coordinates of points J, K, and L. Multiply each coordinate by the similarity ratio $\frac{1}{2}$, or divide the coordinates of each point by 2.

Divide the coordinates of point J ($^-6$, $^-4$) by ■. 2

$^-6 \div 2 =$ ■ $^-3$
$^-4 \div$ ■ $= {^-2}$ 2
$J' = (^-3,$ ■$)$ $^-2$

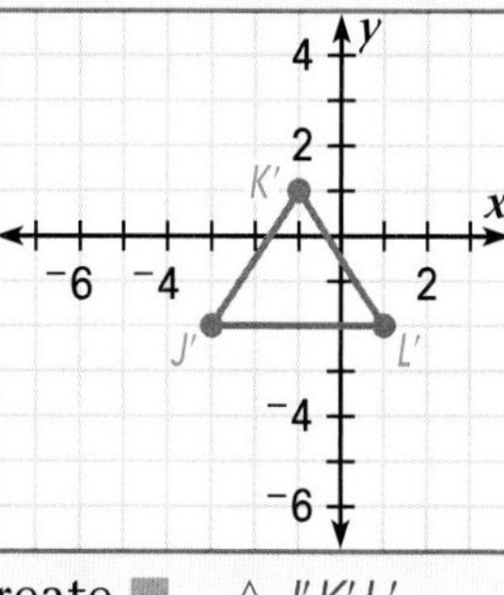

Divide the ■ coordinates of point K ($^-2$, 2) by ■. 2

■ $^-2$ $\div 2 = {^-1}$
$2 \div$ ■ 2 $=$ ■ 1
$K' = ($■$, 1)$ $^-1$

Divide the ■ coordinates of point L (2, $^-4$) by ■. 2

■ 2 $\div 2 = 1$
$^-4 \div 2 =$ ■ $^-2$
$L' = ($■$,$ ■$)$ 1 $^-2$

Graph points J', K', and L'. Connect these points to create ■. $\triangle J'K'L'$

Practice

Use grid paper. Dilate $\triangle DFH$ with each similarity ratio. Draw $\triangle D'F'H'$.
See Additional Answers in the back of this book.

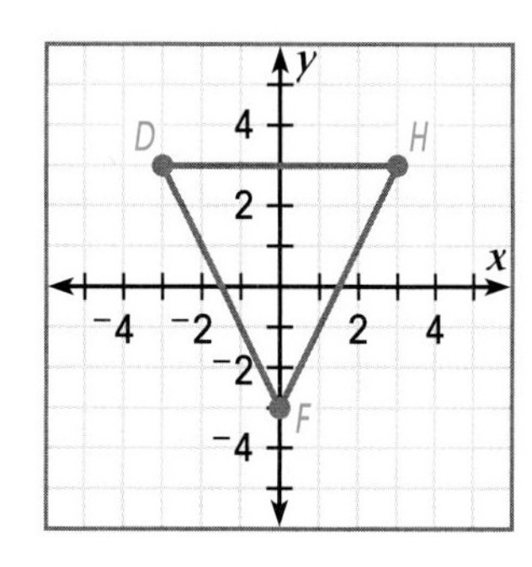

1. similarity ratio of 3
2. similarity ratio of $\frac{1}{3}$
3. similarity ratio of 1

Share Your Understanding

4. Explain to a partner how to find the dilation image of one of the exercises above.
Students should understand that each coordinate must be multiplied by the similarity ratio.
5. **CRITICAL THINKING** Use grid paper and $\triangle DFH$ above. Work with a partner to dilate $\triangle DFH$ with a similarity ratio of $1\frac{1}{3}$.
Check students' work. $D' = (^-4, 4)$, $F' = (0, {^-4})$, $H' = (4, 4)$

More practice is provided in the Workbook, page 83, and in the Classroom Resource Binder, page 135.

12·10 Points in Space

Getting Started
Have volunteers practice drawing the three axes needed to graph a point in space. Use grid paper.

You learned to locate points on the coordinate plane. The coordinate plane has two dimensions. They are the *x*-axis and the *y*-axis.

You can also locate points in space. You will need three axes: the *x*-axis, the *y*-axis, and the *z*-axis. A point in space has three coordinates (x, y, z). The coordinates of point *P* are (3, 4, 5).

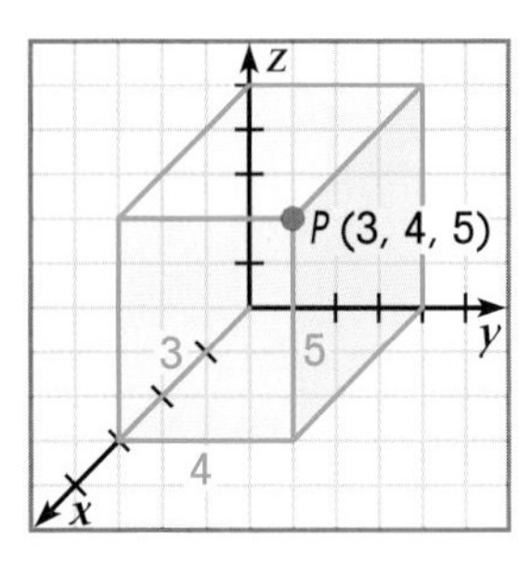

EXAMPLE

Graph point *A* (2, 3, 4).

Use grid paper.
Draw the *x*-, *y*-, and *z*-axes.
Start at the origin.

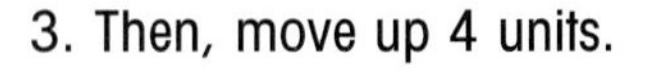

1. Move 2 units on the *x*-axis.
2. Next, move 3 units to the right.

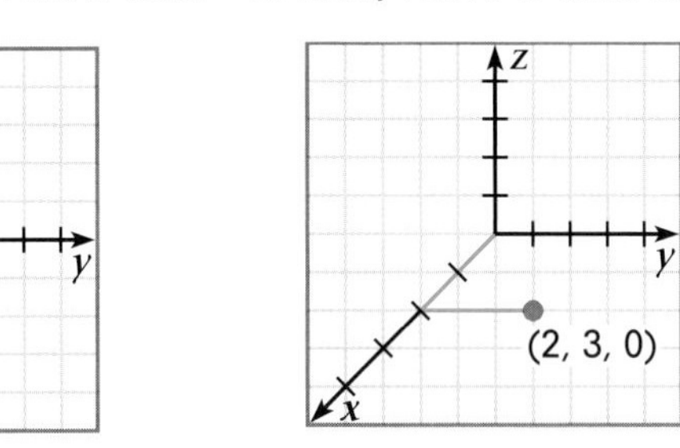

3. Then, move up 4 units.

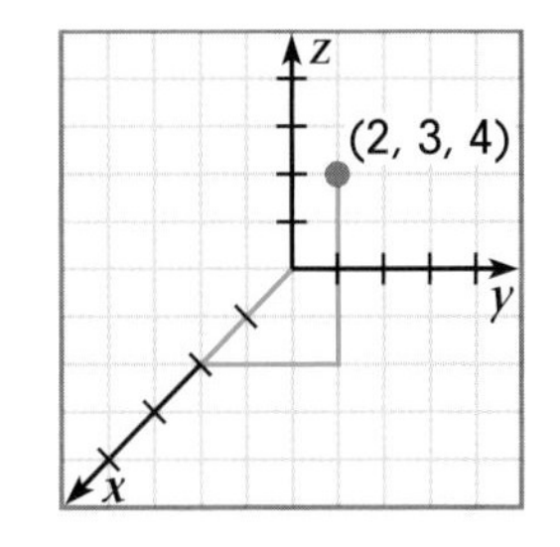

4. Graph and label point *A*.

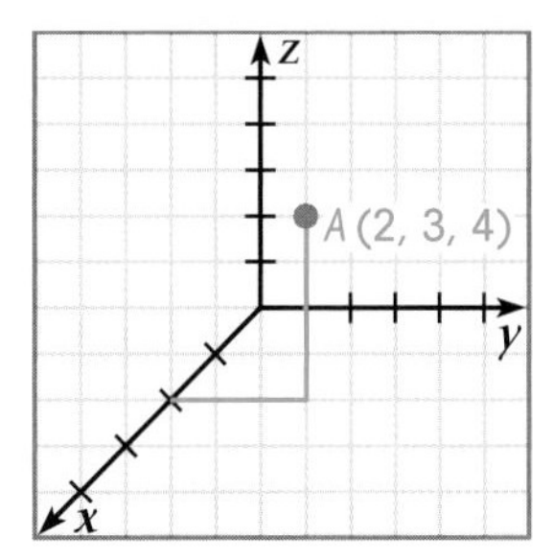

Point *A* is located at (2, 3, 4).

Try This

Graph point B (2, 0, 4).

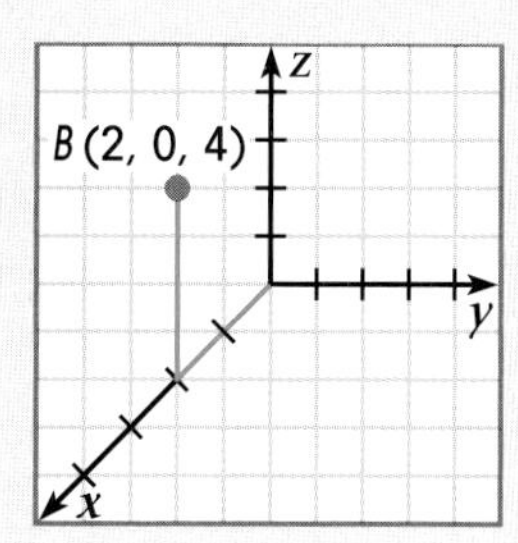

Use grid paper.

Draw the x-, y-, and z-axes

Start at the origin.

Move ■ units on the x-axis. 2

Do not move any units on the y-axis.

Then, move up ■ units. 4

Point B is located at ■. (2, 0, 4)

Practice

Use grid paper. Graph each point on a separate set of axes.

See Additional Answers in the back of this book.

1. A (3, 4, 2) **2.** B (3, 2, 3) **3.** C (3, 3, 3)

4. D (2, 0, 3) **5.** E (3, 0, 4) **6.** F (0, 2, 2)

7. G (0, 3, 2) **8.** H (3, 4, 0) **9.** I (2, 2, 0)

10. J (0, 3, 0) **11.** K (3, 0, 0) **12.** L (0, 0, 3)

Share Your Understanding

13. Explain to a partner how you graphed one of the points above. Have your partner follow your directions and graph the point. Answers will vary. Answers should reflect appropriate movements along the x-axis, the y-axis. and the z-axis.

14. **CRITICAL THINKING** Graph points A (4, 0, 0), B (0, 4, 0), and C (0, 0, 4) on the same set of axes. Connect the points to form a triangle. Find the length of each side.
$AC = 4\sqrt{2}$; $AB = 4\sqrt{2}$; and $BC = 4\sqrt{2}$

More practice is provided in the Workbook, page 83, and in the Classroom Resource Binder, page 135.

12·11 Finding Distance in Space

Getting Started
Plot two points on a coordinate plane. Call on a volunteer to explain how to find the distance between the points. Review the distance formula.

You used a formula to find the distance between two points in the coordinate plane. You can also find the distance between two points in space. You need to use the following formula.

$$d = \sqrt{(x_2 - x_1)^2 + (y_2 - y_1)^2 + (z_2 - z_1)^2}$$

EXAMPLE 1

Find the distance between points A (3, 1, 4) and B (1, −2, 0).

Let (x_1, y_1, z_1) represent one point and (x_2, y_2, z_2) the other.

$A\ (3, 1, 4) = (x_1, y_1, z_1)$
$B\ (1, {}^{-}2, 0) = (x_2, y_2, z_2)$

Write the distance formula. $d = \sqrt{(x_2 - x_1)^2 + (y_2 - y_1)^2 + (z_2 - z_1)^2}$

Substitute the given values. $d = \sqrt{(1 - 3)^2 + ({}^{-}2 - 1)^2 + (0 - 4)^2}$

Simplify.

$d = \sqrt{({}^{-}2)^2 + ({}^{-}3)^2 + ({}^{-}4)^2}$
$d = \sqrt{4 + 9 + 16}$
$d = \sqrt{29}$

The distance between points A and B is $\sqrt{29}$.

EXAMPLE 2

Find the distance between points C (0, 3, 2) and D (4, ${}^{-}1$, 3).

Let (x_1, y_1, z_1) represent one point and (x_2, y_2, z_2) the other.

$C\ (0, 3, 2) = (x_1, y_1, z_1)$
$D\ (4, {}^{-}1, 3) = (x_2, y_2, z_2)$

Write the distance formula. $d = \sqrt{(x_2 - x_1)^2 + (y_2 - y_1)^2 + (z_2 - z_1)^2}$

Substitute the given values. $d = \sqrt{(4 - 0)^2 + ({}^{-}1 - 3)^2 + (3 - 2)^2}$

Simplify.

$d = \sqrt{(4)^2 + ({}^{-}4)^2 + (1)^2}$
$d = \sqrt{16 + 16 + 1}$
$d = \sqrt{33}$

The distance between points C and D is $\sqrt{33}$.

Try This

Find the distance between points E (4, 3, $^-2$) and G (5, 2, 3).

Let (x_1, y_1, z_1) represent one point and (x_2, y_2, z_2) the other.	$E\ (4, 3, {}^-2) = (x_1, y_1, \blacksquare)$ z_1 $G\ (5, 2, 3) = (x_2, \blacksquare, z_2)$ y_2
Write the distance formula.	$d = \sqrt{(x_2 - x_1)^2 + (y_2 - y_1)^2 + (z_2 - z_1)^2}$
Substitute the given values.	$d = \sqrt{(5 - \blacksquare)^2 + (2 - \blacksquare)^2 + (3 - \blacksquare)^2}$ 4, 3, $^-2$
Simplify.	$d = \sqrt{(1)^2 + (\blacksquare)^2 + (5)^2}$ $^-1$ $d = \sqrt{1 + 1 + \blacksquare}$ 25 $d = \blacksquare$ $\sqrt{27}$, or $3\sqrt{3}$

The distance between points E and G is $\blacksquare$. $\sqrt{27}$, or $3\sqrt{3}$

Practice

Find the distance between each pair of points in space.

1. H (2, 4, 0) and J (3, 5, 2) $\sqrt{6}$

2. K (7, 3, 7) and L (4, 1, 5) $\sqrt{17}$

3. M (0, 1, 0) and P ($^-1$, 0, 3) $\sqrt{11}$

4. R (9, 5, 8) and S ($^-10$, $^-1$, 9) $\sqrt{398}$

5. T (7, 2, 6) and U ($^-9$, $^-2$, 8) $\sqrt{276}$ or $2\sqrt{69}$

6. A (2, 3, 4) and B ($^-1$, $^-3$, $^-9$) $\sqrt{214}$

7. C ($^-2$, $^-3$, $^-4$) and D (2, 3, 4) $\sqrt{116}$ or $2\sqrt{29}$

8. E ($^-8$, $^-9$, $^-10$) and G (8, 9, 10) $\sqrt{980}$ or $14\sqrt{5}$

Share Your Understanding

9. Explain to a partner how to find the distance between two points in space. Use the words *substitute* and *square root* in your explanation. Let (x_1, y_1, z_1) represent one point and (x_2, y_2, z_2) the other. Write the distance formula. Substitute the given values and simplify the square root of the sum.

10. **CRITICAL THINKING** Write the coordinates for a point in space. Have your partner write the coordinates for a different point in space. Work together to find the distance between the two points. Answers will vary. Students should demonstrate an understanding of the distance formula for points in space.

More practice is provided in the Workbook, page 84, and in the Classroom Resource Binder, page 135.

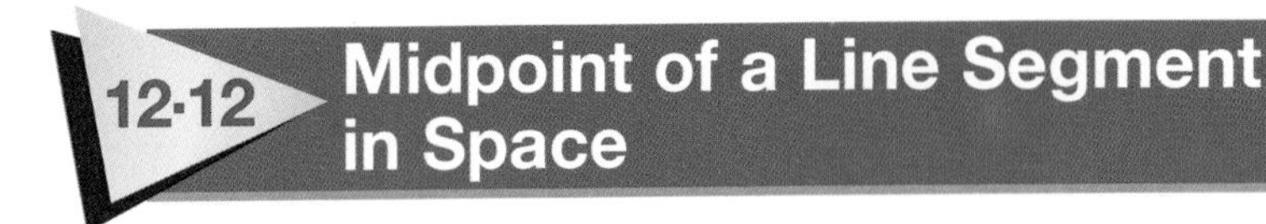

12·12 Midpoint of a Line Segment in Space

Getting Started
Graph two points on a coordinate grid. Call on a volunteer to explain how to find the midpoint of the line segment between them.

You learned how to find the midpoint of a line segment by finding the average of the x-coordinates and of the y-coordinates.

You can also find the midpoint of a line segment in space. Use the following formula.

$$\text{midpoint} = \left(\frac{x_1 + x_2}{2}, \frac{y_1 + y_2}{2}, \frac{z_1 + z_2}{2}\right)$$

EXAMPLE 1

Find the coordinates of the midpoint of $\overline{AB}$ for $A\ (2, 5, 4)$ and $B\ (4, {}^-1, 0)$.

Let (x_1, y_1, z_1) represent one point and (x_2, y_2, z_2) the other.

$$A\ (2, 5, 4) = (x_1, y_1, z_1)$$
$$B\ (4, {}^-1, 0) = (x_2, y_2, z_2)$$

Write the midpoint formula.

$$\text{midpoint} = \left(\frac{x_1 + x_2}{2}, \frac{y_1 + y_2}{2}, \frac{z_1 + z_2}{2}\right)$$

Substitute the given values.

$$= \left(\frac{2 + 4}{2}, \frac{5 + ({}^-1)}{2}, \frac{4 + 0}{2}\right)$$

Simplify.

$$= \left(\frac{6}{2}, \frac{4}{2}, \frac{4}{2}\right) = (3, 2, 2)$$

The midpoint of $\overline{AB}$ is located at (3, 2, 2).

EXAMPLE 2

Find the coordinates of the midpoint of $\overline{PT}$ for $P\ ({}^-2, 4, {}^-7)$ and $T\ ({}^-6, 0, 3)$.

Let (x_1, y_1, z_1) represent one point and (x_2, y_2, z_2) the other.

$$P\ ({}^-2, 4, {}^-7) = (x_1, y_1, z_1)$$
$$T\ ({}^-6, 0, 3) = (x_2, y_2, z_2)$$

Write the midpoint formula.

$$\text{midpoint} = \left(\frac{x_1 + x_2}{2}, \frac{y_1 + y_2}{2}, \frac{z_1 + z_2}{2}\right)$$

Substitute the given values.

$$= \left(\frac{{}^-2 + ({}^-6)}{2}, \frac{4 + 0}{2}, \frac{{}^-7 + 3}{2}\right)$$

Simplify.

$$= \left(-\frac{8}{2}, \frac{4}{2}, -\frac{4}{2}\right) = ({}^-4, 2, {}^-2)$$

The midpoint of $\overline{PT}$ is located at $({}^-4, 2, {}^-2)$.

Try This

Find the coordinates of the midpoint of $\overline{AB}$ for A (4, 5, 2) and B (2, 3, 0).

Let (x_1, y_1, z_1) represent one point and (x_2, y_2, z_2) the other.

$A(4, 5, 2) = (x_1, y_1, z_1)$
$B(2, 3, 0) = (x_2, y_2, z_2)$

Write the ■ formula. midpoint

$$\text{midpoint} = \left(\frac{x_1 + x_2}{2}, \frac{y_1 + y_2}{2}, \frac{z_1 + z_2}{2}\right)$$

Substitute the given values.

$$= \left(\frac{4 + 2}{■}, \frac{■ + 3}{2}, \frac{■ + ■}{2}\right)$$

2 5 2 0

Simplify.

$$= \left(\frac{6}{2}, \frac{8}{2}, \frac{2}{2}\right) = (■, 4, ■)$$

3 1

The midpoint of $\overline{AB}$ is located at ■. (3, 4, 1)

Practice

Find the coordinates of the midpoint of each line segment.

1. A (2, 4, 0) and B (2, 2, 2) (2, 3, 1)
2. C (7, 3, 7) and D (1, 5, 5) (4, 4, 6)
3. E (0, 1, 0) and F (6, 1, 10) (3, 1, 5)
4. G (3, 5, 8) and H (1, 7, 6) (2, 6, 7)
5. L (0, $^-4$, $^-2$) and N ($^-6$, 2, 8) ($^-3$, $^-1$, 3)
6. M (3, 0, 7) and P ($^-1$, $^-6$, $^-7$) (1, $^-3$, 0)
7. R (2, 4, $^-10$) and S ($^-8$, $^-2$, 2) ($^-3$, 1, $^-4$)
8. U (6, $^-2$, 8) and V (0, 0, 0) (3, $^-1$, 4)

9. Let (x_1, y_1, z_1) represent one point and (x_2, y_2, z_2) the other. Write the midpoint formula and substitute the given values. Then simplify.

Share Your Understanding

9. Explain to a partner how to find the midpoint of the line segment in one of the exercises above. See above.

10. **CRITICAL THINKING** Write the coordinates of a point in space. Have a partner write coordinates for another point in space. Work together to find the midpoint of the line segment that connects your points. Check students' work. Students should demonstrate understanding of the midpoint formula for points in space.

12·13 Calculator: Magnitude of a Vector

Getting Started
Give students two legs of a right triangle, 3 and 4. Call on a volunteer to show how to find the length of the hypotenuse.

A **vector** shows both the distance and direction between two points. Vector AB is written $\overrightarrow{AB}$. The length of a vector is the **magnitude**. The magnitude is also the distance between two points. The magnitude of vector AB is written $|\overrightarrow{AB}|$.

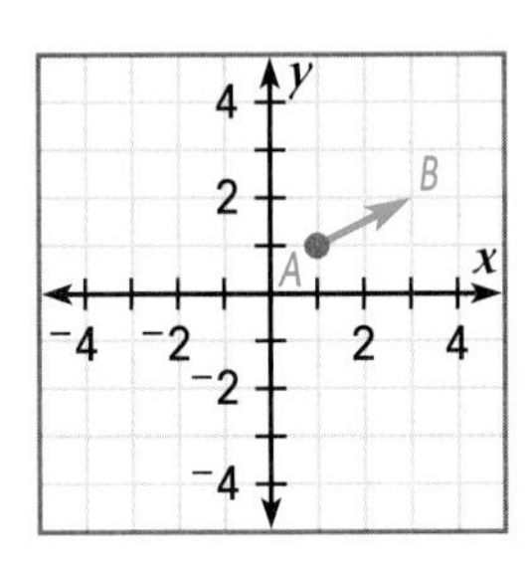

You can use the distance formula to find the magnitude of a vector.

$$|\overrightarrow{AB}| = \sqrt{(x_2 - x_1)^2 + (y_2 - y_1)^2}$$

EXAMPLE

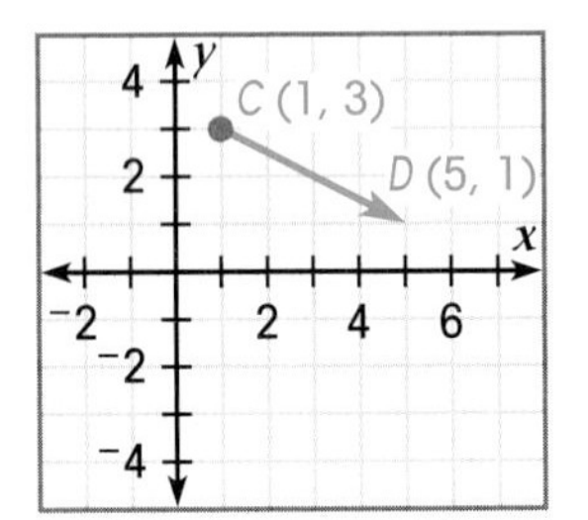

Find the magnitude of $\overrightarrow{CD}$.

Let (x_1, y_1) represent one point and (x_2, y_2) the other.

$C\ (1, 3) = (x_1, y_1)$
$D\ (5, 1) = (x_2, y_2)$

Now, use your calculator.

	Display
PRESS: 5 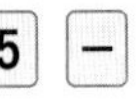 1 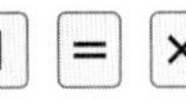 × 4 =	16
Clear the display. PRESS: 1 − 3 = × 2 +/− =	4
Do not clear the display. PRESS: + 1 6 =	20
Do not clear the display. PRESS: √	4.47213

Round the answer to the nearest tenth. $4.47213 \approx 4.5$

The magnitude of $\overrightarrow{CD}$, where $C = (1, 3)$ and $D = (5,1)$, is about 4.5.

Practice

Find the magnitude of each vector. Use your calculator. Round the answer to the nearest tenth if necessary.

1. 10

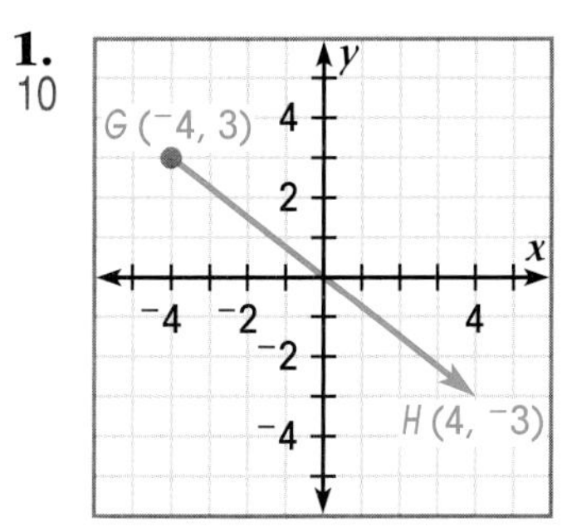

2. 5

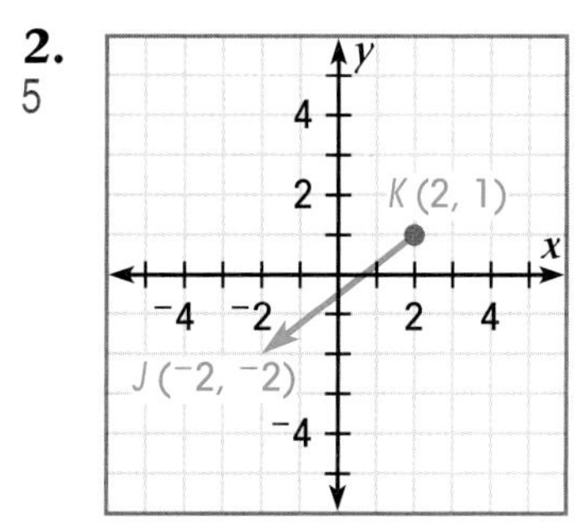

3. 5.4

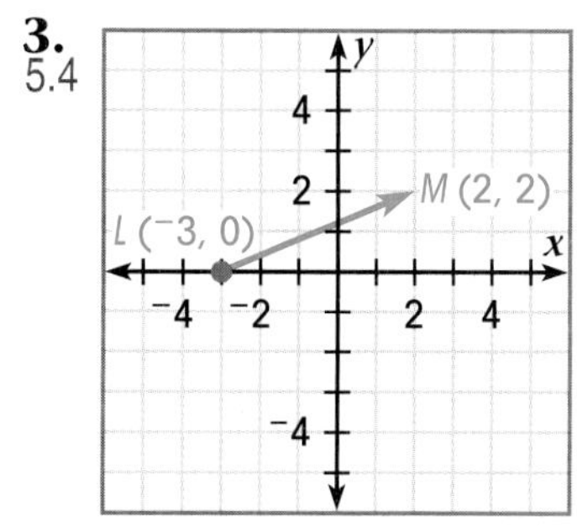

4. 5.7

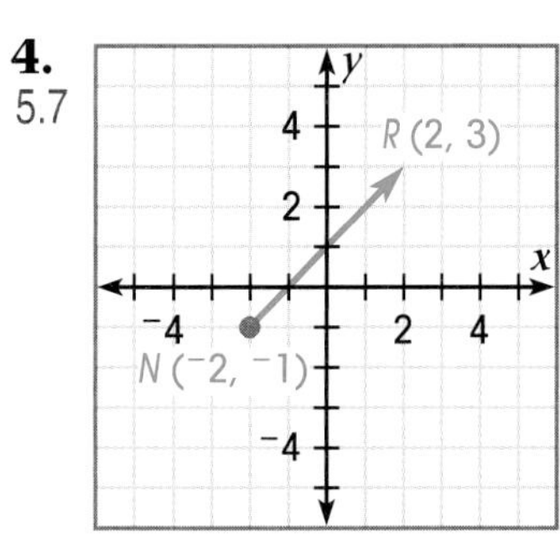

5. 8.5

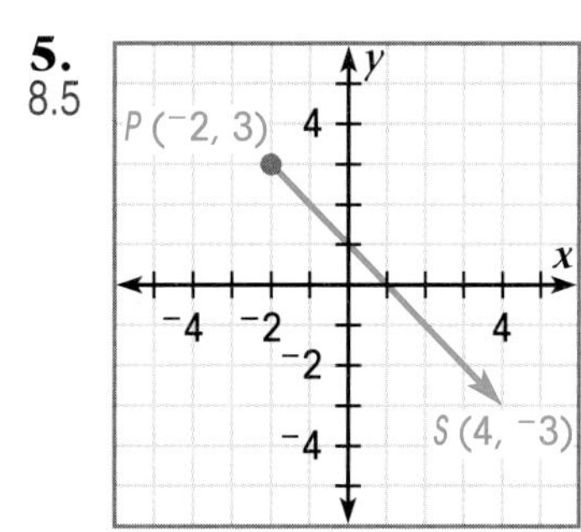

6. 7.2

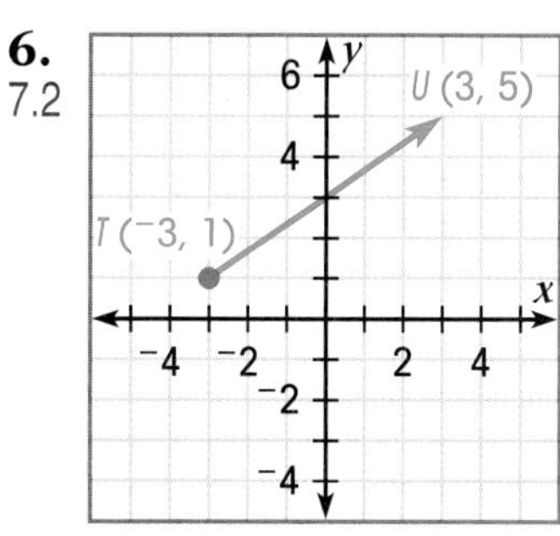

Math In Your Life

NAVIGATION

Coordinate geometry is used to navigate ships and planes from place to place around the earth. The position of a ship or a plane can be given by coordinates of longitude and latitude. These are lines drawn on a globe of the earth.

Lines of longitude are drawn north and south. Lines of latitude are drawn east and west. A ship at 30° W, 15° N is near the Cape Verde Islands.

More practice is provided in the Workbook, page 85, and in the Classroom Resource Binder, page 136.

Problem-Solving Skill: Find the Resultant Vector

Getting Started
Show students a coordinate grid. Ask volunteers to show which way along the grid is north, south, east, and west.

You can add vectors. To add $\overrightarrow{JK}$ and $\overrightarrow{JL}$, draw a parallelogram with the vectors as sides. The sum of $\overrightarrow{JK}$ and $\overrightarrow{JL}$ is the diagonal, $\overrightarrow{JM}$. $\overrightarrow{JM}$ is called the **resultant vector**. The resultant vector is the sum of two vectors.

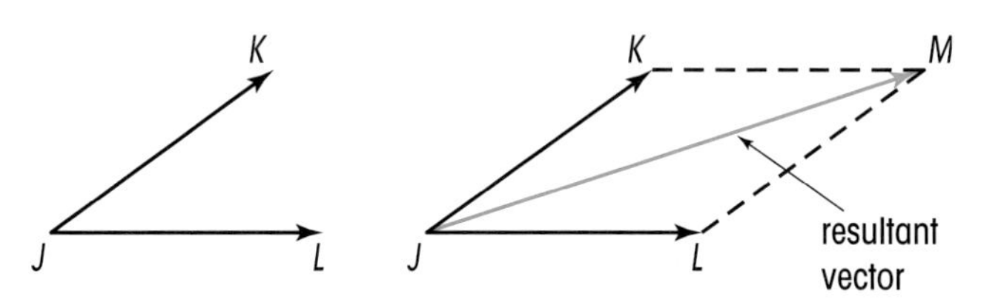

Sometimes the parallelogram formed by the two vectors is a rectangle. Then, the triangle formed by the resultant vector and the sides of the rectangle is a right triangle. You can use what you know about vectors and right triangles to solve problems.

EXAMPLE

A model plane can fly 12 miles per hour in still air. Zian wants to fly his model plane east. There is a wind blowing north at 5 miles per hour. In what direction will the plane actually fly? How fast will it travel?

Draw $\overrightarrow{PQ}$ to show the motion of the model plane. Draw $\overrightarrow{PR}$ to show the motion of the wind. Because the directions are east and north, the vectors form right angles. Draw rectangle *PRSQ*. Draw $\overrightarrow{PS}$ to show the distance and direction the plane will actually fly.

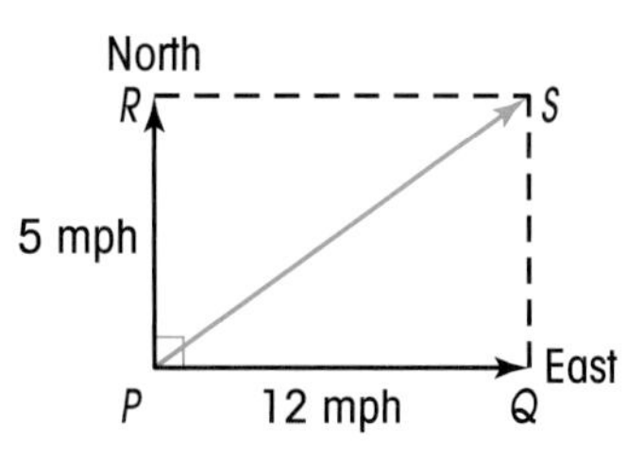

Remember
Opposite sides of a parallelogram are congruent.

Use the Pythagorean Theorem to find the magnitude of $\overrightarrow{PS}$. Substitute the magnitude of $\overrightarrow{PQ}$ for *a*. Substitute the magnitude of $\overrightarrow{PR}$ for *b*.

$$a^2 + b^2 = c^2$$
$$12^2 + 5^2 =$$
$$13 = c$$
$$13 = |\overrightarrow{PS}|$$

The plane flies 13 miles per hour in a northeast direction.

Try This

Azi wants to travel west across Cedar River in his canoe. He can paddle 8 miles per hour in still water. The river flows south at 6 miles per hour. In what direction will the canoe actually travel? How fast will it travel?

Draw vectors to show the motion of the canoe and the river. Draw a rectangle.

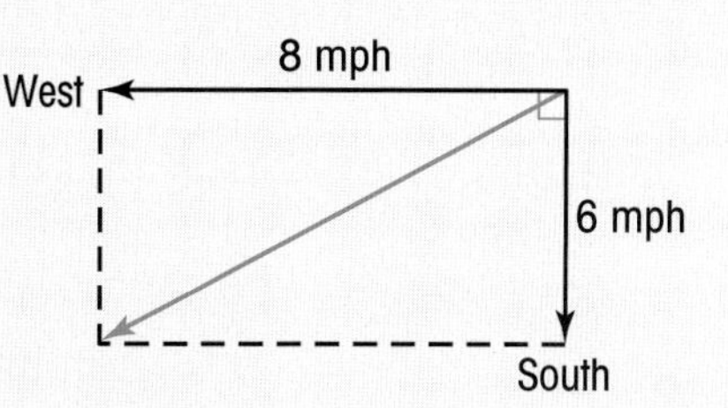

Use the Pythagorean Theorem to find the magnitude of the resultant vector. $a^2 + b^2 = c^2$

Substitute 6 for *a* and 8 for *b*. $6^2 + \blacksquare = c^2$ (8^2)

Simplify. $\blacksquare = c^2$ (100)

$\blacksquare = c$ (10)

Azi travels ■ (10) miles per hour in a ■ (southwest) direction.

Practice

Solve each problem. Draw a diagram if you need to.

1. A ferry heads east across a river. It travels 20 miles per hour in still water. The river flows south at 15 miles per hour. In what direction will the ferry actually travel? How fast will it travel?
southeast; 25 mph

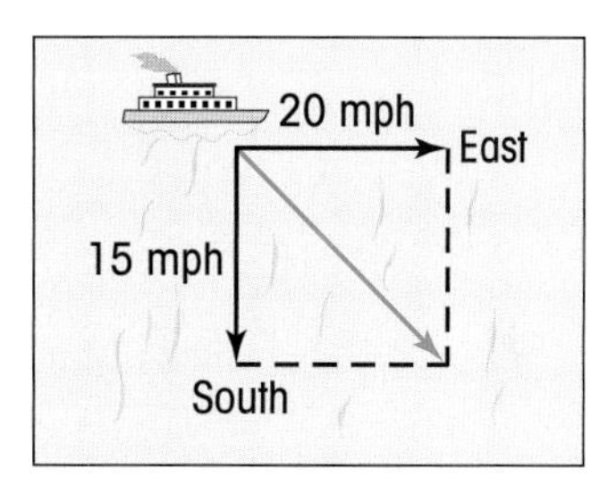

2. An airplane heads north. Its cruising speed is 550 miles per hour in still air. The wind is blowing west at 100 miles per hour. In what direction will the plane actually fly? How fast will it travel? northwest; about 559 mph

3. Carla is canoeing east across a river. She can paddle 6 miles per hour in still water. The river flows north at 4 miles per hour. In what direction will the canoe actually travel? How fast will it travel?
northeast; about 7 mph

More practice is provided in the Workbook, page 86, and in the Classroom Resource Binder, page 137.

Getting Started Review special right triangles. Remind students about the relationships between the sides and the hypotenuse.

12·15 Problem-Solving Application: The Effect of Two Forces

Sometimes you can use what you learned about special right triangles to find the magnitude of the resultant vector.

EXAMPLE

The speed of a model plane in still air is $5\sqrt{3}$ miles per hour. Mari wants to fly the plane east. The wind is blowing north at 5 miles per hour. In what direction will the plane actually fly? How fast will it travel?

Draw $\overrightarrow{AB}$ to show the motion of the model plane. Draw $\overrightarrow{AC}$ to show the motion of the wind. Because the directions are east and north, the vectors form right angles. Draw rectangle *ACDB*. Draw $\overrightarrow{AD}$ to show the direction the plane will actually fly.

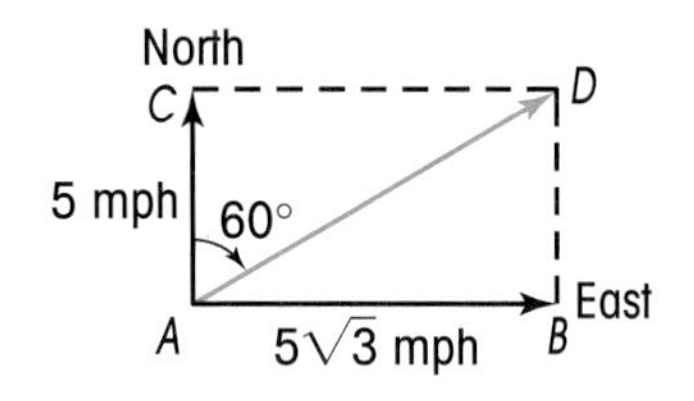

The vectors are the legs of two right triangles. The legs of the triangles are 5 and $5\sqrt{3}$. These are 30°-60°-90° triangles. The hypotenuse of each triangle is the resultant vector.

Remember
The length of the hypotenuse of a 30°-60°-90° triangle is twice the length of the short side.

Use the formula for hypotenuse of a 30°-60°-90° triangle to find the resultant vector. $\text{hypotenuse} = 2 \cdot \text{short leg}$

Substitute the given values. $|\overrightarrow{AD}| = 2 \cdot 5$

$= 10$

The plane flies 10 miles per hour at a direction of 60° east of north.

Try This

The speed of a hot-air balloon in still air is 4 miles per hour. Yuha wants to fly the balloon east. There is a wind blowing north at 4 miles per hour. In what direction will the balloon actually fly? How fast will it travel?

Draw vectors to show the motion of the balloon and the wind. Draw square *GJKH*. Draw *GK* to show the direction the balloon will actually fly.

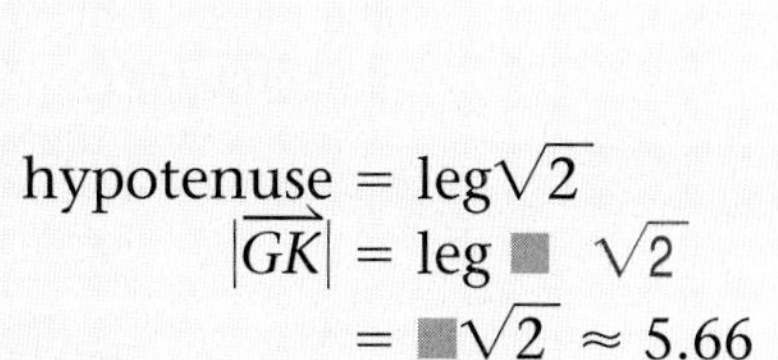

The vectors are the legs of two right triangles. The legs of the triangle are equal. This is a 45°-45°-90° triangle. The hypotenuse of each triangle is the ■ vector. resultant

Use the formula for hypotenuse of a 45°-45°-90° triangle to find the resultant vector.

$$\text{hypotenuse} = \text{leg}\sqrt{2}$$

$$|\overrightarrow{GK}| = \text{leg} \blacksquare \sqrt{2}$$

Substitute the given value.

$$= \blacksquare\sqrt{2} \approx 5.66$$

4

The hot-air balloon flies about ■ miles per hour in a ■ direction. 6; northeast

Practice

Solve each problem. Draw a diagram if you need to.

1. The speed of a boat in still water is $8\sqrt{3}$ miles per hour. The current in the river flows south at 8 miles per hour. The boat heads west across the river. In what direction will the boat travel? How fast will it travel?
southwest; 16 mph

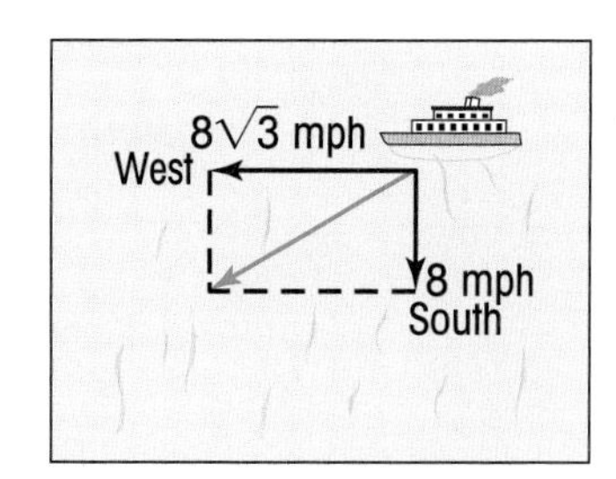

2. The speed of a small plane in still air is 80 miles per hour. Julius flies the plane heading east. The wind blows north at 80 miles per hour. In what direction will the plane travel? How fast will it travel? northeast; about 113 mph

An alternate two-column proof lesson is provided on page 417 of the student book.

12-16 Proof: Coordinate Geometry

Getting Started
Review the formulas for slope and distance.

You learned that the slopes of perpendicular lines are negative reciprocals. You can use this fact to prove that the diagonals of a square are perpendicular. First, you need to draw a square on the coordinate plane.

EXAMPLE

You are given: ▱*ABCD* is a square.

Prove: The diagonals of a square are perpendicular.

Draw a square on a coordinate plane. Place one vertex at (0, 0). You know that the sides are the same length. Let the length of each side be *s*. Then, write the coordinates of each vertex.

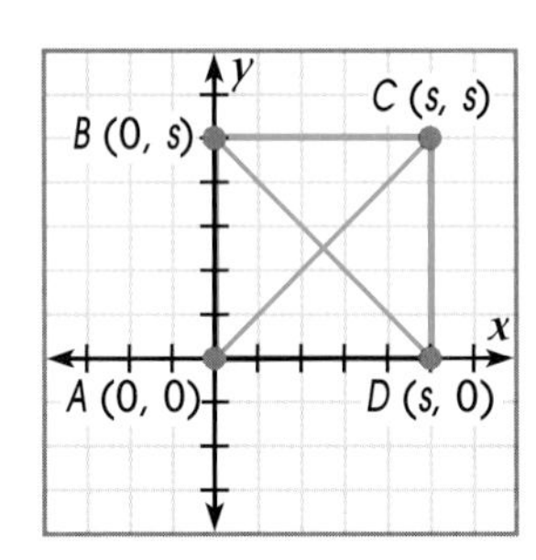

You know that the slopes of perpendicular lines are negative reciprocals. Find the slopes of $\overline{AC}$ and $\overline{BD}$.

$$\text{slope} = \frac{y_2 - y_1}{x_2 - x_1} \qquad \text{slope} = \frac{y_2 - y_1}{x_2 - x_1}$$

Let $A(0, 0) = (x_1, y_1)$ and $C(s, s) = (x_2, y_2)$ Let $D(s, 0) = (x_1, y_1)$ and $B(0, s) = (x_2, y_2)$

$$\text{slope of } \overline{AC} = \frac{s - 0}{s - 0} = \frac{s}{s} = 1 \qquad \text{slope of } \overline{BD} = \frac{0 - s}{s - 0} = \frac{^-s}{s} = {}^-1$$

The numbers 1 and $^-1$ are negative reciprocals. Then, $\overline{AC} \perp \overline{BD}$.

So, you prove that the diagonals of a square are perpendicular. ✓

Try This

Copy and complete the proof.

You are given: ▱*EFGH* is a rectangle.

Prove: The diagonals of a rectangle are congruent.

Draw a rectangle on a coordinate plane. Place one vertex at ■. You know that the opposite 0, 0 sides are the same length. Let the width be w and the length be l. Then, write the coordinates of each vertex.

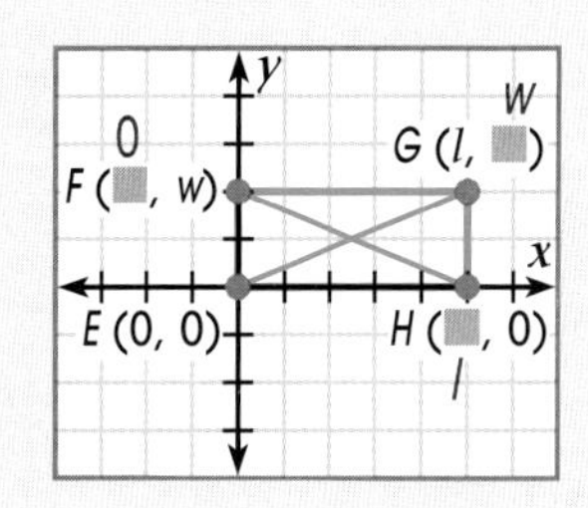

You can use the distance formula to find the length of $\overline{EG}$ and $\overline{FH}$.

$$d = \sqrt{(x_2 - x_1)^2 + (y_2 - y_1)^2} \qquad d = \sqrt{(x_2 - x_1)^2 + (y_2 - y_1)^2}$$

$$EG = \sqrt{(l - 0)^2 + (w - 0)^2} \qquad FH = \sqrt{(0 - l)^2 + (w - 0)^2}$$

$$= \sqrt{l^2 + (■)^2} \; w \qquad = \sqrt{(^-l)^2 + (■)^2} \; w$$

$$= \sqrt{l^2 + w^2} \qquad = \sqrt{l^2 + w^2}$$

$\overline{EG}$ and ■ have the same length. Then, $\overline{EG} \cong$ ■. $\overline{FH}$
$\overline{FH}$

So, you prove that the diagonals of a rectangle are congruent. ✓

Practice See Additional Answers in the back of this book.

Write a proof for each of the following.

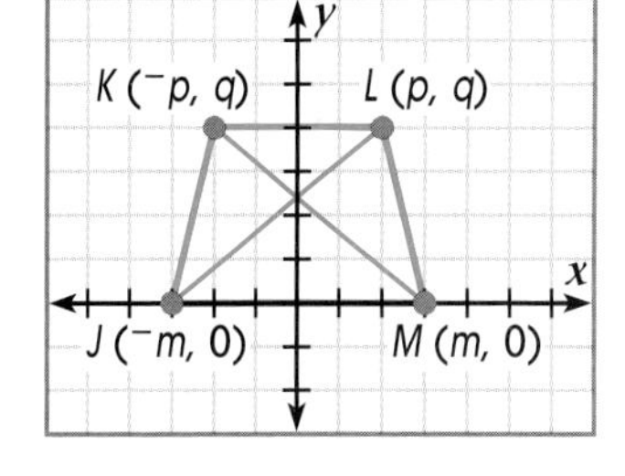

1. You are given:
isosceles trapezoid *JKLM*
$\overline{JK} \cong \overline{ML}$

Prove: $\overline{JL} \cong \overline{MK}$

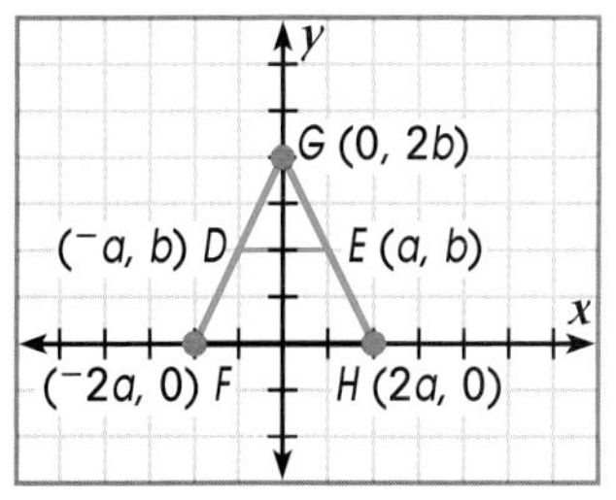

2. You are given:
$\triangle FGH$
$\overline{DE}$ is the midsegment.

Prove: $DE = \frac{1}{2}FH$

More Practice is provided in the Classroom Resource Binder.

Chapter 12 Review

Summary

You can graph points on a coordinate plane.
Use the formula $(\frac{x_1 + x_2}{2}, \frac{y_1 + y_2}{2})$ to find the midpoint of a line segment.
Use the formula $\frac{y_2 - y_1}{x_2 - x_1}$ to find slope and identify parallel lines and perpendicular lines.
Translations, reflections, rotations, and dilations are all transformations in the plane.
You can graph points in space. Use the x-, y-, and z-axes.
The distance between two points in space is $d = \sqrt{(x_2 - x_1)^2 + (y_2 - y_1)^2 + (z_2 - z_1)^2}$
A calculator can be used to find the magnitude of a vector.
You can use vectors to solve problems.
You can use coordinate planes to prove statements.

More vocabulary review is provided in the Classroom Resource Binder.

image
magnitude
resultant vector
slope
transformation
vector
coordinate plane
preimage
center of rotation

Vocabulary Review

Complete the sentences with the words from the box.

1. A change in the position or size of a figure is called a ____. transformation
2. The ____ is a figure resulting from a transformation. image
3. The ____ of a line is the ratio of vertical change to horizontal change. slope
4. The length of a vector is called its ____. magnitude
5. A grid with two axes where points are graphed is called a ____. coordinate plane
6. The sum of two vectors is called the ____. resultant vector
7. A quantity that has size and direction is called a ____. vector
8. The ____ is the original figure before a transformation. preimage
9. The ____ is the point about which a figure is turned. center of rotation

Chapter Quiz

More assessment is provided in the Classroom Resource Binder.

Draw a coordinate plane on grid paper. Graph each point.

1. $A\ (3, {}^{-}4)$
   See Additional Answers in the back of this book.
2. $B\ ({}^{-}5, 2)$

Use formulas for exercises 3–5.

3. Find the distance between $C\ (1, 5)$ and $D\ ({}^{-}2, 1)$. 5
4. $E\ (6, 0)$ and $F\ (4, {}^{-}2)$ are endpoints of $\overline{EF}$. Find the midpoint. $(5, {}^{-}1)$
5. $G\ ({}^{-}2, 6)$ and $H\ (1, 0)$ are points on a line. Find the slope. ${}^{-}2$

Decide if the lines are parallel or perpendicular.

6. Line 1 points: $(0, {}^{-}1)$ and $(2, 0)$
   Line 2 points: $({}^{-}2, 3)$ and $(1, {}^{-}3)$
   Perpendicular
7. Line 1 points: $({}^{-}3, {}^{-}2)$ and $(2, 3)$
   Line 2 points: $({}^{-}2, {}^{-}3)$ and $(2, 1)$
   Parallel

Use grid paper. Draw the image of $\triangle ABC$ for each transformation.

8–12. See Additional Answers in the back of this book.

8. $T\ (x, y) \longrightarrow (x - 8, y - 1)$
9. Rotate $\triangle ABC$ 90° clockwise about the origin.
10. Reflect $\triangle ABC$ over the x-axis.
11. Dilate $\triangle ABC$ with a similarity ratio of $\frac{1}{2}$.

Use what you know about points in space for exercises 12–14.

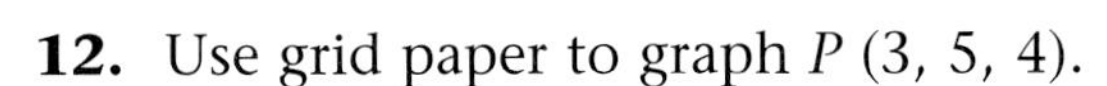

12. Use grid paper to graph $P\ (3, 5, 4)$.
13. Find the distance between $M\ (1, 0, 1)$ and $N\ (6, 4, 4)$. $5\sqrt{2}$
14. $Q\ (2, 1, 5)$ and $R\ (4, 3, 1)$ are endpoints of $\overline{QR}$. Find the midpoint.
    $(3, 2, 3)$

Write About Math No. The transformation image would be the result of a translation horizontally to the left.

Problem Solving

Solve. Show your work.

15. The speed of a boat in still water is 20 mph. The water current flows south at 12 mph. The boat heads east across the river. In what direction will the boat travel? How fast will the boat travel?
    Southeast; $4\sqrt{34}$ mph, or about 23 mph

> **Write About Math**
> Would a reflection of $\triangle ABC$ over the y-axis in exercise 8 be the same as $T\ (x, y) \longrightarrow (x - 8, y)$? Explain.

Additional Practice for this chapter is provided on p. 404 of this text.

Chapter 13 Right Triangle Trigonometry

Waves rolling across the ocean imitate a math function called a sine curve. Where are some other places you can find waves?

Learning Objectives

- Write trigonometric ratios.
- Read the Table of Trigonometric Ratios.
- Use sine, cosine, and tangent ratios to find measures of angles and lengths of sides in a right triangle.
- Use a calculator to evaluate trigonometric ratios.
- Use trigonometric ratios to solve problems.
- Apply concepts and skills to real-life problems involving right triangles.
- Prove trigonometric identities.

ESL/ELL Note Point out that the word *tangent* has two very different meanings in mathematics. A tangent can be a line that touches a circle. A tangent can also be a ratio.

Words to Know

trigonometry the study of the relationships between the sides and angles of a right triangle

trigonometric ratio the ratio of the length of two sides of a right triangle

sine the ratio of the length of the side opposite an angle to the length of the hypotenuse in a right triangle

cosine the ratio of the length of the side adjacent to an angle to the length of the hypotenuse in a right triangle

tangent the ratio of the length of the side opposite an angle to the length of the adjacent side in a right triangle

clinometer a tool used to measure angles

angle of elevation the angle formed by a horizontal line and the line of sight when looking up

angle of depression the angle formed by a horizontal line and the line of sight when looking down

trigonometric identity a trigonometric equation that is always true

Clinometer Project

Make a clinometer like the one on page 381. Use it to help you find the height of your school building. Go outside and stand 20 feet from the school building. At this distance, use the clinometer to find the angle of elevation from your eye to the top of the school.

Use the tangent ratio to find the height of the building from your eye level to the top of the building. Finally, add your height to your answer to find the total height of the school.

Project Students can begin work on the project after completing Lesson 13.7. See the Classroom Resource Binder for a scoring rubric to assess this project.

More practice is provided in the Workbook, page 87, and in the Classroom Resource Binder, page 142.

Trigonometric Ratios

Getting Started
Have students draw three similar 30°-60°-90° right triangles. For each triangle, have them find the ratio of the short leg to the hypotenuse and discuss the results. [$\frac{1}{2}$]

Trigonometry is the study of the relationship of the sides and angles of a right triangle.

A **trigonometric ratio** is a ratio of the lengths of two sides of a right triangle. There are trigonometric ratios for each acute angle. Three of the ratios are called **sine**, **cosine**, and **tangent**.

Trigonometric ratios of $\triangle BCA$ are shown below.

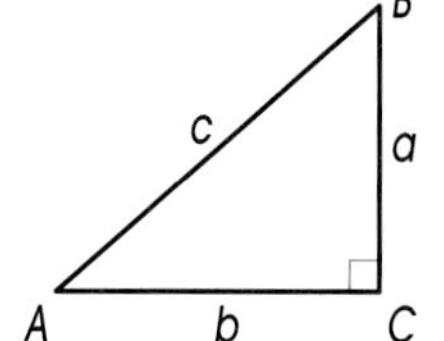

sine $\angle A = \dfrac{\text{length of side opposite } \angle A}{\text{length of hypotenuse}} = \dfrac{a}{c}$	**sine** $\angle B = \dfrac{\text{length of side opposite } \angle B}{\text{length of hypotenuse}} = \dfrac{b}{c}$
cosine $\angle A = \dfrac{\text{length of side adjacent to } \angle A}{\text{length of hypotenuse}} = \dfrac{b}{c}$	**cosine** $\angle B = \dfrac{\text{length of side adjacent to } \angle B}{\text{length of hypotenuse}} = \dfrac{a}{c}$
tangent $\angle A = \dfrac{\text{length of side opposite } \angle A}{\text{length of side adjacent to } \angle A} = \dfrac{a}{b}$	**tangent** $\angle B = \dfrac{\text{length of side opposite } \angle B}{\text{length of side adjacent to } \angle B} = \dfrac{b}{a}$

If you know the lengths of the sides, you can write the trigonometric ratios of the angles.

EXAMPLE

Find three trigonometric ratios of $\angle W$.

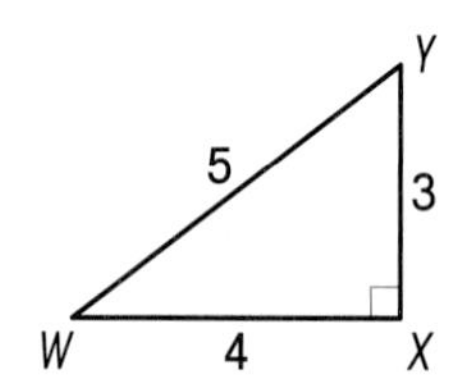

Write the sine ratio of $\angle W$.
You can write *sin* for sine.
$$\sin W = \frac{\text{opposite}}{\text{hypotenuse}} = \frac{3}{5}$$

Write the cosine ratio of $\angle W$.
You can write *cos* for cosine.
$$\cos W = \frac{\text{adjacent}}{\text{hypotenuse}} = \frac{4}{5}$$

Write the tangent ratio of $\angle W$.
You can write *tan* for tangent.
$$\tan W = \frac{\text{opposite}}{\text{adjacent}} = \frac{3}{4}$$

Three trigonometric ratios of $\angle W$ are $\frac{3}{5}$, $\frac{4}{5}$, and $\frac{3}{4}$.

Try This

Find three trigonometric ratios of $\angle K$.

K, 5, 13, L, 12, M

Write the sine ratio of $\angle K$. $\sin K = \frac{\text{opposite}}{■} = \frac{■}{13}$ (hypotenuse; 12)

Write the cosine ratio of $\angle K$. $\cos K = \frac{■}{\text{hypotenuse}} = \frac{5}{■}$ (adjacent; 13)

Write the tangent ratio of $\angle K$. $\tan K = \frac{■}{■} = \frac{12}{5}$ (opposite; adjacent)

Three trigonometric ratios of $\angle K$ are $\frac{12}{13}$, ■, and ■. ($\frac{5}{13}$; $\frac{12}{5}$)

1. $\sin D = \frac{8}{17}$; $\cos D = \frac{15}{17}$; $\tan D = \frac{8}{15}$ 2. $\sin D = \frac{1}{\sqrt{2}}$ or $\frac{2}{\sqrt{2}}$; $\cos D = \frac{1}{\sqrt{2}}$ or $\frac{2}{\sqrt{2}}$; $\tan D = 1$

Practice

3. $\sin D = \frac{\sqrt{3}}{2}$; $\cos D = \frac{1}{2}$
$\tan D = \sqrt{3}$

Find three trigonometric ratios of $\angle D$ for each triangle.

1.

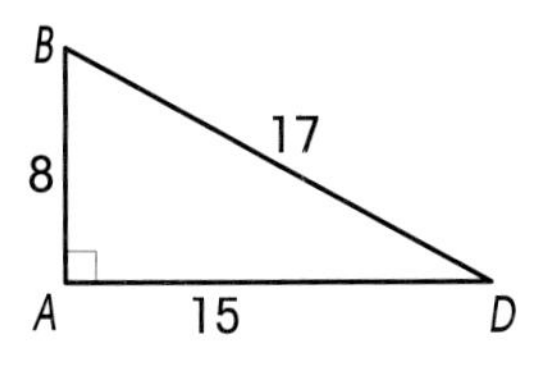

2.

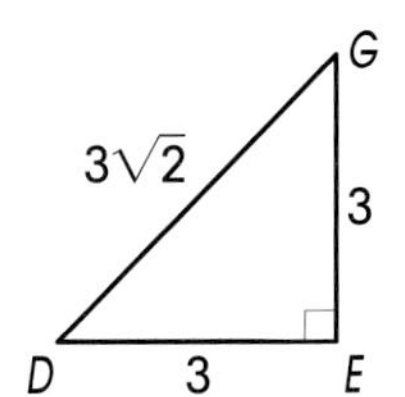

3.

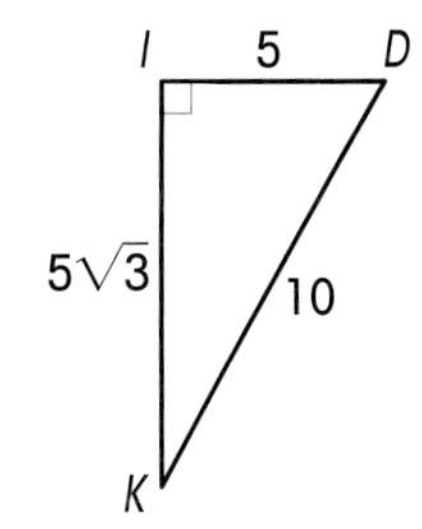

4. To find sine, write the length of the side opposite the angle over the length of the hypotenuse. To find cosine, write the length of the side adjacent to the angle over the length of the hypotenuse. To find tangent, write the length of the side opposite the angle over the length of the adjacent side.

Share Your Understanding

4. Explain to a partner how to find three trigonometric ratios. Use the words *adjacent, opposite,* and *hypotenuse*. See above.

5. **CRITICAL THINKING** Draw three similar 45°-45°-90° right triangles. Make one with legs of 2 cm, a second with legs of 3 cm, and a third with legs of 4 cm. Find the tangent of each acute angle. Will the tangent of a 45° angle always be 1? Why or why not? Yes, a 45°-45°-90° right triangle is also isosceles. So, the opposite side will always be equal to the adjacent side. When the numerator and the denominator are equal, the fraction ratio equals one.

More practice is provided in the Workbook, page 87, and in the Classroom Resource Binder, page 143.

Table of Trigonometric Ratios

Getting Started
Review changing a fraction to a decimal. Have students write $\frac{1}{2}$ as a decimal. [0.5]

If you know the trigonometric ratio of an angle in a right triangle, you can find the angle measure. You can look it up in the Table of Trigonometric Ratios. The ratios in the table are in decimal form. Part of the table is shown below. The entire table is on page 426.

Remember
You can change a ratio into decimal form.
$\frac{3}{4} = 3 \div 4 = 0.75$

Table of Trigonometric Ratios

Angle	Sine	Cosine	Tangent
0°	0.0000	1.0000	0.0000
1°	0.0175	0.9998	0.0175
2°	0.0349	0.9994	0.0349
3°	0.0523	0.9986	0.0524

Here are three trigonometric ratios of a 2° angle.
$\sin 2° = 0.0349$, $\cos 2° = 0.9994$, and $\tan 2° = 0.0349$

EXAMPLE

Write three trigonometric ratios of a 54° angle. Use the Table of Trigonometric Ratios on page 426.

First, find 54° in the table. Look across to the column labeled *sine*. Write the value. $\sin 54° = 0.8090$

Next, look across to the column labeled *cosine*. Write the value. $\cos 54° = 0.5878$

Now, look across to the column labeled *tangent*. Write the value. $\tan 54° = 1.3764$

Three trigonometric ratios of a 54° angle are
$\sin 54° = 0.8090$; $\cos 54° = 0.5878$; $\tan 54° = 1.3764$.

You can find the measure of an angle when you know one of the trigonometric ratios. Sometimes the measure is an approximation because the exact decimal for the ratio may not be in the table.

Try This

Cosine $\angle A = 0.4375$. Find the measure of $\angle A$ to the nearest degree.

Table of Trigonometric Ratios			
Angle	Sine	Cosine	Tangent
63°	0.8910	0.4540	1.9626
64°	0.8988	0.4384	2.0503
65°	0.9063	0.4226	2.1445

Look at the *cosine* column in the Table of Trigonometric ■ on page 426. Ratios
Look for 0.4375 in that column.

Because 0.4375 is not in that column, find the closest decimal. $0.4384 \approx 0.4375$

Identify the angle for this decimal. $\cos 64° = 0.4384$

So, $\angle A$ is about ■. 64°

1. sin 88° = 0.9994 cos 88° = 0.0349 tan 88° = 28.6363
2. sin 23° = 0.3907 cos 23° = 0.9205 tan 23° = 0.4245
3. sin 67° = 0.9205 cos 67° = 0.3907 tan 67° = 2.3559
4. sin 49° = 0.7547 cos 49° = 0.6561 tan 49° = 1.1504
5. sin 16° = 0.2756 cos 16° = 0.9613 tan 16° = 0.2867

Practice

Write three trigonometric ratios for each angle. Use the Table of Trigonometric Ratios on page 426.

1. 88° **2.** 23° **3.** 67° **4.** 49° **5.** 16°

Find the measure of each angle to the nearest degree. Use the table on page 426.

6. $\cos B = 0.0955$ 85°

7. $\tan C = 10.0999$ 84°

8. $\sin D = 0.4856$ 29°

9. $\tan E = 1.8440$ 62°

10. $\sin F = 0.8907$ 63°

11. $\cos G = 0.6998$ 46°

12. $\sin H = 0.0995$ 6°

13. $\cos I = 0.6323$ 51°

14. $\tan J = 2.6075$ 69°

Share Your Understanding

15. Explain to a partner how to find the measure of an angle whose ratio is not in the Table of Trigonometric Ratios. Use the words *closest* and *decimal.* Look at the column in the table for the decimal ratio you know. Find the decimal that is closest to the given decimal. Identify the corresponding angle.

16. **CRITICAL THINKING** Sin 20° is equal to 0.3420. What angle has a cosine equal to 0.3420? 70°

More practice is provided in the Workbook, page 88, and in the Classroom Resource Binder, page 144.

13-3 Tangent Ratio

Getting Started
Have students identify the sides opposite and adjacent to each acute angle in a right triangle.

You know that in a right triangle the tangent of an acute angle is a ratio.

$$\tan A = \frac{a}{b} \qquad \tan B = \frac{b}{a}$$

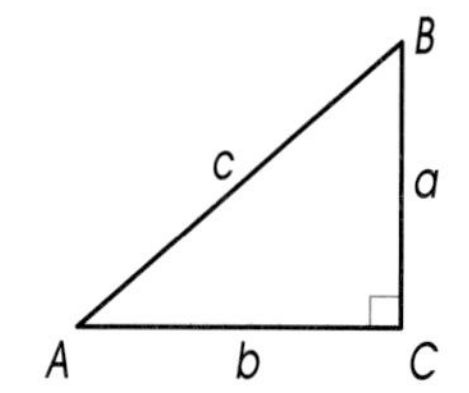

Sometimes you know the lengths of the two legs of a right triangle. You can use the tangent ratio to find the measure of an acute angle.

EXAMPLE 1

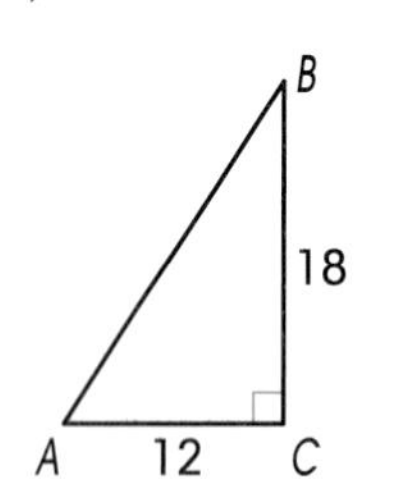

Find the measure of $\angle A$ to the nearest degree.

Write the tangent ratio.	$\tan A = \frac{\text{opposite}}{\text{adjacent}}$
Substitute the given values. Simplify.	$= \frac{18}{12} = 1.5$
Look in the *tangent* column of the table on page 426. Find the closest decimal to 1.5. Identify the angle.	$1.4826 \approx 1.5$ $\tan 56° = 1.4826$

So, $\angle A$ is about 56°.

You can find the unknown length of a leg of a right triangle if you know the measure of one acute angle and the length of the other leg. Use the tangent ratio.

EXAMPLE 2

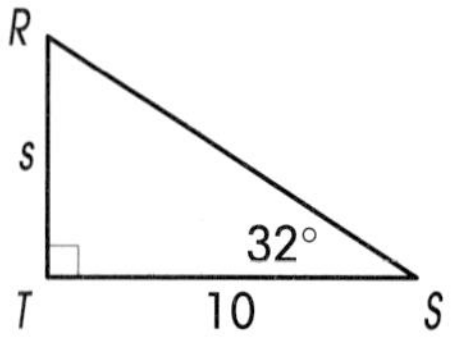

Find the value of s. Round the value to the nearest whole number.

Write the tangent ratio.	$\tan S = \frac{\text{opposite}}{\text{adjacent}}$
Substitute the given values.	$\tan 32° = \frac{s}{10}$
Find 32° in the table on page 426. Look across to the *tangent* column. Substitute.	$0.6249 = \frac{s}{10}$
Solve for s.	$10(0.6249) = s$ $\approx s$

The value of s is about 6.

Try This

Find the value of p. Round the value to the nearest whole number.

Write the tangent ratio.	$\tan K = \frac{\text{opposite}}{■}$ adjacent
Substitute the given values.	$\tan 58° = \frac{■}{p}$ 12
Find 58° in the table on page 426. Look across to the *tangent* column. Substitute.	$1.6003 = \frac{12}{■}$ p
Solve for ■. p	$(1.6003)\,p = ■$ 12
	$p \approx 7$

P, 12, 58°, K, p, M

The value of p is about ■. 7

7. Write the tangent ratio. Substitute the measure of the angle and the length of the given side. Find the tangent of that angle in the table. Substitute the value. Then, simplify to find the unknown length of the side.

Practice

Find the measure of $\angle W$ to the nearest degree. Use the table on page 426.

1. 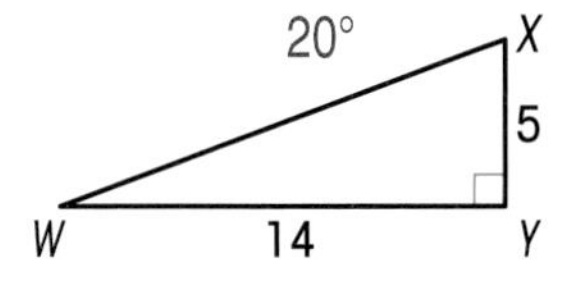

2. 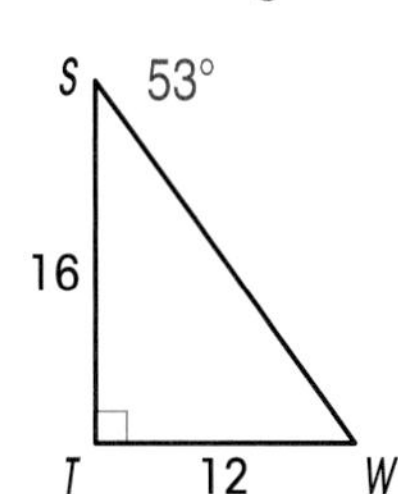

3. 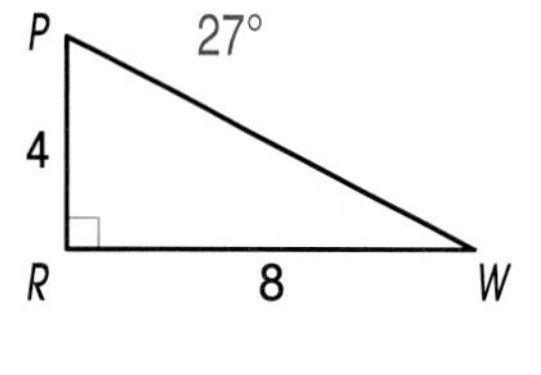

Find the value of a in each diagram. Use the table on page 426. Round the value to the nearest whole number.

4. 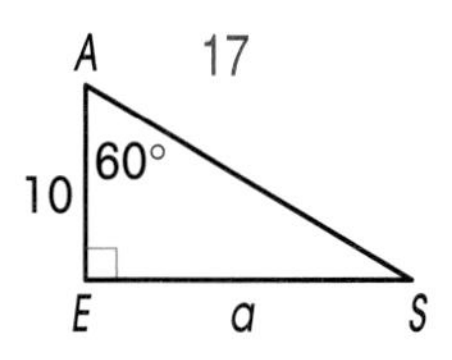

5. 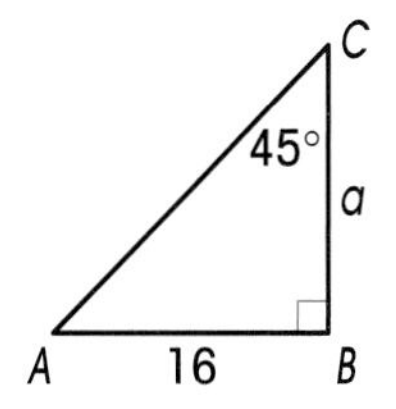

6. 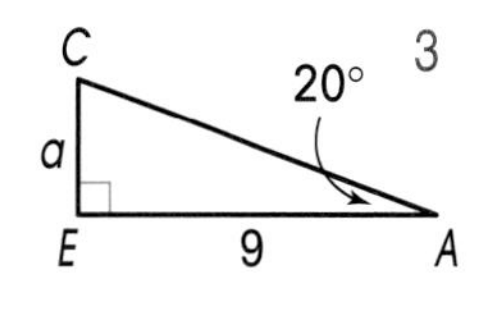

Share Your Understanding

7. Explain to a partner how to use the tangent ratio to find the unknown length of a side in a right triangle. See above.

8. **CRITICAL THINKING** Draw a 30°-60°-90° triangle. The short leg should be 2 cm long. Show that tan 60° equals $\sqrt{3}$.

$\tan 60° = \frac{\text{opposite}}{\text{adjacent}} = \frac{2\sqrt{3}}{2} = \sqrt{3}$

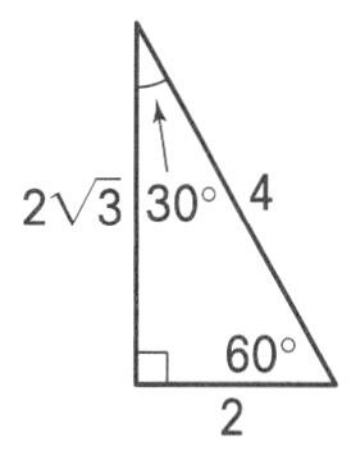

More practice is provided in the Workbook, page 89, and in the Classroom Resource Binder, page 145.

13·4 Sine Ratio

Getting Started
Have students identify the hypotenuse and the side opposite each acute angle in a right triangle.

You know that in a right triangle, the sine of an acute angle is a ratio.

$\sin A = \frac{a}{c} \qquad \sin B = \frac{b}{c}$

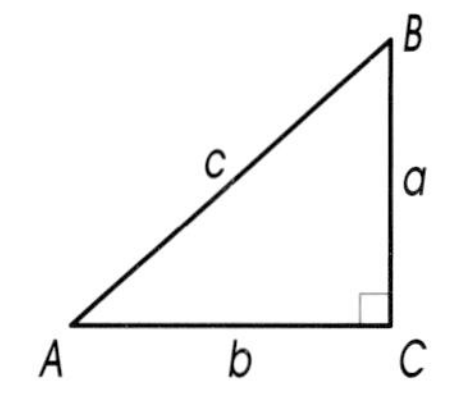

Sometimes you know the lengths of one leg and the hypotenuse of a right triangle. You can use the sine ratio to find the measure of an acute angle.

EXAMPLE 1

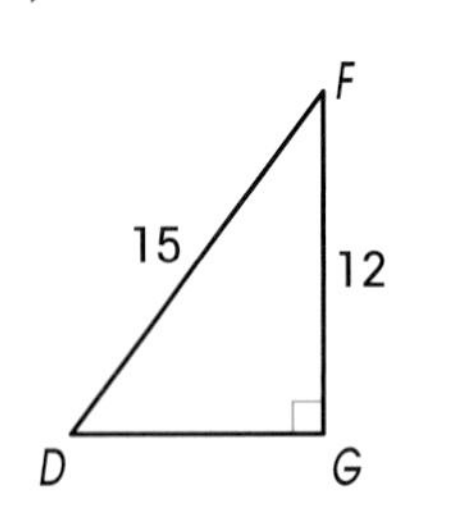

Find the measure of $\angle D$ to the nearest degree.

Write the sine ratio. $\sin D = \frac{\text{opposite}}{\text{hypotenuse}}$

Substitute the given values. Simplify. $= \frac{12}{15} = 0.8$

Look at the *sine* column in the table on page 426. Find the closest decimal to 0.8. Write the measure of the angle.
$0.8 \approx 0.7986$
$\sin 53° = 0.7986$

So, $\angle D$ is about 53°.

You can find the unknown length of a leg of a right triangle if you know the measure of one acute angle and the length of the hypotenuse. Use the sine ratio.

EXAMPLE 2

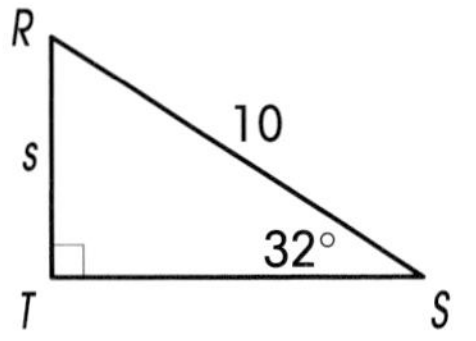

Find the value of s. Round the value to the nearest whole number.

Write the sine ratio. $\sin S = \frac{\text{opposite}}{\text{hypotenuse}}$

Substitute the given values. $\sin 32° = \frac{s}{10}$

Find 32° in the table on page 426. Look across to the *sine* column. Substitute. $0.5299 = \frac{s}{10}$

Solve for s.
$10(0.5299) = s$
$5 \approx s$

The value of s is about 5.

Try This

Find the value of h. Round the value to the nearest whole number.

Write the sine ratio. $\sin G = \frac{\text{opposite}}{■}$ hypotenuse

Substitute the given values. $\sin 48° = \frac{5}{■}$ h

Find 48° in the table on page 426. Look across to the *sine* column. Substitute the value. $0.7431 = \frac{5}{■}$ h

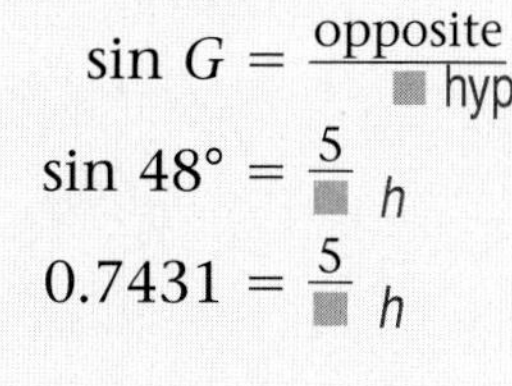

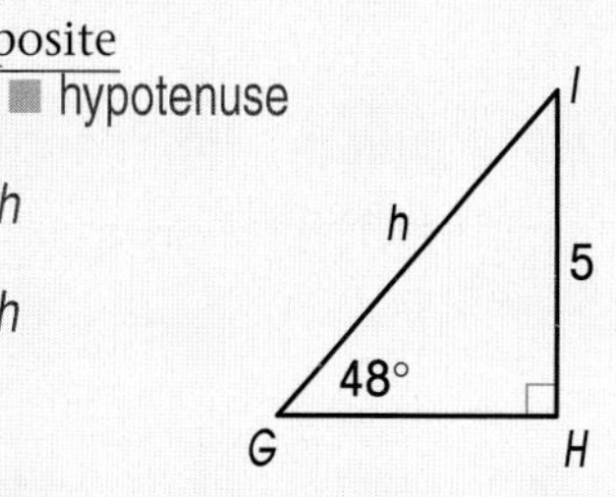

Solve for ■. h $(0.7431)h = 5$

$h \approx$ ■ 7

The value of h is about ■. 7

Practice

Find the measure of $\angle F$ to the nearest degree. Use the table on page 426.

1. 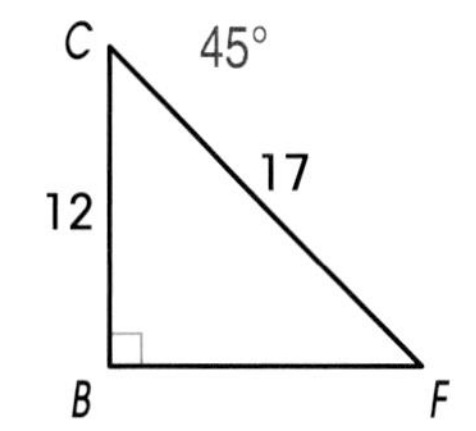

2. 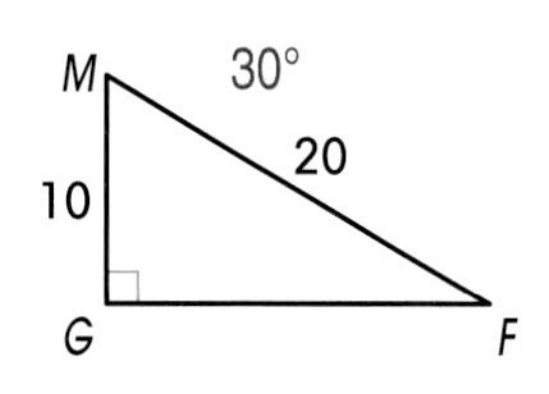

3. 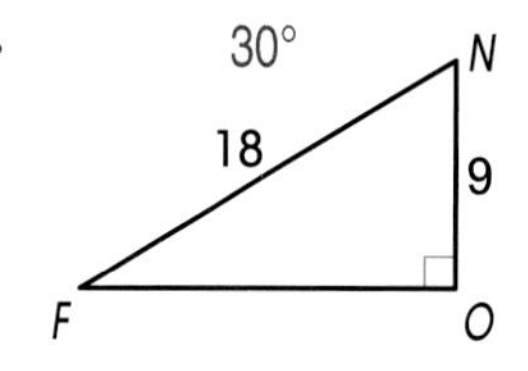

Find the value of t in each diagram. Use the table on page 426. Round the value to the nearest whole number.

4. 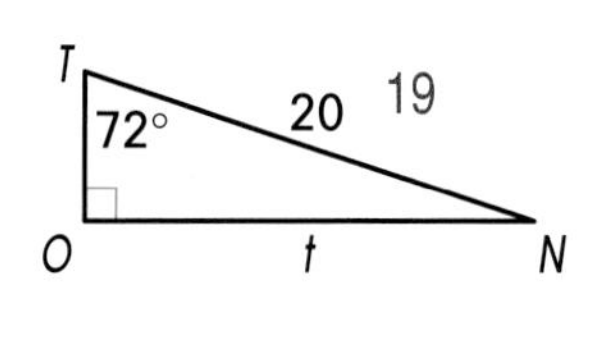

5. 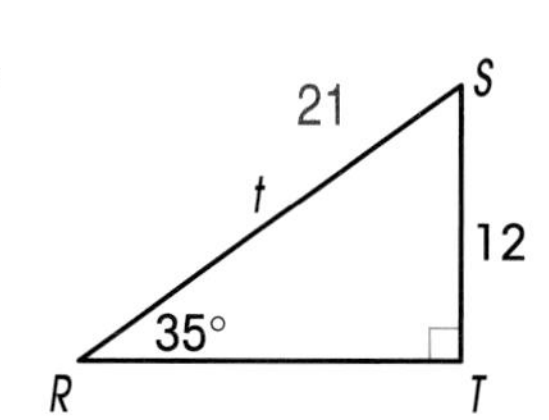

6. 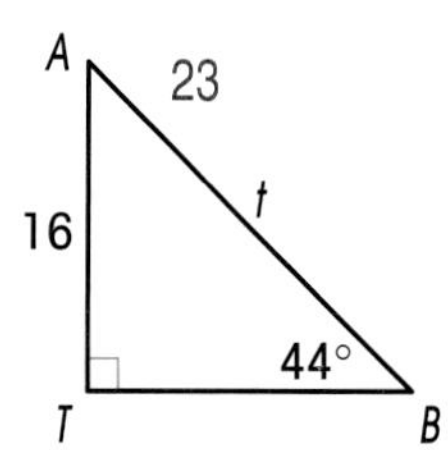

7. Write the formula for sine. Substitute the measure of the angle and the length of the given side. Find the sine of that angle in the table. Substitute the value. Then, simplify to find the unknown length.

Share Your Understanding

7. Explain to a partner how to use the sine ratio to find the unknown length of a side of a right triangle. See above.

8. **CRITICAL THINKING** Why is the sine ratio always less than 1?

8. The sine is the length of the side opposite the given angle divided by the longest side, the hypotenuse. Dividing by a value greater than the dividend gives a quotient less than 1.

More practice is provided in the Workbook, page 90, and in the Classroom Resource Binder, page 145.

13·5 Cosine Ratio

Getting Started
Have students identify the hypotenuse and the side adjacent to each acute angle in a right triangle.

You know that in a right triangle, the cosine of an acute angle is a ratio.

$\cos A = \frac{b}{c}$ $\cos B = \frac{a}{c}$

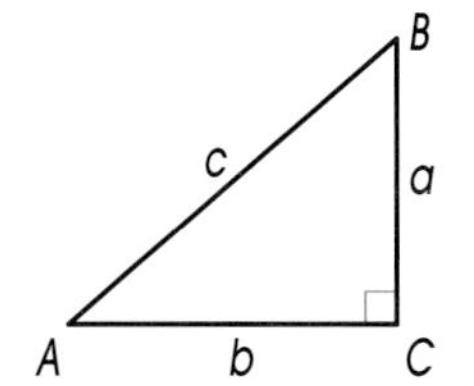

Sometimes you know the lengths of one leg and the hypotenuse of a right triangle. You can use the cosine ratio to find the measure of an acute angle.

EXAMPLE 1

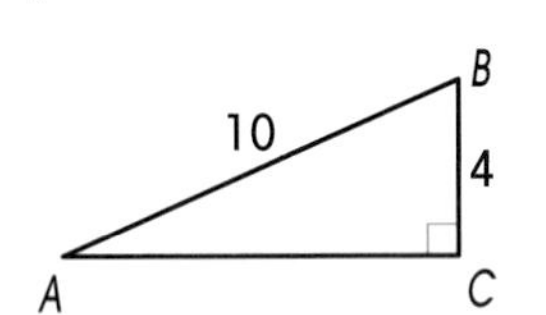

Find the measure of $\angle B$ to the nearest degree.

Write the cosine ratio. $\cos B = \frac{\text{adjacent}}{\text{hypotenuse}}$

Substitute the given values. Simplify. $= \frac{4}{10} = 0.4$

Look in the *cosine* column in the table on page 426. Find the closest decimal to 0.4. Write the measure of the angle. $0.4 \approx 0.4067$ $\cos 66° = 0.4067$

So, $\angle B$ is about 66°.

You can find the unknown length of a leg of a right triangle if you know the measure of one acute angle and the length of the hypotenuse. Use the cosine ratio.

EXAMPLE 2

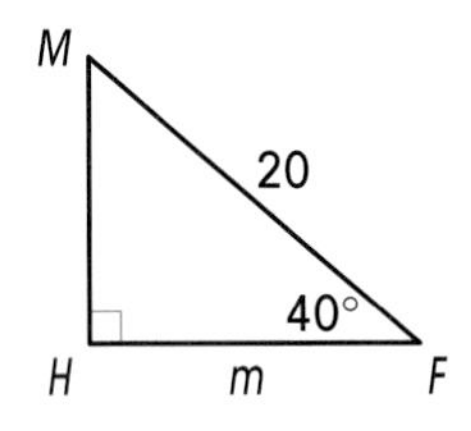

Find the value of m. Round the value to the nearest whole number.

Write the cosine ratio. $\cos F = \frac{\text{adjacent}}{\text{hypotenuse}}$

Substitute the given values. $\cos 40° = \frac{m}{20}$

Find 40° in the table on page 426. Look across to the *cosine* column. Substitute. $0.7660 = \frac{m}{20}$

Solve for m. $20(0.7660) = m$

$15.32 = m$

The value of m is about 15.

Try This

Find the length of p. Round the value to the nearest whole number.

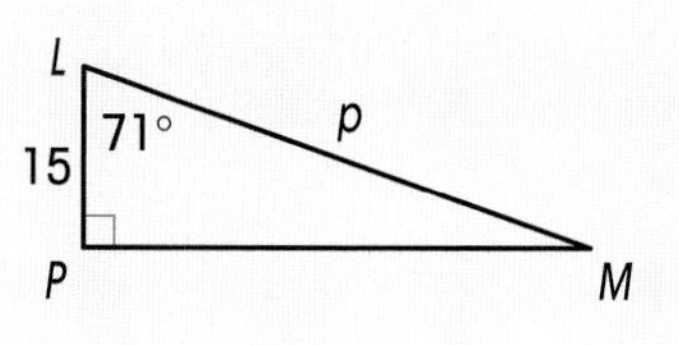

Write the cosine ratio.	$\cos L = \frac{\text{adjacent}}{\text{hypotenuse}}$
Substitute the given values.	$\cos 71° = \frac{15}{■}$ p
Find 71° in the table on page 426. Look across to the *cosine* column. Substitute.	$0.3256 = \frac{15}{■}$ p
Solve for ■. p	$(0.3256)p = 15$
	$p \approx$ ■ 46

The length of p is about ■. 46

Practice

Find the measure of $\angle T$ to the nearest degree. Use the table on page 426.

1. 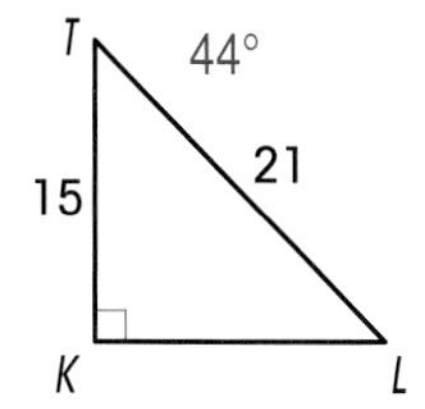

2. 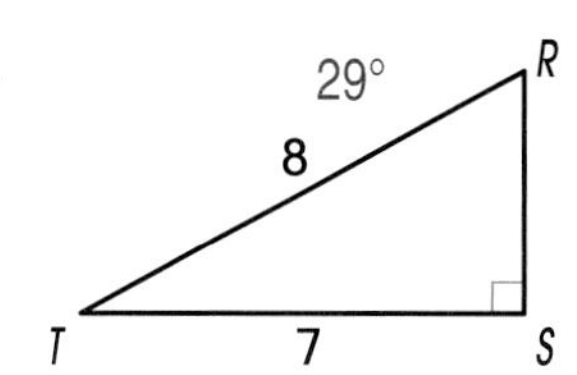

3. 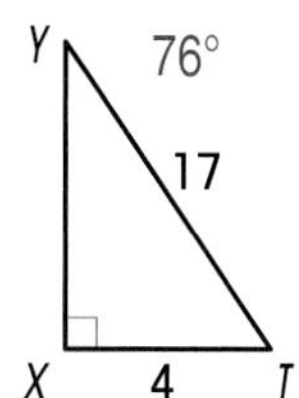

Find the value of m in each diagram. Use the table on page 426. Round the value to the nearest whole number.

4. 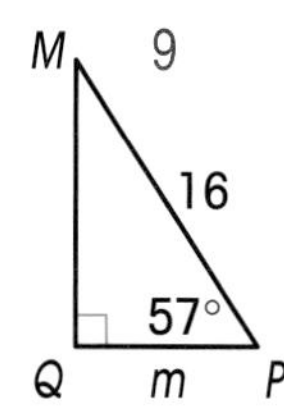

5. 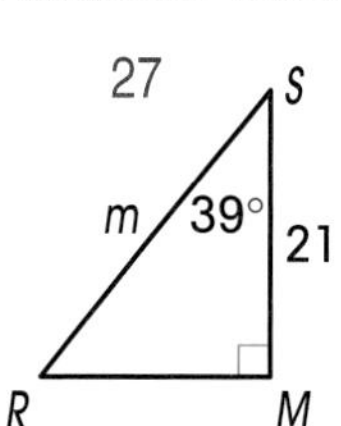

6. 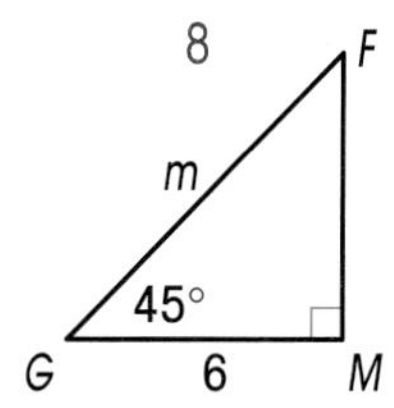

7. Write the cosine ratio. Substitute the measure of the angle and the length of the given side. Find the cosine of the given angle in the table. Substitute the value. Then, simplify to find the unknown length.

Share Your Understanding

7. Explain to a partner how to use the cosine ratio to find the unknown length of a side of a right triangle. See above.

8. **CRITICAL THINKING** Explain why cos 60° is 0.5. Draw a diagram.
Cos 60° is 0.5 because in a 30°-60°-90° triangle, the side adjacent to the 60° angle is half the length of the hypotenuse. Cos $A = \frac{x}{2x} = \frac{1}{2}$

B
2x
60°
A x C

13·6 Calculator: Finding the Missing Side

Getting Started
Have students locate the keys for the trigonometric ratios on their scientific calculators.

You know how to use ratios for sine, cosine, and tangent to find the lengths of sides of right triangles. You also learned how to find angle measures. Now you will learn how to work with trigonometric ratios on a scientific calculator.

EXAMPLE 1

Find tan 56°. Round to four decimal places.

Use your scientific calculator to find tan 56°. **Display**

PRESS: [5] [6] [tan] 1.482560969

So, tan 56° = 1.4826, rounded to four decimal places.

EXAMPLE 2

The cosine of $\angle A$ is 0.8192. Find the measure of $\angle A$. Round the measure to the nearest degree.

Using your scientific calculator, follow these steps. **Display**

PRESS: [.] [8] [1] [9] [2] [2nd] [cos] 34.99520932

So, $\angle A$ is about 35°.

EXAMPLE 3

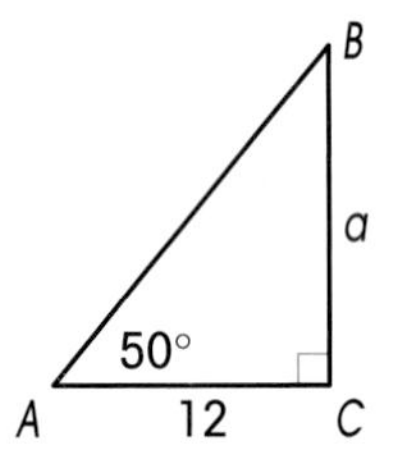

Find the value of a. Round the value to the nearest whole number.

You know the measure of $\angle$A and the length of the side adjacent to it. You need to find the length of the side opposite it. Write the tangent ratio. $\tan A = \frac{\text{opposite}}{\text{adjacent}}$

Substitute the given values. $\tan 50° = \frac{a}{12}$

Solve for a. $12(\tan 50°) = a$

Now, use your calculator to find the value of a.

Display

PRESS: [1] [2]    [×] [5] [0] [tan] [=] 14.30104311

The value of a is about 14.

Practice

Use a calculator. Find the measure of $\angle M$ in each diagram. Round the measure to the nearest degree.

1. 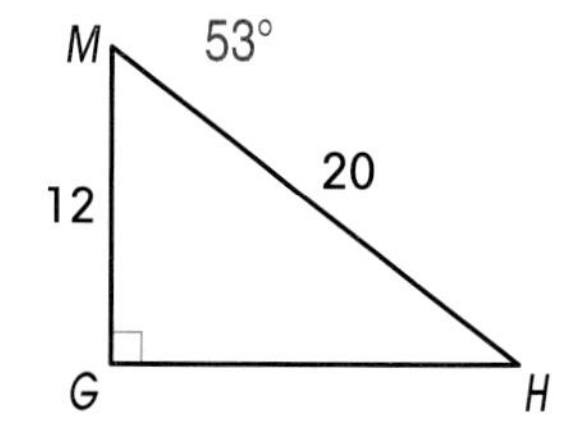

2. 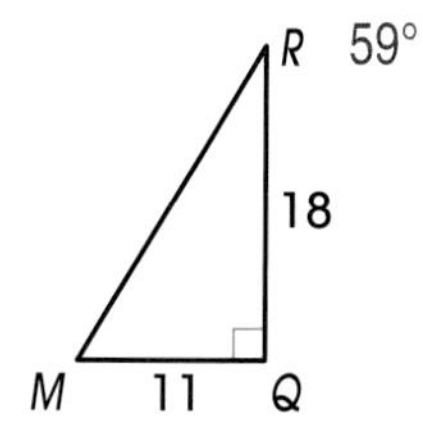

3. 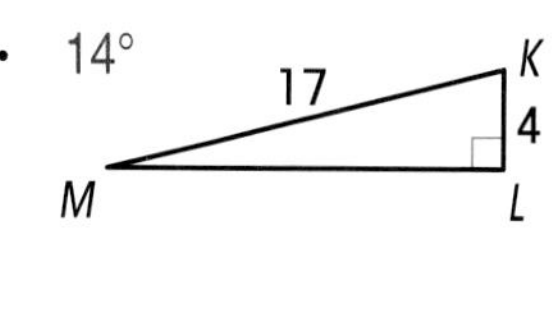

Use a calculator. Find the value of a in each diagram. Round the value to the nearest whole number.

4. 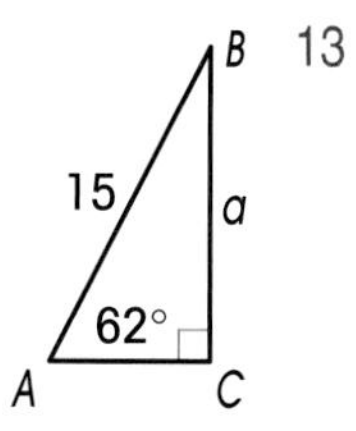

5. 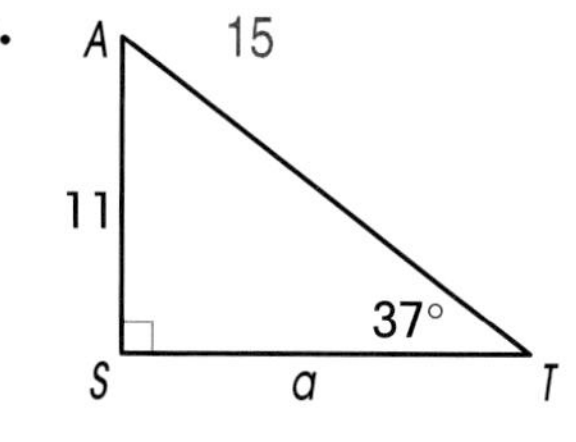

6. 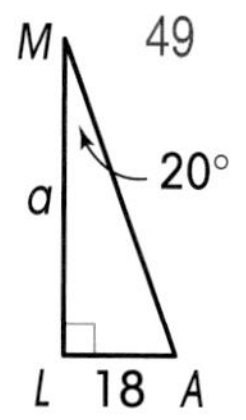

Math In Your Life

MAKING A CLINOMETER

A **clinometer** is a tool that can be used to measure an angle. It can help you make indirect measurements.

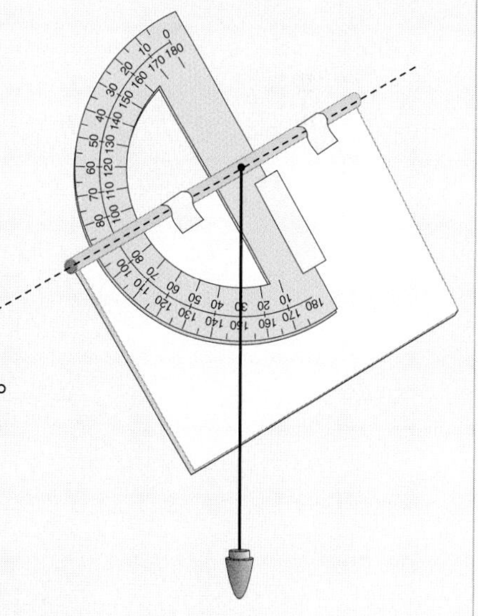

You can make a clinometer. You need a protractor, string, tape, cardboard, a small weight, and a straw. Tape the protractor to the cardboard as shown. Next, tape the straw along the top of the cardboard through the 90° mark. Attach the string to the center of the protractor. Attach the weight to the string.

To use the clinometer, hold it as shown. Look through the straw to the top of the object you want to measure. Have someone read the angle marked by the string on the protractor.

More practice is provided in the Workbook, page 91, and in the Classroom Resource Binder, page 146.

13-7 Problem-Solving Skill: Angles of Elevation and Depression

Getting Started
Draw two parallel lines cut by a transversal. Remind students that alternate interior angles are congruent. These angles are similar to angles of elevation and depression.

Joya looks up at Malia at a 40° angle above a horizontal line. This angle is an **angle of elevation.** Malia looks down at Joya at a 40° angle below a horizontal line. This angle is an **angle of depression.**

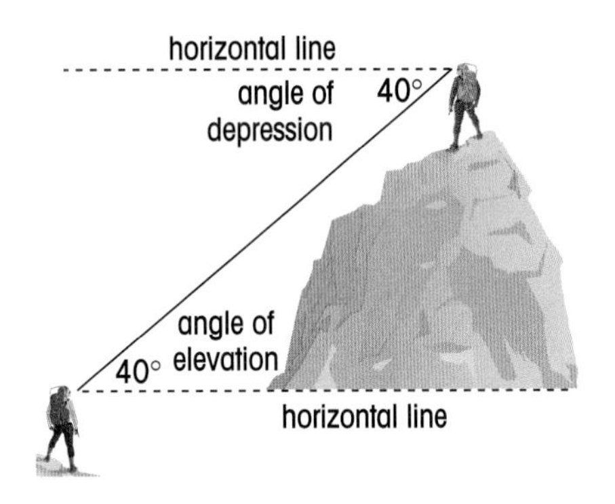

Knowing an angle of elevation or an angle of depression can help you solve a problem about distance.

EXAMPLE

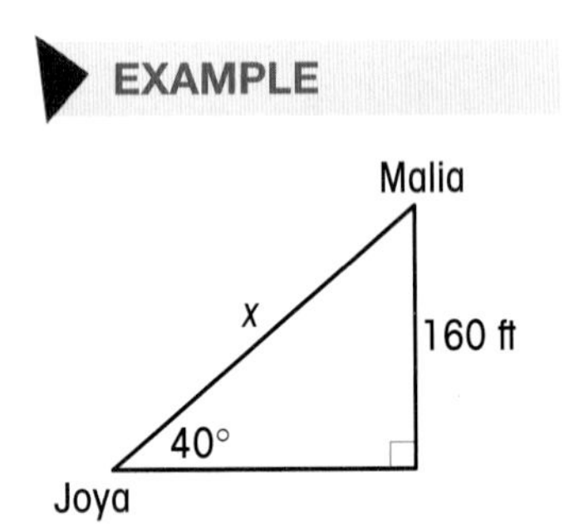

In the diagram above, the distance between the two horizontal lines is 160 feet. What is the distance between Joya and Malia? Round your answer to the nearest foot.

First, draw a triangle to show the problem. Label the parts you can use to find the unknown distance. Let $x =$ the distance between Joya and Malia.

Then, decide which trigonometric ratio to use. You know the measure of an angle and the length of the side opposite it. You need to find the length of the hypotenuse. Use the sine ratio.

Write the sine ratio. $\sin A = \frac{\text{opposite}}{\text{hypotenuse}}$

Substitute the given values. $\sin 40° = \frac{160}{x}$

Use the decimal value of sin 40° in the ratio. $0.6428 = \frac{160}{x}$

Solve for x. Round to the nearest whole number. $0.6428x = 160$
$x \approx 249$

The distance between Joya and Malia is about 249 feet.

Try This

Aaron flies a kite at a 65° angle of elevation. The kite string is 300 feet long. Aaron is 5 feet tall. How high off the ground is the kite, to the nearest foot?

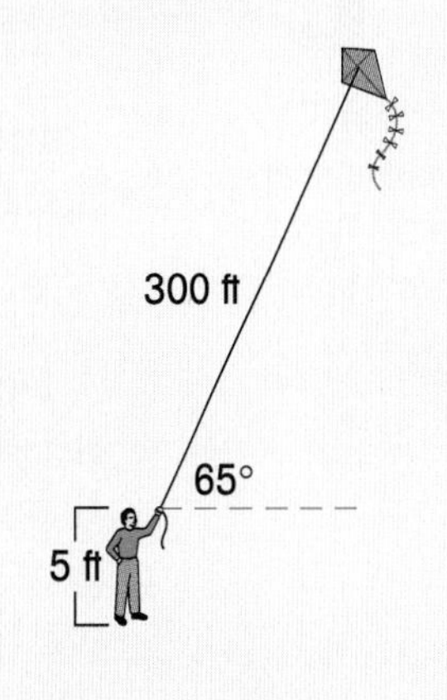

First, draw a triangle to show the problem. Let x = the distance from the kite to the horizontal line.

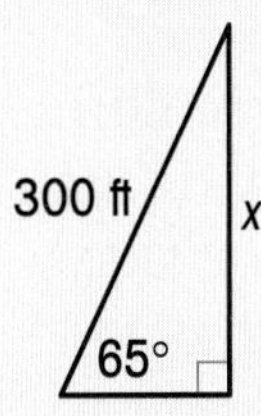

Then, decide which trigonometric ratio to use. $\sin A = \frac{\text{opposite}}{■}$ hypotenuse

Substitute the given values. $\sin 65° = \frac{x}{■}$ 300

Use the decimal value of sin 65° in the formula. 0.9063 $■ = \frac{x}{300}$

Solve for x. Round to the nearest whole number. $■ \cdot 300 = x$ 0.9063

$272 \approx x$

Now, add Aaron's height. $272 + 5 = ■$ 277

The kite is about ■ off the ground. 277 ft

Practice

Solve each problem. Round your answer to the nearest whole number. Draw a diagram if you need to.

1. Kate looks up at a house on top of a hill. The angle of elevation is 50°. The height of the hill is 32 feet. What is the distance between Kate and the house? about 42 ft

2. An airplane flies 8 miles above the ground. The horizontal distance from the airplane to the start of the runway is 20 miles. What is the angle of depression the airplane must use to land on the runway? about 22°

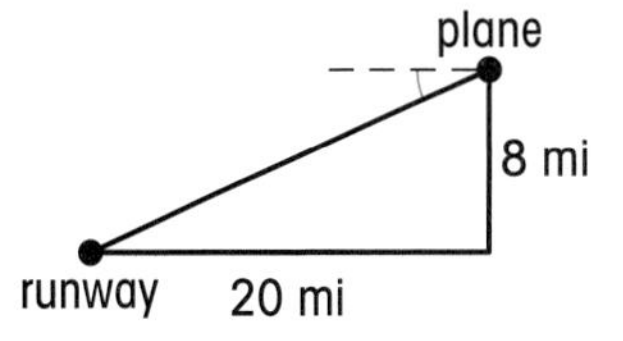

More practice is provided in the Workbook, page 92, and in the Classroom Resource Binder, page 147.

13·8 Problem-Solving Application: Using Trigonometric Ratios

Getting Started
Draw right triangle *ABC*. Ask students what information they would need to find m∠*A*, using the sine ratio.

Sometimes you cannot use ordinary tools to measure distances and angles. You can, however, use what you have learned about sine, cosine, and tangent.

EXAMPLE

A ladder is 26 feet long. It is leaning against a building. It makes a 67° angle with the ground. How far up the side of the building does the ladder reach? Round your answer to the nearest foot.

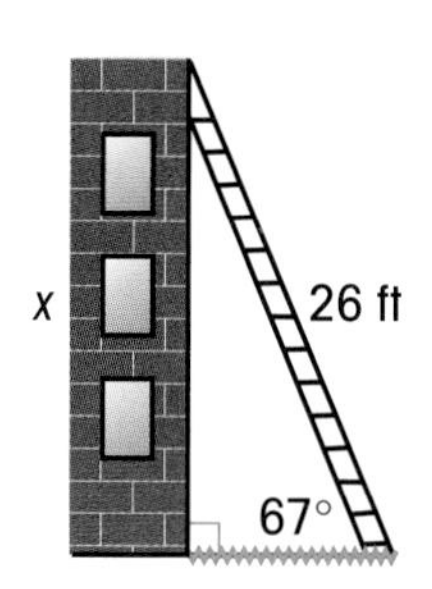

Draw a triangle. Label the triangle with the given values. Let x = the distance the ladder reaches.

You know the measure of an angle and the length of the hypotenuse. You need to find the length of the side opposite the angle. Use the sine ratio. $\sin 67° = \frac{\text{opposite}}{\text{hypotenuse}}$

Substitute the given values. $\sin 67° = \frac{x}{26}$

Find sin 67° in the table on page 426. Substitute the value. $0.9205 = \frac{x}{26}$

Solve for x.

$$0.9205(26) = x$$

$$23.933 = x$$

The ladder reaches about 24 feet up the side of the building.

Try This

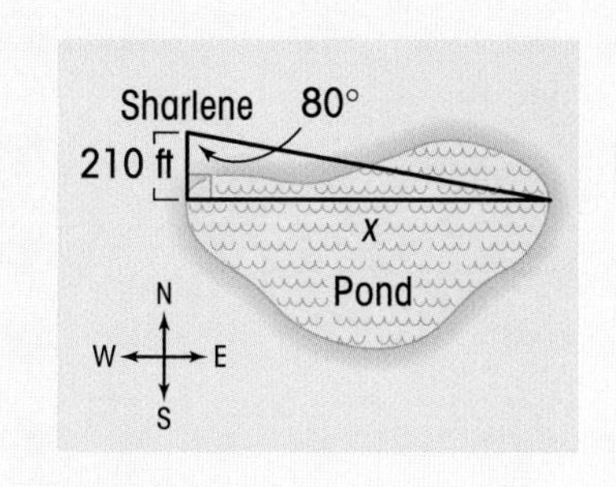

Sharlene wants to measure the length of a pond. She stood at a point 210 feet north of one end of the pond. The angle from where she stood to the other end of the pond is 80°. How long is the pond, to the nearest foot?

You know the measure of an angle and the length of the side adjacent to it. You need to find the length of the side opposite the angle. Let x = the length. Use the tangent ratio. $\tan 80° = \frac{■}{\text{adjacent}}$ opposite

Substitute the given values. $\tan 80° = \frac{x}{210}$

Find ■ (tan 80°) in the table on page 426. Substitute the value. $5.6713 = \frac{x}{210}$

Solve for x.

$$210(5.6713) = x$$
$$1{,}191 \approx x$$

The pond is about ■ long. 1,191 ft

Practice

Solve each problem. Use trigonometric ratios. Round your answer to the nearest whole number.

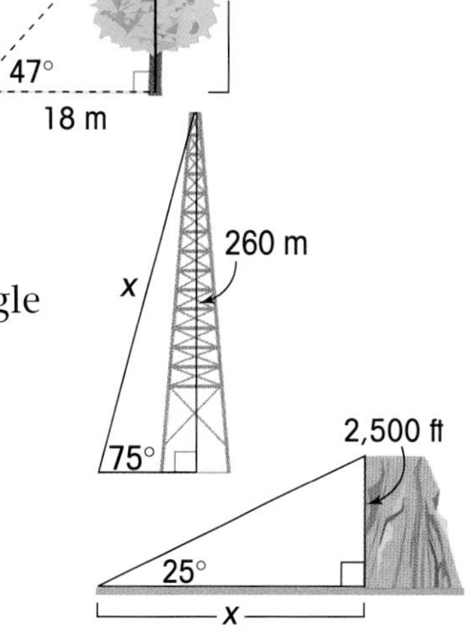

1. A tree casts a shadow 18 meters long. The angle of elevation is 47°. What is the height of the tree?
   about 19 m

2. A cable is used to support a television tower that is 260 meters tall. The cable makes a 75° angle with the ground. How long is the cable?
   about 269 m

3. A surveyor needs to find out how far she is from a cliff that is 2,500 feet high. The angle of elevation is 25°. How far is she from the cliff?
   about 5,361 ft

13-9 Proof: Trigonometric Identities

Getting Started
Review the Pythagorean Theorem.

In this chapter, you learned how to find the sine, cosine, and tangent of the acute angles of a right triangle.

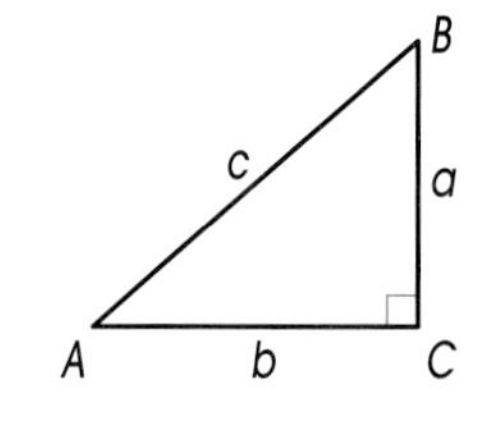

$\sin A = \frac{a}{c}$ $\qquad$ $\sin B = \frac{b}{c}$

$\cos A = \frac{b}{c}$ $\qquad$ $\cos B = \frac{a}{c}$

$\tan A = \frac{a}{b}$ $\qquad$ $\tan B = \frac{b}{a}$

You can use the definitions of sine, cosine, and tangent to prove trigonometric identities. A **trigonometric identity** is a trigonometric equation that is always true.

EXAMPLE

You are given:

$\triangle ABC$ is a right triangle.

Prove: $\sin^2 A + \cos^2 A = 1$

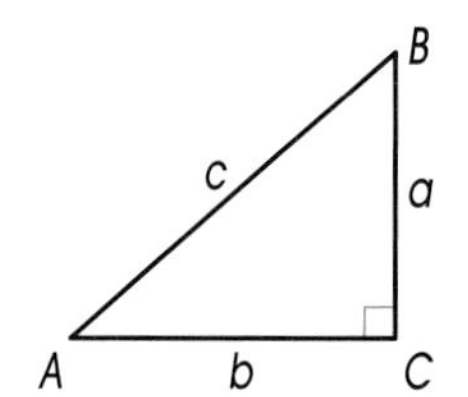

To prove the identity is true, change the left side of the equation to show that it equals 1.

You know that $\sin A = \frac{a}{c}$ *and* $\cos A = \frac{b}{c}$.

You can use these ratios to prove the equation is true.

Write the equation.	$\sin^2 A + \cos^2 A = 1$
Substitute the ratios for sin A and cos A.	$\left(\frac{a}{c}\right)^2 + \left(\frac{b}{c}\right)^2 = 1$
Simplify and add.	$\frac{a^2}{c^2} + \frac{b^2}{c^2} = 1$
	$\frac{a^2 + b^2}{c^2} = 1$
Substitute c^2 *for* $a^2 + b^2$.	$\frac{c^2}{c^2} = 1$
	$1 = 1$

So, you prove that $\sin^2 A + \cos^2 A = 1$. ✓

Try This

Copy and complete the proof. You are given:

$\triangle ABC$ is a right triangle.

Prove: $\tan A = \frac{\sin A}{\cos A}$

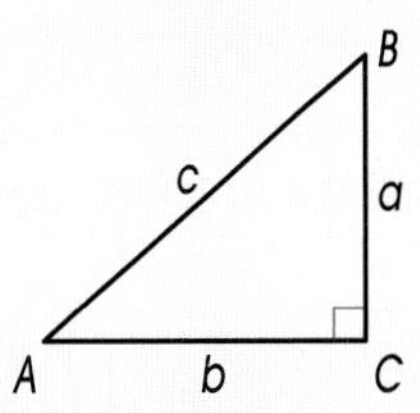

Change the right side of the equation to show that it equals tan A.

You know that $\sin A = \frac{a}{c}$, $\cos A = \frac{b}{c}$, *and* $\tan A = \frac{a}{b}$.

You can use these ratios to prove the equation is true.

Write the equation. $\tan A = \frac{\sin A}{\cos A}$

Substitute the ratios for sin A and cos A. $\tan A = \frac{a}{c} \div \frac{b}{c}$

Do the division. Remember, dividing by $\frac{b}{c}$ *is the same as multiplying by* $\frac{c}{b}$. $\tan A = \frac{a}{c} \cdot \frac{c}{b} = \frac{a}{b}$

Substitute tan A for $\frac{a}{b}$. $\tan A = \tan A$

So, you prove that $\tan A = \frac{\sin A}{\cos A}$. ✓

Practice

Write a proof for each of the following. Use the diagram below for each proof. See Additional Answers in the back of this book.

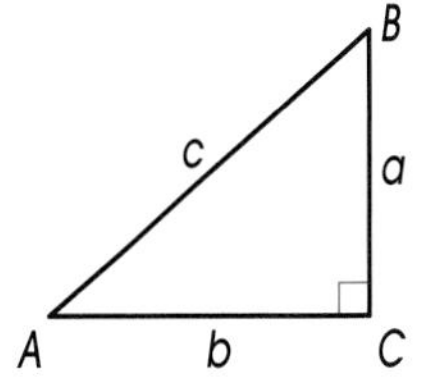

1. You are given:

$\triangle ABC$ is a right triangle.

Prove: $\cos A = \frac{\sin A}{\tan A}$

2. You are given:

$\triangle ABC$ is a right triangle.

Prove: $\sin A = \frac{\cos A}{\tan B}$

3. You are given:

$\triangle ABC$ is a right triangle.

Prove: $\tan A \cdot \cos A = \sin A$

More Practice is provided in the Classroom Resource Binder.

Chapter 13 Review

Summary

You can use the Table of Trigonometric Ratios to find the sine, cosine, and tangent of an angle.
In right triangle *ABC* where $\angle C$ is the right angle, tan $A = \frac{a}{b}$ and tan $B = \frac{b}{a}$.
In right triangle *ABC* where $\angle C$ is the right angle, sin $A = \frac{a}{c}$ and sin $B = \frac{b}{c}$.
In right triangle *ABC* where $\angle C$ is the right angle, cos $A = \frac{b}{c}$ and cos $B = \frac{a}{c}$.
You can use sine, cosine, and tangent to find the lengths of sides and angle measures.
You can use trigonometric ratios to solve problems.
You can use what you know about trigonometry to prove trigonometric identities.

More vocabulary review is provided in the Classroom Resource Binder, page 141.

trigonometry
trigonometric ratios
sine
cosine
tangent
angle of elevation
angle of depression

Vocabulary Review

Complete the sentences with the words from the box.

1. The tangent of an acute angle in a right triangle is the ratio of the length of the side opposite the angle and the length of the side adjacent to the angle.
2. Trigonometry is the study of the relationship between sides and angles of a right triangle.
3. In a right triangle, the ratio of the length of the side opposite an acute angle and the length of the hypotenuse is the sine of that acute angle.
4. When looking up, the angle formed by a horizontal line and the line of sight is the angle of elevation.
5. The ratios of the lengths of two sides of a right triangle are called trigonometric ratios.
6. The cosine of an acute angle in a right triangle is the ratio of the length of the side adjacent to the angle and the length of the hypotenuse.
7. When looking down, the angle formed by a horizontal line and the line of sight is the angle of depression.

Chapter Quiz

More assessment is provided in the Classroom Resource Binder.

Use the diagram on the right.

1. Find three trigonometric ratios of $\angle A$.

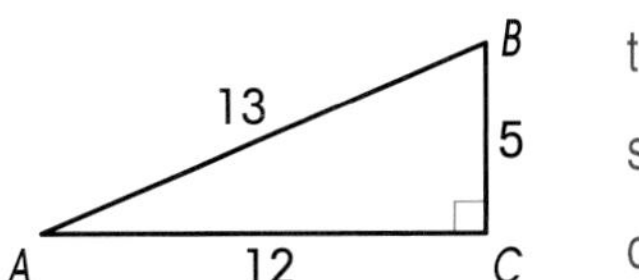

$\tan A = \frac{5}{12}$

$\sin A = \frac{5}{13}$

$\cos A = \frac{12}{13}$

Write three trigonometric ratios for each angle. Use the Table of Trigonometric Ratios on page 426.

2. 30° tan = 0.5774; cos = 0.8660; sin = 0.5000

3. 45° tan = 1.0000; cos = 0.7071; sin = 0.7071

4. 60° tan = 1.7321; cos = 0.5000; sin = 0.8660

Find the measure of $\angle D$ to the nearest degree. Use the table on page 426.

5.

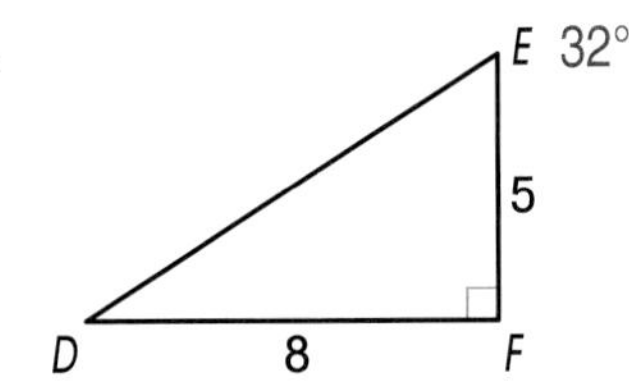

E 32°

6.

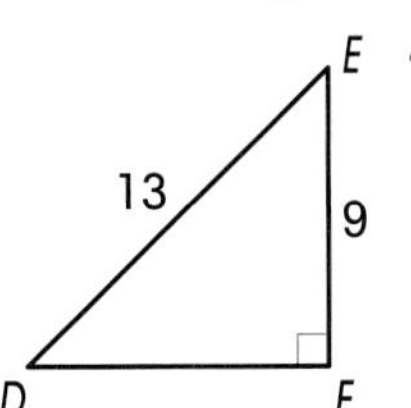

E 44°

7.

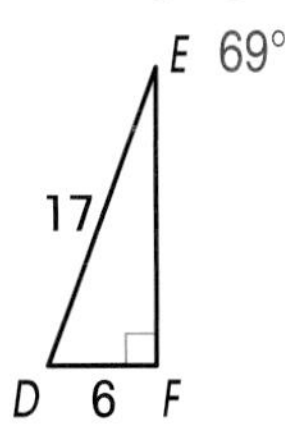

E 69°

Find the value of *a* in each diagram. Use the table on page 426. Round the value to the nearest whole number.

8.

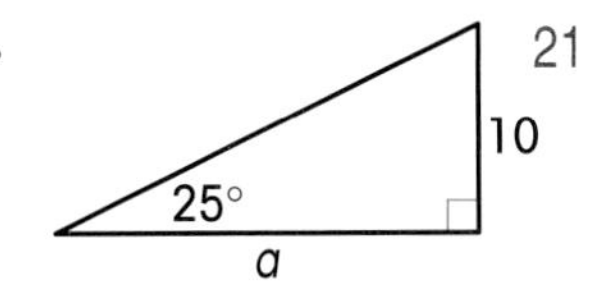

21

9.

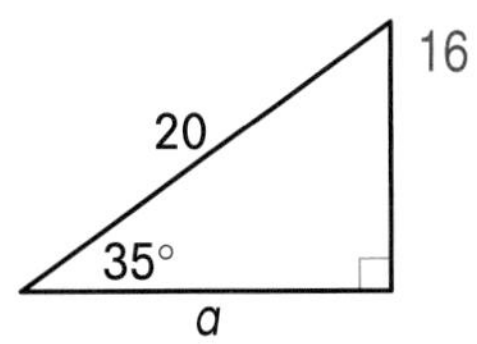

16

10.

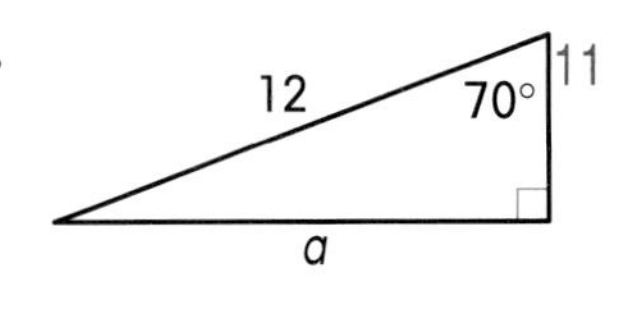

11

Problem Solving

Solve each problem. Show your work.

The sine ratio of $\angle A$ is $\frac{a}{c}$, where a is the length of the side opposite the angle and c is the length of the hypotenuse.

B
2x
60°
A x C

11. Allison stands on a balcony, looking for Garry. She spots him 100 feet away from her building. The balcony is 45 feet above the ground. What is the angle of depression from the balcony to Garry? about 24°

12. Emily flies a kite at the beach. She lets out 95 feet of string. Emily is 5 ft tall. The kite's angle of elevation from the ground is 57°. How high is the kite? about 85 feet

Write About Math

Explain the sine ratio of an acute angle of a right triangle. Draw a diagram. See above.

Additional Practice for this chapter is provided on page 405 of the student book.

Unit 4 Review

Standardized Test Preparation This unit review follows the format of many standardized tests. A Scantron sheet is provided in the Classroom Resource Binder.

Write the letter of the correct answer.

1. The edge of a cube is 5 cm. Find the surface area.
A. 25 cm^2
(B.) 150 cm^2
C. 300 cm^2
D. 625 cm^2

2. Find the volume of the cylinder.
A. 40π
B. 80π
C. 100π
(D.) 160π

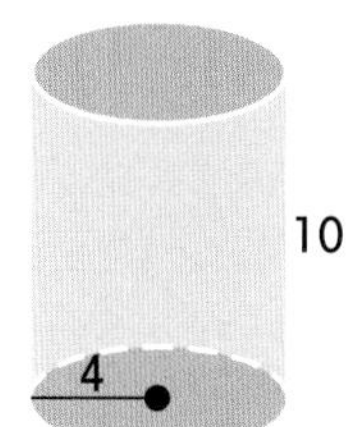

3. Find the distance between point A ($^-2$, 1) and point B (3, $^-1$).
A. 4
B. $\sqrt{5}$
(C.) $\sqrt{29}$
D. $\sqrt{7}$

4. A square has one side on one axis of the coordinate plane. P (0, 3) is one corner of the square. Which could not be another corner?
(A.) (3, 5)
B. (0, 0)
C. (3, 0)
D. ($^-3$, 0)

5. The angle of elevation from the ground to the top of a tree is 48°. This is measured from a point 50 m from the tree. About how high is the tree?
A. 33 m
B. 37 m
C. 45 m
(D.) 56 m

6. A 12-ft ladder leaning against a building makes a 60° angle with the ground. About how far from the building is the ladder?
A. 4 ft
(B.) 6 ft
C. 10 ft
D. 15 ft

Critical Thinking

The diagonals of a rhombus are 6 and 8. What trigonometric ratio can you use to find angles in the rhombus? tangent

CHALLENGE Find the measure of the larger angle to the nearest degree. 106°

Critical Thinking

See the Classroom Resource Binder for a scoring rubric to assess the Critical Thinking Question.

Student Handbook Contents

ADDITIONAL PRACTICE

Chapter 1: Basic Geometric Concepts

Name each figure.

1. N O

ray NO

2. D E

line segment DE

3. R S T

Possible answer: plane RST

Add or subtract.

4. $^-7 + (^-4)$ $^-11$ **5.** $16 + (^-2)$ 14 **6.** $6 - (^-4)$ 10 **7.** $^-5 - 9$ $^-14$

Find the distance between each set of points.

8. Points O and M 4

9. Points K and L 8

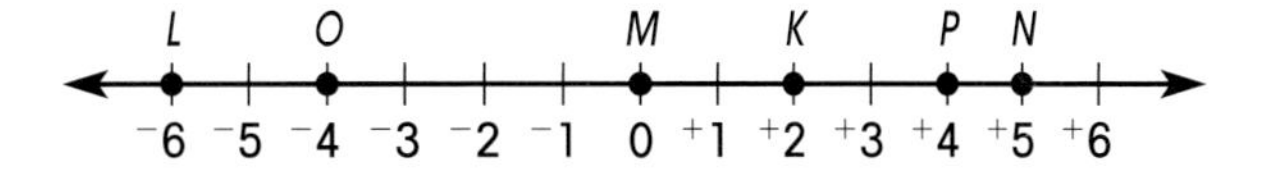

Use the diagrams on the right for exercises 10–12.

10. Solve for x. $x = 1$

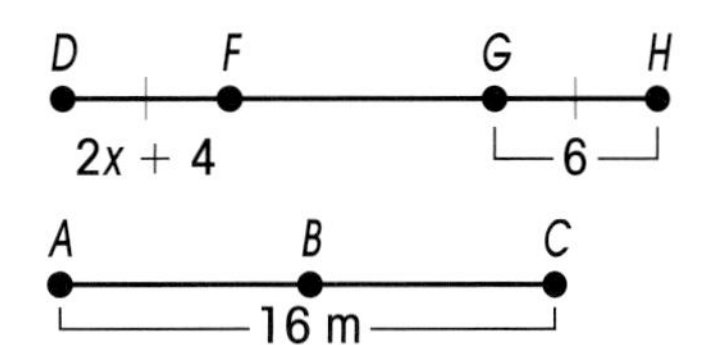

11. Point B is the midpoint of $\overline{AC}$. Find the length of $\overline{AB}$. 8 m

12. Find the length of $\overline{MN}$. 6 cm

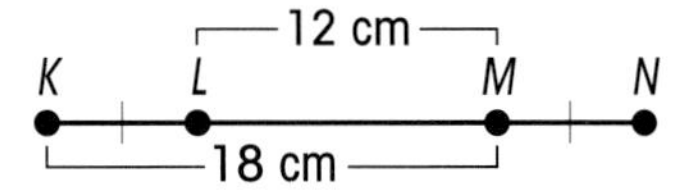

Problem Solving

Solve each problem. Show your work.

13. Points A, R, and Z are collinear. The points are not in that order. The distance from point A to point R is 6 meters. The distance from point R to point Z is 34 meters. Find the distance from point A to point Z. 40 m or 28 m

14. A wire 120 cm long is cut into two pieces. One piece is twice as long as the other piece. How long is the shorter piece of wire?
40 cm

Write About Math

Do the symbols $\overrightarrow{PD}$ and $\overrightarrow{DP}$ both name the geometric figure below? Explain why or why not.

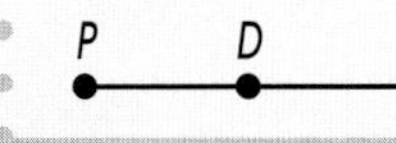

No. Only $\overrightarrow{PD}$ names the figure. A ray must be named by two points, with the endpoint named first.

ADDITIONAL PRACTICE
Chapter 2: Angles

Draw and name an angle for each measure. Then, classify the angle. Write *acute, right, obtuse,* or *straight*. Check students' work.

1. 25° acute **2.** 110° obtuse **3.** 90° right **4.** 180° straight

Solve for x.

5. $4x - (2x - 4) = 90$
$x = 43$

6. $7x = 2x + 180$
$x = 36$

7. $90 - (x + 15) = 4x$
$x = 15$

8. Find the angle that is complementary to a 41° angle. 49°

9. Find the angle that is supplementary to a 41° angle. 139°

Use the diagram on the right for exercises 10 and 11.

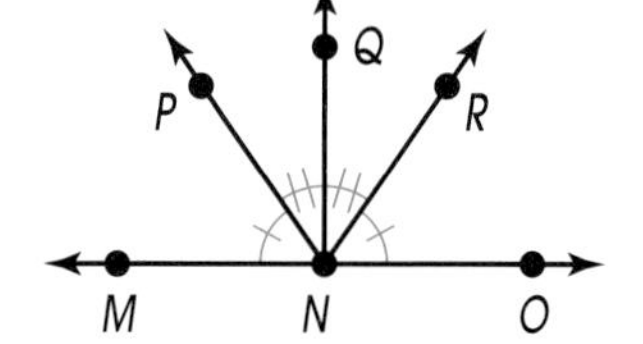

10. $\angle QNR \cong$ ■ $\angle QNP$ or $\angle PNQ$

11. If $\angle RNO$ is 55°, then find the measure of $\angle MNP$. 55°

Find the measure of each angle.

12. $\overrightarrow{KJ}$ bisects $\angle LKM$. $\angle LKM$ is 124°. Find the measure of $\angle JKL$. 62°

13. Find the measure of $\angle TUX$. 79°

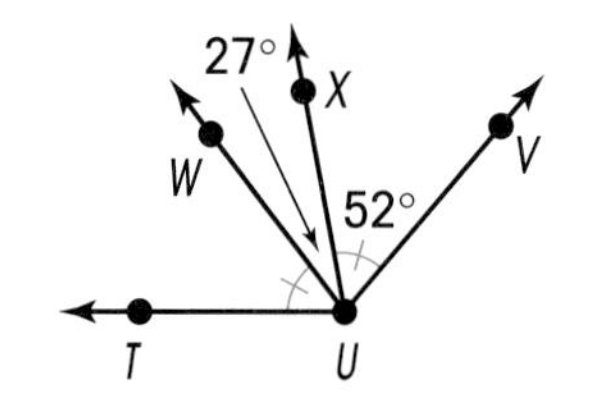

Yes. $\angle A$ plus $\angle B$ must be equal to 180°. $\angle A$ can be greater than 90° if $\angle B$ is less than 90°.

Problem Solving

Solve the problem. Show your work.

14. Walter has to stretch his arm from 68° to 112°. He wants to straighten his elbow by about 11° every 3 days. How many days will it take to get to 112°? 12 days

Write About Math

$\angle A$ and $\angle B$ are supplementary angles. Can $\angle A$ be greater than 90°? Explain why or why not.

ADDITIONAL PRACTICE

Chapter 3: Reasoning and Proofs

Continue each pattern.

1. 3, 6, 4, 7, 5, ____, ____ 8, 6

2. 

Write a conclusion for each conditional statement.

3. If $\angle D \cong \angle E$, then ____. $\angle E \cong \angle D$

4. If $JK = LM$, then ____. $LM = JK$

5. If E is between D and F, then ____. $DE + EF = DF$

6. If $\overrightarrow{AB}$ bisects $\angle CAD$, then ____. $\angle BAC \cong \angle BAD$

Match the statement on the left with the property on the right.

7. If $m\angle P = m\angle Q$, then $m\angle Q = m\angle P$. C

8. If $x = 5$, then you can use 5 for x. So, $x + y = 5 + y$. D

9. If $\angle 5 \cong \angle 6$ and $\angle 6 \cong \angle 7$, then $\angle 5 \cong \angle 7$. A

10. If $a = b$, then $a - 3 = b - 3$. B

A. Transitive Property

B. Subtraction Property

C. Symmetric Property

D. Substitution Property

Write About Math You would use conclusions and conditional statements when making plans that are dependent on conditions such as weather, money, and health. For example, if it rains, we will go to the movies. If it is sunny, we will go to the park.

Problem Solving

Write a proof for each of the following.

11. Write an indirect proof.
You are given: n is an odd number.
Prove: $n + n$ is an even number.
(Hint: $2n$ is an even number.)

Write About Math
When would you use conclusions and conditional statements in everyday life?

12. Write a proof. Choose the form you like best.
You are given: Points A, B, and C are collinear.
$\overline{AB}$ is congruent to $\overline{BC}$.
Prove: Point B is the midpoint of $\overline{AC}$.

12. Proofs should show understanding of a midpoint and congruent line segments.

11. You are given that n is an odd number. Assume $n + n$ is an odd number. You can combine the like terms to write $n + n = 2n$. However, you know that $2n$ is an even number because $2n$ is divisible by 2. This contradicts the assumption. So, $n + n$ is an even number.

ADDITIONAL PRACTICE

Chapter 4: Perpendicular and Parallel Lines

Decide if the lines or line segments are perpendicular. Write *yes* or *no*. If they are perpendicular, use the $\perp$ symbol to write a statement.

1.

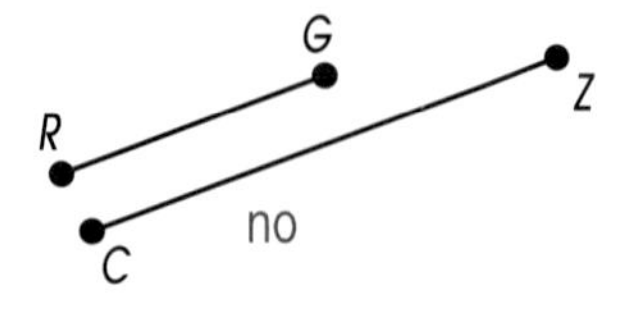

2.

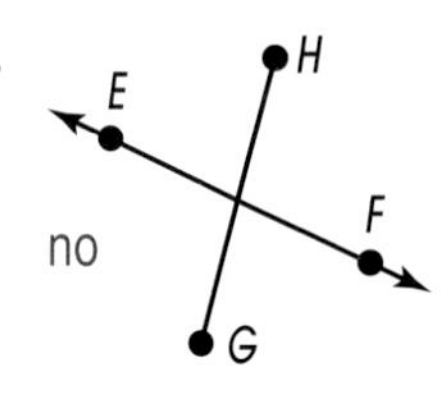

3.

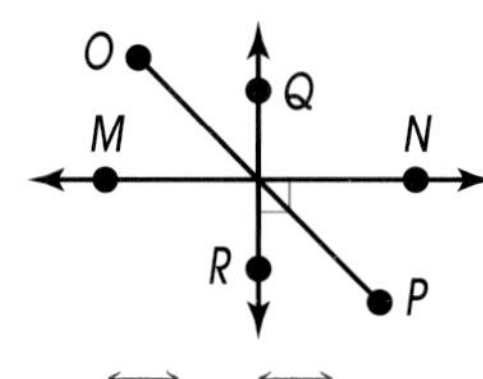

yes; $\overleftrightarrow{QR} \perp \overleftrightarrow{MN}$

Use the diagram below for exercises 4–10. In the diagram, $m \parallel n$.

Name the following angle pairs.

4. same-side interior angles
$\angle 4$ and $\angle 6$; $\angle 3$ and $\angle 5$

5. alternate interior angles
$\angle 3$ and $\angle 6$; $\angle 4$ and $\angle 5$

6. corresponding angles
$\angle 1$ and $\angle 5$; $\angle 2$ and $\angle 6$; $\angle 3$ and $\angle 7$; $\angle 4$ and $\angle 8$

If $\angle 2 = 124°$, find the measure of each angle.

7. $\angle 5$ 56°   **8.** $\angle 1$ 56°   **9.** $\angle 7$ 124°   **10.** $\angle 8$ 56°

Problem Solving

Solve each problem. Show your work.

11. Create a one-point perspective drawing of a cube. Check students' work.

12. Look at the map on page 114. How many routes can a taxicab take from 32nd Street and 7th Avenue to 36th Street and 4th Avenue? 3 routes

Write About Math
If two lines do not intersect, must they be parallel? Explain why or why not.

No. Parallel lines lie in the same plane and do not intersect. If they are not coplanar and do not intersect, they are not parallel.

ADDITIONAL PRACTICE
Chapter 5: Triangles

Classify each triangle by its angles and sides.

1. 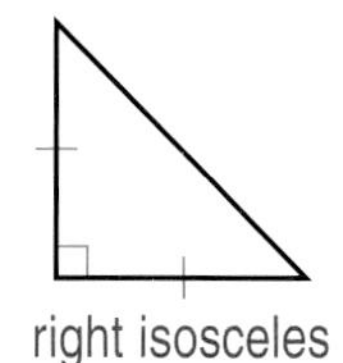

right isosceles

2. 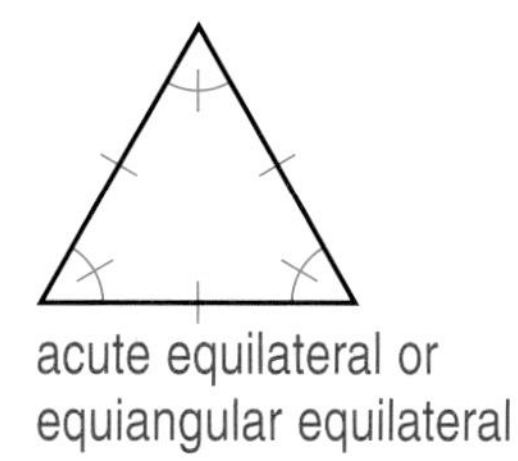

acute equilateral or equiangular equilateral

3. 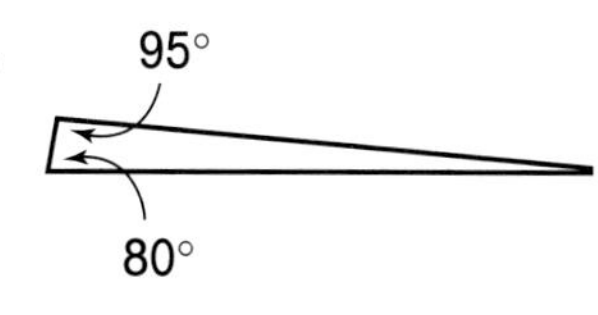

obtuse scalene

Find the value of x.

4. 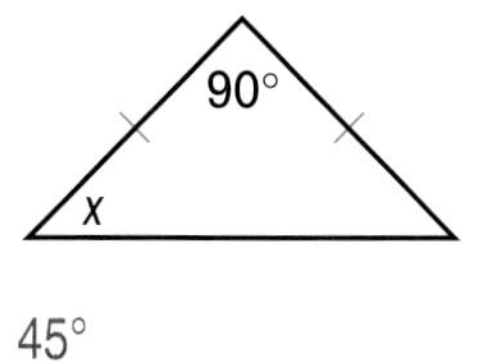

45°

5. 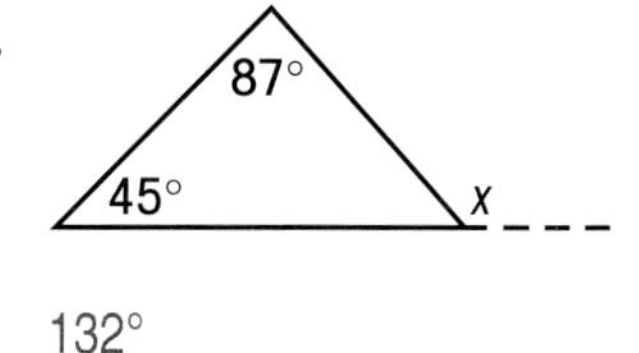

132°

6. 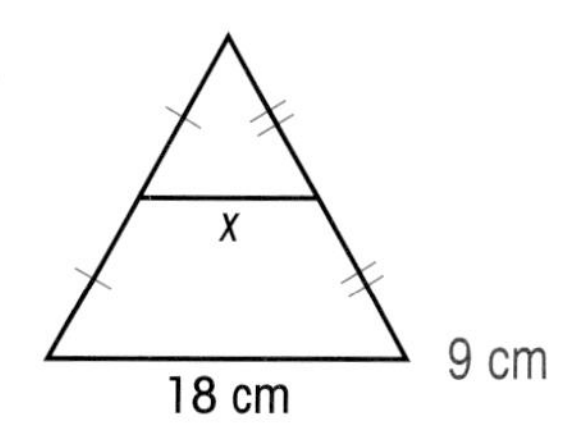

9 cm

7. Can segments 12 cm, 16 cm, and 30 cm long form a triangle? Write *yes* or *no*. If your answer is *no*, tell why.
No; $12 + 16 < 30$

Write a congruence statement for each pair of triangles. Name the postulate or theorem that you used.

8. 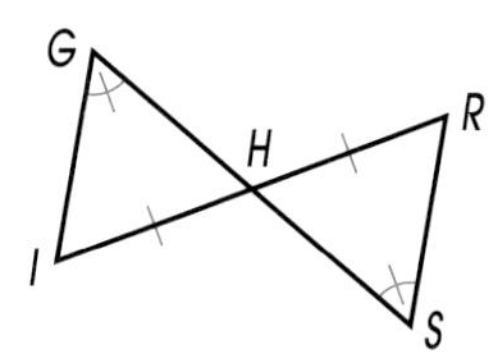

$\triangle GHI \cong \triangle SHR$;
Angle-Angle-Side

9. 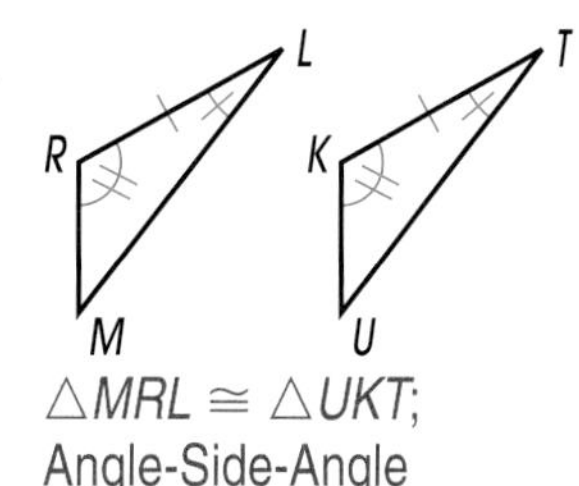

$\triangle MRL \cong \triangle UKT$;
Angle-Side-Angle

10. 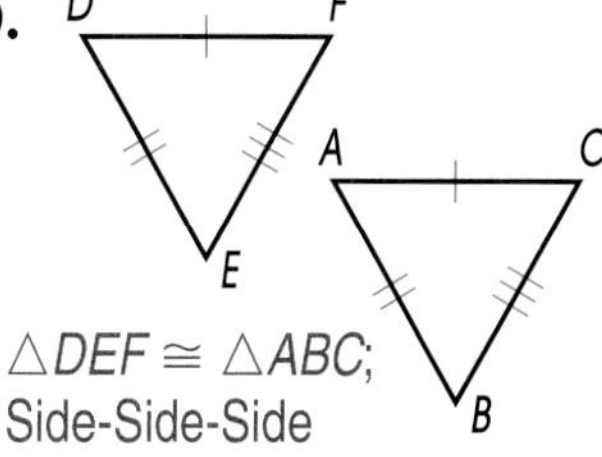

$\triangle DEF \cong \triangle ABC$;
Side-Side-Side

Problem Solving

Solve the problem. Show your work.

11. $\overline{QR}$ is 16 inches long. $\overline{SQ}$ is a median of $\triangle PSR$. What is the length of $\overline{PR}$? 32 inches

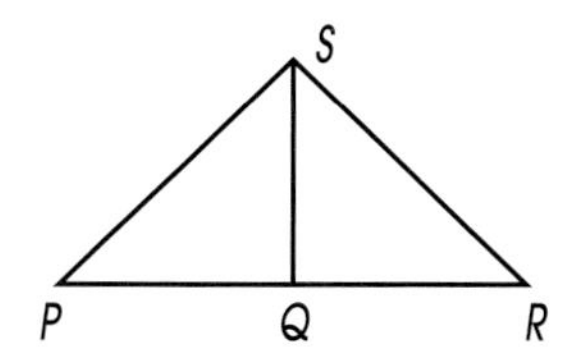

Write About Math

Can an equiangular triangle also be a right triangle? Explain?

No. An equiangular triangle has three 60° angles. Since it does not have a right angle, it cannot also be a right triangle.

ADDITIONAL PRACTICE

Chapter 6: Right Triangles

Simplify each expression.

1. 15^2 225

2. $\sqrt{121}$ 11

3. $\sqrt{52}$ $2\sqrt{13}$

4. $\sqrt{98}$ $7\sqrt{2}$

Find the unknown lengths of the sides.

5.

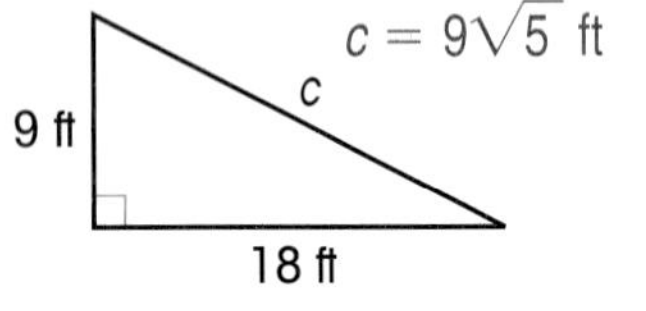

$c = 9\sqrt{5}$ ft

6.

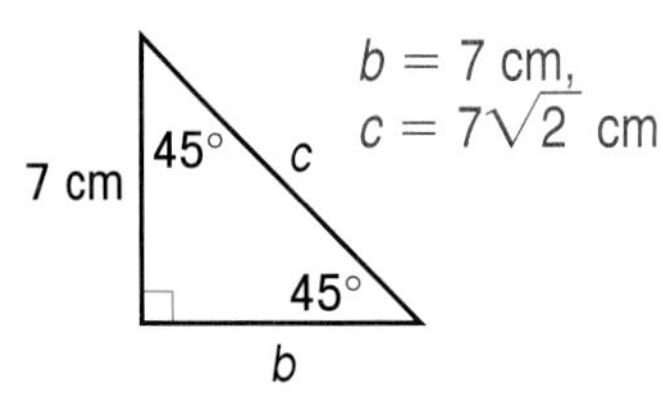

$b = 7$ cm, $c = 7\sqrt{2}$ cm

7.

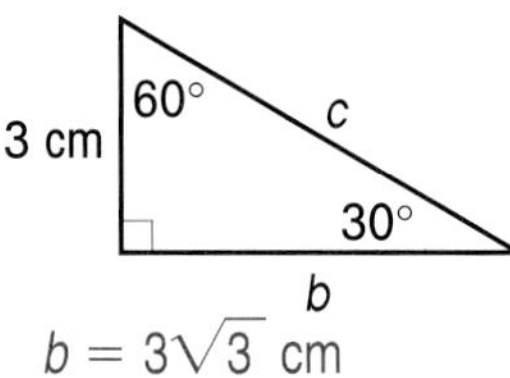

$b = 3\sqrt{3}$ cm
$c = 6$ cm

8.

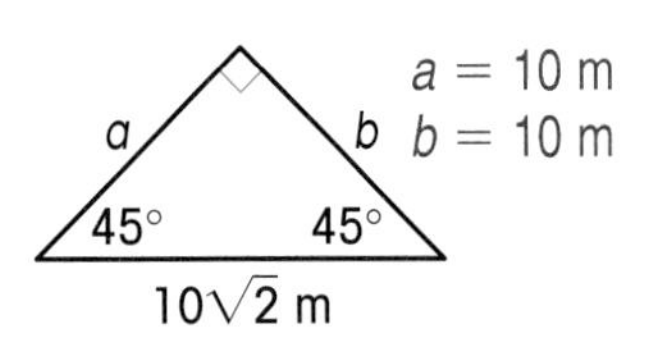

$a = 10$ m
$b = 10$ m

9.

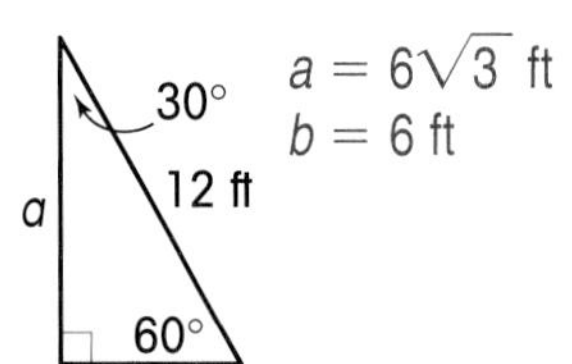

$a = 6\sqrt{3}$ ft
$b = 6$ ft

10.

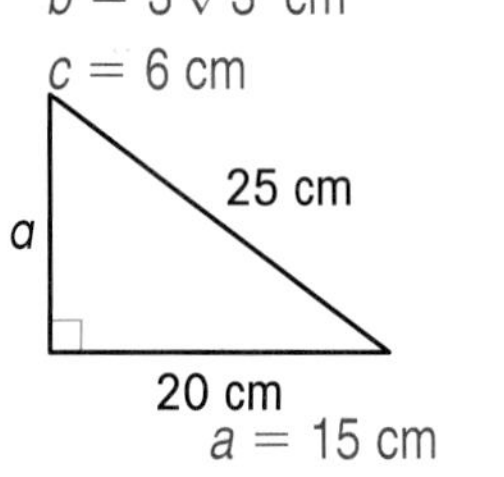

$a = 15$ cm

Tell whether the three lengths can form the sides of a right triangle. Write *yes* or *no*.

11. 11 ft, 15 ft, 18 ft no

12. 12 cm, 16 cm, 20 cm yes

Problem Solving

Solve each problem. Show your work.

13. A ladder is 15 feet long. It reaches 12 feet up the side of a building. How far from the building is the bottom of the ladder? 9 ft

14. Katie drives from her house to school. She drives 4 miles south, then she drives 10 miles west. What is the shortest distance from Katie's house to school?
$2\sqrt{29}$ miles

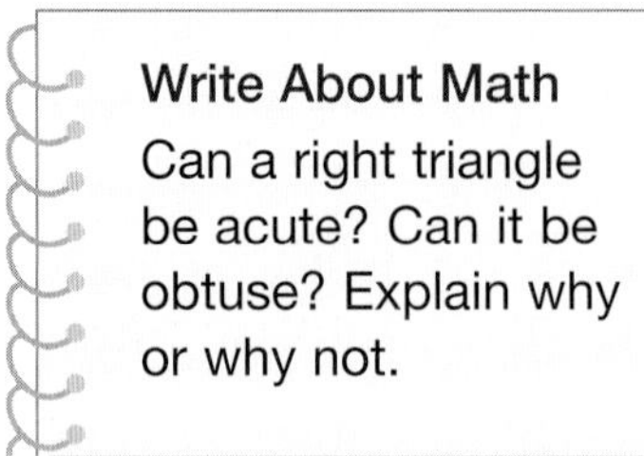

Write About Math

Can a right triangle be acute? Can it be obtuse? Explain why or why not.

A right triangle cannot be acute because the right angle is equal to 90°. All the angles must be less than 90° in an acute triangle. A right triangle cannot be obtuse because one angle in an obtuse triangle must be greater than 90°. In a right triangle, the 90° angle is the largest angle because the sum of the angles in any triangle must be 180°.

ADDITIONAL PRACTICE

Chapter 7: Quadrilaterals and Polygons

1. Name a polygon with 7 sides. heptagon

Complete each statement.

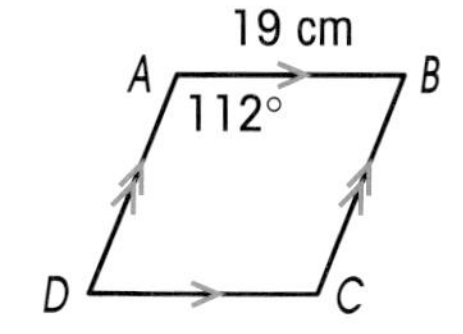

2. $\angle A \cong$ ■ $\angle C$

3. $\overline{AD} \cong$ ■ $\overline{BC}$

4. $\overline{DC} \parallel$ ■ $\overline{AB}$

5. $m\angle B =$ ■ $m\angle D$, or 68

6. A parallelogram with four congruent sides and four right angles is a ____. square

▱RSUT is a rectangle. RX is 12 cm. Find the length of each line segment and the measure of each angle.

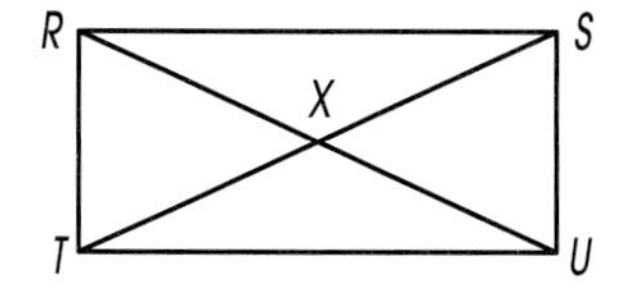

7. $\overline{RU}$ 24 cm

8. $\overline{TS}$ 24 cm

9. $\angle RTU$ 90°

10. $\overline{XU}$ 12 cm

Trapezoid *DEGF* is isosceles. *DF* is 11 meters. Find the length of each line segment and the measure of each angle.

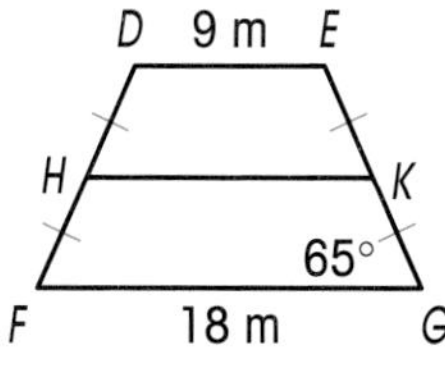

11. $\overline{HK}$ 13.5 m

12. $\angle F$ 65°

13. $\angle E$ 115°

14. $\overline{EG}$ 11 m

Problem Solving

Solve each problem. Show your work.

15. Find the measure of an exterior angle of a regular nonagon. 40°

16. Find the measure of an interior angle of a regular hexagon. 120°

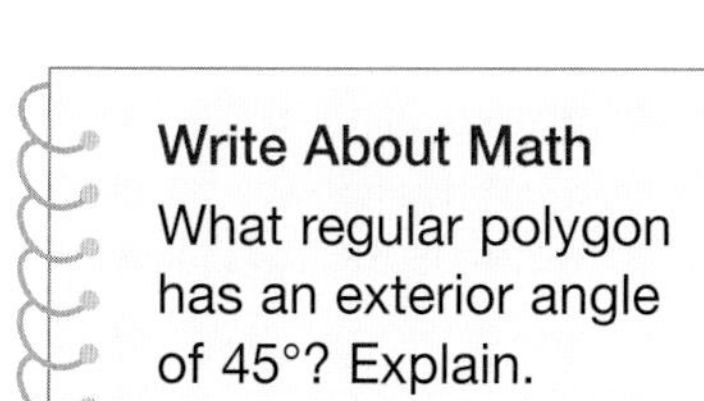

An octagon has an exterior angle of 45°. Possible explanation: $360 \div 45 = 8$. So, the polygon has 8 sides.

ADDITIONAL PRACTICE

Chapter 8: Perimeter and Area

Find the perimeter of each figure.

1.

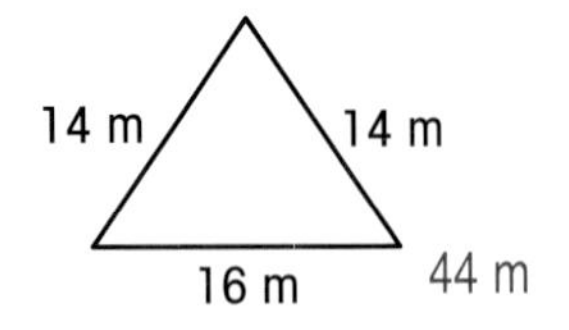

44 m

2.

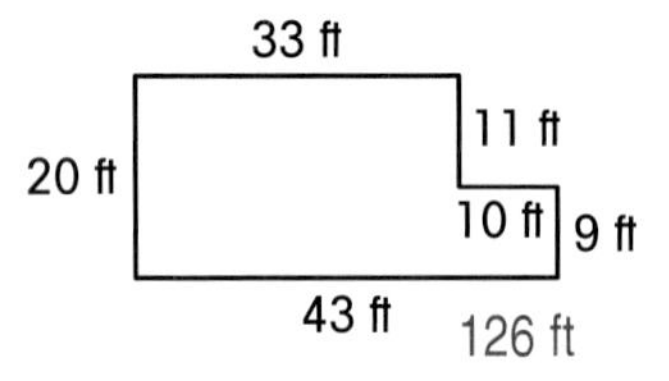

126 ft

3.

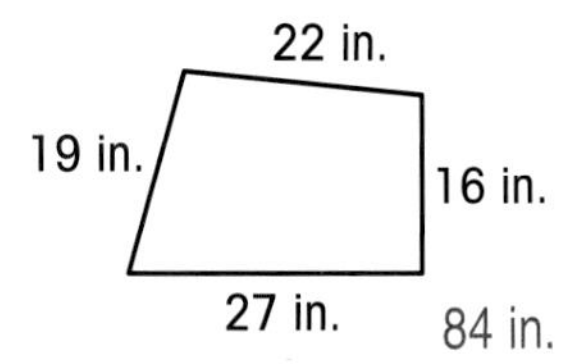

84 in.

Find the area of each figure.

4.

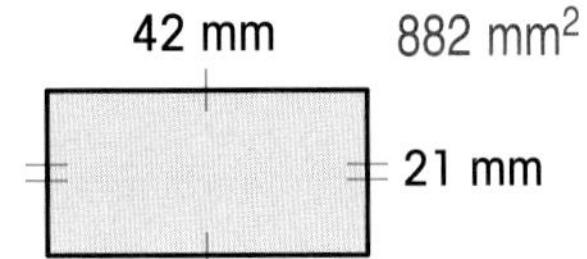

882 mm^2

5.

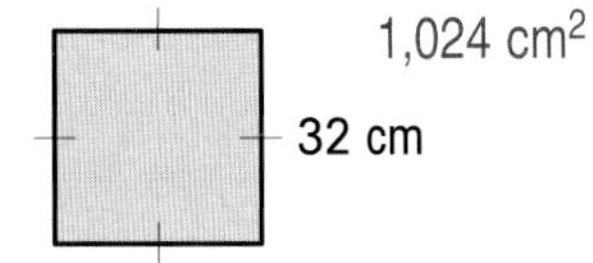

1,024 cm^2

6.

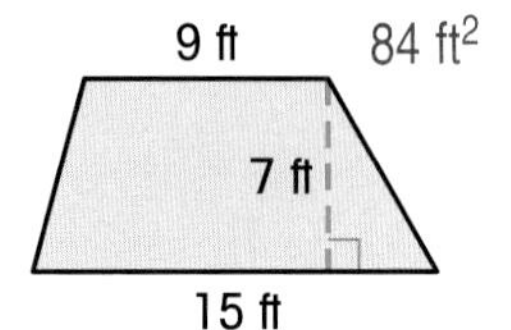

84 ft^2

Find the value of *x* in each diagram.

7.

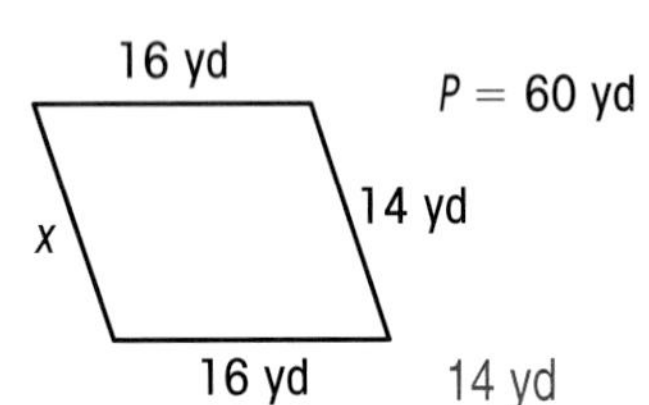

$P = 60$ yd

14 yd

8.

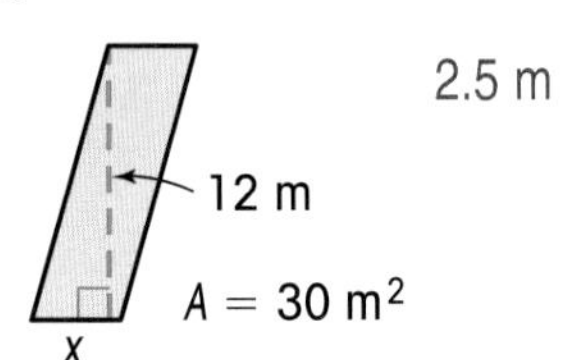

$A = 30$ m^2

2.5 m

9.

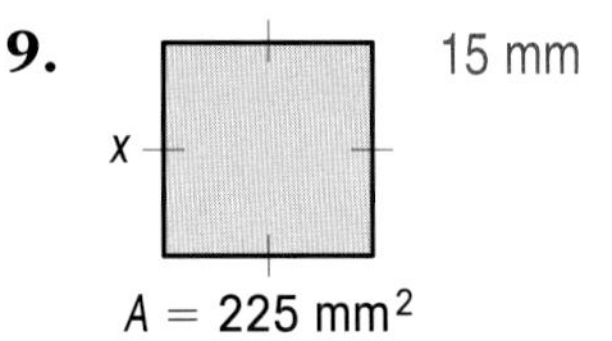

$A = 225$ mm^2

15 mm

Problem Solving

Solve the problem. Show your work.

10. Agnes is carpeting the floor of a closet. The shape of the floor is a trapezoid. The bases measure 72 inches and 90 inches. The height of the trapezoid is 40 inches. What is the area of the carpet she needs? (Hint: Draw a diagram.) 3,240 in.^{2}

Write About Math

If you know the perimeter of a square, can you find its area? Explain why or why not.

Yes. If you know the perimeter of a square, you can determine the area because all sides are congruent. Divide the perimeter by the number of sides, 4. Then, square the length of one side to find the area.

ADDITIONAL PRACTICE

Chapter 9: Similar Polygons

Use the diagram on the right.

1. Write the ratio of blue squares to all squares.
3 to 7

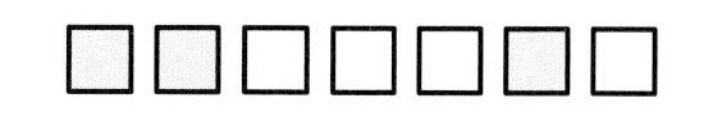

Solve each proportion.

2. $\frac{1}{3} = \frac{25}{x}$ 75 **3.** $\frac{x}{32} = \frac{1}{8}$ 4 **4.** $\frac{3}{x} = \frac{x}{27}$ 9

Each pair of triangles is similar. Find the value of *x*.

5.

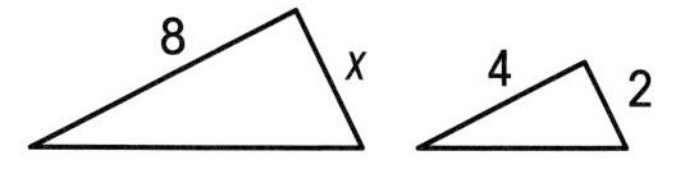

4

6.

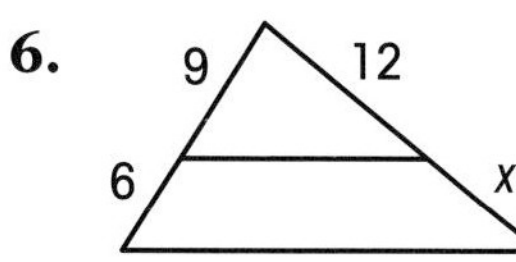

8

7.

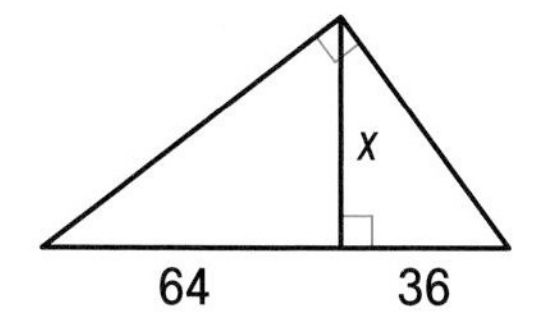

48

8.

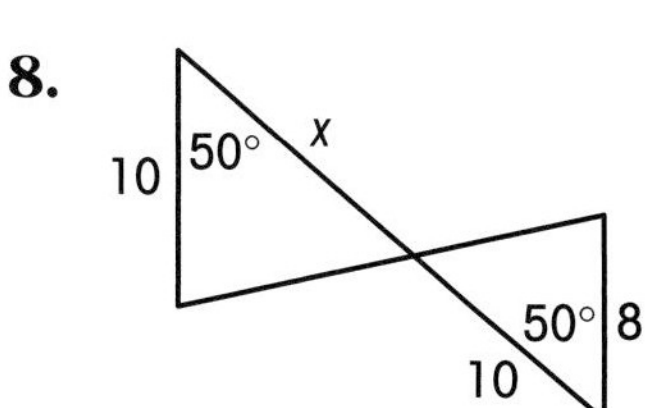

12.5

Find the perimeter or the area.

9. ▱*ABCD* ~ ▱*EFGH*. The ratio of a pair of corresponding sides is 4:5. The perimeter of ▱*ABCD* is 32 cm. Find the perimeter of ▱*EFGH*. 40 cm

10. △*JKL* ~ △*MNP*. The ratio of a pair of corresponding sides is 2 to 10. The area of △*JKL* is 12 cm^2. Find the area of △*MNP*. 300 cm^2

Write About Math In any triangle, there can only be at most one right angle. Therefore both triangles have angles that measure 90°, 45°, and 45°. By the Angle-Angle Similarity Postulate, isosceles right triangles are similar.

Problem Solving

Solve each problem. Show your work.

11. A 5-foot-tall boy casts an 8-foot shadow. At the same time, a tree casts an 18-foot shadow. How tall is the tree? 11.25 ft

12. The scale on a scale drawing is $\frac{1}{8}$ inch = 1 foot. A room on the drawing is $1\frac{3}{4}$ inches. long. What is the actual length of the room? 14 ft

Write About Math
Two right triangles are isosceles. Are the triangles similar? Explain why or why not. See above.

ADDITIONAL PRACTICE
Chapter 10: Circles

Find the circumference and area of each circle. Use 3.14 for π. Round your answers to the nearest whole number.

1.

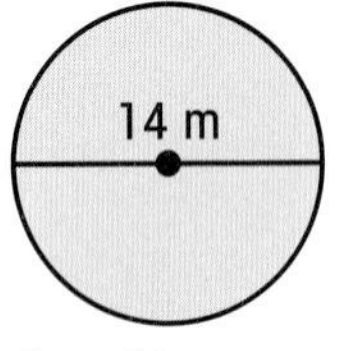

$C \approx 44$ m;
$A \approx 154$ m^2

2.

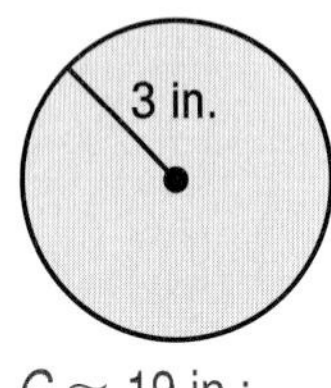

$C \approx 19$ in.;
$A \approx 28$ in.^{2}

3.

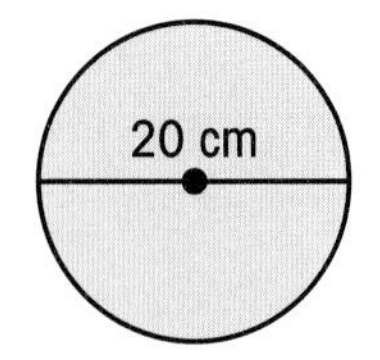

$C \approx 63$ cm;
$A \approx 314$ cm^2

Find the value of x in each figure.

4. 140°

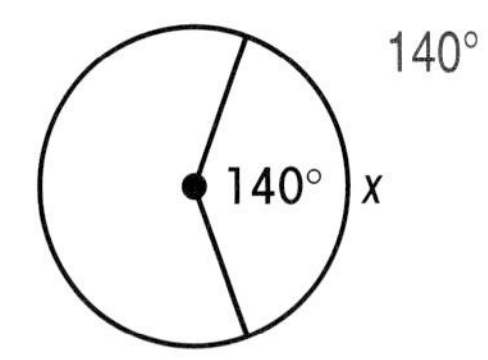

5. 110°

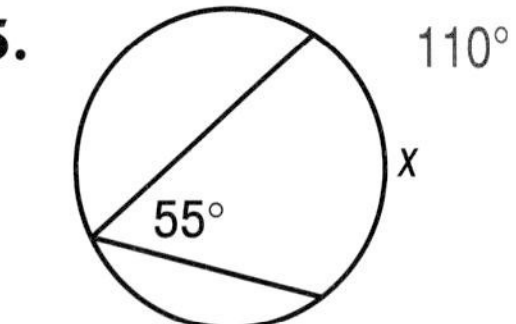

6.

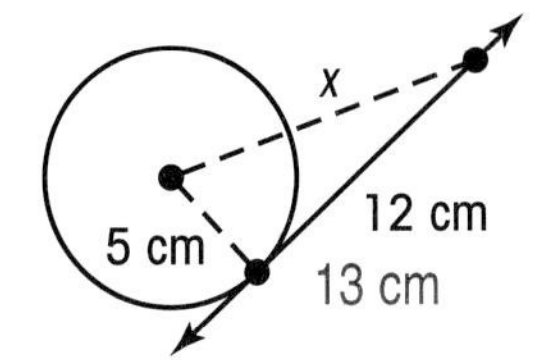

13 cm

7. 40°

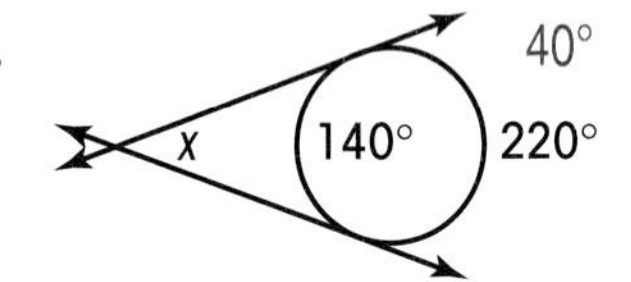

8. 12°

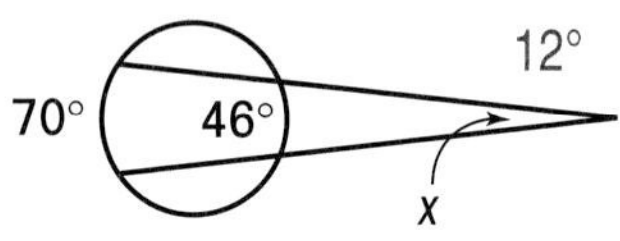

9. 4

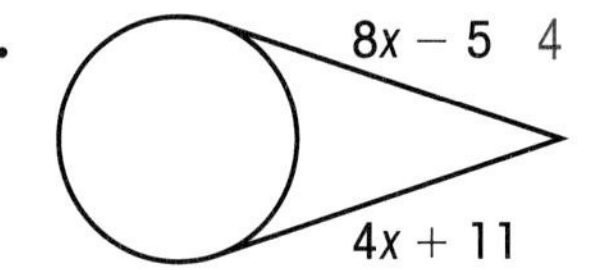

Write About Math. The smaller tire will make more revolutions per mile because the formula for determining the number of revolutions is distance ÷ circumference. The smaller the circumference, the greater the quotient.

Problem Solving

Solve each problem. Show your work. Round your answer to the nearest whole number.

10. Gerard's truck has tires with a diameter of 33 inches. He moved the truck forward 1,200 feet. About how many revolutions did one tire make? about 139 revolutions

11. A circle with a diameter of 12 inches is inscribed in a square. What is the area of the circle? What is the area of the square?
Area of the circle ≈ 113 in.^{2} Area of the square $= 144$ in.^{2}

Write About Math

A vehicle has tires that are two different sizes. Which tire will make more revolutions per mile traveled, the larger tire or the smaller tire? Explain.

ADDITIONAL PRACTICE

Chapter 11: Surface Area and Volume

Name the space figure you can make from each net.

1.

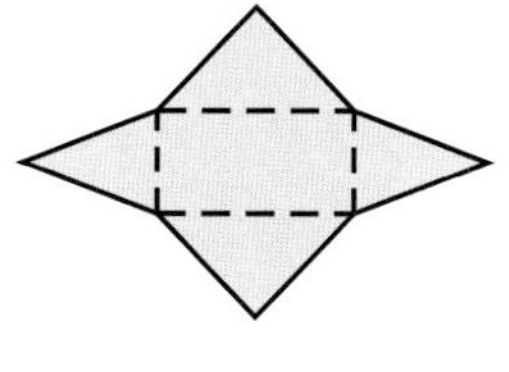

rectangular pyramid

2.

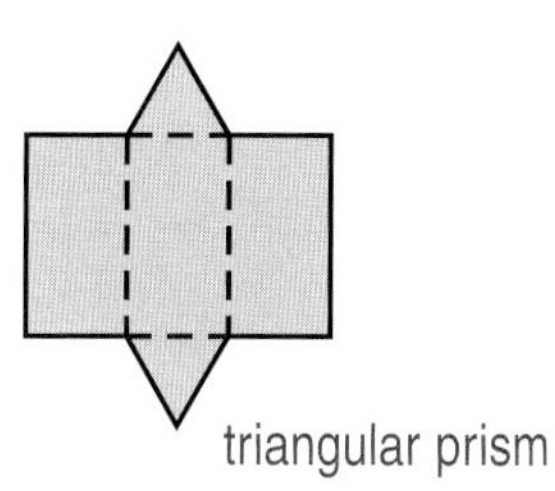

triangular prism

3.

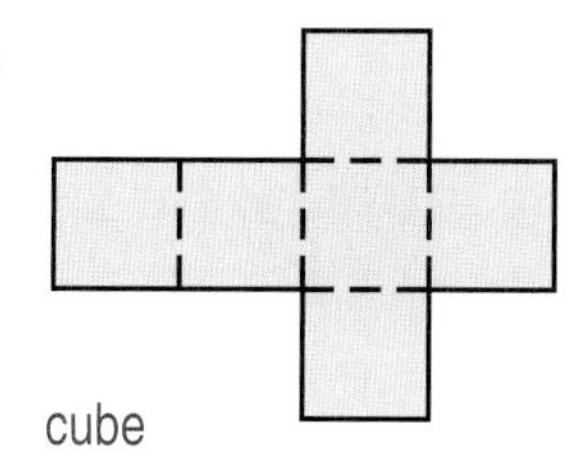

cube

Find the surface area of each space figure. Use 3.14 for π as needed.

4.

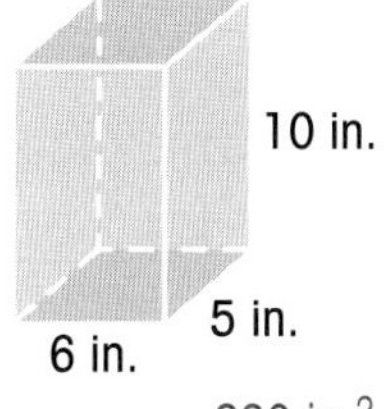

280 in.^{2}

5.

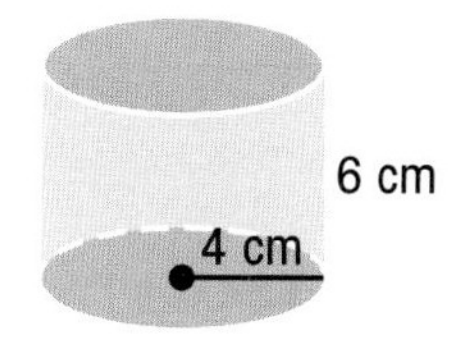

about 251.2 mm^2

6.

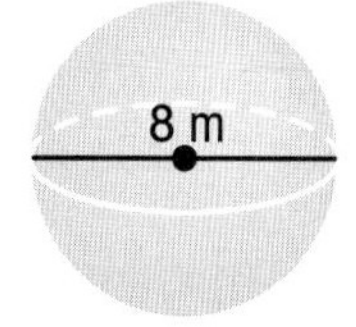

about 201 m^2

Find the volume of each space figure. Use 3.14 for π as needed.

7.

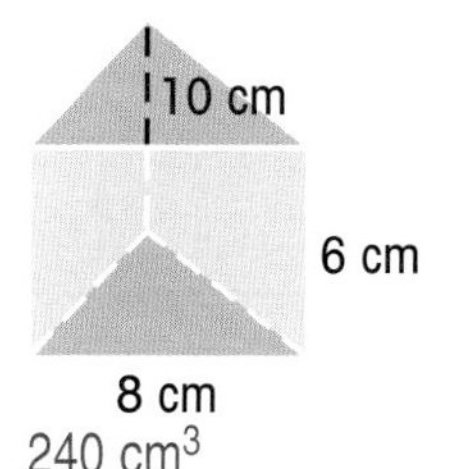

240 cm^3

8.

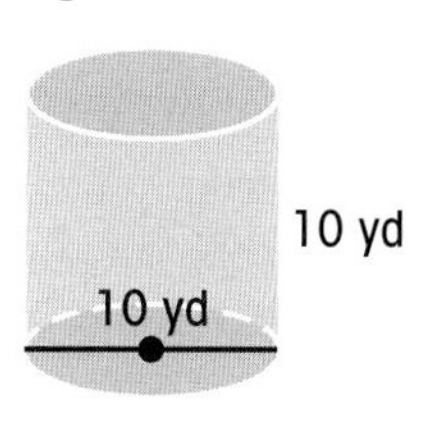

about 785 yd^3

9.

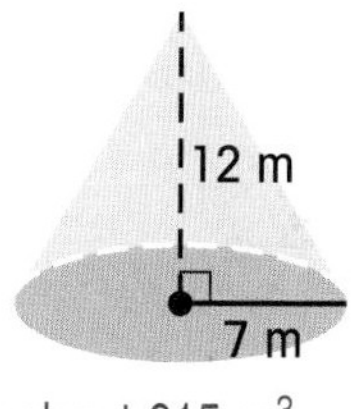

about 615 m^3

Problem Solving

Solve each problem. Show your work.

10. Two cylinders have a similarity ratio of 1 to 3. The volume of the larger cylinder is 270 ft^3. What is the volume of the smaller one? 10 ft^3

11. How much paper is needed to cover all the sides of a cube that measures 7 inches on a side? 294 in.^{2}

Write About Math
Which has a larger volume, a cube with 8-cm sides or a sphere with an 8-cm diameter? Explain.

Write Abouth Math The cube has a larger volume. Because a side of the cube equals the diameter of the sphere, the sphere can fit inside the cube. However, because the sphere is round, it does not fill the space inside the cube completely.

ADDITIONAL PRACTICE
Chapter 12: Coordinate Geometry and Transformations

Draw a coordinate plane on grid paper. Graph each point. See Additional Answers.

1. $A\ (2, {}^-1)$ **2.** $B\ ({}^-4, 3)$

Use formulas for exercises 3–5.

3. Find the distance between $C\ (5, 1)$ and $D\ ({}^-2, 4)$. $\sqrt{58}$

4. $E\ (4, 1)$ and $F\ (0, {}^-3)$ are endpoints of $\overline{EF}$. Find the midpoint. $(2, {}^-1)$

5. $G\ (0, 2)$ and $H\ ({}^-1, 6)$ are points on a line. Find the slope. ${}^-4$

Decide if the lines are parallel or perpendicular.

6. Line 1: points $({}^-1, {}^-2)$ and $({}^-2, 0)$
Line 2: points $({}^-2, 2)$ and $(0, 3)$
perpendicular

7. Line 1: points $({}^-1, 0)$ and $(3, 5)$
Line 2: points $(0, {}^-3)$ and $(4, 2)$
parallel

Use grid paper. Draw the image of $\triangle ABC$ for each transformation. See Additional Answers.

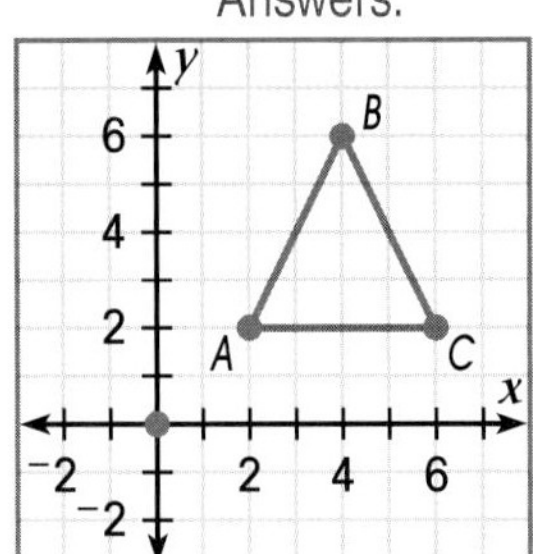

8. $T\ (x, y) \longrightarrow (x - 4, y + 1)$
$A'\ ({}^-2, 3)$, $B'\ (0, 7)$, $C'\ (2, 3)$

9. Rotate $\triangle ABC$ 180° counterclockwise about the origin.
$A'\ ({}^-2, {}^-2)$, $B'\ ({}^-4, {}^-6)$, $C'\ ({}^-6, {}^-2)$

10. Reflect $\triangle ABC$ over the y-axis.
$A'\ ({}^-2, 2)$, $B'\ ({}^-4, 6)$, $C'\ ({}^-6, 2)$

11. Dilate $\triangle ABC$ with a similarity ratio of $\frac{3}{2}$.
$A'\ (3, 3)$, $B'\ (6, 9)$, $C'\ (9, 3)$

Use what you know about points in space for exercises 12–14.

12. Use grid paper to graph $P\ (1, 5, 2)$.
See Additional Answers.

13. Find the distance between $M\ (4, 2, 0)$ and $N\ (2, 6, 6)$.
$2\sqrt{14}$

14. $Q\ (1, 6, 4)$ and $R\ (1, 2, 2)$ are endpoints of $\overline{QR}$. Find the midpoint.
$(1, 4, 3)$

Problem Solving

Solve the problem. Show your work.
southwest; 17 mph

15. The speed of a boat in still water is 8 mph. The water current flows south at 15 mph. The boat heads west across the river. In which direction will the boat travel? How fast will the boat travel?

Write About Math
If two points have the same y-coordinate, what is the slope of the line between them? How do you know?

Write About Math The slope will always be 0. Points with the same y-coordinate will always form a horizontal line with zero slope. Slope is the difference in y-coordinates divided by the difference in x-coordinates and $y_2 - y_1$ will always be 0 when $y_1 = y_2$. Zero divided by any number is 0.

ADDITIONAL PRACTICE
Chapter 13: Right Triangle Trigonometry

Use diagram on the right.

1. Find three trigonometric ratios for $\angle A$.

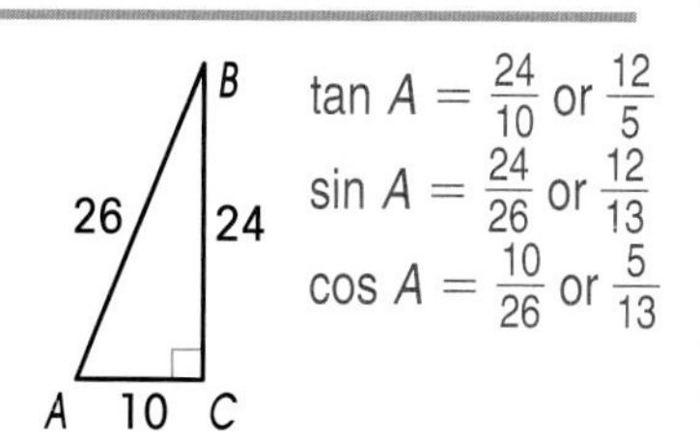

$\tan A = \frac{24}{10}$ or $\frac{12}{5}$
$\sin A = \frac{24}{26}$ or $\frac{12}{13}$
$\cos A = \frac{10}{26}$ or $\frac{5}{13}$

Write three trigonometric ratios for each angle. Use the Table of Trigonometric Ratios on page 426.

2. 12° tan = 0.2126; cos = 0.9781; sin = 0.2079

3. 43° tan = 0.9325; cos = 0.7314; sin = 0.6820

4. 90° tan = undefined; cos = 0.0000; sin = 1.00000

Find the measure of $\angle D$ to the nearest degree. Use the table on page 426.

5. 42°

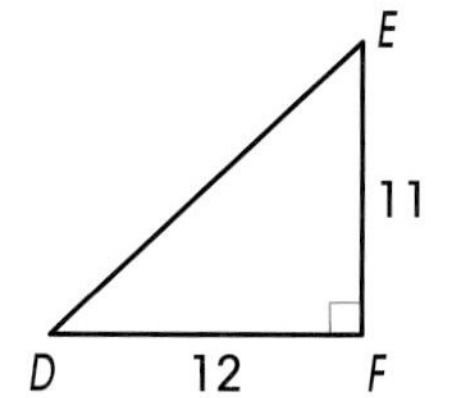

6. 34°

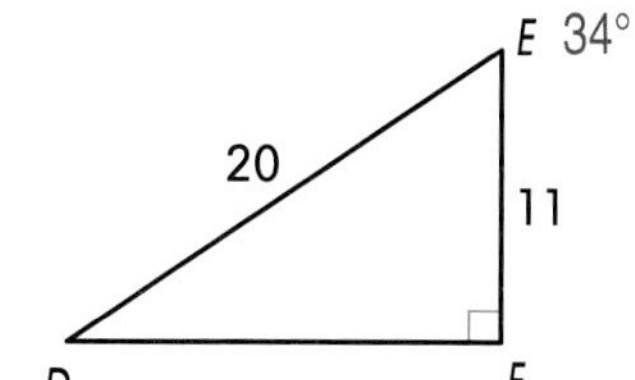

7. 48°

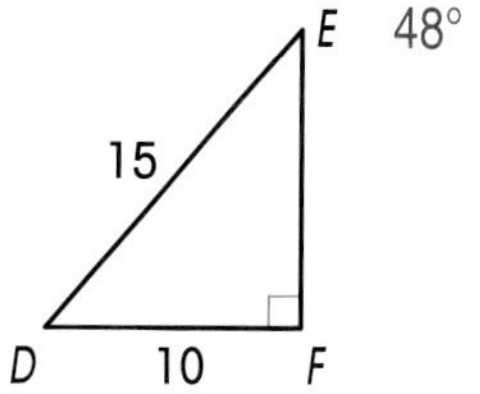

Find the value of *a* in each diagram. Use the table on page 426. Round the value to the nearest whole number.

8. 17

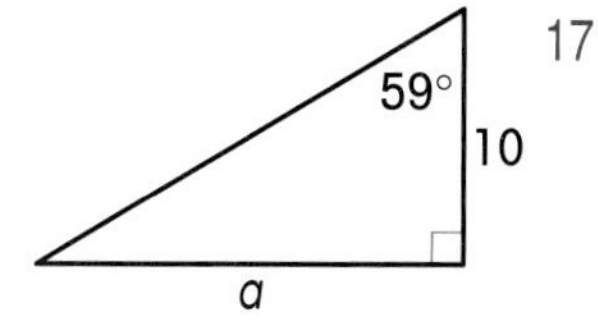

9. 12

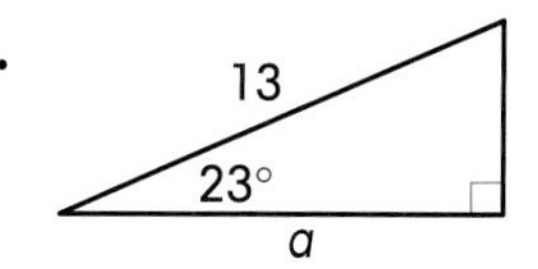

10. 11

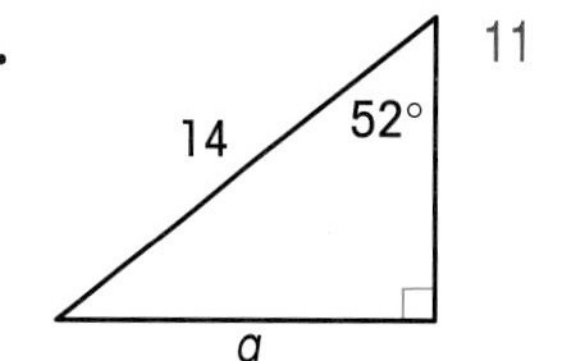

Problem Solving

Solve each problem. Show your work.

11. Alice stands at a window looking for Lewis. She spots him 20 feet away from the house, on a patio 25 feet below her. What is the angle of depression as Alice looks at Lewis? 51°

12. Urbano stands at the top of a 100-foot-tall cliff. He spots a boat 600 feet off shore. What is the angle of elevation from the boat to Urbano? 10°

> **Write About Math**
> Can a 30°-60°-90° triangle have legs that are lengths of 6 cm and 8 cm long? Explain.

Write About Math No. Possible answer: The length of the longer leg of a 30°-60°-90° triangle is the length of the shorter leg multiplied by $\sqrt{3}$. Because $6\sqrt{3} \neq 8$, the triangle cannot have side lengths of 6 and 8.

TWO-COLUMN PROOF: Lesson 1·10

Line Segments *pages 26–27*

THEOREM 2
Midpoint Theorem

The midpoint M of $\overline{AB}$ divides $\overline{AB}$ in half so that $AM = \frac{1}{2}AB$.

Getting Started Review the definition of *midpoint* with the students.

You are given:

M is the midpoint of $\overline{AB}$.

Prove: $AM = \frac{1}{2}AB$

Statements	Reasons
1. M is the midpoint of $\overline{AB}$.	1. Given
2. $\overline{AM} \cong \overline{MB}$	2. Definition of a midpoint
3. $AM = MB$	3. Definition of congruent line segments
4. $AB = AM + MB$	4. Segment Addition Postulate
5. $AB = AM + AM$	5. Substitution Property
6. $AB = 2AM$	6. Addition
7. $AM = \frac{1}{2}AB$	7. Divide both sides of the equation by 2.

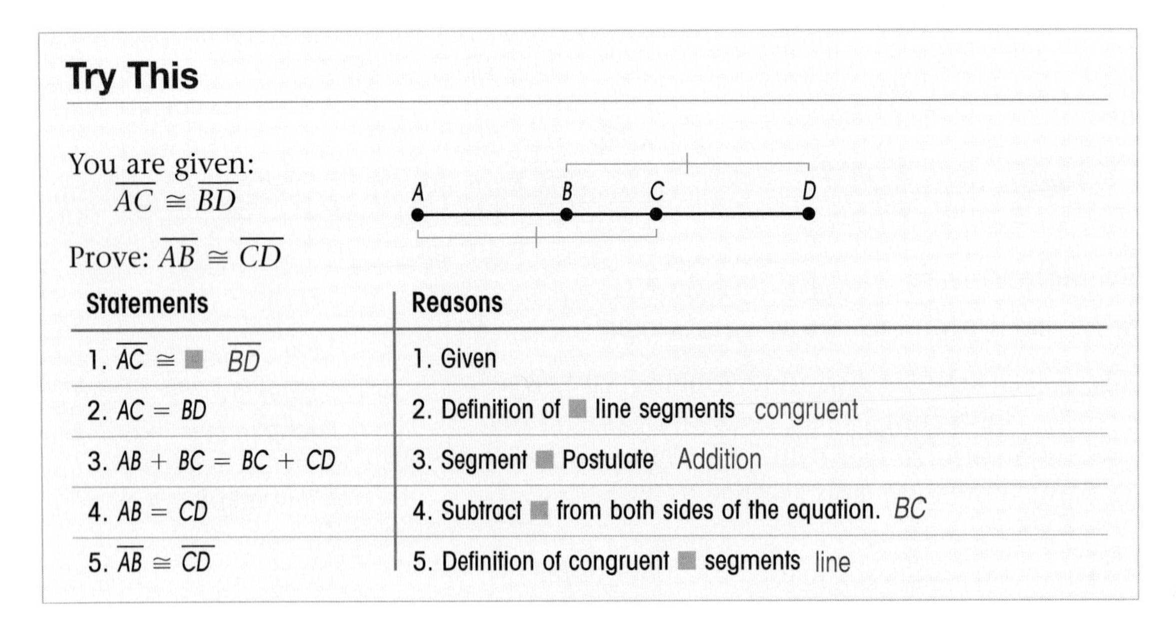

Try This

You are given:

$\overline{AC} \cong \overline{BD}$

Prove: $\overline{AB} \cong \overline{CD}$

Statements	Reasons
1. $\overline{AC} \cong$ ■ $\overline{BD}$	1. Given
2. $AC = BD$	2. Definition of ■ line segments congruent
3. $AB + BC = BC + CD$	3. Segment ■ Postulate Addition
4. $AB = CD$	4. Subtract ■ from both sides of the equation. BC
5. $\overline{AB} \cong \overline{CD}$	5. Definition of congruent ■ segments line

Practice is provided on page 27 of the student book. Have students write the practice proof in two-column form. **More Practice** is provided in the Classroom Resource Binder.

TWO-COLUMN PROOF: Lesson 2·12

Angles *pages 58–59*

THEOREM 4 Angle Bisector Theorem

The bisector $\overrightarrow{BD}$ of $\angle ABC$ divides $\angle ABC$ in half so that $m\angle ABD = \frac{1}{2}m\angle ABC$.

Getting Started Review the proofs in Chapter 1.

You are given:

$\overrightarrow{BD}$ is the bisector of $\angle ABC$.

Prove: $m\angle ABD = \frac{1}{2}m\angle ABC$

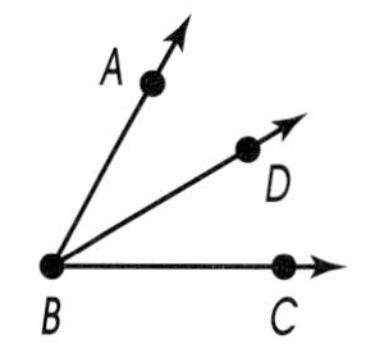

Statements	Reasons
1. $\overrightarrow{BD}$ bisects $\angle ABC$	1. Given
2. $\angle ABD \cong \angle DBC$	2. Definition of an angle bisector
3. $m\angle ABD = m\angle DBC$	3. Definition of congruent angles
4. $m\angle ABC = m\angle ABD + m\angle DBC$	4. Angle Addition Postulate
5. $m\angle ABC = m\angle ABD + m\angle ABD$	5. Substitution Property
6. $m\angle ABC = 2m\angle ABD$	6. Addition
7. $m\angle ABD = \frac{1}{2}m\angle ABC$	7. Divide both sides of the equation by 2.

Try This

You are given:

$\angle ABE \cong \angle DBC$

Prove: $\angle ABD \cong \angle EBC$

Statements	Reasons
1. $\angle ABE \cong$ ■ $\angle DBC$	1. Given
2. $m\angle ABE = m\angle DBC$	2. Definition of ■ angles congruent
3. $m\angle ABD + m\angle DBE = m\angle DBE + m\angle EBC$	3. Angle ■ Postulate Addition
4. $m\angle ABD = m\angle EBC$	4. Subtract ■ from both sides of the equation. $m\angle DBE$
5. $\angle ABD \cong \angle EBC$	5. Definition of congruent ■ angles

Practice is provided on page 59 of the student book. Have students write the practice proof in two-column form. **More Practice** is provided in the Classroom Resource Binder.

TWO-COLUMN PROOF: Lesson 3·11

Angles and Line Segments *pages 84–85*

THEOREM 6

If two angles are supplementary to the same angle, then they are congruent.

Getting Started Review supplementary and complementary angles.

You are given: $\angle A$ and $\angle B$ are supplementary.
$\angle C$ and $\angle B$ are supplementary.

Prove: $\angle A \cong \angle C$

Statements	Reasons
1. $\angle A$ and $\angle B$ are supplementary. $\angle C$ and $\angle B$ are supplementary.	1. Given
2. $m\angle A + m\angle B = 180$ $m\angle C + m\angle B = 180$	2. Definition of supplementary angles
3. $m\angle A + m\angle B = m\angle C + m\angle B$	3. Transitive Property
4. $m\angle A = m\angle C$	4. Subtraction Property
5. $\angle A \cong \angle C$	5. Definition of congruent angles

Try This

You are given:
$\overleftrightarrow{AB}$ intersects $\overleftrightarrow{CD}$.

Prove: $\angle 1 \cong \angle 3$

C B 2 1 3 4 A D

Statements	Reasons
1. $\overleftrightarrow{AB}$ ■ $\overleftrightarrow{CD}$. intersects	1. Given
2. $\angle 1$ and $\angle 2$ form a straight angle. $\angle 2$ and $\angle 3$ form a straight angle.	2. Given in the diagram
3. $m\angle 1 + m\angle 2 = 180$ $m\angle 2 + m\angle 3 =$ ■ 180	3. Definition of a ■ angle straight
4. $\angle 1 \cong \angle 3$	4. Two angles supplementary to the same angle are ■. congruent

Practice is provided on page 85 of the student book. **More Practice** is provided in the Classroom Resource Binder.

TWO-COLUMN PROOF: Lesson 4-11
Proving Lines Are Parallel *pages 116–117*

POSTULATE 8

If two lines are cut by a transversal so that the alternate interior angles are congruent, then the lines are parallel.

Getting Started Point out that the above postulate begins with the angles. Since the alternate interior angles are congruent, the lines are parallel.

You are given:
$\angle 3$ and $\angle 5$ are supplementary angles.

Prove: $a \parallel b$

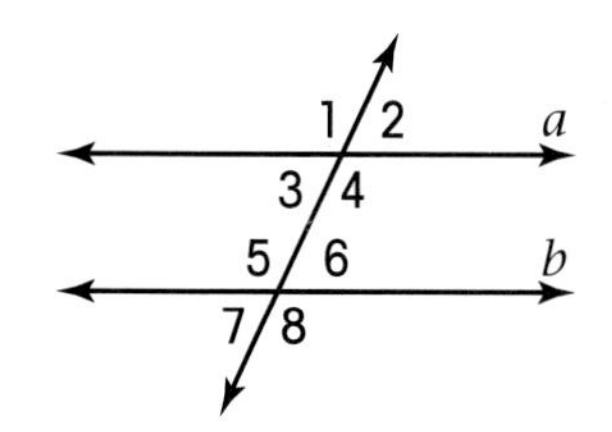

Statements	Reasons
1. $\angle 3$ and $\angle 5$ are supplementary.	1. Given
2. $m\angle 3 + m\angle 5 = 180$	2. Definition of supplementary angles
3. $\angle 5$ and $\angle 6$ form a straight angle.	3. Given in diagram
4. $m\angle 5 + m\angle 6 = 180$	4. Definition of a straight angle
5. $m\angle 3 + m\angle 5 = m\angle 5 + m\angle 6$	5. Transitive Property
6. $m\angle 3 = m\angle 6$	6. Subtraction Property
7. $\angle 3 \cong \angle 6$	7. Definition of congruent angles
8. $a \parallel b$	8. Postulate 8

Try This

You are given:
$a \perp c$ and $b \perp c$

Prove: $a \parallel b$

Statements	Reasons
1. $a \perp c$ and $b \perp$ ■ c	1. Given
2. $\angle 3$ and $\angle 5$ are right angles.	2. Perpendicular lines form ■ angles. right
3. $m\angle 3 = 90$ and $m\angle 5 =$ ■ 90	3. Definition of a ■ angle right
4. $\angle 3$ and $\angle 5$ are ■ supplementary	4. Definition of supplementary angles
5. $a \parallel b$	5. Proof above

Practice is provided on page 117 of the student book. **More Practice** is provided in the Classroom Resource Binder.

TWO-COLUMN PROOF: Lesson 5·16
CPCTC *pages 154–155*

Math Fact
Corresponding parts of congruent triangles are congruent. This statement is known as CPCTC.

Getting Started Review the corresponding parts of two congruent triangles.

You are given: $\overline{AB} \parallel \overline{CD}$
E is the midpoint of $\overline{AD}$.

Prove: $\overline{AB} \cong \overline{DC}$

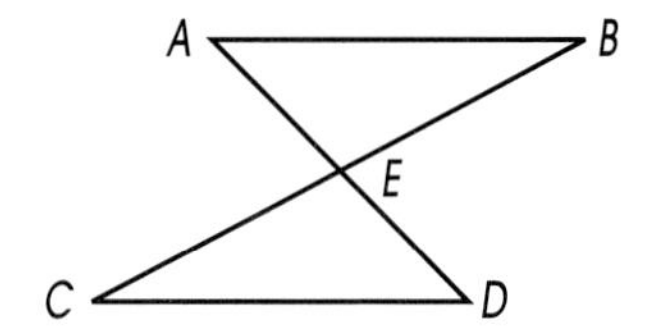

Statements	Reasons
1. $\overline{AB} \parallel \overline{CD}$	1. Given
2. $\angle BAE \cong \angle CDE$	2. Alternate interior angles are congruent.
3. E is the midpoint of $\overline{AD}$.	3. Given
4. $\overline{AE} \cong \overline{DE}$	4. Definition of a midpoint
5. $\angle AEB \cong \angle DEC$	5. Vertical angles are congruent.
6. $\triangle AEB \cong \triangle DEC$	6. ASA
7. $\overline{AB} \cong \overline{DC}$	7. CPCTC

Try This

You are given: $\triangle JKL$ is isosceles.
$\overline{JK} \cong \overline{LK}$
$\overline{KM}$ bisects $\angle K$.

Prove: $\overline{JM} \cong \overline{LM}$

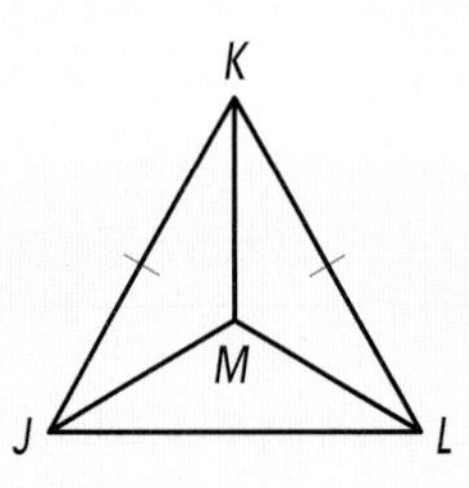

Statements	Reasons
1. $\overline{JK} \cong \overline{LK}$	1. Given
2. $\overline{KM}$ bisects ■. $\angle K$	2. Given
3. $\angle JKM \cong \angle LKM$	3. Definition of an ■ bisector angle
4. $\overline{KM} \cong \overline{KM}$	4. ■ Property Reflexive
5. $\triangle JKM \cong$ ■ $\triangle LKM$	5. SAS
6. $\overline{JM} \cong \overline{LM}$	6. CPCTC

Practice is provided on page 155 of the student book. **More Practice** is provided in the Classroom Resource Binder.

TWO-COLUMN PROOF: Lesson 6·11

Hypotenuse–Leg Theorem *pages 180–181*

THEOREM 20
Hypotenuse–Leg Theorem

If the hypotenuse and one leg of one right triangle are congruent to the hypotenuse and one leg of another right triangle, then the triangles are congruent.

Getting Started Review congruent triangles.

You are given:

$\angle B$ and $\angle D$ are right angles.
$\overline{AB} \cong \overline{CD}$

Prove: $\triangle ABC \cong \triangle CDA$

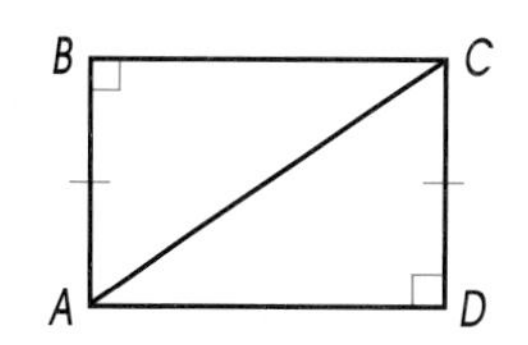

Statements	Reasons
1. $\angle B$ and $\angle D$ are right angles.	1. Given
2. $\triangle ABC$ and $\triangle CDA$ are right triangles.	2. Definition of a right triangle
3. $\overline{AB} \cong \overline{CD}$	3. Given
4. $\overline{AC} \cong \overline{CA}$	4. Reflexive Property
5. $\triangle ABC \cong \triangle CDA$	5. Hypotenuse–Leg Theorem

Try This

You are given: $\angle H$ and $\angle K$ are right angles.
$\overline{HK}$ bisects $\overline{GJ}$.
$\overline{GH} \cong \overline{JK}$

Prove: $\triangle GHI \cong \triangle JKI$

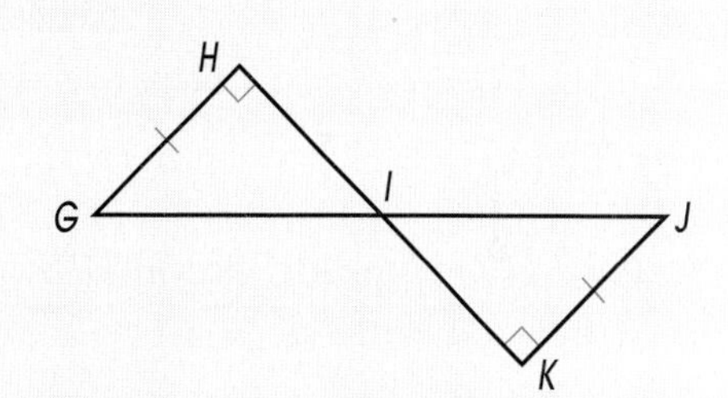

Statements	Reasons
1. $\angle H$ and $\angle K$ are ■ angles. right	1. Given
2. $\triangle GHI$ and $\triangle JKI$ are right triangles.	2. Definition of a ■ triangle right
3. $\overline{GH} \cong$ ■ $\overline{JK}$	3. Given
4. $\overline{HK}$ bisects $\overline{GJ}$.	4. ■ Given
5. $\overline{GI} \cong \overline{JI}$	5. Definition of a ■ bisector
6. $\triangle GHI \cong$ ■ $\triangle JKI$	6. Hypotenuse–Leg Theorem

Practice is provided on page 181 of the student book. **More Practice** is provided in the Classroom Resource Binder.

TWO-COLUMN PROOF: Lesson 7·10

Proving a Quadrilateral Is a Parallelogram *pages 204–205*

THEOREM 24

If a quadrilateral has diagonals that bisect each other, then the quadrilateral is a parallelogram.

Getting Started Review the properties of a parallelogram.

You are given:

quadrilateral $ABCD$
$\overline{AC}$ bisects $\overline{BD}$.
$\overline{BD}$ bisects $\overline{AC}$.

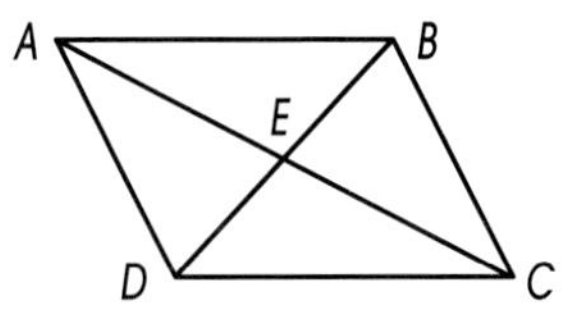

Prove: Quadrilateral $ABCD$ is a parallelogram.

Statements	Reasons
1. $\overline{AC}$ bisects $\overline{BD}$. $\overline{BD}$ bisects $\overline{AC}$.	1. Given
2. $\overline{BE} \cong \overline{DE}$. $\overline{AE} \cong \overline{CE}$	2. Definition of a bisector
3. $\angle BEA \cong \angle DEC$	3. Vertical angles are congruent.
4. $\triangle BEA \cong \triangle DEC$	4. SAS
5. $\angle BAE \cong \angle DCE$	5. CPCTC
6. $\overline{AB} \parallel \overline{CD}$ Similarly, $\overline{BC} \parallel \overline{DA}$.	6. If alternate interior angles are congruent, then the lines are parallel.
7. Quadrilateral $ABCD$ is a parallelogram.	7. Definition of a parallelogram

Try This

You are given: quadrilateral $HIJK$
$\angle H \cong \angle J$ and $\angle K \cong \angle I$

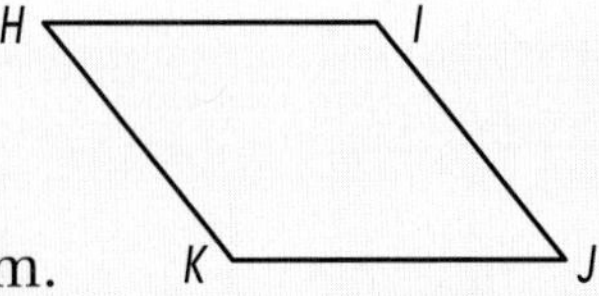

Prove: Quadrilateral $HIJK$ is a parallelogram.

Statements	Reasons
1. $\angle H \cong$ ■; $\angle K \cong$ ■ $\angle J$; $\angle I$	1. Given
2. $m\angle H =$ ■; $m\angle K =$ ■ $m\angle J$; $m\angle I$	2. Definition of congruent angles
3. $2m\angle H + 2m\angle K = 360$	3. Sum of the interior angles of a quadrilateral
4. $m\angle H + m\angle K =$ ■ 180	4. Division Property of Equality
5. $\overline{HI} \parallel \overline{JK}$ Similar, $\overline{HK} \parallel \overline{IJ}$.	5. If same-side ■ angles are supplementary, then the lines are parallel. interior
6. Quadrilateral $HIJK$ is a parallelogram.	6. Definition of a parallelogram

Practice is provided on page 205 of the student book. **More Practice** is provided in the Classroom Resource Binder.

TWO-COLUMN PROOF: Lesson 8·9

Equal Areas *pages 228–229*

THEOREM 25

Two triangles have equal areas if the heights are congruent and the bases are congruent.

Getting Started Review the triangle area formula.

You are given: $\overline{BD} \parallel \overline{AC}$

Prove: Area $\triangle ABC$ = Area $\triangle ADC$

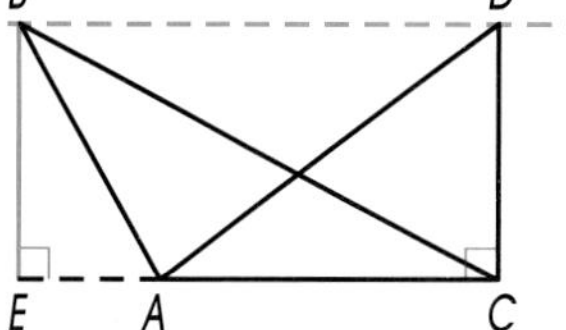

Statements	Reasons
1. $\overline{BD} \parallel \overline{AC}$	1. Given
2. $BE = DC$	2. Definition of parallel lines
3. $AC = AC$	3. Reflexive Property
4. $\frac{1}{2}(AC)(BE) = \frac{1}{2}(AC)(DC)$	4. Substitution and Multiplication Properties
5. Area $\triangle ABC$ = Area $\triangle ADC$	5. Definition of area of a triangle

Try This

You are given:

K is the midpoint of $\overline{HI}$.

$\overline{JL} \parallel \overline{GI}$ and $\overline{GH} \parallel \overline{LI}$

Prove: Area $\triangle GHI$ = Area ▱$GJLI$

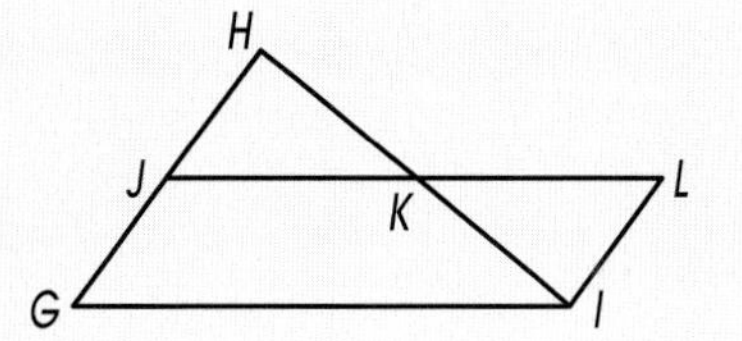

Statements	Reasons
1. K is the ■ of $\overline{HI}$. midpoint	1. Given
2. $\overline{KH} \cong \overline{KI}$	2. Definition of a ■ midpoint
3. $\angle HKJ \cong$ ■ $\angle IKL$	3. Vertical angles are congruent.
4. $\overline{GH} \parallel \overline{LI}$	4. ■ Given
5. $\angle HJK \cong \angle ILK$	5. Alternate interior angles are congruent.
6. $\triangle JHK \cong \triangle$■ LIK	6. AAS
7. Area $\triangle JHK$ = Area $\triangle$■ $\angle LIK$	7. Congruent triangles have equal areas.
8. Area $\triangle JHK$ + Area trapezoid $GJKI$ = Area $\triangle LIK$ + Area trapezoid $GJKI$	8. Addition Property of Equality
9. Area $\triangle GHI$ = Area ▱$GJLI$	9. Substitution Property

Practice is provided on page 229 of the student book. Have students write the practice proof in two-column form. **More Practice** is provided in the Classroom Resource Binder.

TWO-COLUMN PROOF: Lesson 9·14
Angle-Angle Similarity Postulate *pages 260–261*

POSTULATE 12
Angle-Angle Similarity Postulate

If two angles of a triangle are congruent to two angles of another triangle, then the two triangles are similar.

Getting Started Have students list postulates and theorems about triangles that they already know.

You are given: $\overline{AD} \parallel \overline{CB}$
$\overline{AB}$ intersects $\overline{CD}$.

Prove: $\triangle ADE \sim \triangle BCE$

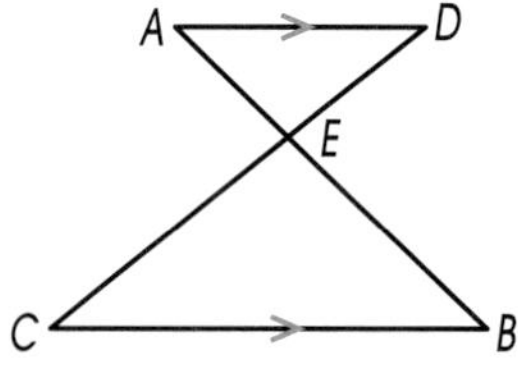

Statements	Reasons
1. $\overline{AD} \parallel \overline{CB}$	1. Given
2. $\angle A \cong \angle B$ $\angle D \cong \angle C$	2. Alternate interior angles of parallel lines are congruent.
3. $\triangle ADE \cong \triangle BCE$	3. Angle-Angle Similarity Postulate

Try This

You are given: $\triangle LMN$ is a right triangle.
$\overline{MP} \perp \overline{LN}$

Prove: $\triangle LMN \sim \triangle LPM$

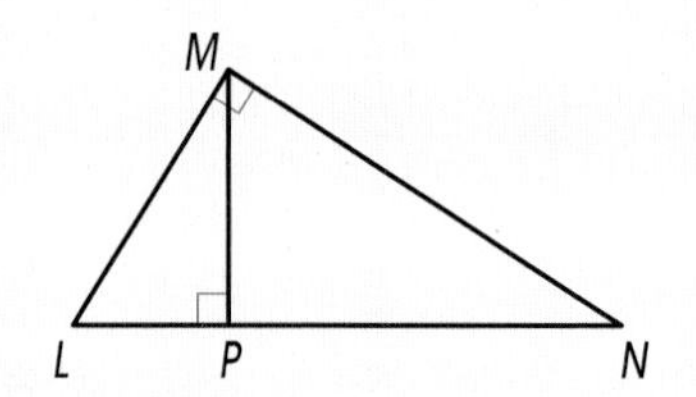

Statements	Reasons
1. $\overline{MP} \perp \overline{LN}$	1. ■ Given
2. $\angle MPL$ is a right angle.	2. Definition of ■ perpendicular lines
3. $\angle LMN$ is a right angle.	3. ■ Given
4. $\angle MPL \cong \angle LMN$	4. All right angles are ■. congruent
5. $\angle L \cong$ ■ $\angle L$	5. Reflexive Property
6. $\triangle LMN \sim \triangle LPM$	6. Angle-Angle ■ Postulate Similarity

Practice is provided on page 261 of the student book. **More Practice** is provided in the Classroom Resource Binder.

TWO-COLUMN PROOF: Lesson 10·15

Circles *pages 294–295*

THEOREM 34

If two inscribed angles of a circle intercept the same arc, then the angles are congruent.

Getting Started Ask students how an inscribed angle in a circle is formed. [by 2 chords with a common point on the circle]

You are given: $\angle DEF$ intercepts $\widehat{DF}$.
$\angle DGF$ intercepts $\widehat{DF}$.

Prove: $\angle DEF \cong \angle DGF$

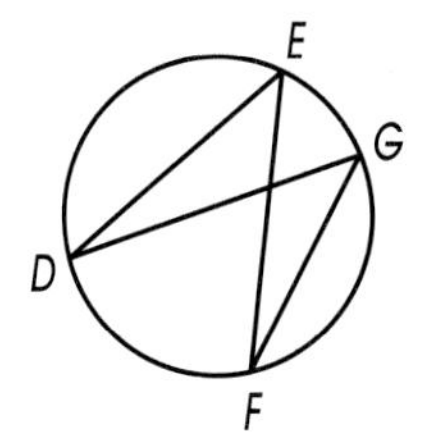

Statements	Reasons
1. $\angle DEF$ intercepts $\widehat{DF}$. $\angle DGF$ intercepts $\widehat{DF}$.	1. Given
2. $m\angle DEF = \frac{1}{2}m\widehat{DF}$ $m\angle DGF = \frac{1}{2}m\widehat{DF}$	2. Definition of an inscribed angle
3. $m\angle DEF = m\angle DGF$	3. Transitive Property
4. $\angle DEF \cong \angle DGF$	4. Definition of congruent angles

Try This

You are given: $\overline{HI} \cong \overline{JK}$

Prove: $\overline{LP} \cong \overline{MP}$

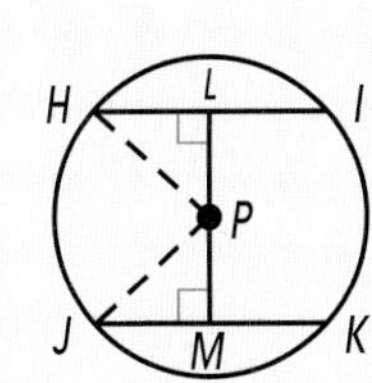

Statements	Reasons
1. $\overline{HI} \cong \overline{JK}$	1. ■ Given
2. $\overline{LM}$ bisects $\overline{HI}$ and $\overline{JK}$.	2. A line segment from the center and perpendicular to a chord ■ the chord. bisects
3. $\overline{HL} \cong \overline{JM}$	3. Congruent segments that are bisected have congruent parts.
4. $\overline{HP} \cong \overline{JP}$	4. Radii of the same circle are ■. congruent
5. $\triangle HLP \cong$ ■ $\triangle JMP$	5. Hypotenuse–Leg Theorem
6. $\overline{LP} \cong$ ■ $\overline{MP}$	6. CPCTC

Practice is provided on page 295 of the student book. **More Practice** is provided in the Classroom Resource Binder.

TWO-COLUMN PROOF: Lesson 11·14

Volume of Figures *pages 328–329*

Getting Started Have students compare a rectangular prism with a cube.

You are given:

a cube with side s

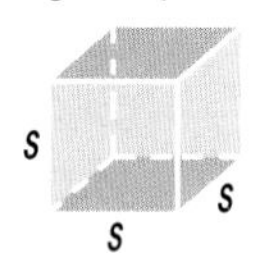

Prove: Volume $= s^3$

Statements	Reasons
1. Cube with side s	1. Given
2. A cube is a prism.	2. Definiton of a cube
3. $V = Bh$	3. Volume of a prism
4. $B = s^2$	4. Area of a square
5. $h = s$	5. The height is a side of the cube.
6. $V = s^2 \bullet s$	6. Substitution
7. $V = s^3$	7. By multiplication

Try This

You are given: a hemisphere with radius r

Prove: Volume $= \frac{2}{3}\pi r^3$

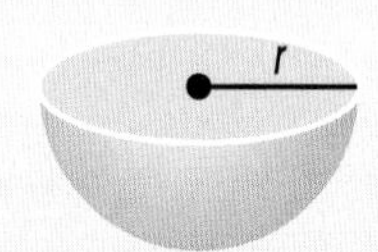

Statements	Reasons
1. A hemisphere with radius r	1. ■ Given
2. A hemisphere is one-half a ■. sphere	2. Definition of a hemisphere
3. $V = \frac{1}{2}(\frac{4}{3}\pi r^3)$	3. Volume of a ■ hemisphere
4. $V = \frac{2}{3}\pi r^3$	4. By ■ multiplication

Practice is provided on page 329 of the student book. **More Practice** is provided in the Classroom Resource Binder.

TWO-COLUMN PROOF: Lesson 12·16

Coordinate Proofs *pages 364–365*

Getting Started Review the formulas for slope and distance.

You are given:

▱*ABCD* is a square.

Prove: The diagonals of a square are perpendicular.

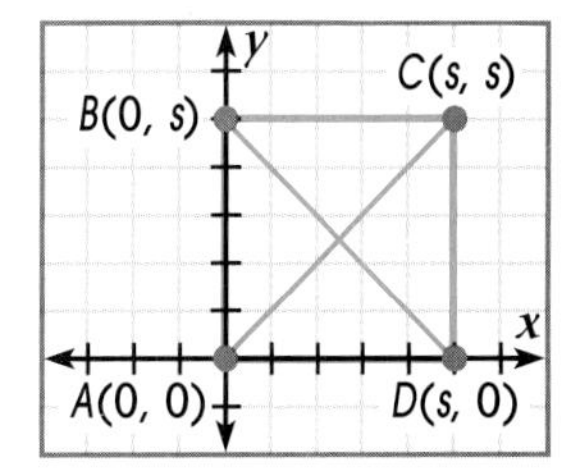

Statements	Reasons
1. ▱*ABCD* is a square.	1. Given
2. Slope of $\overline{AC} = \frac{s-0}{s-0} = 1$ Slope of $\overline{BD} = \frac{s-0}{0-s} = {}^{-}1$	2. Definition of slope
3. $\overline{AC} \perp \overline{BD}$	3. Their slopes are negative reciprocals of each other.

Try This

You are given:

▱*EFGH* is a rectangle.

Prove: The diagonals of a rectangle are congruent.

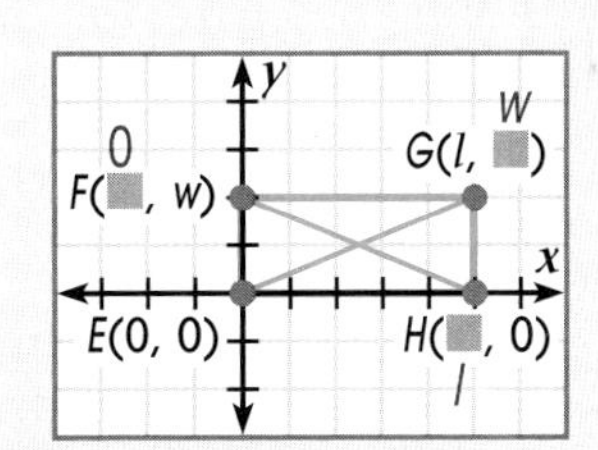

Statements	Reasons
1. ▱*EFGH* is a rectangle.	1. ■ Given
2. $EG = \sqrt{(l-0)^2 + (w-0)^2} = \sqrt{l^2 + w^2}$ $FH = \sqrt{(0-l)^2 + (w-0)^2} = \sqrt{l^2 + w^2}$	2. ■ formula Distance
3. $EG = FH$	3. Transitive ■ Property
4. $\overline{EG} \cong$ ■ $\overline{FH}$	4. Definition of congruent line segments

Practice is provided on page 365 of the student book. **More Practice** is provided in the Classroom Resource Binder.

POSTULATES AND THEOREMS

Chapter 1 Basic Geometric Concepts

POSTULATE 1 Two points determine exactly one line. *(p. 4)*

POSTULATE 2 Three points not on a single line determine exactly one plane. *(p. 4)*

POSTULATE 3 The distance between points C and D is $|c - d|$ or $|d - c|$. *(p. 10)*

THEOREM 1 **Common Segment Theorem**
If $AB = CD$, then $AC = BD$. If $AC = BD$, then $AB = CD$. *(p. 20)*

A B C D

POSTULATE 4 **Segment Addition Postulate**
If three points are on the same straight line and point B is between points A and C, then $AB + BC = AC$. *(p. 24)*

THEOREM 2 **Midpoint Theorem**
The midpoint M of $\overline{AB}$ divides $\overline{AB}$ in half so that $AM = \frac{1}{2}AB$. *(p. 26)*

Chapter 2 Angles

POSTULATE 5 **Angle Addition Postulate**
If point D is in the interior of $\angle ABC$, then $m\angle ABD + m\angle DBC = m\angle ABC$. *(p. 38)*

THEOREM 3 Vertical angles are congruent. *(p. 46)*

THEOREM 4 **Angle Bisector Theorem**
The bisector $\overrightarrow{BD}$ of $\angle ABC$ divides $\angle ABC$ in half so that $m\angle ABD = \frac{1}{2} m\angle ABC$. *(p. 48 and p. 58)*

THEOREM 5 **Common Angle Theorem**
Points D and E are in the interior of $\angle ABC$. If $m\angle ABD = m\angle EBC$, then $m\angle ABE = m\angle DBC$. If $m\angle ABE = m\angle DBC$, then $m\angle ABD = m\angle EBC$. *(p. 52)*

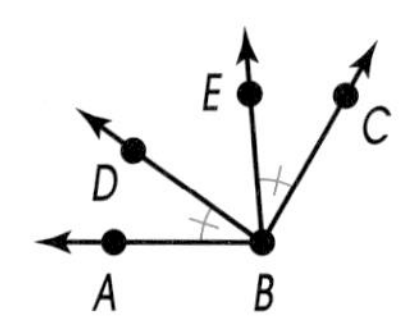

Chapter 3 Reasoning and Proofs

THEOREM 6 If two angles are supplementary to the same angle, then they are congruent. *(p. 84)*

Chapter 4 Perpendicular and Parallel Lines

THEOREM 7 Perpendicular lines form four congruent right angles. *(p. 90)*

THEOREM 8 A point on the perpendicular bisector of a segment is equidistant from the endpoints of the segment. *(p. 94)*

POSTULATE 6 **Alternate Interior Angles Postulate**
If parallel lines are cut by a transversal, then the alternate interior angles are congruent. *(p. 102)*

THEOREM 9 If parallel lines are cut by a transversal, then the same-side interior angles are supplementary. *(p. 104)*

THEOREM 10 **Corresponding Angles Theorem**
If parallel lines are cut by a transversal, then each pair of corresponding angles is congruent. *(p. 106)*

POSTULATE 7 **Parallel Postulate**
Through a given point P not on a line, exactly one line may be drawn parallel to the line. *(p. 109)*

POSTULATE 8 If two lines are cut by a transversal so that the alternate interior angles are congruent, then the lines are parallel. *(p. 116)*

Chapter 5 Triangles

THEOREM 11 **Angle Sum Theorem for Triangles**
The sum of the angles of a triangle is 180°. *(p. 126)*

THEOREM 12 **Exterior Angle Theorem**
The measure of an exterior angle of a triangle equals the sum of the measures of the two remote interior angles. *(p. 128)*

THEOREM 13 **Triangle Inequality Theorem**
The sum of the lengths of any two sides of a triangle must be greater than the length of the third side. *(p. 132)*

THEOREM 14 **Isosceles Triangle Theorem**
If two sides of a triangle are congruent, then the angles opposite those sides are also congruent. *(p. 134)*

THEOREM 15 **Opposite Side-Angle Theorem**
The longest side of a triangle is opposite the largest angle. *(p. 136)*

POSTULATE 9 **Side-Side-Side Postulate (SSS)**
If three sides of one triangle are congruent to the three sides of another triangle, then the triangles are congruent. *(p. 140)*

POSTULATE 10 **Side-Angle-Side Postulate (SAS)**
If two sides and the included angle of one triangle are congruent to the corresponding parts of another triangle, then the triangles are congruent. *(p. 140)*

POSTULATE 11 **Angle-Side-Angle Postulate (ASA)**
If two angles and the included side of one triangle are congruent to the corresponding parts of another triangle, then the triangles are congruent. *(p. 142)*

THEOREM 16 **Angle-Angle-Side Theorem (AAS)**
If two angles and a nonincluded side of one triangle are congruent to the corresponding parts of another triangle, then the two triangles are congruent. *(p. 142)*

THEOREM 17 **Triangle Midsegment Theorem**
If a segment joins the midpoints of two sides of a triangle, then the segment is parallel to the third side and half its length. *(p. 146)*

Chapter 6 Right Triangles

THEOREM 18 **Pythagorean Theorem**
In a right triangle, the sum of the squares of the legs is equal to the square of the hypotenuse. $a^2 + b^2 = c^2$ *(p. 166)*

THEOREM 19 **Converse of the Pythagorean Theorem**
If a, b, and c are the lengths of the sides of a triangle so that $a^2 + b^2 = c^2$, then the triangle is a right triangle. *(p. 174)*

THEOREM 20 **Hypotenuse-Leg Theorem**
If the hypotenuse and one leg of one right triangle are congruent to the hypotenuse and one leg of another right triangle, then the triangles are congruent. *(p. 180)*

Chapter 7 Quadrilaterals and Polygons

THEOREM 21 **Trapezoid Midsegment Theorem**
The length of the midsegment of a trapezoid is half the sum of the lengths of both bases. *(p. 194)*

THEOREM 22 **Exterior-Angle Sum Theorem for Polygons**
The sum of the measures of the exterior angles of a polygon is 360°. *(p. 198)*

THEOREM 23 **Interior-Angle Sum Theorem for Polygons**
The sum of the measures of the interior angles of a polygon with n sides is $(n - 2) \cdot 180$. *(p. 200)*

THEOREM 24 If a quadrilateral has diagonals that bisect each other, then the quadrilateral is a parallelogram. *(p. 204)*

Chapter 8 Perimeter and Area

THEOREM 25 Two triangles have equal areas if the heights are congruent and the bases are congruent. *(p. 228)*

Chapter 9 Similar Polygons

POSTULATE 12 **Angle-Angle Similarity Postulate**
If two angles of one triangle are congruent to two angles of another triangle, then the two triangles are similar. *(p. 240 and p. 260)*

THEOREM 26 **Side-Splitter Theorem**
If a line is parallel to one side of a triangle and intersects the other two sides, then it divides the two sides proportionally. *(p. 246)*

THEOREM 27 If two polygons are similar, then the ratio of their perimeters is equal to the ratio of any pair of corresponding sides. *(p. 250)*

THEOREM 28 If two polygons are similar, then the ratio of their areas is equal to the ratio of the squares of the lengths of any pair of corresponding sides. *(p. 252)*

Chapter 10 Circles

THEOREM 29 The measure of an angle inscribed in a circle is equal to one-half the measure of its intercepted arc. *(p. 274)*

THEOREM 30 If a line is tangent to a circle, then it is perpendicular to the radius drawn to the point of tangency. *(p. 276)*

THEOREM 31 **Tangent-Tangent Angle Theorem**
The measure of the angle formed by two tangents is one-half the difference of the degree measures of the intercepted arcs. *(p. 278)*

THEOREM 32 **Tangent-Segment Theorem**
If two segments from the same exterior point are tangent to a circle, then they are congruent. *(p. 280)*

THEOREM 33 The measure of an angle formed by two intersecting chords is one-half the sum of the intercepted arcs. *(p. 284)*

THEOREM 34 If two inscribed angles of a circle intercept the same arc, then the angles are congruent. *(p. 294)*

Chapter 11 Surface Area and Volume

THEOREM 35 The surface area of a prism is the sum of the lateral area and the area of each of the two bases. *(p. 306)*

THEOREM 36 The surface area of a cylinder is the sum of the area of each base and the lateral surface. *(p. 308)*

THEOREM 37 If the similarity ratio of the sides of two similar figures is $a{:}b$, then the ratio of the volumes of the figures is $a^3{:}b^3$. *(p. 320)*

SYMBOLS

$=$	is equal to
$\neq$	is not equal to
$>$	is greater than
$<$	is less than
$\approx$	is approximately equal to
$\sim$	is similar to
$\cong$	is congruent to
$\lvert x \rvert$	absolute value of x
$\sqrt{x}$	square root of x
$\overleftrightarrow{AB}$	line AB
$\overline{AB}$	line segment AB
$\overrightarrow{AB}$	ray AB
$\angle ABC$	angle ABC
$m\angle ABC$	measure of angle ABC
$^\circ$	degrees
	right angle
	congruent angles
	congruent line segments
$\perp$	is perpendicular to
$\parallel$	is parallel to
	parallel line segments
$\triangle ABC$	triangle ABC
▱$ABCD$	parallelogram $ABCD$
π	pi, about $\frac{22}{7}$ or 3.14
$\odot M$	circle M
$\overset{\frown}{AB}$	minor arc AB
$m\overset{\frown}{AB}$	measure of arc AB
$\overset{\frown}{ABC}$	major arc ABC
$\overrightarrow{AB}$	vector AB
(a, b)	ordered pair with x-coordinate a and y-coordinate b
$\sin A$	sine A
$\cos A$	cosine A
$\tan A$	tangent A

FORMULAS

Rectangle

Perimeter $= 2l + 2w$

Area $= lw$

Square

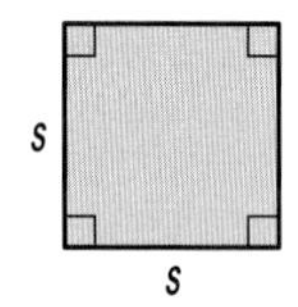

Perimeter $= 4s$

Area $= s^2$

Triangle

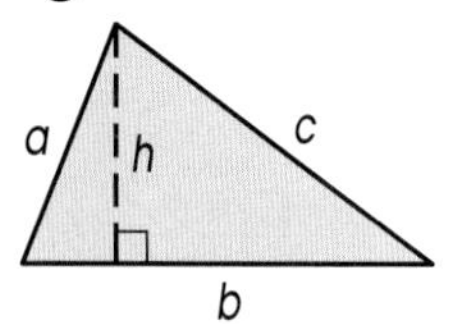

Perimeter $= a + b + c$

Area $= \frac{1}{2}bh$

Parallelogram

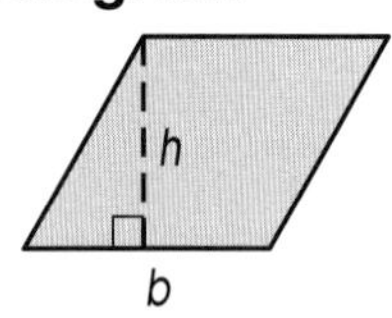

Area $= bh$

Trapezoid

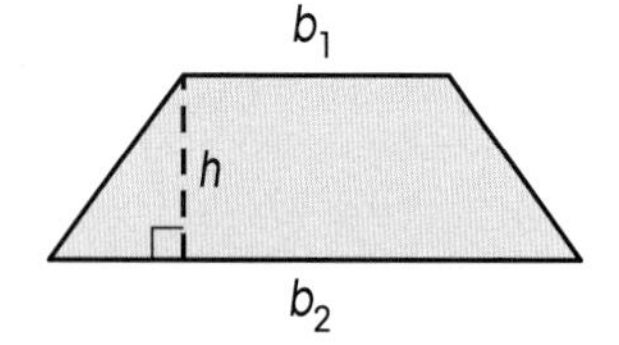

Area $= \frac{1}{2}h(b_1 + b_2)$

Regular Polygon

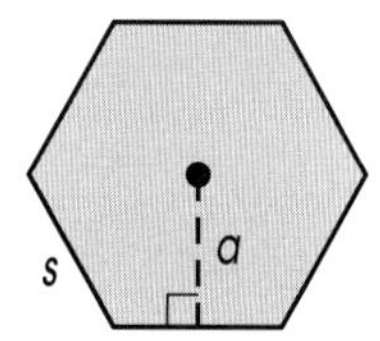

Perimeter $= ns$
  for $n =$ number of sides

Area $= \frac{1}{2}ap$
  for $p =$ perimeter

Circle

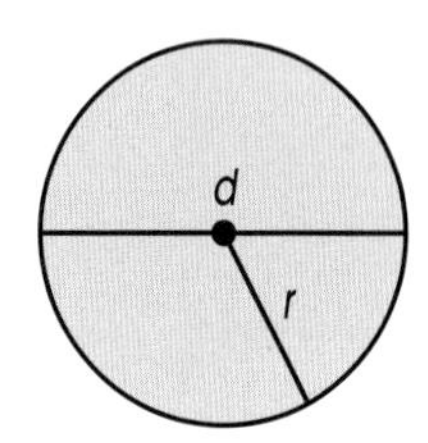

Circumference $= \pi d = 2\pi r$

Area $= \pi r^2$

Sphere

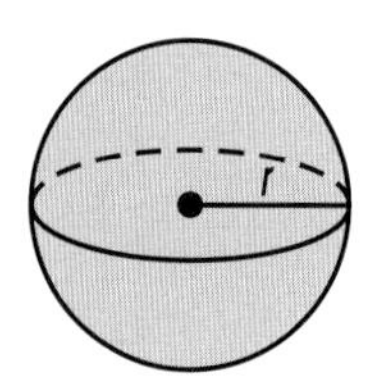

Surface Area $= 4\pi r^2$

Volume $= \frac{4}{3}\pi r^3$

Prism

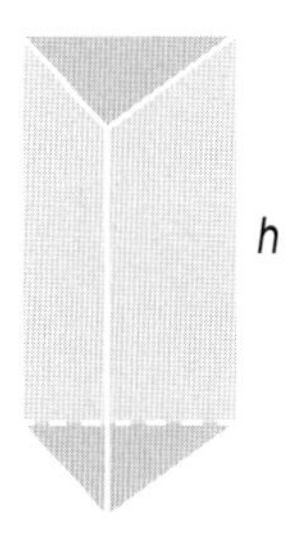

Volume = Bh
for B = area of the base

Rectangular Prism

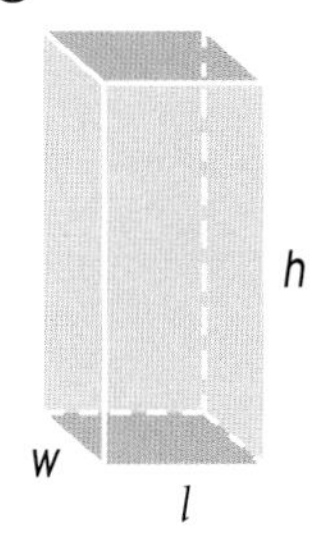

Volume = lwh

Cube

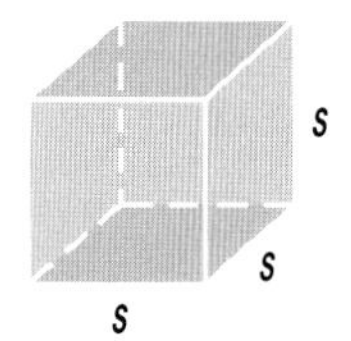

Volume = s^3

Cylinder

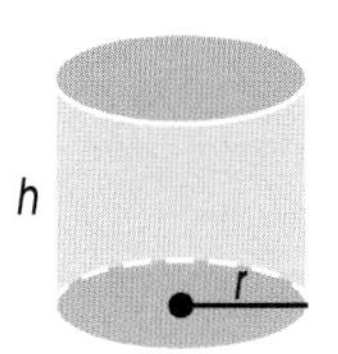

Volume = $\pi r^2 h$

Pyramid

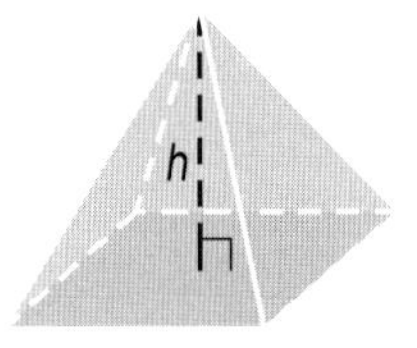

Volume = $\frac{1}{3}Bh$
for B = area of the base

Cone

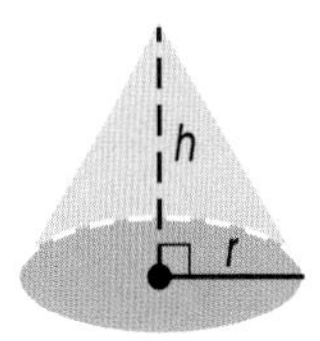

Volume = $\frac{1}{3}\pi r^2 h$

Coordinate Plane

The length of $\overline{AB}$, or distance

$$\text{Distance} = \sqrt{(x_1 - x_2)^2 + (y_1 - y_2)^2}$$

The coordinates of midpoint M

$$\text{Midpoint} = \left(\frac{x_1 + x_2}{2}, \frac{y_1 + y_2}{2}\right)$$

The slope of $\overline{AB}$

$$\text{Slope} = \frac{y_2 - y_1}{x_2 - x_1}$$

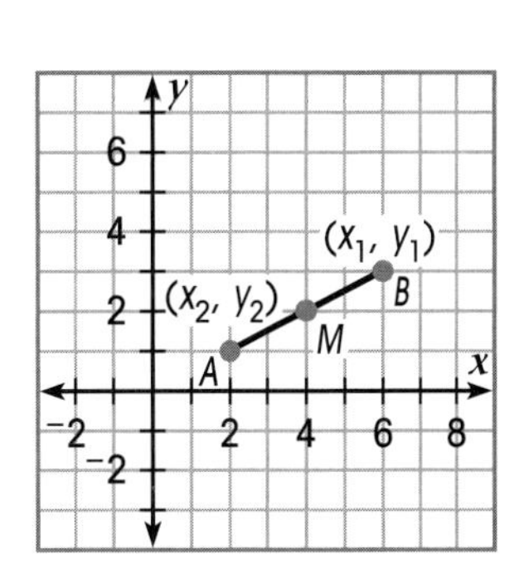

TABLE OF TRIGONOMETRIC RATIOS

Angle	Sine	Cosine	Tangent
1°	0.0175	0.9998	0.0175
2°	0.0349	0.9994	0.0349
3°	0.0523	0.9986	0.0524
4°	0.0698	0.9976	0.0699
5°	0.0872	0.9962	0.0875
6°	0.1045	0.9945	0.1051
7°	0.1219	0.9925	0.1228
8°	0.1392	0.9903	0.1405
9°	0.1564	0.9877	0.1584
10°	0.1736	0.9848	0.1763
11°	0.1908	0.9816	0.1944
12°	0.2079	0.9781	0.2126
13°	0.2250	0.9744	0.2309
14°	0.2419	0.9703	0.2493
15°	0.2588	0.9659	0.2679
16°	0.2756	0.9613	0.2867
17°	0.2924	0.9563	0.3057
18°	0.3090	0.9511	0.3249
19°	0.3256	0.9455	0.3443
20°	0.3420	0.9397	0.3640
21°	0.3584	0.9336	0.3839
22°	0.3746	0.9272	0.4040
23°	0.3907	0.9205	0.4245
24°	0.4067	0.9135	0.4452
25°	0.4226	0.9063	0.4663
26°	0.4384	0.8988	0.4877
27°	0.4540	0.8910	0.5095
28°	0.4695	0.8829	0.5317
29°	0.4848	0.8746	0.5543
30°	0.5000	0.8660	0.5774
31°	0.5150	0.8572	0.6009
32°	0.5299	0.8480	0.6249
33°	0.5446	0.8387	0.6494
34°	0.5592	0.8290	0.6745
35°	0.5736	0.8192	0.7002
36°	0.5878	0.8090	0.7265
37°	0.6018	0.7986	0.7536
38°	0.6157	0.7880	0.7813
39°	0.6293	0.7771	0.8098
40°	0.6428	0.7660	0.8391
41°	0.6561	0.7547	0.8693
42°	0.6691	0.7431	0.9004
43°	0.6820	0.7314	0.9325
44°	0.6947	0.7193	0.9657
45°	0.7071	0.7071	1.0000

Angle	Sine	Cosine	Tangent
46°	0.7193	0.6947	1.0355
47°	0.7314	0.6820	1.0724
48°	0.7431	0.6691	1.1106
49°	0.7547	0.6561	1.1504
50°	0.7660	0.6428	1.1918
51°	0.7771	0.6293	1.2349
52°	0.7880	0.6157	1.2799
53°	0.7986	0.6018	1.3270
54°	0.8090	0.5878	1.3764
55°	0.8192	0.5736	1.4281
56°	0.8290	0.5592	1.4826
57°	0.8387	0.5446	1.5399
58°	0.8480	0.5299	1.6003
59°	0.8572	0.5150	1.6643
60°	0.8660	0.5000	1.7321
61°	0.8746	0.4848	1.8040
62°	0.8829	0.4695	1.8807
63°	0.8910	0.4540	1.9626
64°	0.8988	0.4384	2.0503
65°	0.9063	0.4226	2.1445
66°	0.9135	0.4067	2.2460
67°	0.9205	0.3907	2.3559
68°	0.9272	0.3746	2.4751
69°	0.9336	0.3584	2.6051
70°	0.9397	0.3420	2.7475
71°	0.9455	0.3256	2.9042
72°	0.9511	0.3090	3.0777
73°	0.9563	0.2924	3.2709
74°	0.9613	0.2756	3.4874
75°	0.9659	0.2588	3.7321
76°	0.9703	0.2419	4.0108
77°	0.9744	0.2250	4.3315
78°	0.9781	0.2079	4.7046
79°	0.9816	0.1908	5.1446
80°	0.9848	0.1736	5.6713
81°	0.9877	0.1564	6.3138
82°	0.9903	0.1392	7.1154
83°	0.9925	0.1219	8.1443
84°	0.9945	0.1045	9.5144
85°	0.9962	0.0872	11.4301
86°	0.9976	0.0698	14.3007
87°	0.9986	0.0523	19.0811
88°	0.9994	0.0349	28.6363
89°	0.9998	0.0175	57.2900
90°	1.0000	0.0000	Undefined

GLOSSARY

30°-60°-90° triangle a right triangle whose acute angles measure 30° and 60° (*p. 172*)

45°-45°-90° triangle a right triangle whose acute angles both measure 45° (*p. 170*)

acute angle an angle that is greater than 0° and less than 90° (*p. 34*)

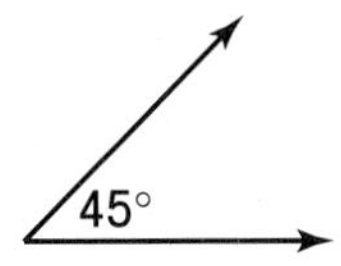

acute triangle a triangle with three acute angles (*p. 124*)

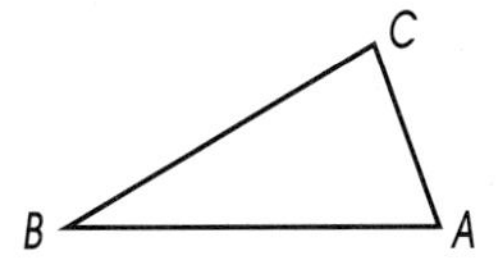

adjacent angles two angles that have a common vertex and a common ray (*p. 38*)

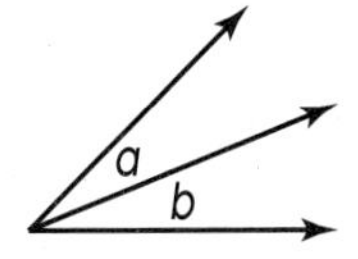

alternate interior angles interior angles that are on opposite sides of a transversal (*p. 100*)

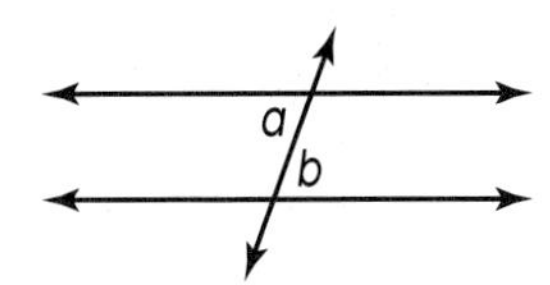

altitude a line segment that joins a vertex to the line containing the opposite side and that is perpendicular to that side (*p. 144*)

angle a geometric figure formed by two rays with a common endpoint (*p. 32*)

angle bisector **1.** a ray that divides an angle into two congruent angles (*p. 48*) **2.** a line segment that joins a vertex of a triangle to the opposite side and that bisects the angle (*p. 144*)

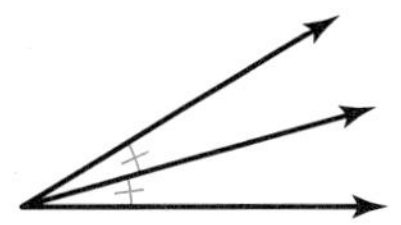

angle of depression the angle formed by a horizontal line and the line of sight when looking down (*p. 382*)

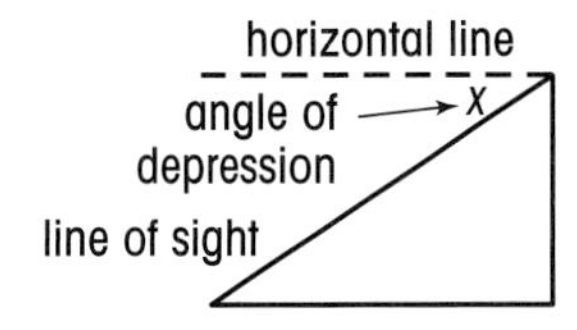

angle of elevation the angle formed by a horizontal line and the line of sight when looking up (*p. 382*)

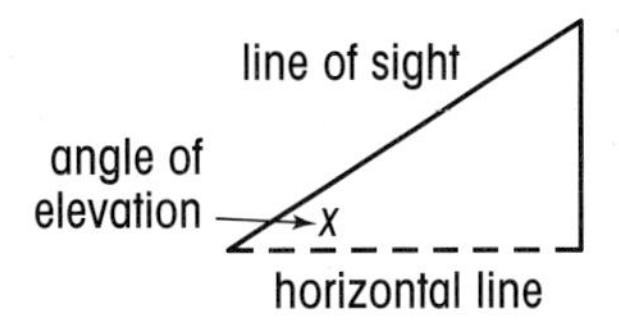

apothem a perpendicular line segment from the center of a regular polygon to any base (*p. 222*)

arc a curve between any two points on a circle (*p. 270*)

area the number of square units needed to cover a surface (*p. 214*)

base **1.** a face of a space figure (*p. 302*) **2.** side of a polygon to which an altitude is drawn (*p. 216*)

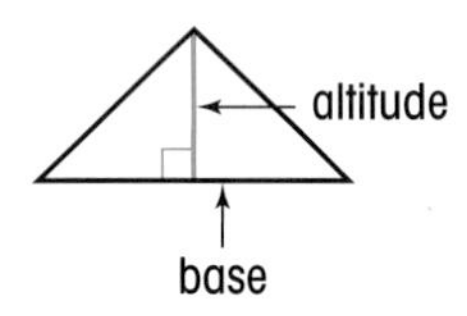

bisector a line that intersects a line segment at its midpoint (*p. 16*)

center the point inside a circle that is the same distance from each point on the circle (*p. 266*)

center of rotation the point about which a figure is turned (*p. 348*)

central angle an angle with its vertex at the center of the circle and sides that are radii (*p. 270*)

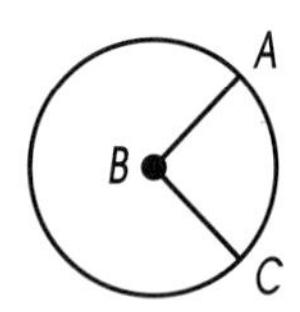

centroid the point where the three medians of a triangle meet (*p. 150*)

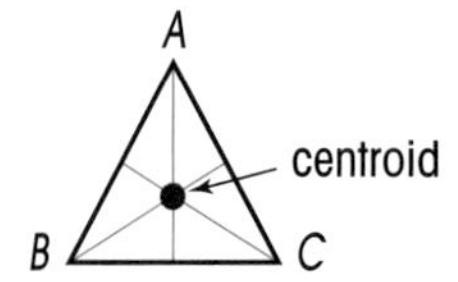

chord a line segment with endpoints on a circle (*p. 282*)

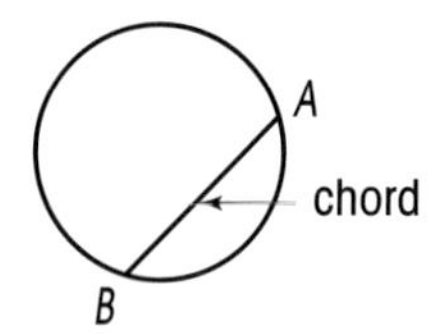

circle a plane figure made up of points all the same distance from a point called the center (*p. 266*)

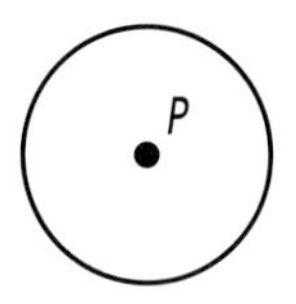

circumference the distance around a circle (*p. 266*)

clinometer a tool used to measure angles (*p. 381*)

collinear points points on the same line (*p. 22*)

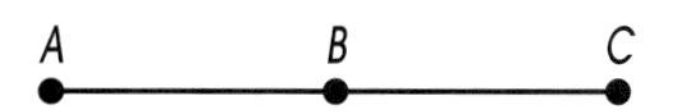

complementary angles a pair of angles whose measures have a sum of 90° (*p. 40*)

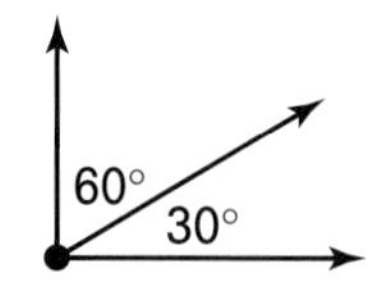

conclusion the "then" part of a conditional statement (*p. 68*)

conditional statement a statement that uses the words *if* and *then* (*p. 68*)

cone a space figure with one circular base and one vertex (*p. 302*)

congruent angles angles that have the same measure (*p. 42*)

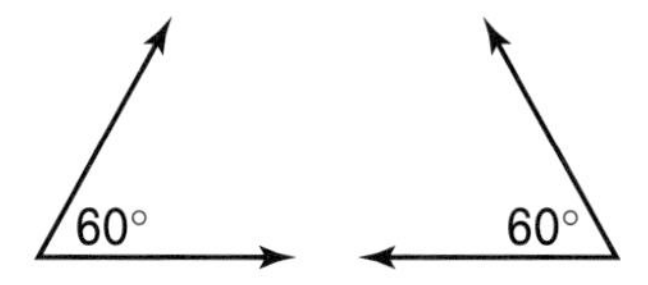

congruent line segments line segments that have the same length (*p. 12*)

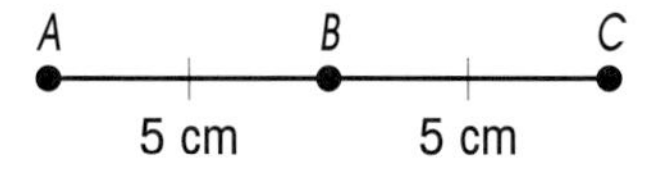

coordinate plane a grid with two axes where points are graphed (*p. 334*)

coordinates the ordered numbers that locate a point on a plane or in space (*p. 334*)

corresponding angles angles on the same side of a transversal; one is an interior angle and one is an exterior angle (*p. 106*)

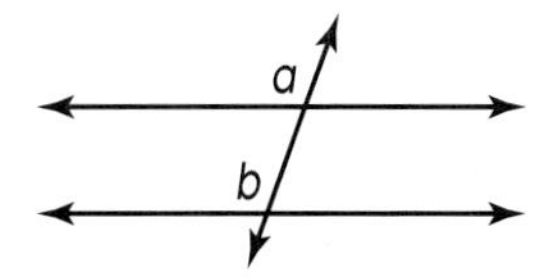

cosine the ratio of the length of the side adjacent to an angle to the length of the hypotenuse in a right triangle (*p. 370*)

counterexample an example that shows a statement is false (*p. 78*)

cross products in a proportion, if $\frac{a}{b} = \frac{c}{d}$, then $a \bullet d = b \bullet c$ (*p. 236*)

cube a prism with six faces that are all congruent squares (*p. 302*)

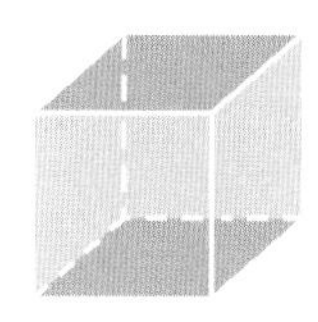

cubic unit a cube with sides one unit long that is used to measure volume (*p. 312*)

cylinder a space figure with two congruent, parallel bases that are circles (*p. 302*)

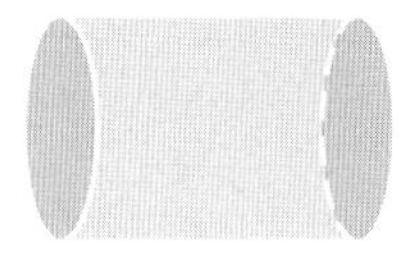

deductive reasoning a way to reach a conclusion based on known facts (*p. 66*)

diagonal a line segment that connects one vertex to another nonconsecutive vertex (*p. 192*)

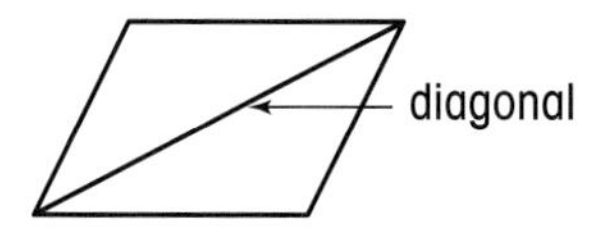

diameter a line segment that passes through the center of a circle with endpoints on the circle (*p. 266*)

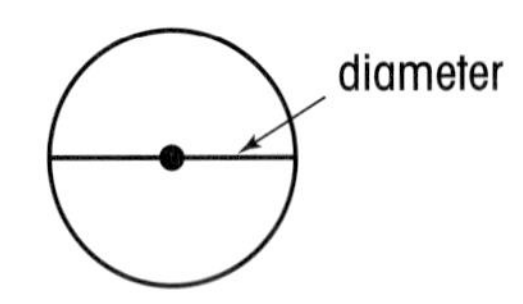

dilation a transformation in which the image is similar to the original figure (*p. 350*)

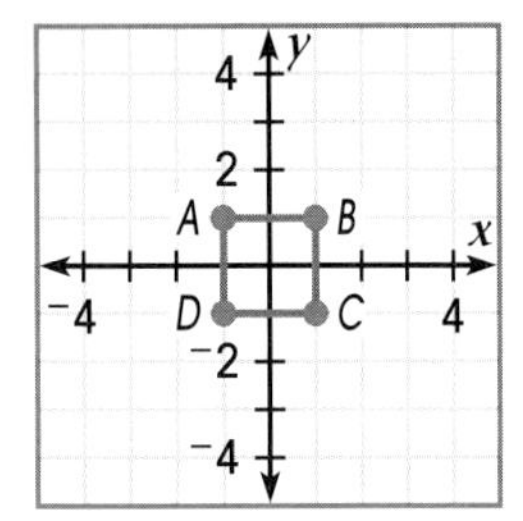

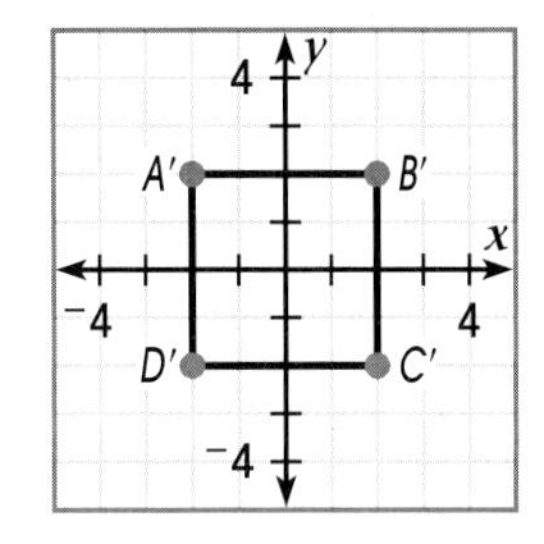

equiangular triangle a triangle with all angles congruent (*p. 124*)

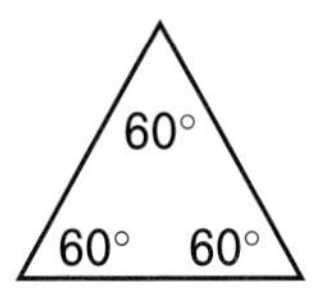

equilateral triangle a triangle with all sides congruent (*p. 130*)

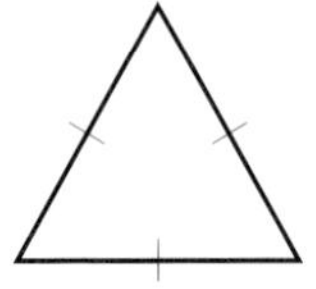

exterior angle the angle formed by extending one side of a triangle (*p. 128*)

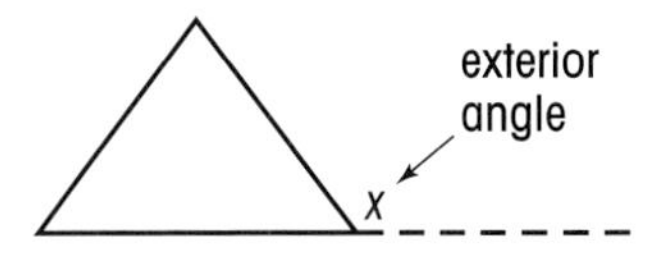

extremes the first and last terms in the ratios of a proportion (*p. 236*)

$$\frac{6}{3} = \frac{4}{2}$$

formula a rule that uses letters to represent measures (*p. 212*)

geometric mean the square root of the product of two positive numbers; in the proportion $\frac{a}{x} = \frac{x}{b}$, the geometric mean of a and b is x (*p. 236*)

height **1.** the length of the altitude of a polygon (*p. 216*) **2.** the length of the altitude of a space figure (*p. 312*)

horizon line the level of the viewer's eyes as the viewer looks across a distance in a drawing (*p. 112*)

hypotenuse the side opposite the right angle of a right triangle (*p. 164*)

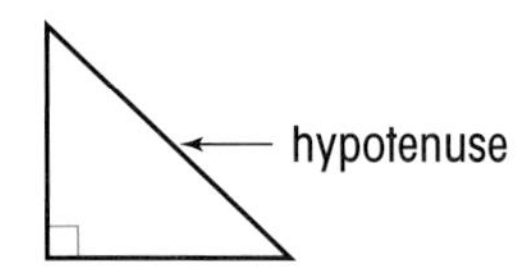

hypothesis the "if" part of a conditional statement (*p. 68*)

image the figure that results from a transformation (*p. 344*)

indirect proof a proof that can be used when there are only two possibilities; if one possibility is false, the other must be true (*p. 80*)

inductive reasoning a way to reach a conclusion based on a pattern (*p. 64*)

inscribed angle an angle with its vertex on the circle and sides that are chords (*p. 274*)

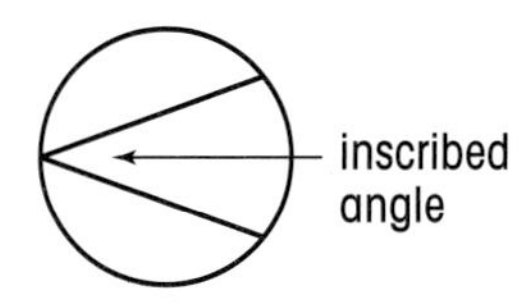

isosceles trapezoid a special trapezoid with congruent legs; each pair of base angles is congruent (*p. 196*)

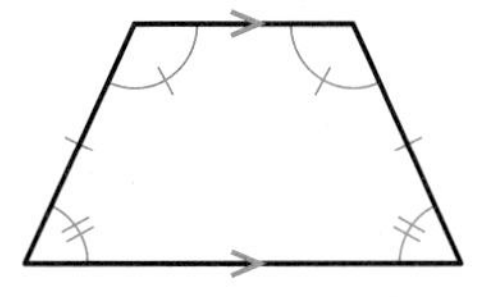

isosceles triangle a triangle with two sides congruent (*p. 130*)

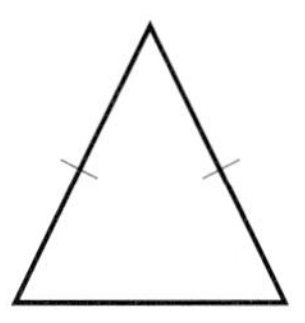

leg **1.** each of the sides that forms the right angle of a right triangle (*p. 164*)
2. each of the sides of a trapezoid that are not parallel (*p. 194*)

line a geometric figure made up of infinitely many points; it extends endlessly in two directions (*p. 4*)

line of reflection the line over which a figure is flipped in a reflection (*p. 346*)

line segment a part of a line; it has two endpoints (*p. 4*)

magnitude the length of a vector (*p. 358*)

means the two middle terms in the ratios of a proportion (*p. 236*)

$$\frac{6}{3} = \frac{4}{2}$$

median a line segment that joins a vertex of a triangle to the midpoint of the opposite side (*p. 144*)

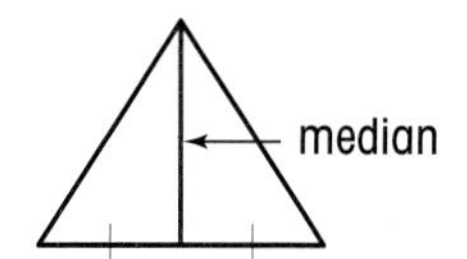

midpoint the point that divides a line segment into two congruent parts (*p. 16*)

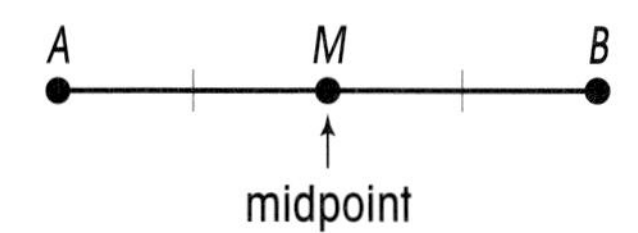

midsegment a line segment that joins the midpoints of two sides of a triangle (*p. 146*)

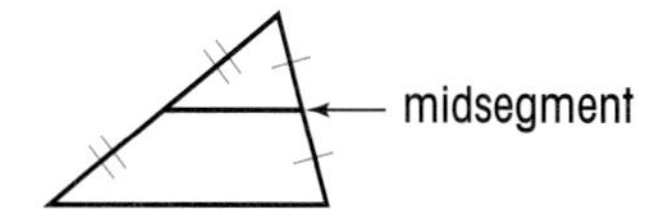

net a two-dimensional pattern that can be folded to create a space figure (*p. 304*)

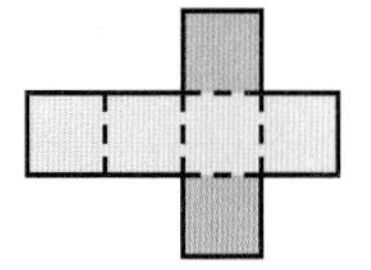

obtuse angle an angle that is greater than 90° and less than 180° (*p. 34*)

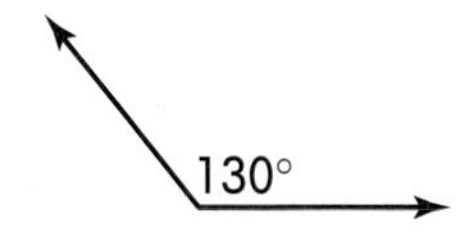

obtuse triangle a triangle with one obtuse angle (*p. 124*)

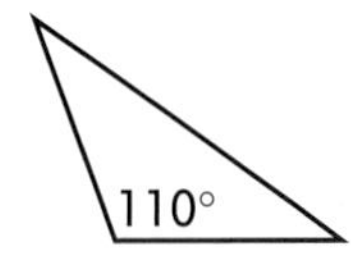

one-point perspective a way to draw real objects on a flat surface as they appear in real life (*p. 112*)

ordered pair two numbers that locate a point on a coordinate plane (*p. 334*)

origin the point where the *x*- and *y*-axes meet (*p. 334*)

parallel lines lines that lie in the same plane and do not intersect (*p. 98*)

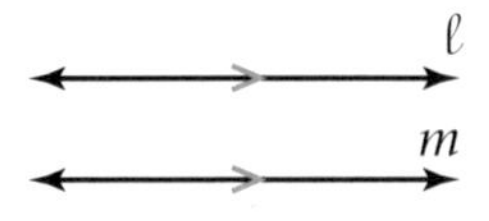

parallelogram a special quadrilateral with both pairs of opposite sides parallel (*p. 188*)

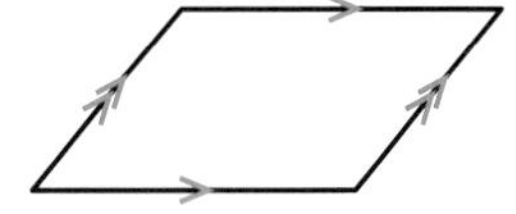

perimeter the distance around a polygon (*p. 212*)

perpendicular bisector a line that bisects a line segment and is perpendicular to the line segment (*p. 94*)

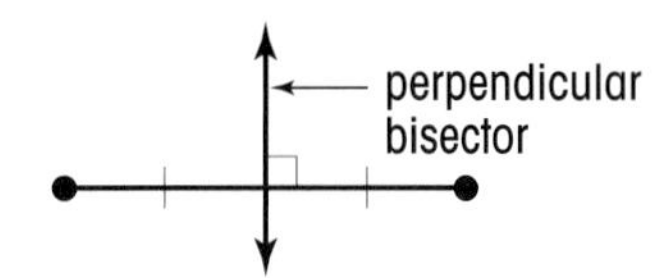

perpendicular lines two lines that intersect to form a right angle (*p. 90*)

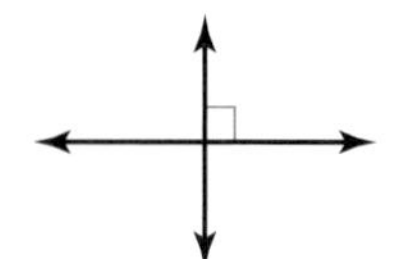

plane a flat surface; it has no thickness; it extends endlessly in all directions (*p. 4*)

point a location in space (*p. 4*)

polygon a closed geometric figure made up of at least three line segments or sides (*p. 186*)

postulate a statement that is accepted without proof (*p. 4*)

preimage the original figure before a transformation (*p. 344*)

prism a space figure with faces that are polygons and two bases that are congruent and parallel (*p. 302*)

proof a way to reach a conclusion using reasoning (*p. 74*)

properties of congruence properties used to prove statements about geometric figures, such as line segments and angles (*p. 72*)

properties of equality properties used to solve equations (*p. 70*)

proportion an equation that states that two ratios are equal (*p. 236*)

protractor a tool used to measure angles (*p. 32*)

pyramid a space figure with one base and triangular lateral faces that meet at a vertex (*p. 302*)

Pythagorean Theorem a formula for finding the length of a side of a right triangle when you know the lengths of the other sides; $\text{leg}^2 + \text{leg}^2 = \text{hypotenuse}^2$ or $a^2 + b^2 = c^2$ (*p. 166*)

radical a number with a square root symbol (*p. 162*)

$$\sqrt{24}$$

radius a line segment from the center to any point on the circle (*p. 266*)

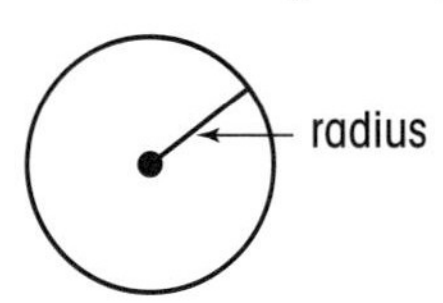

ratio a comparison of two quantities (*p. 234*)

ray a part of a line; it has one endpoint (*p. 4*)

rectangle a parallelogram with four right angles (*p. 190*)

reflection a transformation in which a figure is flipped over a line (*p. 346*)

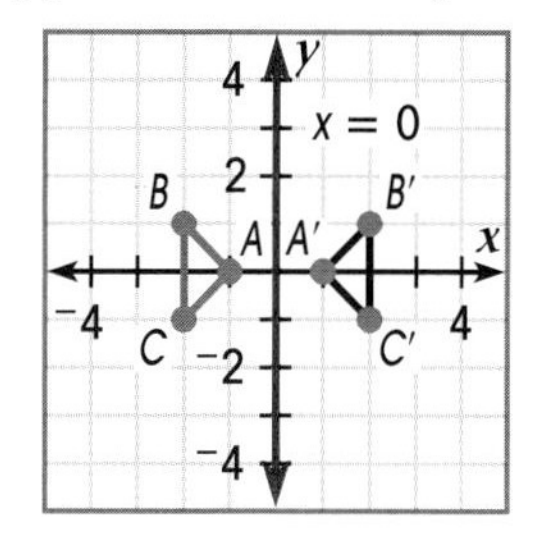

regular polygon a polygon that has all congruent sides and all congruent angles (*p. 198*)

remote interior angle an interior angle of a triangle not adjacent to the given exterior angle (*p. 128*)

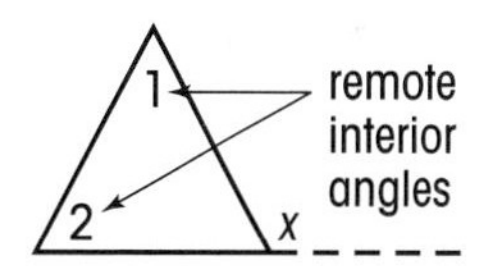

resultant vector the sum of two vectors (*p. 360*)

rhombus a parallelogram with four congruent sides (*p. 190*)

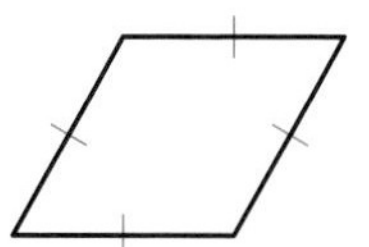

right angle an angle that is equal to 90° (*p. 34*)

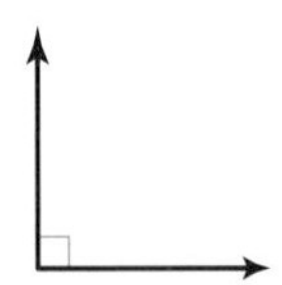

right triangle a triangle with one right angle (*p. 124*)

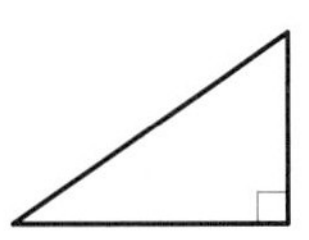

rotation a transformation in which a figure is turned about a point (*p. 348*)

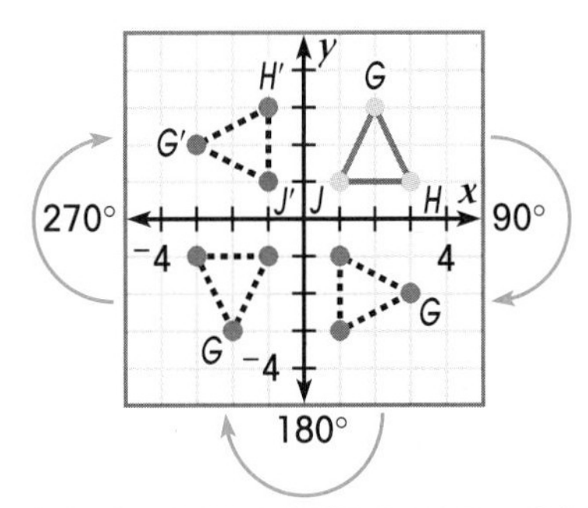

clockwise rotation 270° about the origin

same-side interior angles interior angles that are on the same side of a transversal (*p. 100*)

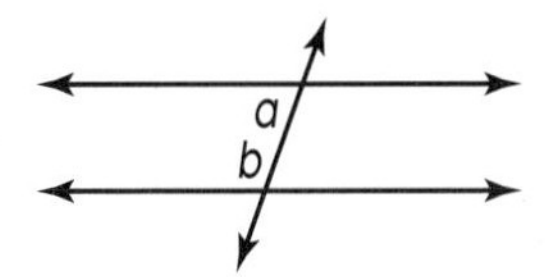

scale drawing a drawing showing dimensions in proportion to the object it represents; a length on the drawing is proportional to the object's actual length (*p. 258*)

scalene triangle a triangle with no congruent sides (*p. 130*)

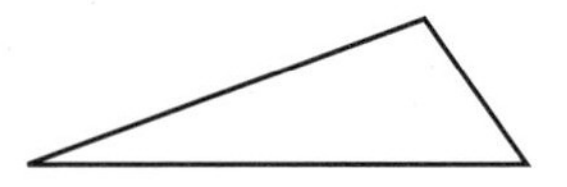

secant a line that intersects a circle at two points (*p. 278*)

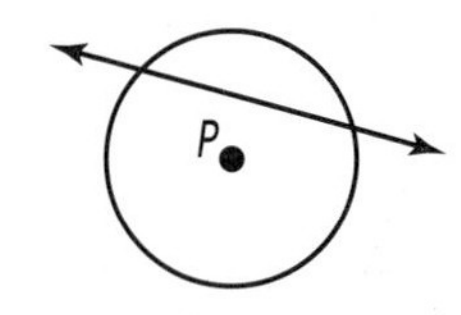

sector a region formed by two radii and the intercepted arc (*p. 272*)

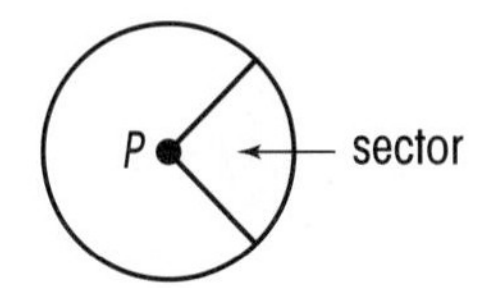

similarity ratio the ratio of the lengths of the corresponding sides of similar polygons (*p. 238*)

similar triangles triangles that have the same shape but that may or may not be the same size (*p. 238*)

sine the ratio of the length of the side opposite an angle to the length of the hypotenuse in a right triangle (*p. 370*)

slope the steepness of a line; the ratio of vertical change to horizontal change (*p. 340*)

space figure a three-dimensional figure (*p. 302*)

sphere a space figure with all points the same distance from the center (*p. 302*)

square **1.** the product of multiplying a number by itself (*p. 160*) **2.** a parallelogram with four right angles and four congruent sides (*p. 190*)

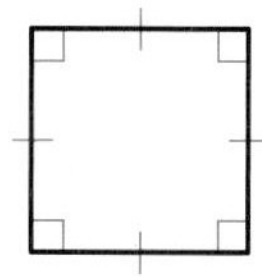

square root the number that was squared (*p. 161*)

straight angle an angle that is equal to 180° (*p. 34*)

supplementary angles a pair of angles whose measures have a sum of 180° (*p. 40*)

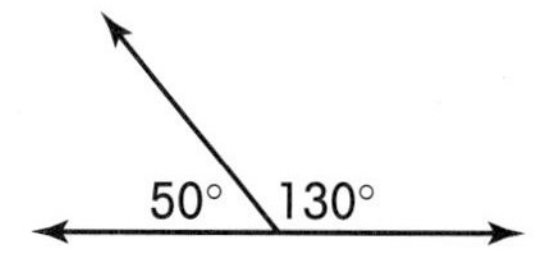

surface area the sum of the areas of all the surfaces of a space figure (*p. 306*)

tangent **1.** the ratio of the length of the side opposite an angle to the length of the adjacent side in a right triangle (*p. 370*) **2.** a line that touches a circle at only one point (*p. 276*)

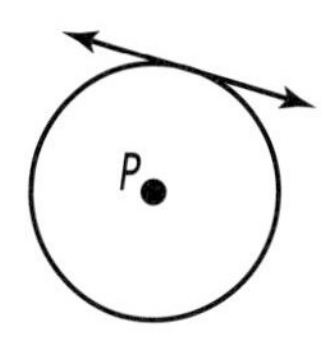

theorem a statement that can be proved (*p. 26*)

transformation a change in the position or size of a figure (*p. 344*)

translation a transformation in which a figure slides to a new position (*p. 344*)

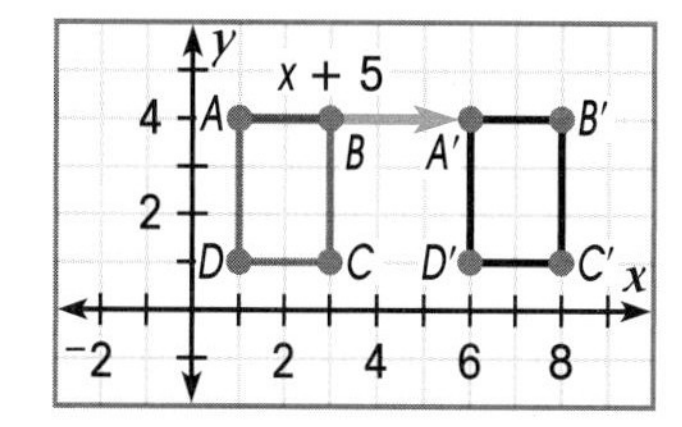

transversal a line that intersects two different lines at two different points (*p. 100*)

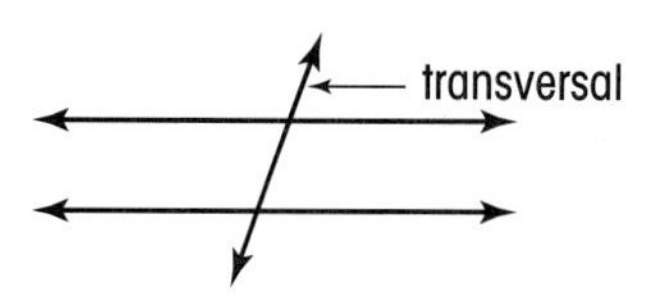

trapezoid a quadrilateral with only one pair of parallel sides (*p. 194*)

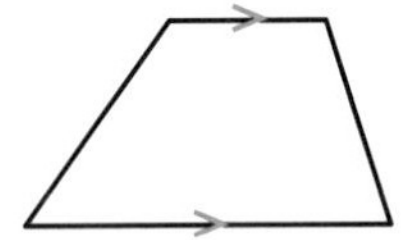

trigonometric ratio the ratio of the length of two sides of a right triangle (*p. 370*)

trigonometric identity a trigonometric equation that is always true (*p. 386*)

trigonometry the study of the relationship between the sides and angles of a right triangle (*p. 370*)

vanishing point a point on the horizon where parallel lines appear to meet (*p. 112*)

vector any quantity that has size and direction (*p. 358*)

vertical angles angles formed by intersecting lines (*p. 46*)

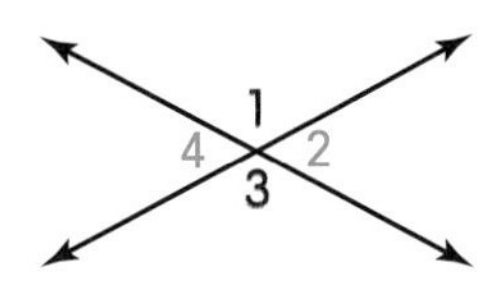

volume the number of cubic units that fill a space figure (*p. 312*)

SELECTED ANSWERS

Selected Answers are given for some of the Practice exercises.

UNIT ONE

Chapter 1 Basic Geometric Concepts

Lesson 1.1 *(Page 5)*

Practice

1. ray *CB* **7.** • *C*

Lesson 1.2 *(Page 7)*

Practice

1. 8 **5.** 1

Lesson 1.3 *(Page 9)*

Practice

1. $5 + (^-4)$ **5.** 1

Lesson 1.4 *(Page 11)*

Practice

1. 2 **4.** 3

Lesson 1.5 *(Pages 13–14)*

Practice

1. $\overline{AB}$ and $\overline{GH}$ **2.** 16
8. $\overline{NO}$ **12.** 1 cm

Lesson 1.6 *(Pages 17–18)*

Practice

1. midpoint **4.** 8 cm
10. 10 cm
13. 12 cm **17.** 12 mm

Lesson 1.7 *(Page 21)*

Practice

1. 24.98

Lesson 1.8 *(Page 23)*

Practice

1. 6 miles

Lesson 1.9 *(Page 25)*

Practice

1. 30 miles

Chapter 2 Angles

Lesson 2.1 *(Page 33)*

Practice

1. 30°

Lesson 2.2 *(Page 35)*

Practice

1. obtuse

Lesson 2.3 *(Page 37)*

Practice

1. 13

Lesson 2.4 *(Page 39)*

Practice

1. 120°

Lesson 2.5 *(Page 41)*

Practice

1. supplementary, because the sum of their measures equals 180°
2. 70° **10.** 160°

Lesson 2.6 *(Pages 43–44)*

Practice

1. $\angle T$ and $\angle V$; $\angle S$ and $\angle R$
2. 82
6. $\angle KHL$ **8.** 50°

Lesson 2.7 *(Page 47)*

Practice

1. 65° **4.** 30°

Lesson 2.8 *(Pages 49–50)*

Practice

1. 30° **7.** 19

Lesson 2.9 *(Page 53)*

Practice

1. 65°

Lesson 2.10 *(Page 55)*

Practice

1. 3 weeks

Lesson 2.11 *(Page 57)*

Practice

1. 1 time

Chapter 3 Reasoning and Proofs

Lesson 3.1 *(Page 65)*

Practice

1. 41, 44

Lesson 3.2 *(Page 67)*

Practice

1. $m\angle N = 180$
7. $m\angle ABD = m\angle DBC = 90$

Lesson 3.3 *(Page 69)*

Practice

1. you work for pay; you will earn money
5. $\angle R \cong \angle S$

Lesson 3.4 *(Page 71)*

Practice

1. substitution

Lesson 3.5 *(Page 73)*

Practice

1. Symmetric

5. $x = y$; Transitive

Lesson 3.8 *(Page 79)*

Practice

1. $n = 11$

Lesson 3.9 *(Page 81)*

Practice

1. $\overrightarrow{JK}$ bisects $\overline{GH}$

Chapter 4 Perpendicular and Parallel Lines

Lesson 4.1 *(Pages 91–92)*

Practice

1. yes; $\overline{AB} \perp \overline{CD}$

4. 90° **10.** 90°

Lesson 4.2 *(Pages 95–96)*

Practice

1. No, $\overline{LM}$ does not bisect $\overline{AB}$.

4. True. The distance between any point on the perpendicular bisector and each endpoint is equal.

9. 19

Lesson 4.3 *(Page 99)*

Practice

1. yes; $a \parallel b$

4. $\overleftrightarrow{MN} \parallel \overleftrightarrow{OP}$; $\overleftrightarrow{MO} \parallel \overleftrightarrow{NP}$

Lesson 4.4 *(Page 101)*

Practice

1. line c

4. $\angle 3$ and $\angle 5$; $\angle 4$ and $\angle 6$

Lesson 4.5 *(Page 103)*

Practice

1. $\angle 5 = 50°$; $\angle 6 = 130°$

Lesson 4.6 *(Page 105)*

Practice

1. $\angle 5 = 130°$; $\angle 8 = 130°$

Lesson 4.7 *(Pages 107–108)*

Practice

1. 40° **7.** 115°

Lesson 4.8 *(Page 111)*

Practice

1. $x = 15$; $\angle 1$ is 110°.

Lesson 4.10 *(Page 115)*

Practice

1. Travel east on 36th Street to 5th Avenue. Then, travel south on 5th Avenue to 33rd Street. Then, travel west on 33rd Street to 8th Avenue.

UNIT TWO

Chapter 5 Triangles

Lesson 5.1 *(Page 125)*

Practice

1. equiangular

Lesson 5.2 *(Page 127)*

Practice

1. 50°

Lesson 5.3 *(Page 129)*

Practice

1. 35°

Lesson 5.4 *(Page 131)*

Practice

1. equilateral

4. equiangular equilateral

Lesson 5.5 *(Page 133)*

Practice

1. no; $4 + 5 \not> 16$

9. yes; isosceles

Lesson 5.6 *(Page 135)*

Practice

1. 50° **4.** 6

Lesson 5.7 *(Page 137)*

Practice

1. $\overline{AC}$ **4.** $\angle K$

Lesson 5.8 *(Page 139)*

Practice

1. $\angle P$ **7.** 45

Lesson 5.9 *(Page 141)*

Practice

1. $\triangle YLF \cong \triangle GSW$, using the SSS Postulate.

Lesson 5.10 *(Page 143)*

Practice

1. $\triangle FPB \cong \triangle WMT$, using the SAS Postulate.

Lesson 5.11 *(Page 145)*

Practice

1. angle bisector

4. $\overline{JG}$

Lesson 5.12 *(Page 147)*

Practice

1. 9 cm

Lesson 5.13 *(Page 149)*

Practice

1. 76°

Lesson 5.15 *(Page 153)*

Practice

1. 20 ft; because the base angles are congruent and the opposite sides are congruent

Chapter 6 Right Triangles

Lesson 6.1 *(Page 160)*

Practice

1. 64

Lesson 6.2 *(Page 161)*

Practice

1. 8

Lesson 6.3 *(Page 163)*

Practice

1. $2\sqrt{3}$

Lesson 6.4 *(Page 165)*

Practice

1. 15 ft

Lesson 6.5 *(Pages 167–168)*

Practice

1. 15 cm **7.** 4 ft

Lesson 6.6 *(Page 171)*

Practice

1. $b = 4$ cm; $c = 4\sqrt{2}$ cm

Lesson 6.7 *(Page 173)*

Practice

1. $b = 4\sqrt{3}$ in.; $c = 8$ in.

Lesson 6.8 *(Page 175)*

Practice

1. no

Lesson 6.9 *(Page 177)*

Practice

1. 13 km

Lesson 6.10 *(Page 179)*

Practice

1. 12 ft

Chapter 7 Quadrilaterals and Polygons

Lesson 7.1 *(Page 187)*

Practice

1. triangle *ABC*

Lesson 7.2 *(Page 189)*

Practice

1. 42 m **7.** 58

Lesson 7.3 *(Page 191)*

Practice

1. 20 mm **4.** 8 mm

Lesson 7.4 *(Page 193)*

Practice

1. 16 in. **7.** 9 mm

Lesson 7.5 *(Page 195)*

Practice

1. 24 m **4.** 15 in.

Lesson 7.6 *(Page 197)*

Practice

1. $\overline{SM}$ **7.** 12

Lesson 7.7 *(Page 199)*

Practice

1. 72° **7.** 12-gon

Lesson 7.8 *(Page 201)*

Practice

1. 720° **10.** 120°

13. hexagon

Lesson 7.9 *(Page 203)*

Practice

1. 200

UNIT THREE

Chapter 8 Perimeter and Area

Lesson 8.1 *(Page 213)*

Practice

1. 80 ft **4.** 25 mm

Lesson 8.2 *(Page 215)*

Practice

1. 78 m^2 **4.** 9 m

Lesson 8.3 *(Page 217)*

Practice

1. 35 cm^2 **4.** 6 ft

Lesson 8.4 *(Page 219)*

Practice

1. 72 cm^2 **4.** 14 ft

Lesson 8.5 *(Page 221)*

Practice

1. 25 mm^2

Lesson 8.6 *(Page 223)*

Practice

1. 21.336 $in.^2$

Lesson 8.8 *(Page 227)*

Practice

1. 729 ft^2

Chapter 9 Similar Polygons

Lesson 9.1 *(Page 235)*

Practice

1. 2 to 1, 2:1, $\frac{2}{1}$

Lesson 9.2 *(Page 237)*

Practice

1. 5 **9.** 6

Lesson 9.3 *(Page 239)*

Practice

1. yes; $\triangle LBR \sim \triangle PDK$; The similarity ratio is 2 to 3.

Lesson 9.4 *(Page 241)*

Practice

1. yes; Explanations may vary.
3. $IN = 12$

Lesson 9.5 *(Page 243)*

Practice

1. $EH = 45$; $GE = 75$; $GF = 100$

Lesson 9.6 *(Page 245)*

Practice

1. $HJ = 45$; $IJ = 60$

Lesson 9.7 *(Page 247)*

Practice

1. 30

Lesson 9.8 *(Page 249)*

Practice

1. 27; 3 to 4

Lesson 9.9 *(Page 251)*

Practice

1. 36 cm

Lesson 9.10 *(Page 253)*

Practice

1. 54 cm^2

Lesson 9.11 *(Page 255)*

Practice

1. 3 hours **3.** no

Lesson 9.12 *(Page 257)*

Practice

1. 70 m

Lesson 9.13 *(Page 259)*

Practice

1. 13 ft

Chapter 10 Circles

Lesson 10.1 *(Page 267)*

Practice

1. about 12.56 in.
7. 8 ft

Lesson 10.2 *(Page 269)*

Practice

1. about 12.56 $in.^2$
7. 3 cm

Lesson 10.3 *(Page 271)*

Practice

1. 90°

Lesson 10.4 *(Page 273)*

Practice

1. Length of minor arc $\approx$ 6.28 in.; Area of sector $\approx$ 28.26 $in.^2$

Lesson 10.5 *(Page 275)*

Practice

1. 30°

Lesson 10.6 *(Page 277)*

Practice

1. 6 cm

Lesson 10.7 *(Page 279)*

Practice

1. 90°

Lesson 10.8 *(Page 281)*

Practice

1. 7 **4.** 88 cm

Lesson 10.9 *(Page 283)*

Practice

1. 20 m

Lesson 10.10 *(Page 285)*

Practice

1. 92°

Lesson 10.11 *(Page 287)*

Practice

1. 24

Lesson 10.12 *(Page 289)*

Practice

1. about 1,351.97 cm^2
4. about 151 in.
7. $A \approx 1{,}808.6\ ft^2$; $C \approx 150.7$ ft

Lesson 10.13 *(Page 291)*

Practice

1. 100 cm^2

Lesson 10.14 *(Page 293)*

Practice

1. about 12,560 feet

UNIT FOUR

Chapter 11 Surface Area and Volume

Lesson 11.1 *(Page 303)*

Practice

1. rectangular prism

Lesson 11.2 *(Page 305)*

Practice

1. rectangular prism

Lesson 11.3 *(Page 307)*

Practice

1. 202 ft^2

Lesson 11.4 *(Page 309)*

Practice

1. about 1,256 cm^2

Lesson 11.5 *(Page 311)*

Practice

1. $324\pi\ in.^2$
4. 4 m

Lesson 11.6 *(Page 313)*

Practice

1. 150 in.^{3}

Lesson 11.7 *(Page 315)*

Practice

1. 108π in.^{3}

4. 9 ft

Lesson 11.8 *(Page 317)*

Practice

1. about 615 ft^3

4. 180 cm

Lesson 11.9 *(Page 319)*

Practice

1. about 4,187 ft^3

Lesson 11.10 *(Page 321)*

Practice

1. 13,824 cm^3

Lesson 11.11 *(Page 323)*

Practice

1. 144 in.^{3}

Lesson 11.12 *(Page 325)*

Practice

1. 805 ft^2

Lesson 11.13 *(Page 327)*

Practice

1. 6,480 BTUs

Chapter 12 Coordinate Geometry and Transformations

Lesson 12.1 *(Page 335)*

Practice

1.

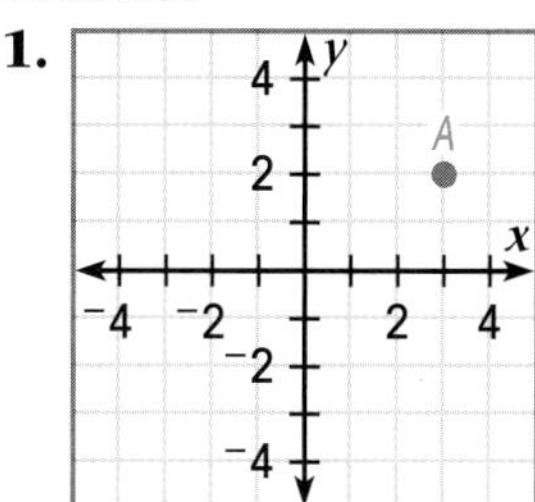

9. (5, 5)

Lesson 12.2 *(Page 337)*

Practice

1. $\sqrt{13}$

Lesson 12.3 *(Page 339)*

Practice

1. (3, 4)

Lesson 12.4 *(Page 341)*

Practice

1. $^-1$

Lesson 12.5 *(Page 343)*

Practice

1. parallel

5. not perpendicular

Lesson 12.6 *(Page 345)*

Practice

1.

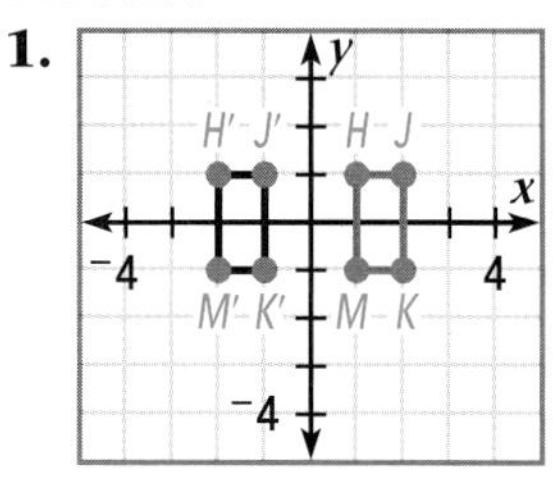

5. $T(x, y) \longrightarrow (x + 4, y - 3)$

Lesson 12.7 *(Page 347)*

Practice

1.

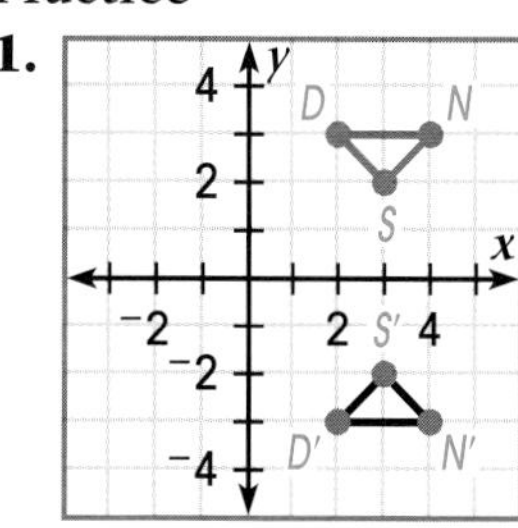

Lesson 12.8 *(Page 349)*

Practice

1.

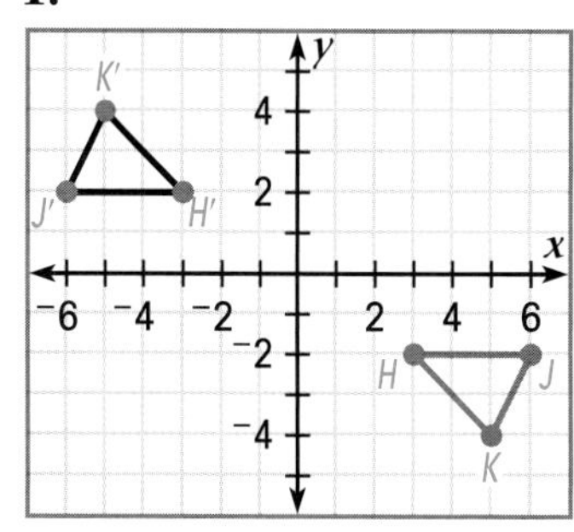

Lesson 12.9 *(Page 351)*

Practice

1.

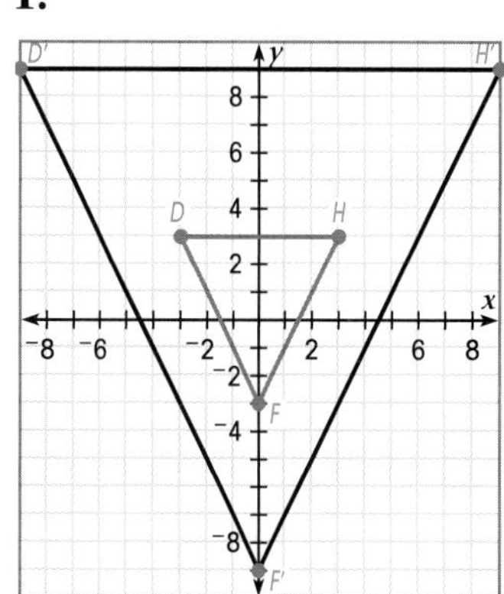

Lesson 12.10 *(Page 353)*

Practice

1.

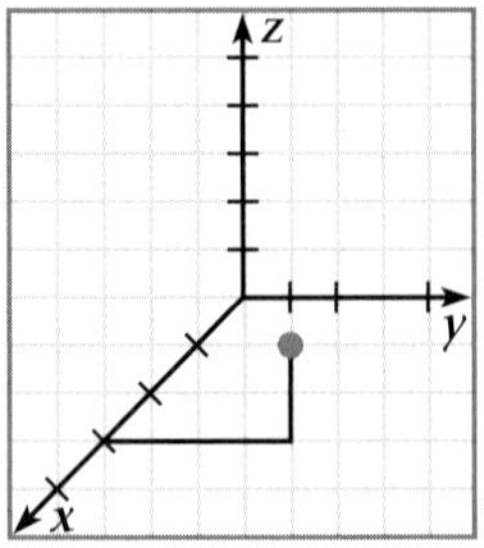

Lesson 12.11 *(Page 355)*

Practice

1. $\sqrt{6}$

Lesson 12.12 *(Page 357)*

Practice

1. (2, 3, 1)

Lesson 12.13 *(Page 359)*

Practice

1. 10

Lesson 12.14 *(Page 361)*

Practice

1. southeast; 25 mph

Lesson 12.15 *(Page 363)*

Practice

1. southwest; 16 mph

Chapter 13 Right Triangle Trigonometry

Lesson 13.1 *(Page 371)*

Practice

1. $\sin D = \frac{8}{17}$;
$\cos D = \frac{15}{17}$;
$\tan D = \frac{8}{15}$

Lesson 13.2 *(Page 373)*

Practice

1. $\sin 88° = 0.9994$;
$\cos 88° = 0.0349$;
$\tan 88° = 28.6363$

6. 85°

Lesson 13.3 *(Page 375)*

Practice

1. 20° **4.** 17

Lesson 13.4 *(Page 377)*

Practice

1. 45° **4.** 19

Lesson 13.5 *(Page 379)*

Practice

1. 44° **4.** 9

Lesson 13.6 *(Page 381)*

Practice

1. 53° **4.** 13

Lesson 13.7 *(Page 383)*

Practice

1. about 42 ft

Lesson 13.8 *(Page 385)*

Practice

1. about 19 m

INDEX

A

B

C

D

E

F

G

H

I

K

L

M

N

O

P

Q

R

S

T

Photo Credits

Cover Credits: M. Getty Images, Inc./PhotoDisc, Inc.; t.r. Stone; b.r. Corbis Digital Stock; b.i. Courtesy of Karen Edmonds; m.i. Rollo Silver.

p. 2: Don Farrall/PhotoDisc, Inc.
p. 30: Andre Jenny/Index Stock Imagery, Inc.
p. 62: Richard Hutchings/PhotoEdit
p. 88: © Tim Davis/Photo Researchers, Inc.
p. 122: Joel Greenstein/Omni-Photo Communications, Inc.
p. 158: Peter Timmermans/Stone
p. 175: Lawrence Sawyer/Index Stock Imagery, Inc.
p. 184: Phil Jason/Stone
p. 210: Adalberto Rios Szalay/Sexto Sol/PhotoDisc, Inc.
p. 232: © Fritz Polking/Peter Arnold, Inc.
p. 255: © Norman Owen Tomalin/Bruce Coleman, Inc.
p. 264: Richard Cummins/Corbis
p. 289: © Martin Dohrn/Science Photo Library/Photo Researchers, Inc.
p. 300: Donovan Reese/Stone
p. 323: Colored French engraving, 1584, The Granger Collection
p. 332: Corbis Digital Stock
p. 359: Kevin Miller/Stone
p. 368: Warren Bolster/Stone

Additional Answers

Answers for proof lessons may vary. Possible answers are given. Answers for proofs in two-column format can be found in the Solution Key.

UNIT ONE

Chapter 3 Reasoning and Proofs

Lesson 3.6, page 75

Practice

1. You are given that $\angle JKM \cong \angle NKL$. Because the measures of congruent angles are equal, $m\angle JKM = m\angle NKL$. Because $\angle MKN$ is congruent to itself, you can add $m\angle MKN$ to both sides of this equation. So, $m\angle JKM + m\angle MKN = m\angle NKL + m\angle MKN$. You can rewrite this equation as $m\angle JKM + m\angle MKN = m\angle MKN + m\angle NKL$. Now, you can use the Angle Addition Postulate to rewrite each side of the equation. Then, $m\angle JKN = m\angle MKL$. By definition of congruent angles, $\angle JKN \cong \angle MKL$.

So, you prove that $\angle JKN \cong \angle MKL$.

2. You are given that $\angle PQT \cong \angle SQR$. Because the measures of congruent angles are equal, $m\angle PQT = m\angle SQR$. Using the Angle Addition Postulate, you can rewrite each side of the equation. So, $m\angle PQS + m\angle SQT = m\angle SQT + m\angle TQR$. You can subtract $m\angle SQT$ from both sides of this equation. Then, $m\angle PQS = m\angle TQR$. By definition of congruent angles, $\angle PQS \cong \angle TQR$.

So, you prove that $\angle PQS \cong \angle TQR$.

Lesson 3.7, page 77

Practice

1.

Statements	Reasons
1. $\angle JKM \cong \angle NKL$	1. Given
2. $m\angle JKM = m\angle NKL$	2. Definition of congruent angles
3. $m\angle MKN = m\angle MKN$	3. Reflexive Property of Equality
4. $m\angle JKM + m\angle MKN = m\angle NKL + m\angle MKN$	4. Addition Property of Equality
5. $m\angle JKM + m\angle MKN = m\angle MKN + m\angle NKL$	5. Commutative Property
6. $m\angle JKN = m\angle JKM + m\angle MKN$ $m\angle MKL = m\angle MKN + m\angle NKL$	6. Angle Addition Postulate
7. $m\angle JKN = m\angle MKL$	7. Substitution
8. $\angle JKN \cong \angle MKL$	8. Definition of congruent angles

2.

Statements	Reasons
1. $\angle PQT \cong \angle SQR$	1. Given
2. $m\angle PQT = m\angle SQR$	2. Definition of congruent angles
3. $m\angle PQT = m\angle PQS + m\angle SQT$ $m\angle SQR = m\angle SQT + m\angle TQR$	3. Angle Addition Postulate
4. $m\angle PQS + m\angle SQT = m\angle SQT + m\angle TQR$	4. Substitution
5. $m\angle PQS + m\angle SQT = m\angle TQR + m\angle SQT$	5. Commutative Property
6. $m\angle PQS = m\angle TQR$	6. Subtraction Property of Equality
7. $\angle PQS \cong \angle TQR$	7. Definition of congruent angles

---

Lesson 3.9, page 81

Share Your Understanding

9. You are given: Line m intersects line n. Prove: Line m and line n intersect at one point.

Assume: Line m and line n intersect at two points.

Draw a diagram.

This contradicts what is given. Line m is a curve. The assumption that line m and line n intersect at two points is false. Then, the prove statement is true.

So, you prove that line m and line n intersect at one point.

Lesson 3.10, page 83

Practice

1.

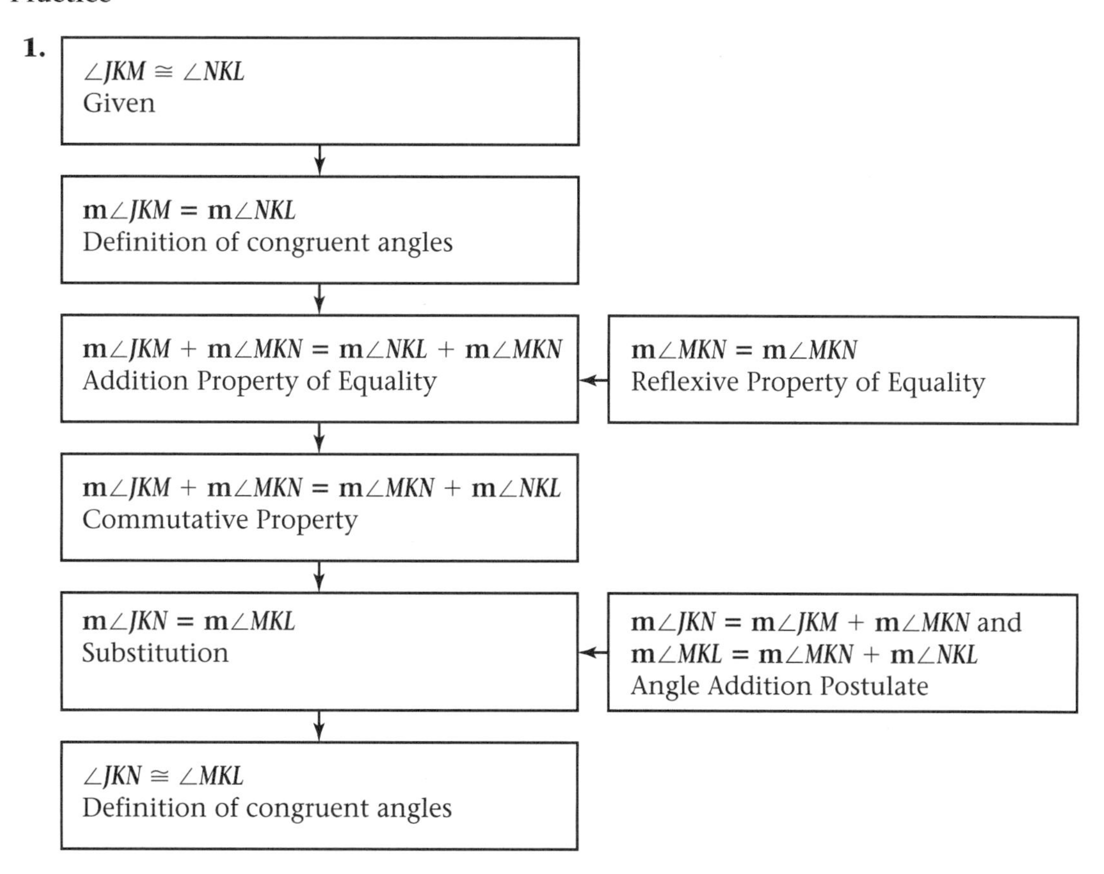

2.

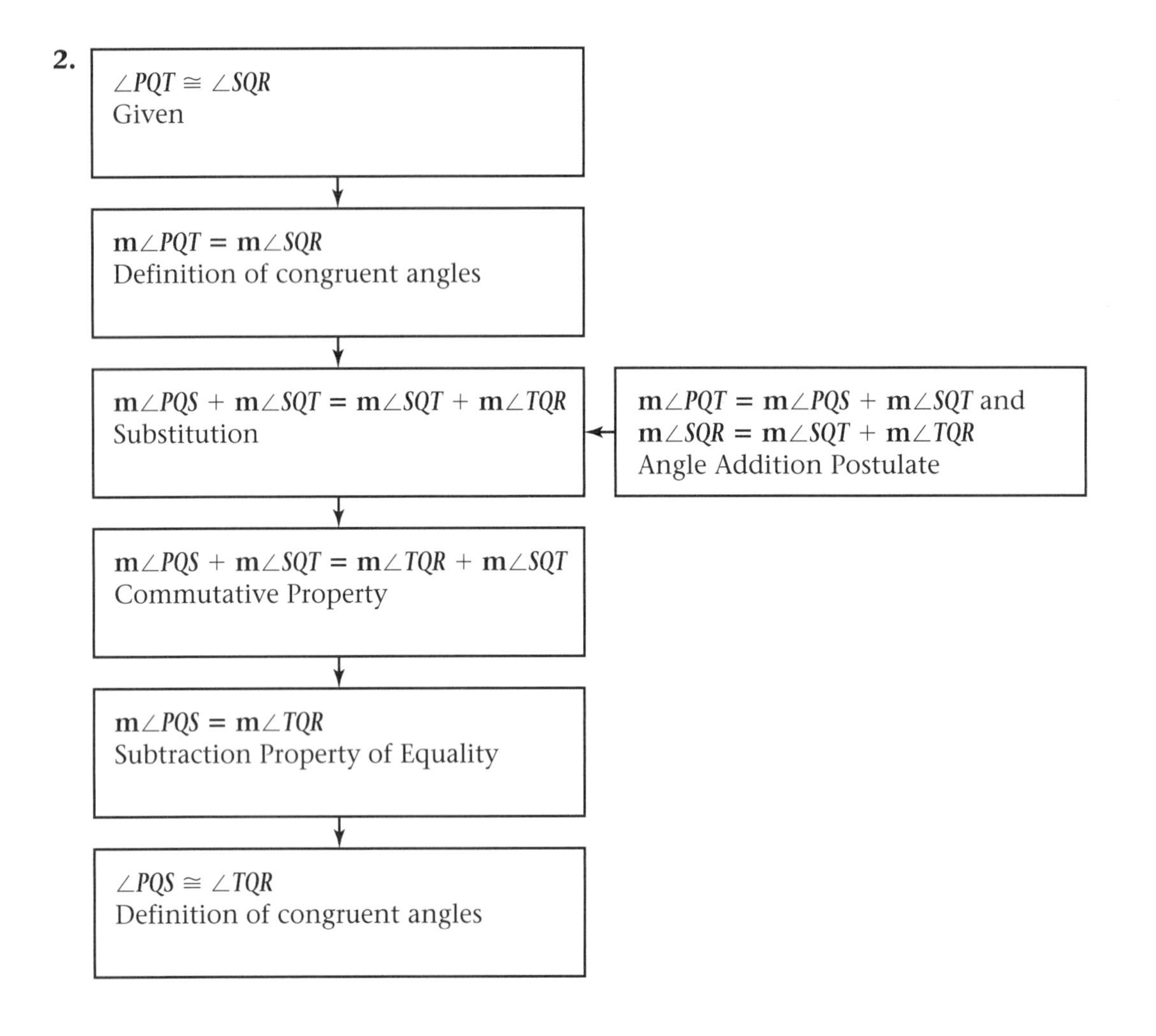

Lesson 3.11, page 85

Practice

1. You are given that $\angle D$ and $\angle E$ are complementary. You are also given that $\angle F$ and $\angle E$ are complementary. Complementary angles are angles whose measures have a sum of 90. So, $m\angle D + m\angle E = 90$ and $m\angle F + m\angle E = 90$. Because both sums equal 90, they are equal to each other.
$m\angle D + m\angle E = m\angle F + m\angle E$
You can subtract $m\angle E$ from both sides of this equation. Then, $m\angle D = m\angle F$. By definition of congruent angles, $\angle D \cong \angle F$.

So, you prove that $\angle D \cong \angle F$.

2. You are given that $\overline{AB} \cong \overline{PQ}$ and $\overline{BC} \cong \overline{QR}$. Congruent line segments have the same length. So, $AB = PQ$ and $BC = QR$. You can use the Segment Addition Postulate to write the equation $AC = AB + BC$. Substitute PQ for AB and QR for BC. So, $AC = PQ + QR$. Then, $AC = PR$. By definition of congruent line segments, $\overline{AC} \cong \overline{PR}$.

   So, you prove that $\overline{AC} \cong \overline{PR}$.

3. You are given that $\overline{JL} \cong \overline{EF}$. You are also given that point L is the midpoint of $\overline{JK}$. A midpoint divides a line segment into two congruent parts. So, $\overline{JL} \cong \overline{LK}$. Because $\overline{JL}$ is congruent to both $\overline{LK}$ and $\overline{EF}$, the line segments are congruent to each other. That is, $\overline{LK} \cong \overline{EF}$.

   So, you prove that $\overline{LK} \cong \overline{EF}$.

Chapter Review, page 87

Chapter Quiz

11. Assume that if n is a prime number, then n^2 is a prime number. Try $n = 5$. You know that 5 is a prime number. However, $5^2 = 25$ and 25 is divisible by 5, 25, and 1. So, 25 is not a prime number. Thus, the assumption is false. The prove statement must be true.

    So, you prove that n^2 is not a prime number.

Chapter 4 Perpendicular and Parallel Lines

Lesson 4.11, page 117

Practice

1. You are given that $\angle 1 \cong \angle 5$. Because $\angle 1$ and $\angle 4$ are vertical angles, $\angle 1 \cong \angle 4$. Because $\angle 4$ and $\angle 5$ are congruent to $\angle 1$ they are congruent to each other. That is, $\angle 4 \cong \angle 5$. $\angle 4$ and $\angle 5$ are alternate interior angles. By Postulate 8, $m \parallel n$.

   So, you prove that $m \parallel n$.

2. You are given that $\angle 1 \cong \angle 8$. Because they are vertical angles, $\angle 1 \cong \angle 4$. Because $\angle 4$ and $\angle 8$ are congruent to $\angle 1$, they are congruent to each other, $\angle 4 \cong \angle 8$. Because $\angle 8$ and $\angle 5$ are vertical angles, $\angle 5 \cong \angle 8$. Because $\angle 5$ and $\angle 4$ congruent to $\angle 8$, they are congruent to each other, $\angle 4 \cong \angle 5$. By Postulate 8, $k \parallel l$.

   So, you prove that $l \parallel k$.

UNIT TWO

Chapter 5 Triangles

Lesson 5.16, page 155

Practice

1. You are given that $\angle Q \cong \angle S$. You are also given that $\overline{QS}$ bisects $\overline{PR}$. This means that $\overline{PT} \cong \overline{RT}$. You know that $\angle QTR \cong \angle STP$ because they are vertical angles. Now, $\triangle QRT \cong \triangle SPT$ by AAS. Then, $\angle P \cong \angle R$ by CPCTC.

   So, you prove that $\angle P \cong \angle R$.

2. You are given that $\overline{DE} \cong \overline{FE}$. You are also given that $\overline{EG}$ bisects $\angle E$. This means that $\angle DEG \cong \angle FEG$. You know that $\overline{EG} \cong \overline{EG}$. Now, $\triangle DEG \cong \triangle FEG$. Then, $\overline{DG} \cong \overline{FG}$ by CPCTC.

   So, you prove that $\overline{DG} \cong \overline{FG}$.

Chapter 6 Right Triangles

Lesson 6.11, page 181

Practice

1. You are given that $\overline{MO} \perp \overline{LO}$ and $\overline{MO} \perp \overline{MN}$. This means that $\angle LOM$ and $\angle NMO$ are right angles. Then, $\triangle LMO$ and $\triangle NOM$ are right triangles. $\overline{LM}$ is the hypotenuse of $\triangle LMO$ and $\overline{NO}$ is the hypotenuse of $\triangle NOM$. You are also given that $\overline{LM} \cong \overline{NO}$. $\overline{MO} \cong \overline{OM}$. Then, by the Hypotenuse–Leg Theorem, the triangles are congruent.

   So, you prove that $\triangle LMO \cong \triangle NOM$.

2. You are given that $\overline{SP}$ is the perpendicular bisector of $\overline{QT}$. This means that $\angle QRP$ and $\angle TRS$ are right angles and $\overline{QR} \cong \overline{TR}$. $\triangle PQR$ and $\triangle STR$ are right triangles. You are given that $\overline{QP} \cong \overline{ST}$. Thus, the hypotenuses of both triangles are congruent. Then, by the Hypotenuse–Leg Theorem, the triangles are congruent.

   So, you prove that $\triangle PQR \cong \triangle STR$.

Chapter 7 Quadrilaterals and Polygons

Lesson 7.10, page 205

Practice

1. You are given that $\overline{CD} \cong \overline{EF}$ and $\overline{CF} \cong \overline{ED}$. You know that $FD \cong DF$. Then, $\triangle FCD \cong \triangle DEF$ by SSS. This means that $\angle CDF \cong \angle EFD$ by CPCTC. These are alternate interior angles. So, $\overline{CD} \parallel \overline{EF}$. You also know that $\angle CFD \cong \angle EDF$ by CPCTC. These are alternate interior angles. So, $\overline{CF} \parallel \overline{ED}$. Because the opposite sides of quadrilateral *CDEF* are parallel, the quadrilateral is a parallelogram.

So, you prove that quadrilateral *CDEF* is a parallelogram.

2. You are given that $\overline{QP} \cong \overline{SR}$. You are also given that $\overline{PQ} \parallel \overline{SR}$. This means that $\angle PQS \cong \angle RSQ$. You know that $\overline{SQ} \cong \overline{QS}$. Then, $\triangle SPQ \cong \triangle QRS$ by SAS. This means that $\angle PSQ \cong \angle RQS$ by CPCTC. These are alternate interior angles. So, $\overline{PS} \parallel \overline{RQ}$. Because the opposite sides of quadrilateral *PQRS* are parallel, the quadrilateral is a parallelogram.

So, you prove that quadrilateral *PQRS* is a parallelogram.

UNIT THREE

Chapter 8 Perimeter and Area

Lesson 8.9, page 229

Practice

1. You are given trapezoid *MNOP*. You know that $\overline{NO} \parallel \overline{MP}$. This means that the heights of $\triangle MNP$ and $\triangle POM$ are congruent. You know that $\overline{MP} \cong \overline{MP}$. This means that the bases are congruent. Because the bases are congruent and the heights are congruent, the areas must be equal.

So, you prove that Area $\triangle MNP$ = Area $\triangle POM$.

2. You are given trapezoid *ABCD* and $\overline{BC} \parallel \overline{AE}$. This means that $\angle CBF \cong \angle DEF$. You know that $\angle BFC \cong \angle EFD$. They are vertical angles. You are given that $\overline{BC} \cong \overline{ED}$. Then, $\triangle BCF \cong \triangle EDF$ by AAS. Because the triangles are congruent, the areas are equal. Then, Area $\triangle BCF$ + Area *ABFD* = Area $\triangle EDF$ + Area *ABFD*.

So, you prove that Area *ABCD* = Area $\triangle ABE$.

Chapter 9 Similar Polygons

Lesson 9.14, page 261

Practice

1. You are given that $\overline{AD}$ intersects $\overline{BE}$. This means $\angle BCA \cong \angle ECD$ because they are vertical angles. Because both $\angle B$ and $\angle E$ are right angles, they are also congruent. By the Angle-Angle Similarity Postulate, the two triangles are similar.

   So, you prove that $\triangle ABC \sim \triangle DEC$.

2. You are given that triangle FGH is a right triangle with right angle G. You are also given that $\overline{GI}$ is an altitude. Then, $\angle GIH$ is a right angle. Because both $\angle G$ and $\angle GIH$ are right angles, they are congruent. You know that $\angle H \cong \angle H$. By the Angle-Angle Similarity Postulate, the two triangles are similar.

   So, you prove that $\triangle FGH \sim \triangle GIH$.

3. You are given that $\overline{MN} \parallel \overline{JL}$. $\overline{KJ}$ is a transversal for the two parallel lines. $\angle KMN$ and $\angle KJL$ are the corresponding angles for these parallel lines, $\angle KMN$ and $\angle KJL$. You know that $\angle K \cong \angle K$. By the Angle-Angle Similarity Postulate, the two triangles are similar.

   So, you prove that $\triangle MKN \sim \triangle JKL$.

Chapter 10 Circles

Lesson 10.15, page 295

Practice

1. $\angle RQS$ and $\angle TQU$ are central angles. The measures of the central angles are equal to the degree measures of their arcs. So, $m\angle RQS = m\overset{\frown}{RS}$ and $m\angle TQU = m\overset{\frown}{TU}$. Because $m\angle RQS \cong m\angle TQU$, $m\overset{\frown}{RS} = m\overset{\frown}{TU}$.

   So, you prove that $\overset{\frown}{RS} \cong \overset{\frown}{TU}$.

2. You are given that $\overline{AB} \cong \overline{CD}$. $\overline{OA}$, $\overline{OB}$, $\overline{OC}$, and $\overline{OD}$ are radii of the circle. Because radii of the same circle are congruent, $\overline{AO} \cong \overline{DO} \cong \overline{CO} \cong \overline{BO}$. Then, $\triangle ABO \cong \triangle DCO$ by SSS. By CPCTC, $\angle AOB \cong \angle DOC$. Because the central angles are congruent, their arcs are congruent.

   So, you prove that $\overset{\frown}{AB} \cong \overset{\frown}{CD}$.

3. You are given that $\overline{EF} \parallel \overline{GH}$. $\angle EFG$ and $\angle FGH$ are alternate interior angles. This means that $\angle EFG \cong \angle FGH$. $\angle EFG$ and $\angle FGH$ are inscribed angles. Because the inscribed angles are congruent, the intercepted arcs are congruent.

   So, you prove that $\overset{\frown}{EG} \cong \overset{\frown}{FH}$.

UNIT FOUR

Chapter 11 Surface Area and Volume

Lesson 11.14, page 329

Practice

1. You know that the base of the cylinder is a circle with radius r. So, the area of the base is πr^2. The height of the cylinder is h. The volume is the area of the base times the height, $V = Bh$. Using substitution, the volume of the cylinder is $\pi r^2 h$.

   So, you prove that Volume $= \pi r^2 h$.

2. You know that the base of the pyramid is a square with side s. So, the area of the base is s^2. The height of the pyramid is h. The volume is $\frac{1}{3}$ the area of the base times the height, $V = \frac{1}{3}Bh$. Using substitution, $V = \frac{1}{3}s^2 h$.

   So, you prove that Volume $= \frac{1}{3}s^2 h$.

Lesson 12.6, page 345

Practice

1.

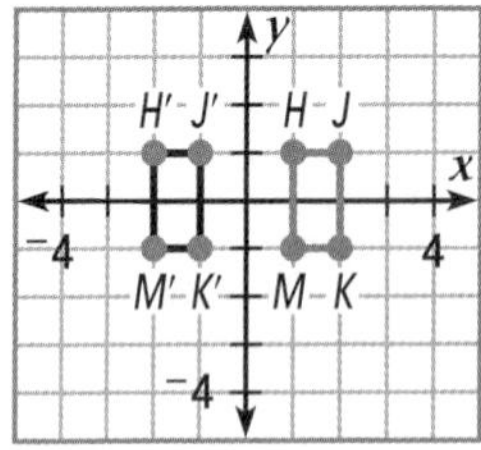

2.

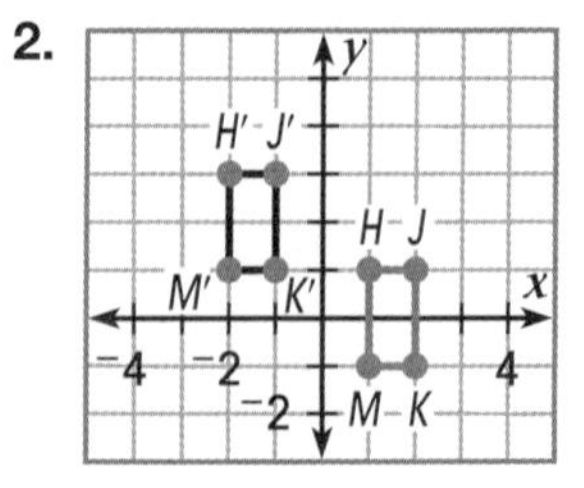

3.

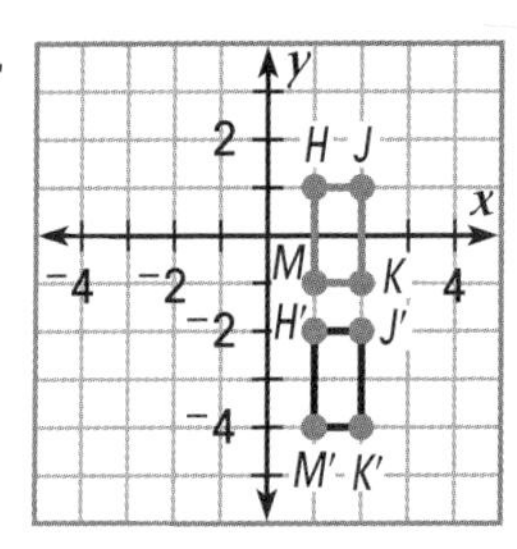

4.

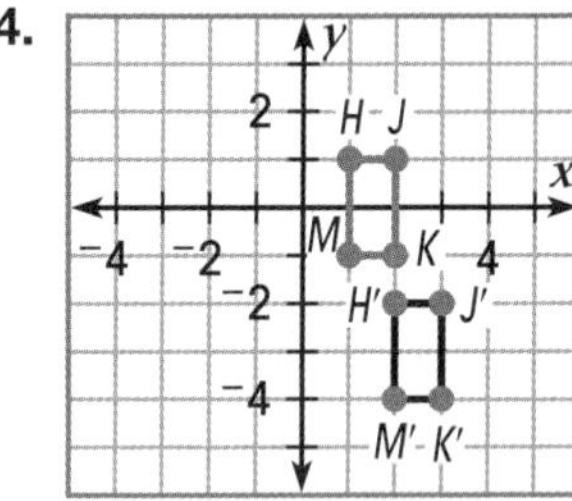

Lesson 12.7, page 347

Practice

1.

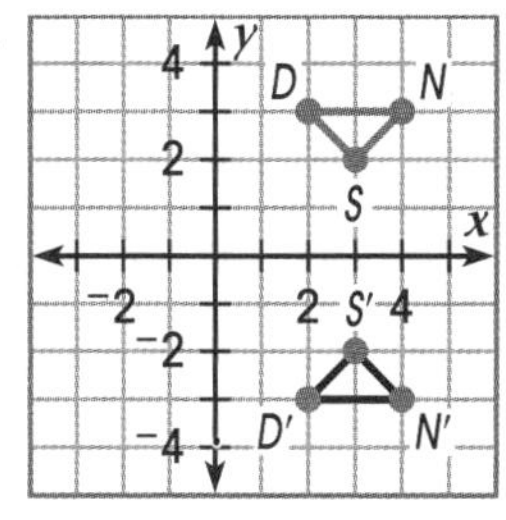

2.

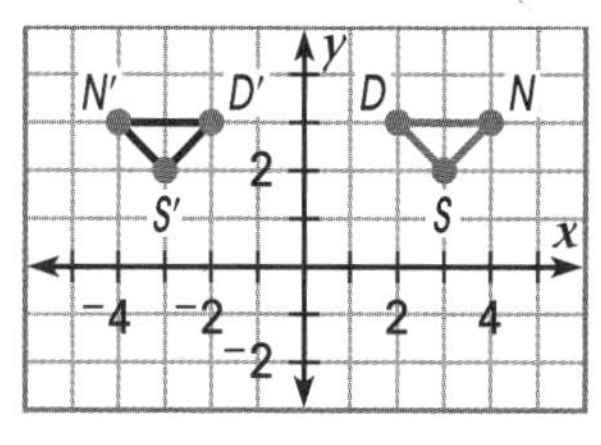

3.

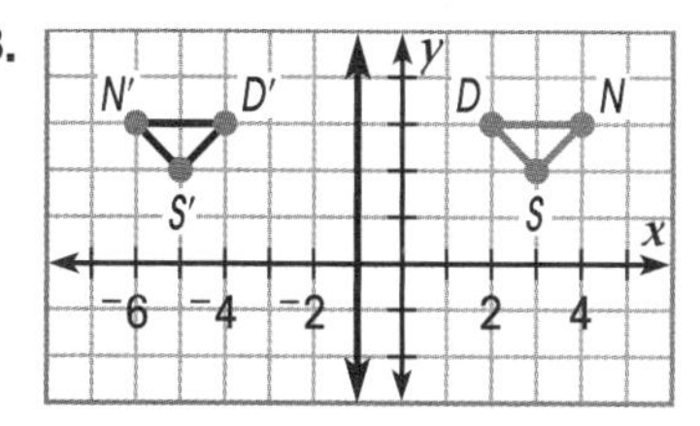

4.

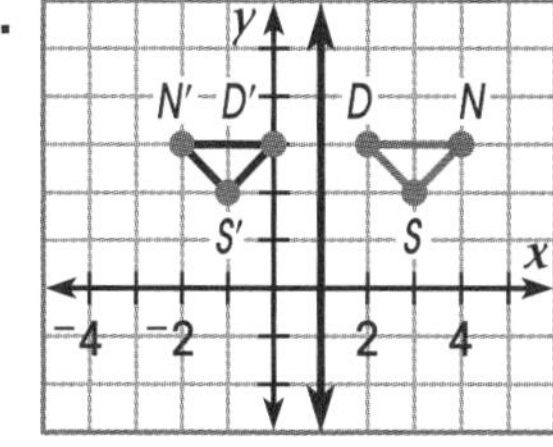

5.

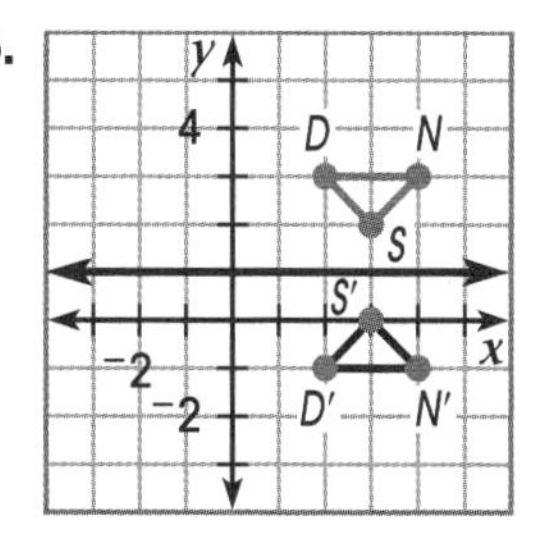

6.

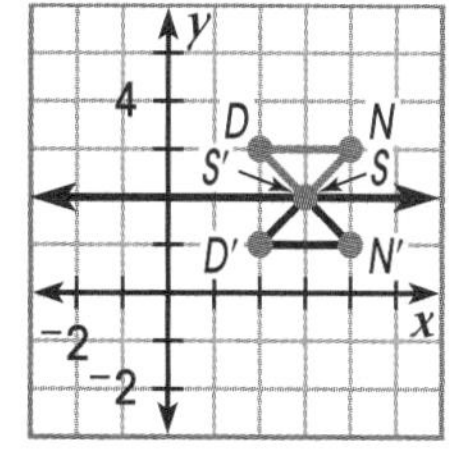

Lesson 12.8, page 349

Practice

1.

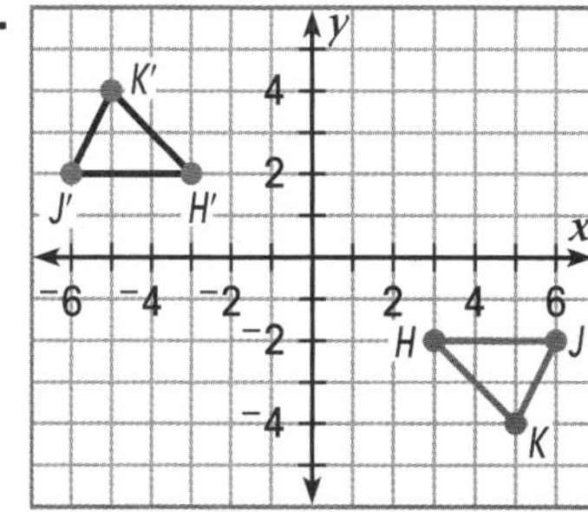

2.

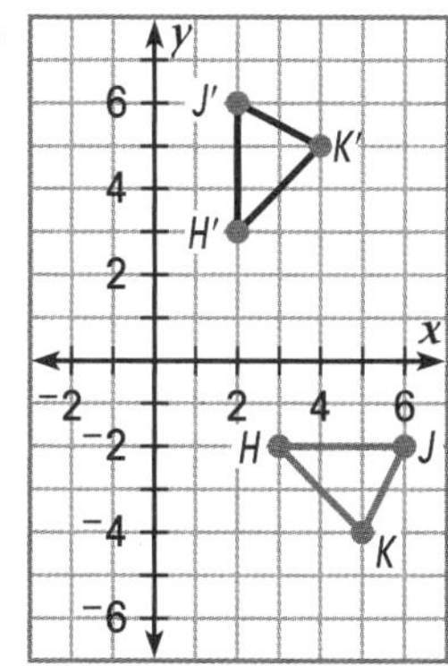

3.

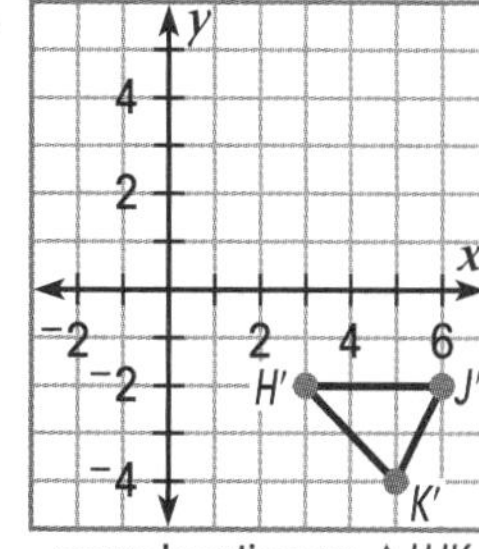

same location as $\triangle HJK$

Lesson 12.9, page 351

Practice

1.

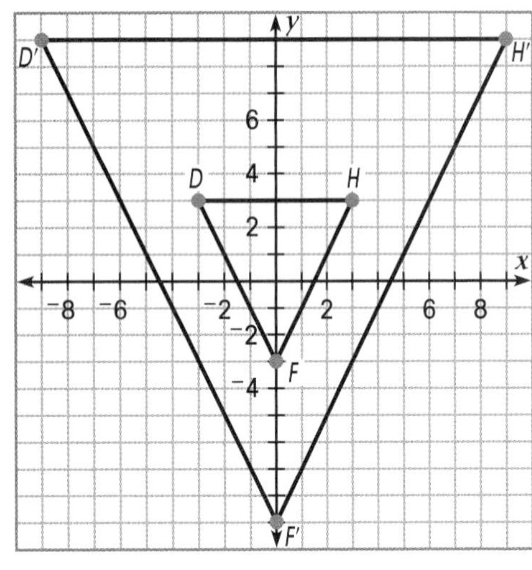

2.

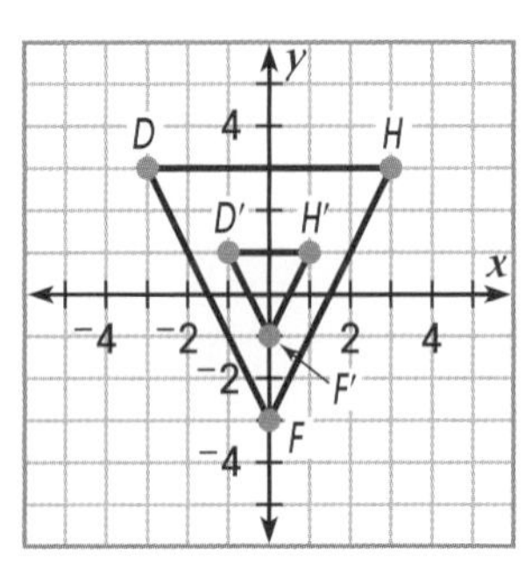

3.

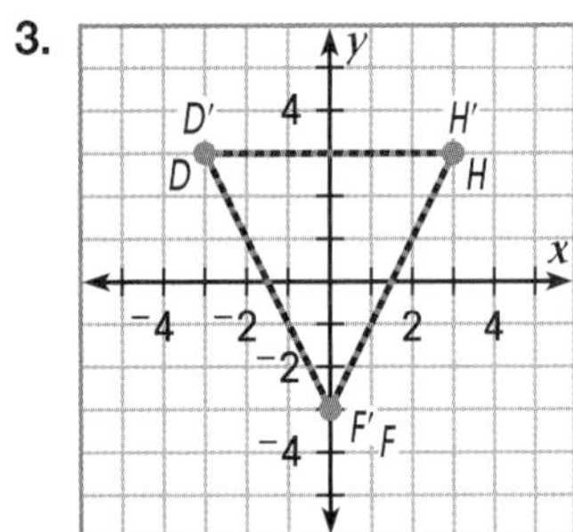

Lesson 12.10, page 353

Practice

1.

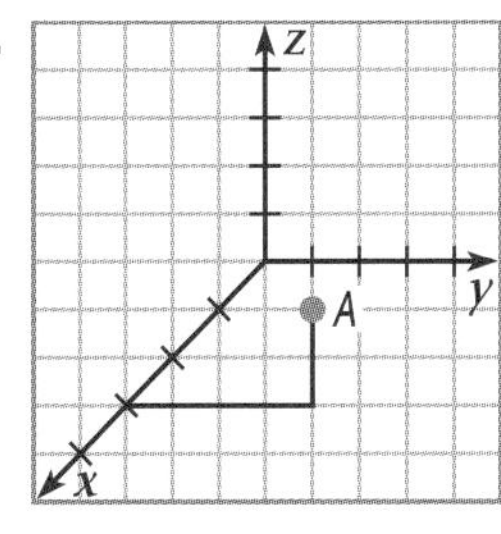

2.

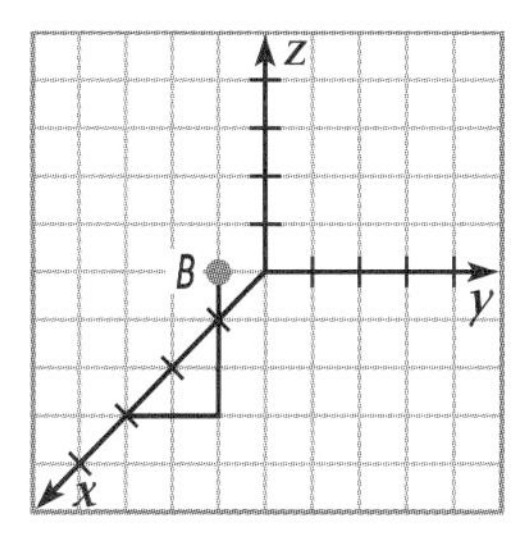

3.

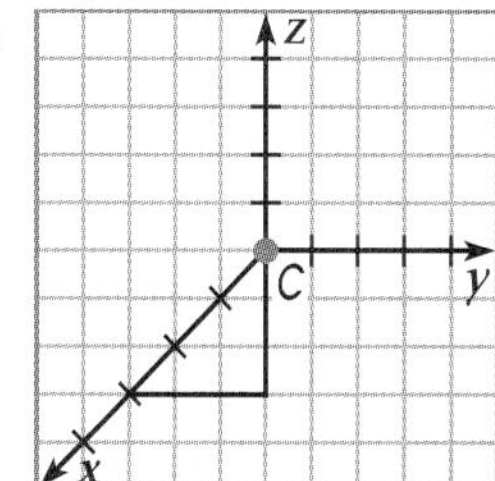

4.

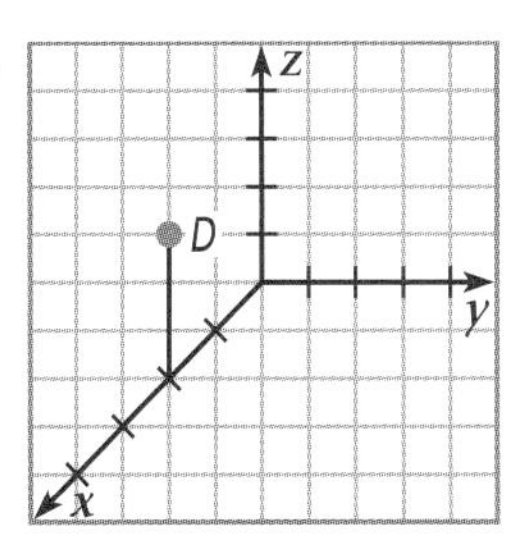

5.

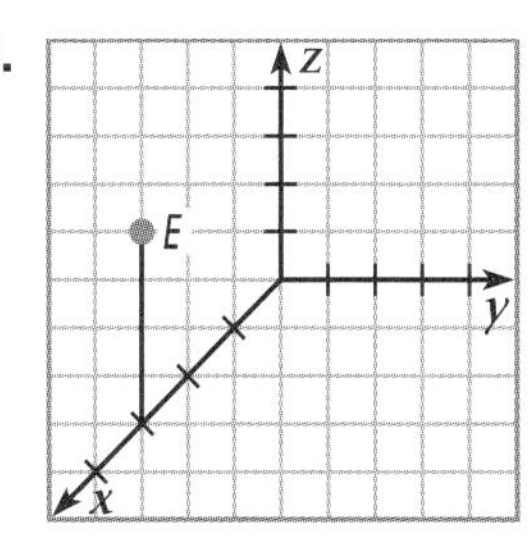

6.

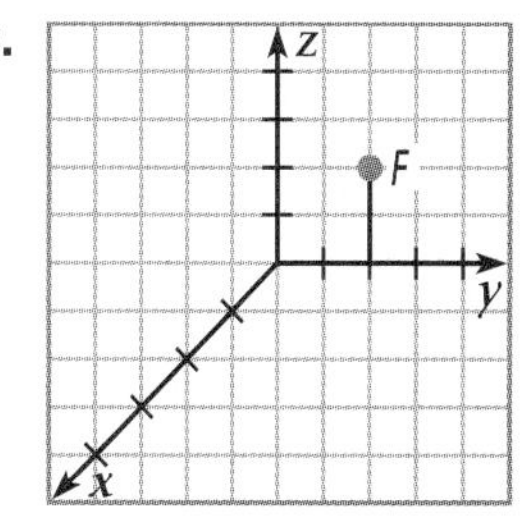

7.

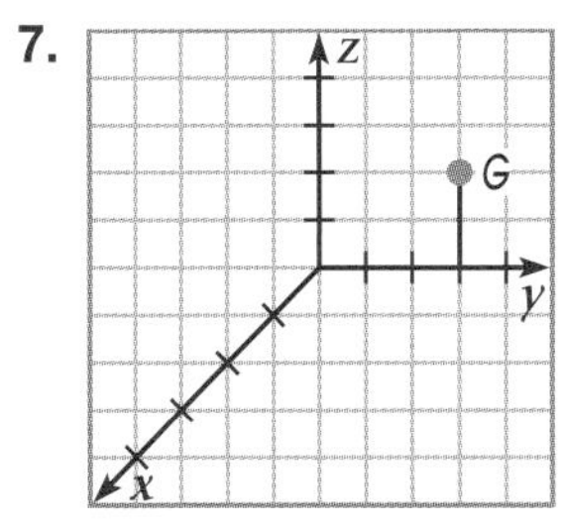

8.

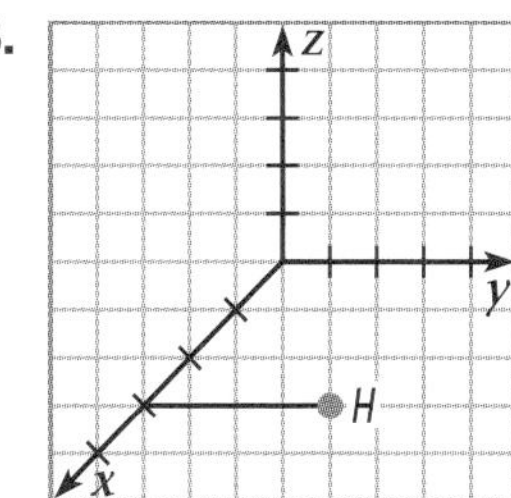

9.

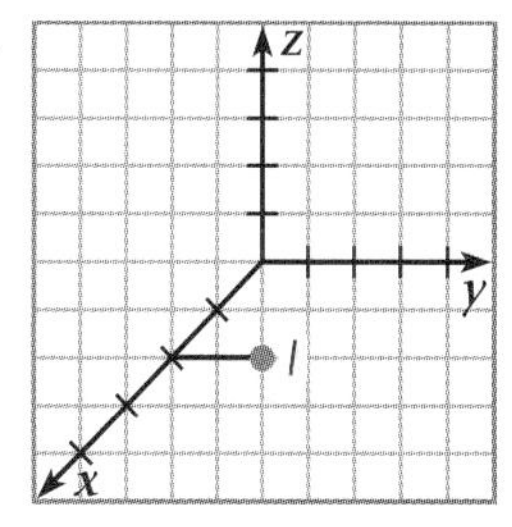

10.

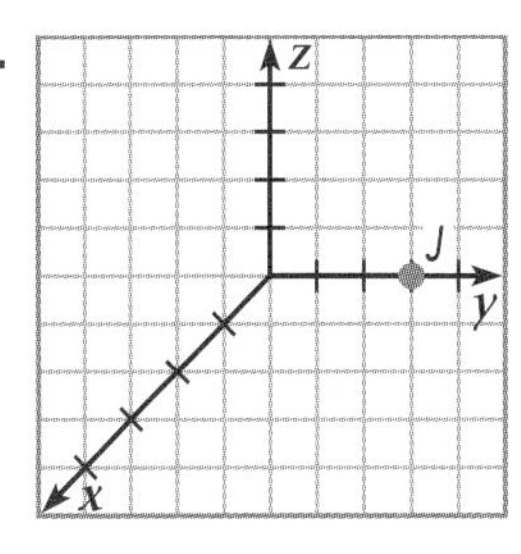

11.

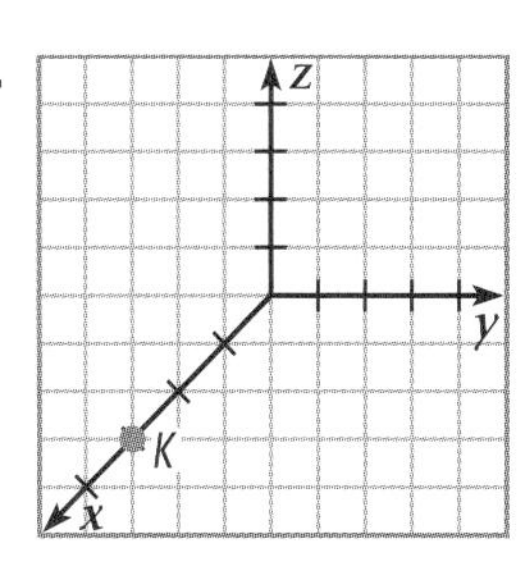

12.

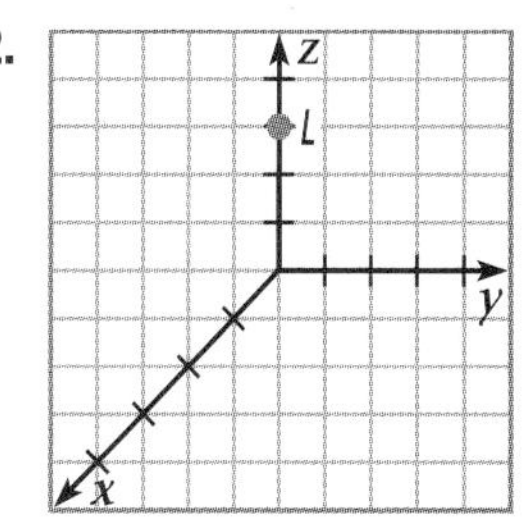

Lesson 12.16, page 365

Practice

1. You can use the distance formula to find the length of $\overline{JL}$ and $\overline{MK}$.

$$JL = \sqrt{[p - (-m)]^2 + (q - 0)^2} \qquad MK = \sqrt{[m - (-p)]^2 + (0 - q)^2]}$$
$$= \sqrt{(p + m)^2 + (q)^2} \qquad = \sqrt{(m + p)^2 + (q)^2}$$
$$= \sqrt{(m + p)^2 + q^2}$$

$\overline{JL}$ and $\overline{MK}$ have the same length. Then, $\overline{JL} \cong \overline{MK}$.

So, you prove that $\overline{JL} \cong \overline{MK}$.

2. You can use the distance formula to prove that the length of $\overline{DE}$ is half the length of $\overline{FH}$.

$$DE = \sqrt{[a - (-a)]^2 + (b - b)^2} \qquad FH = \sqrt{[2a - (-2a)]^2 + (0 - 0)^2}$$
$$= \sqrt{(2a)^2 + 0^2} \qquad = \sqrt{(4a)^2}$$
$$= \sqrt{4a^2} \qquad = \sqrt{16a^2}$$
$$= 2a \qquad = 4a$$

The length of $\overline{DE}$ is $2a$. The length of $\overline{FH}$ is $4a$.

So, you prove that $DE = \frac{1}{2}\overline{FH}$.

Chapter Quiz, page 367

Exercises 1 – 2

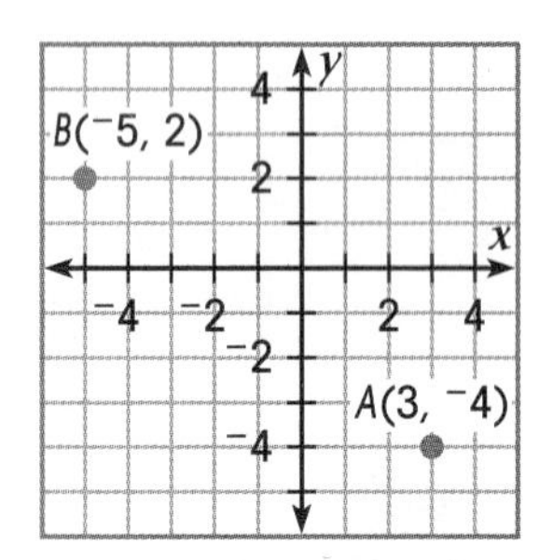

Exercises 8 – 11

8. 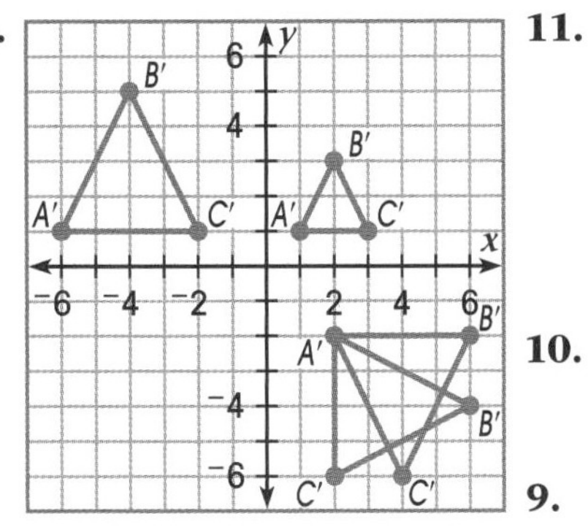

8. $A'\ (^-6, 1), B'\ (^-4, 5), C'\ (^-2, 1)$

9. $A'\ (2, ^-2), B'\ (6, ^-4), C'\ (2, ^-6)$

10. $A'\ (2, ^-2), B'\ (4, ^-6), C'\ (6, ^-2)$

11. $A'\ (1, 1), B'\ (2, 3), C'\ (3, 1)$

12. 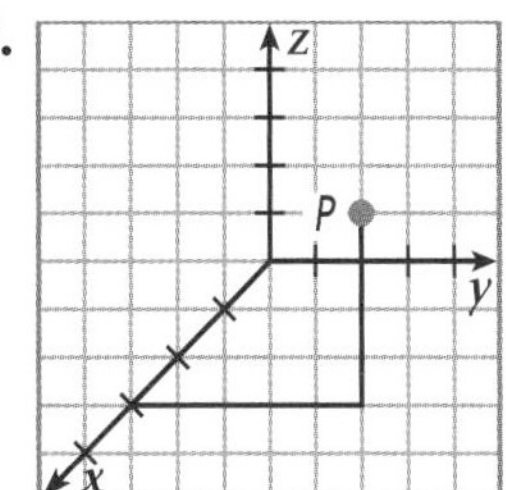

Chapter 13 Right Triangle Trigonometry

Lesson 13.9, page 387

Practice

1. $\cos A = \frac{\sin A}{\tan A}$

$\cos A = \frac{a}{c} \div \frac{a}{b}$

$\cos A = \frac{a}{c} \bullet \frac{b}{a}$

$\cos A = \frac{b}{c}$

$\cos A = \cos A$

So, you prove that $\cos A = \frac{\sin A}{\tan A}$. ✓

2. $\sin A = \frac{\cos A}{\tan B}$

$\sin A = \frac{b}{c} \div \frac{b}{a}$

$\sin A = \frac{b}{c} \bullet \frac{a}{b}$

$\sin A = \frac{a}{c}$

$\sin A = \sin A$

So, you prove that $\sin A = \frac{\cos A}{\tan B}$. ✓

3. $\tan A \bullet \cos A = \sin A$

$\frac{a}{b} \bullet \frac{b}{c} = \sin A$

$\frac{a}{c} = \sin A$

$\sin A = \sin A$

So, you prove that
$\tan A \bullet \cos A = \sin A$. ✓

Additional Practice

Chapter 12, page 404

Exercises **1 – 2**

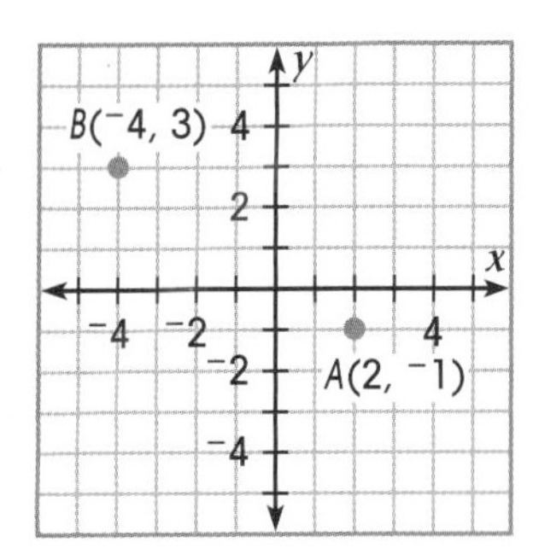

Exercises **8 – 11**

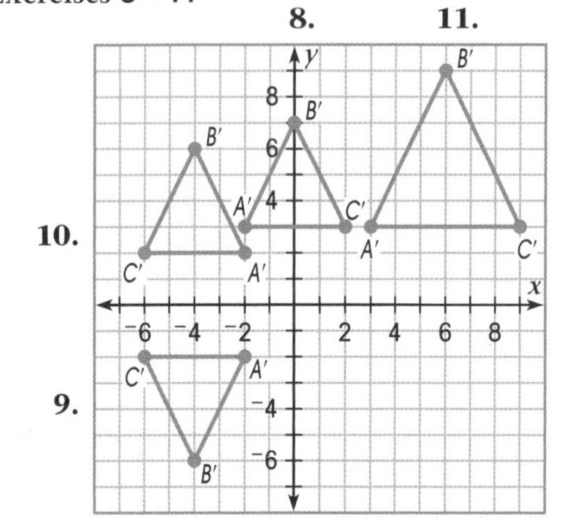

8. A' ($^-2$, 3), B' (0, 7), C' (2, 3)

9. A' ($^-2$, $^-2$), B' ($^-4$, $^-6$), C' ($^-6$, $^-2$)

10. A' ($^-2$, 2), B' ($^-4$, 6), C' ($^-6$, 2)

11. A' (3, 3), B' (6, 9), C' (9, 3)

12.

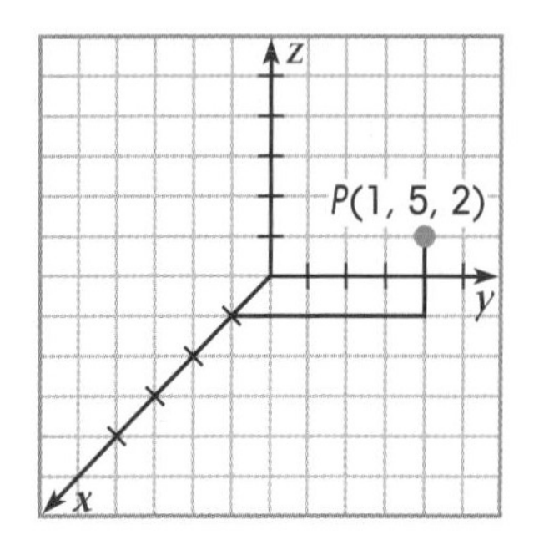